FOUNDATIONS OF MULTIATTRIBUTE UTILITY

Many of the complex problems faced by decision-makers involve uncertainty as well as multiple conflicting objectives. This book provides a complete understanding of the types of objective functions that should be used in multiattribute decision-making. By using tools such as preference, value, and utility functions, readers will learn state-of-the-art methods to analyze prospects to guide decision-making and will develop a process that guarantees a defensible analysis to rationalize choices.

Summarizing and distilling classical techniques and providing extensive coverage of recent advances in the field, the author offers practical guidance on how to make good decisions in the face of uncertainty. This text will appeal to graduate students and practitioners alike in systems engineering, operations research, business, management, government, climate change, energy, and health care.

Ali E. Abbas is Professor of Industrial and Systems Engineering and Professor of Public Policy and Governance at the University of Southern California. He directs the Center for Decisions and Ethics (DECIDE) and previously directed the Center for Risk and Economic Analysis of Terrorism Events (CREATE). His research covers all aspects of decision-making under uncertainty broadly defined. He has written two books and edited two others, and has more than 100 refereed publications in the field of decision analysis. He is editor of numerous professional journals and has extensive media coverage on all aspects of decision-making and risk.

FOUNDATIONS OF MULTIATTRIBUTE UTILITY

ALI E. ABBAS

University of Southern California, Los Angeles

CAMBRIDGE
UNIVERSITY PRESS

University Printing House, Cambridge CB2 8BS, United Kingdom

One Liberty Plaza, 20th Floor, New York, NY 10006, USA

477 Williamstown Road, Port Melbourne, VIC 3207, Australia

314–321, 3rd Floor, Plot 3, Splendor Forum, Jasola District Centre, New Delhi – 110025, India

79 Anson Road, #06-04/06, Singapore 079906

Cambridge University Press is part of the University of Cambridge.

It furthers the University's mission by disseminating knowledge in the pursuit of education, learning, and research at the highest international levels of excellence.

www.cambridge.org
Information on this title: www.cambridge.org/9781107150904
DOI: 10.1017/9781316596739

First published 2018

Printed in the United States of America by Sheridan Books, Inc.

A catalogue record for this publication is available from the British Library.

ISBN 978-1-107-15090-4 Hardback

To my mother and family

Contents

Preface

Decisions are the only means you have to change your life. What we encounter in life is shaped by the decisions that we make. Whenever you make a decision, you get an outcome, and you live your life with the outcome of this decision, and with the decision that you have made. There is always uncertainty about what might happen following our decisions. We are not guaranteed the best outcome as we make good decisions, but by using the approach presented in this book, we are guaranteed a process and a logic that can guide us and help us make the right decision despite the uncertainties that prevail. We are also guaranteed defensible analysis that can help us rationalize our choice regardless of the outcome that we get.

When thinking about what might happen following a decision, it is useful to use the term "prospects" of a decision instead of "consequences." A prospect highlights that we are thinking not only about the result of a past decision but also about our future life following the decision and the outcomes of the uncertainties that have been revealed. A prospect is characterized by the decisions that we make, the outcomes of any uncertainties following the decisions, and our future life with the decisions that we have made and with their outcomes.

If there is no uncertainty about what will happen, then the prospect you get will be determined by the decision that you make. You can determine the best decision by simply ranking the prospects you face from most preferred to least preferred. Ranking the prospects is a simple task if the prospects are few, but if they are numerous, then we will need a structured way of ranking them. A function that helps us rank the prospects is known as a "preference function." The higher the score on the preference, the more preferred is the prospect.

Sometimes we care about more than just the ordering the prospects. We might care about the value of the prospect that we get, and how much we would need to be paid (or pay) to receive one prospect instead of another. We can then think about a function that associates a value with each prospect to help with this task. This function is known as a "value function."

Many of the prospects that we face in our decisions include more than one preference criterion (or attribute) that contributes to the value of the prospect. For example, in a medical decision, we might consider attributes such as health and wealth. In a job selection decision, we might consider salary, vacation days, advancement, intellectual stimulation, and the overall collegial environment. When multiple attributes are present, it is helpful to have a function that can rank the multiattribute prospects. This is known a "multiattribute preference function."

When there are multiple attributes of value in the decision, as well as uncertainty about the prospect that you will get, it becomes essential to rely on a sound decision-making process. In this case, you will need to assign a "utility" to each prospect, and then choose the alternative that has the highest expected utility. The term "utility of a prospect" is widely misused in various contexts, and we shall explain its meaning in more detail in this book. We can also think of a function that returns a utility for each prospect. We call this a "utility function."

The purpose of this book is to help you make quality decisions by thinking about your preference, value, and utility for the prospects of a decision, and particularly when prospects are characterized by multiple attributes. Characterizing preferences for prospects is one of the most fundamental steps in decision analysis. Other elements of making good decisions include choosing the appropriate frame for the decision; characterizing the information that you know through a joint probability distribution; selecting the appropriate alternatives, and having a committed decision-maker that uses sound logic to make the decision. A good reference on this general topic of decision analysis is Howard and Abbas (2015).

WHY I WROTE THIS BOOK

Many researchers and practitioners in industry and in the government face decisions with uncertainty and multiple objectives. Many of the analyses of such decisions immediately rush into assuming that there is some weighed combination of some selected metrics that needs to be constructed, and that the best decision should be determined by this weighted combination. They immediately talk about "the weights" in the decision as if every decision requires some multi-objective function of the form of some weighted combination. Of course, this is not the case and this message will be propagated throughout this book. In fact, weighted combinations of metrics could be very remote from the decision-maker's preferences, and the resulting score based on those chosen metrics could lead to poor decision-making. One main motivation for writing this book was to provide a proper understanding of the types of objective functions that should be used in multiattribute decision-making, and to highlight that these functions need not necessarily be some weighted combination. Another motivation was to highlight the arbitrariness in several existing approaches, and to explain how to determine the errors that might result when using arbitrary forms of objective functions in multiattribute decisions.

I have also found that people confuse the terms "preference," "value," and "utility," and so I felt that it is important to have a reference that clarifies the difference between these terms, to remove ambiguity about their use in the field.

When writing this book, I also wanted to highlight that many engineering and physical phenomena should be incorporated in the construction of preference, value, and utility functions, using a "structural model." The structural model often relies on physical principles and domain knowledge, and it is not merely about assessing weights and univariate functions for the attributes. To illustrate, think about the trade-offs between widths of pillars in a bridge. If you want to reduce the width of one pillar at the expense of increasing the width of another, then there are structural and engineering principles that should be used to determine this new design. Simply assessing weights and preference functions over the diameters of the pillars and then combining them into some arbitrary combination is a recipe for disaster. Capturing physical and engineering phenomena into the construction of preference, value, and utility functions has often been overlooked in much of the literature, particularly in engineering design.

I included a section titled "How People Get It Wrong" in many of the chapter appendices to explain some common misuses of utility theory. I have deliberately avoided citations to such misuses. The purpose is to present this message to as broad an audience as possible, and to influence the way many applications are modelled without causing anybody to feel defensive about their work, or about learning a different approach. I also included a section titled "Dialogue between Instructor and Student" in numerous chapters to answer some of the misconceptions people have about the field of multiattribute utility in a less formal way. These conversations also help explain some of the mathematical portions of the book.

While writing this book, I interacted with many people in leadership positions in the federal government and at large enterprises. I realized that in order to have an impact on the modeling of their decisions, there is a need for a reference on preference, value, and utility functions that can present the foundational concepts in a way that is accessible to a broad audience, and one that highlights the implications of using inappropriate or overly simplistic modeling methods. The content and exposition of this book have been designed to achieve this purpose. I kept some of the mathematical portions because I also felt it is important for organizations to realize that there is a science and a skill that needs to be learned if you wish to build models at the enterprise, public policy, or societal level, and that a simple "weight and rate" analysis will not necessarily lead to good decision-making.

When writing this book, I also wanted to highlight an important distinction that relates to many engineering phenomenon: the difference between problem solving and decision-making. Problem solving is about reacting to a problem and finding a solution, such as how to decrease the mass of an engine for space applications because of limited power. Decision-making is about making a decision even if you do not have a problem, and, more importantly, it requires you to think about preferences. At what cost does decreasing the engine mass come with? How do we decide on the best alternative? Do we need to reduce the mass or are there other decision alternatives that can be considered? This difference is essential in almost all aspects of our modern era.

Another motivation for writing this book was for students and practitioners in the field: to show them that there are many methods for constructing multiattribute functions. In the past fifty years, the field of multiattribute utility theory has often been classified by what is known as either the "Stanford Approach" using utility functions over value functions, or the "Harvard Approach" constructing multiattribute utility functions by assessing individual utility functions over the attributes and then combining them using independence assertions. Both of these approaches have had tremendous impact on the field. There are, however, many other methods that have been developed in recent years, and my purpose was to explain these new methods to a broader audience. I hope this work will provide the foundations of multiattribute utility theory for graduate students, researchers, and practitioners who wish to understand the fundamentals and the latest developments in the field. I also hope that it will inspire new research in many different directions.

REFERENCE

Howard, R. A. and A. E. Abbas. 2015. *Foundations of Decision Analysis*. Pearson, New York.

Acknowledgments

It goes without saying that I owe special gratitude to Ronald A. Howard, who has helped me get my thinking straight about the field of decision analysis since my early days of graduate school at Stanford University. Ron is one of the deepest thinkers and clear-minded people that I have ever met. Ron has influenced a lot of my learning, particularly in the first parts of this book covering structural models, value functions, and utility functions over value functions. Jim Matheson has also helped advocate the concept of utility transversality and assigning utility functions over value in some of our joint papers. I have included historic documents from the early days of decision analysis at Stanford Research Institute (SRI) to show the clarity they had about these concepts since the late 1960s. Ralph Keeney is also a special friend who has influenced my understanding of many concepts of utility independence. David Bell has also been a great collaborator and thinker about concepts of utility theory and one-switch independence.

Basic Structure of the Book

This book is divided into several parts.

PART I: INTRODUCTION

This part of the book explains the basic definitions of preference, value, and utility of a prospect of a decision, because there is much confusion about these terms. The term "utility" is used with numerous meanings. A second objective of this part is to provide the rules of decision-making under uncertainty that motivate the need for determining the preference, value, and utility of the prospects of decision. This section also presents flawed methods of decision-making that are widely used to explain why it is important to follow a rigorous approach, and that simplicity is not an excuse for using a bad decision-making method.

PART II: DECISIONS WITH NO UNCERTAINTY

This part of the book focuses on ordering deterministic prospects of a decision and then assigning a value measure to them. In deterministic decisions, the preference and value of a prospect determine the best decision alternative because each alternative corresponds to only one prospect. This part highlights that sometimes there is no need for a model before ordering the prospects, and that the mere visualization of the prospects could be sufficient. In other cases, particularly when the prospects are numerous and are characterized by multiple measurable attributes, a preference or a value function can be assigned. The part distinguishes between prospects characterized by a single attribute, such as money, and prospects characterized by multiple attributes, such as money and health. This part also emphasizes that quite often engineering or accounting principles are needed to determine preference and value functions, and that problems might arise if they are constructed arbitrarily.

PART III: DECISIONS WITH UNCERTAINTY USING VALUE MEASURES

This part of the book introduces utility functions that are needed when uncertainty is present. It first describes how to assess a utility function over a value measure using simple assessments, and then shows how multiattribute utility functions can be constructed by assigning a one-dimensional utility function over the value function or over an attribute of the preference function.

PART IV: PROPERTIES OF SINGLE-ATTRIBUTE UTILITY FUNCTIONS OVER VALUE MEASURES

Because of the important role that single-attribute utility functions play in the construction of multiattribute utility functions (both in assigning a utility function over value or by combining individual single-attribute utility functions to construct a multiattribute utility function), this part provides a rigorous treatment of the properties and implications of single-attribute utility functions. This section uses functional equations to derive exciting new formulations in utility theory, but no prior knowledge of functional equations is required. Readers who are unfamiliar with the concept will find a detailed explanation throughout the part, as well as an additional Appendix at the end of the book.

PART V: CONSTRUCTING MULTIATTRIBUTE UTILITY FUNCTIONS WITHOUT PREFERENCE OR VALUE FUNCTIONS

This part of the book presents methods to construct multiattribute utility functions without preference or value functions. A general expansion theorem of multiattribute utility functions in terms of conditional utility assessments is presented, as well as methods to simplify the expansion such as attribute dominance, utility independence, boundary independence, and interpolation independence conditions. New concepts of multiattribute utility functions and graphical representations of utility functions are also introduced.

PART VI: CONSTRUCTING MULTIATTRIBUTE UTILITY FUNCTIONS USING COPULA STRUCTURES

This part of the book presents methods to construct multiattribute utility functions using single-attribute utility assessments and new constructs in utility theory called utility copula functions. The main idea is to construct a utility surface that matches conditional utility assessments made at the boundary values of the domain of the attributes.

How to Use This Book

ENGINEERING COURSE ON SYSTEM DESIGN

If you are interested in a course on systems design using utility theory that may be applicable to undergraduate or first-year graduate engineering courses, then appropriate chapters would be those of Parts I, II, and III, in which the focus is on explaining a rigorous approach to decision-making, identifying flaws in some widely used methods of decision-making, using structural models, and identifying the value of various engineering design parameters using appropriate value functions. The use of value functions is particularly important when physical or engineering connections relate the attributes of the problem.

PRACTITIONER INTERESTED IN ADVANCES IN MULTIATTRIBUTE UTILITY THEORY

If you are a practitioner interested in advanced models for constructing multiattribute utility functions, then Parts I, II, III, V, and VI can be relevant, in which new approaches such as expansion theorems and graphical representations are presented.

GRADUATE CLASS IN UTILITY THEORY

If you are interested in a course on utility theory that provides advanced topics on properties of utility functions and their characterizations, then the entire book will be relevant. Part IV deals with a rigorous treatment of the mathematical foundations of single-attribute utility functions, which may appeal to an advanced audience. Parts V and VI present new approaches for constructing multiattribute utility functions and various independence conditions.

PART I

FOUNDATIONS OF PREFERENCE, VALUE, AND UTILITY

The purpose of this part is to explain the meaning of three concepts – preference, value, and utility of a prospect of a decision – and to illustrate their use in decision-making. This part also explains the rationale for using the expected utility criterion as a basis for sound decision-making. It also discusses other methods of decision-making that are motivated by their simplicity but can lead to errors in decision-making.

CHAPTER 1

Preference, Value, and Utility

Chapter Concepts

- Decision
- Decision alternative
- Prospect of a decision
- A prospect is deterministic; there is no uncertainty about what it entails.
- When uncertainty is present, a decision alternative may have multiple prospects
- Preference, value, and utility of a prospect

1.1 CHARACTERIZING PROSPECTS OF A DECISION

We start with the main building blocks of a decision, and in so doing we distinguish between a decision, a decision alternative, and a prospect of a decision.

Decision vs. Decision Alternative vs. Prospect of a Decision

> **Definition**
> **Decision:** A decision is a choice between two or more alternatives that involves an irrevocable allocation of resources. The term "irrevocable" implies that you will not be in the same state after making the decision.
>
> **Decision Alternative:** A decision alternative is one of possible feasible actions that you can do in a decision you are facing. You must own your alternatives. They must be feasible.

Whenever you make a decision, you face the consequences of the decision that you have made. If you choose to wear blue jeans to a black-tie cocktail party, then you might very well end up with the consequence of being the only person in blue jeans in the party. Think about the consequence of this decision. Imagine yourself in this situation.

There is always uncertainty about what might happen following our decisions. For example, there is a possibility that people will appreciate your courage for showing up this way, and there is also a possibility that they will not let you in through the front door of the cocktail party in the first place. You might then miss the party if you make this "blue jeans" choice, or you might choose to go to purchase appropriate clothes and come back to attend the remainder of the party.

Decisions are the only means you have to affect your future. From here on we shall use the term "**prospect**" of a decision instead of "consequence" of a decision because it highlights the importance of thinking not only about consequences of past decisions but also about your future life following the decisions that you have made.

Definition
Prospect of a Decision: A prospect of a decision is one of the possible (deterministic) states of the world that might occur following a decision alternative (or a sequence of decision alternatives) that you have chosen.

EXAMPLE 1.1 Uncertainty Exists Even in Simple Decisions

The prospects we encounter in life are shaped by the decisions that we make. Think of the simple decision of either cooking dinner at home or going out to watch a movie and eating at a restaurant.

If you decide to stay at home and cook dinner, then one <u>possible</u> prospect is "eating a nice dinner at home." Taking uncertainty into account, can you think of other possible prospects for this decision alternative?

If you choose to go out to watch a movie and eat dinner at a restaurant instead, then a <u>possible</u> prospect could be "watching a movie and having a nice dinner somewhere else." Can you think of other prospects for this alternative?

Each of the alternatives discussed above is a feasible alternative, but the uncertainties that are associated with the alternative enable a number of possible prospects. This is why we say "*one possible prospect*" of this decision is "eating a nice dinner at home" instead of asserting that there will be only one prospect of "eating a nice dinner." If you choose to stay at home and cook dinner, there is a possibility that the food will burn, and so you will have another decision to make: cook another meal, go out for dinner, order pizza, or just forget about dinner and munch on some snacks.

If you choose to go to the movie theater, then there is an uncertainty about whether or not you will enjoy the movie. If you do not enjoy it, then you might have another decision to make about whether you will leave in the middle of the movie or stay till the end. You might also have an uncertainty about whether you will be sitting in a good seat (with a good view? With a noisy neighbor?) … etc.

The uncertainties about this simple decision continue to grow. At the movie theater, you might bump into an old friend while parking, and so you might have another decision to make: whether to stay at the movie theater or to go somewhere else with your friend where you can talk and catch up on old times.

Uncertainties continue to surround us as we make decisions. The art of modeling decisions requires capturing the important uncertainties that can affect the alternative that you choose and the important aspects of your preferences.

If there is no *uncertainty* in the decision, then the alternative you choose determines the prospect that you will get. When uncertainty is present, each alternative can result in a number of possible prospects.

Note: While there might be uncertainty about which prospect you will get, following a decision, each of these prospects is deterministic: It should be described clearly before the decision is made, and so there is no uncertainty about what a prospect entails. You just don't know which one you will get.

Characterizing a Prospect of a Decision

The characterization of each prospect requires us to think about:

1. The decision(s) that we make;
2. The outcomes of any uncertainties that may characterize this prospect, and
3. Our future life with this prospect having made those decisions and having received their corresponding outcomes.

Because a prospect is deterministic, you should be able to visualize your life with this prospect to help you better characterize your preferences for it. If you are uncertain about a particular aspect of a prospect, such as whether or not you will like your colleagues in a new job, then you should further divide the prospect of a new job into additional prospects. For example, you might consider two prospects: one where you are in your new job and you like your colleagues, and another where you are in your new job and you do not like your colleagues. Each of these prospects should be deterministic. If they are not, then you can continue to create more prospects by thinking about the different aspects of the prospects of the job.

The main philosophy of this book is to help you think about the best decision alternative (even when uncertainty is present) by reducing the problem into thinking about the deterministic prospects of the decision, your preferences for these prospects, and their likelihood. The bulk of our discussion, therefore, will focus on preferences for prospects, while keeping in mind that the ultimate goal of this discussion is to use these prospects to determine the best decision alternative under uncertainty. We shall discuss how to use the preference, value, and utility of a prospect to determine the best decision alternative when there is uncertainty. We shall not discuss probability-encoding techniques or other aspects of a decision such as framing or generating alternatives, but refer the reader to Howard and Abbas (2015) for a detailed discussion on characterizing the uncertainty about the decision, as well as framing of a decision and other elements and applications of the foundations of decision analysis.

Note: As we continue with our endeavor throughout this book, we must remember that we are not guaranteed to get our most preferred prospect using decision analysis, but we are guaranteed a process and a logic that can guide us and help us make the right decision even when uncertainty prevails.

This observation often leads people to question the validity of decision analysis, saying things like "Well, if I am not guaranteed a good outcome, then why should I use it?" As a result many people might use arbitrary methods of decision-making justifying them by their simplicity. There is no method of decision-making that will guarantee you a good outcome when faced with uncertainty that is out of your control. Everybody wants a good outcome, but using an arbitrary method of decision-making is not the way to get there. It is because you are not guaranteed a good outcome that you need to rely on a sound decision-making method, especially if you are asked to justify your decision in the case of a bad outcome.

Before we can think about the best decision alternative, it is important to characterize the prospects of the decision alternatives that we are facing. The following examples illustrate how to characterize such prospects.

Example 1.2 Characterizing Prospects of Purchasing a Lottery Ticket

The characterization of the prospects of a decision to purchase a lottery ticket may include the decision to purchase the ticket (and the price that you will pay for it) in addition to the uncertainty about whether or not you will win the lottery. At first, you might consider only three prospects:

> Prospect A: (Buy the lottery ticket, Win the lottery)
> Prospect B: (Buy the lottery ticket, Do not win the lottery)
> Prospect C: (Do not buy the lottery ticket)

Figure 1.1. depicts these prospects graphically using a decision tree. A square (or rectangular) node represents a decision among alternatives and a circle (or oval) represents an uncertainty.

While these three prospects might be sufficient to provide clarity about this decision for some people, others might wish to further characterize them before they can decide. Winning the lottery might affect your lifestyle, the relations that you have with friends and family, as well as many other factors that you might not have thought about at first. If you have uncertainty about how the prospect of winning the lottery could affect your future life, and if this characterization can affect your decision, then you should create distinctions to help you think about each possible scenario.

The prospect of buying the lottery ticket and winning the lottery can be further divided into two prospects:

> (Buying the lottery ticket, winning the lottery, Friendships remain the same)
> (Buying the lottery ticket, winning the lottery, Friends only care about my money)

Now we have a new characterization of the decision situation as shown in Figure 1.2.

This process can go on. For example, you might add another distinction about whether you will enjoy the publicity or not, as well as many other distinctions. The process ends when you have sufficient clarity to help you make a decision visualize your life with these prospects without the need for further classification.

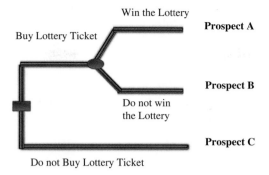

Figure 1.1. A decision tree characterizing three prospects of a lottery ticket purchase decision. The square node represents a decision, and the circle node represents an uncertainty.

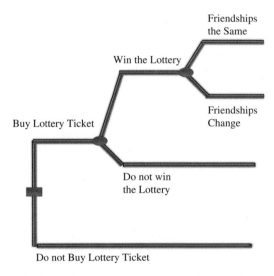

Figure 1.2. Further characterization of the prospects of the lottery ticket purchase decision.

We have not defined what we mean by preference, value, and utility of a deterministic prospect, so let us define them in more detail.

1.2 PREFERENCE FOR A DETERMINISTIC PROSPECT

Let us now clarify what we mean by a preference statement.

Definition
Preference for deterministic prospects: Preference is a statement about the order of a set of deterministic prospects. This preference order is determined by the decision-maker. Making a preference statement for a set of prospects requires you to order

the prospects in a list (with ties allowed) such that the higher the prospect is in the list, the more preferred it is. The preference statement is also referred to as an ordinal rank of the prospects.

For example, you might be considering three different houses for purchase and you consider several factors such as the view, the cost, the size of the house, and the location. After some thought, you might then say:

> *I prefer to purchase house A to house B to house C to not purchasing any of these houses.*

This statement takes into account the whole purchase prospect for each house and your future life when you own it. In other words, it might express something like "given what I have to pay for the house; the view I will get; my use of the house, and the convenience of the location and the commute time, taking all that into account, I choose to purchase House *A* over House *B* over House *C*."

It will be convenient to use the notation "≻" to make a preference statement. For example, we can write:

> *House A≻House B≻House C≻Not purchasing any of these houses.*

Note: If there is no uncertainty about the decision or the prospect that you will get, then this statement is sufficient to help you determine the best decision alternative: you simply purchase House A.

We have not discussed how you come up with a preference statement. We shall provide more discussion on making such preference statements in future chapters. If you give this preference statement to an agent, he can make this purchase decision on your behalf if the price of each house remains the same. But sometimes we need more than just preference statements. We cannot tell from this preference statement, for example, by how much you prefer to purchase House *A* to House *C*. If the owner of a house, say House *B*, offers a major reduction in price, then the agent would not know whether you would still prefer House *A* to House *B* at this discounted price. This is why preference statements alone are not always sufficient, and where the need for a value statement comes into play.

1.3 VALUE OF A DETERMINISTIC PROSPECT

Having defined the preference for a set of prospects, and illustrated some of the limitations of having a preference statement alone, particularly when monetary purchases are made, let us now define the value of a prospect. First, we recall that alternatives under consideration must be feasible. Therefore, you own the alternatives under consideration, and a prospect is one of possible states of the world that you could own following your decision.

> **Definition**
> **Value of a deterministic prospect:** The value of a deterministic prospect is the amount that makes you just indifferent to giving up that prospect if you owned it. The value of a prospect is the personal indifference selling price of the prospect.

A value statement for a prospect requires some value measure that is defined on an absolute scale. The value statement is usually expressed in dollars, but it can also be expressed in terms of any other value measure where more is strictly preferred to less.

> **Note: Value of an Item within the Prospect:** Sometimes an outcome of a decision could be getting an item, such as a free ticket to an opera. When we talk about the personal indifference selling price, we may simply talk about the personal indifference selling price of the ticket. By that we mean the price that makes you just indifferent to the prospect of having the ticket with your current life, and the prospect of not having the ticket with your current life, in addition to some monetary amount that is equal to the personal indifference selling price of the ticket. The personal indifference buying price of the ticket is the amount of money that makes you just indifferent to buying back the ticket given your current life situation.
> Around a cycle of ownership (where you buy and sell an item at your indifference buying and selling prices), the personal indifference buying price is equal to the personal indifference selling price of a prospect.

An example of a value statement for an item is

I value prospect A, of owning a mountain bike at my current life situation, by $800.

This does not mean that you value the prospect of your life, health, and wealth with the bike by $800. It means that if you made a decision, and you ended up with prospect A of having the mountain bike at your current state, then you would be just indifferent to keeping it or giving it up and receiving $800 with all else being the same except that you have no bike. Furthermore, if you sold the bike at $800, then you would be just indifferent to buying it back immediately for $800. This is referred to as the cycle of ownership.

Another example of a value statement for a house is

I value house A by $1 million, house B by $600,000, and house C by $400,000.

This statement is the value of each house to you at your current state. It means that if you owned house A, then you are just indifferent to keeping it or selling it for $1 million. It also means that if you sold house A for $ 1 million, then you are just indifferent to

1. Purchasing house A back and having $ 1 million less in your bank account, or
2. Not purchasing house A and keeping the $1 million.

The value statement also determines the amount by which you prefer one prospect over another. For example, this value statement also means that if you owned House B, then you would be just indifferent to paying an extra $400,000 and giving up the house in exchange for getting House A. Similarly, if you get House A for $1 million, then you would be indifferent to getting House B and receiving $400,000 (the difference). Value statements can of course change with time, wealth, or information. For example, you might learn that House A is in a flood zone and so you will need to evacuate it frequently or buy excessive flood insurance. Your value for House A might then decrease.

Note: A value statement is stronger than a preference statement, because you can infer the preference order of the prospects from their value: The higher the value, the more preferred is the prospect. But, as we discussed, you cannot infer value from a preference statement. You can merely assert that one has a higher value than another from a preference statement, and if two prospects are equally preferred, then they must have the same value.

With a value statement, the agent can now answer on your behalf deterministic questions about the house for a given price. For example, if the seller offers House A for $1 million and House B for $600,000, the agent will know that you are just indifferent between the two deals. But if the seller offers House B for $400,000, then the agent will know to purchase House B on your behalf. The situation is now clear for the agent, provided there is no uncertainty.

1.4 UTILITY OF A DETERMINISTIC PROSPECT

When uncertainty is present, value statements might not be sufficient to determine the best decision. For example, suppose that the agent pays $1 million on your behalf to get House A, and then the seller offers him an exotic deal:

"For $1 million, you can either keep House A, or you can get an investment."

With the sudden perplexed look on your agent's face, the seller explains the investment:

"The investment will pay either $3 million or $100,000 with equal probability."

The seller says that a decision needs to be made instantaneously. Should the agent choose to invest in this deal on your behalf or keep house A?

Figure 1.3 depicts this decision graphically.

The agent cannot make this decision on your behalf using only the value statement you have provided because he does not know your taste for risk (your risk attitude). He would like to win the $3 million to surprise you, but he is worried that he might get the $100,000 instead (which would be an unpleasant surprise). To help with this decision, we need to determine the utility values of the prospects involved.

The utility of a prospect only has a meaningful interpretation when it is expressed in terms of two other prospects, one that is more preferred and another that is less preferred than the prospect under consideration.

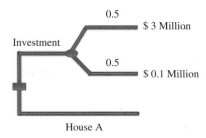

Figure 1.3. Choice between receiving House A or an investment after paying $ 1 million.

> **Definition**
> **Utility of a deterministic prospect:** Given three deterministic prospects, A, B, and C, with strict preference $A \succ B \succ C$, the utility of prospect B in terms of prospects A and C is the probability, p, that makes you indifferent between receiving prospect B for certain or receiving a binary deal that provides prospect A with probability p and prospect C with probability 1-p.

Let us refer back to the case of House A that has a value of $1 million, or the investment that pays either $3 million or $100,000. It is clear we can make the following preference statement:

> ***Receiving $3 million is preferred to receiving house A, which is preferred to receiving $100,000***

We know that because the purchaser has made a value statement that he values House A at $1 million and we assume that he prefers more money to less. So in fact we are merely saying that

> ***Receiving $3 million is preferred to receiving $1 million, which is preferred to receiving $100,000***

To express the utility of a deterministic prospect, we need to think about the notion of indifference between this prospect, and a binary deal that yields either a more preferred prospect or a less preferred one. For example, suppose we would like to express the utility of House A with respect to $3 million and $100,000. To do this we need the indifference probability, p, that makes the decision-maker indifferent between the two deals of Figure 1.4: either getting House A for certain, or getting a binary deal having a p chance of getting $3 million and (1-$p$) chance of getting $100,000. The binary deal is shown in Figure 1.4 together with the indifference probability. While the deal in Figure 3.1 shows the items of a prospect at your current state, the indifference probability must take into account your current life and the important circumstances you are currently facing in addition to the house purchase.

This indifference probability must be between zero and one because $p = 0$ implies indifference between your life with House A and your life without House A but with $100,000, which is a contradiction to the value statement of House A. Similarly, $p = 1$ implies indifference between your life with House A and your life with $3 million without house A, another contradiction to the value statement.

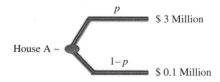

Figure 1.4. Indifference probability (utility) of House A expressed in terms of a better and worse prospect.

We may use the following notation to express the utility of the deterministic prospect of our life with House A in terms of our life with the prospects of the binary deal,

$$U(House\ A)\,|_{\$0.1\,Million}^{\$3\,Million} = p.$$

If a decision-maker states that his indifference probability is 0.7, for example, then he is implying that he is indifferent between receiving House A or trading House A for a binary deal that pays him \$3 million with a probability 0.7 and \$100,000 with a probability 0.3. We write this as

$$U(House\ A)\,|_{\$0.1\,Million}^{\$3\,Million} = 0.7$$

This indifference probability has nothing to do with the probability of getting \$3M in the investment that the agent is offering (that probability was 0.5). The indifference probability is related to the preference for the deterministic consequences and not their probabilities of occurrence.

From the previous value statements, we can also infer that

$$U(House\ B)\,|_{\$0.1\,Million}^{\$3\,Million} < 0.7$$

because the decision-maker prefers House A to House B and therefore would be willing to accept a lower chance of the \$3 million for the trade. Similarly, we can infer that

$$U(House\ C)\,|_{\$0.1\,Million}^{\$3\,Million} < U(House\ B)\,|_{\$0.1\,Million}^{\$3\,Million} < U(House\ A)\,|_{\$0.1\,Million}^{\$3\,Million} = 0.7$$

Note: The more preferred a prospect is, the higher is its indifference probability (utility) when compared to the same best and worst prospects.

We can infer preference statements from utility (indifference probability) statements. But we cannot infer value statements from indifference probability statements. Knowing that $U(House\ A)\,|_{\$0.1\,Million}^{\$3\,Million} = 0.7$ tells us that the value of House A is less than \$3 million and higher than \$100,000, but it does not tell us about the exact value of the house.

The indifference probability of 0.7 stated above for House A only makes sense for the binary deal of \$3 million and \$100,000. We do not know the decision-maker's indifference probability for House A in terms of prospects \$4 million and \$100,000, for example. We do know, however, that if $U(House\ A)\,|_{\$0.1\,Million}^{\$3\,Million} = 0.7$, then

$$U(House\ A)\,|_{\$0.1\,Million}^{\$4\,Million} < 0.7$$

This is because, if the decision-maker prefers more money to less and if he is willing to give up House A for a 0.7 chance of getting $3 million and 0.3 of getting $100,000, then he should be willing to give up House A for a lower chance in a binary deal that gives $4 million and $100,000 because it is a better deal.

If the agent knows your utility of House A, he can now answer questions about trading the house for the investment. If the investment provides a 50/50 chance of $3 million and $100,000, then he knows you will not accept the trade because you need at least a 0.7 chance of $3 million. On the other hand, if the investment pays 0.9 chance of $3 million and 0.1 chance of $100,000, then the agent will know that you would prefer the investment to the house.

Dialogue between Instructor (I) and Student (S)

I: Is everyone clear on the difference between preference, value, and utility of a prospect?

S: Yes, but let us just recap. So preference is merely a statement about the order of the prospects.

I: Correct. It also helps you determine the best decision if there is no uncertainty. We often refer to this order as "ordinal preference" or "ordinal ranking."

S: But preference alone does not help you quantify the dollar value of a prospect, or the willingness to pay to give up one prospect for a better one.

I: Correct. You need value statements for that. Value statements provide a monetary (dollar) equivalent for each prospect and so you can determine the dollars needed to give up one prospect for another. The dollar value represents your indifference selling price for that prospect.

S: So you can infer preference statements from value statements. Correct?

I: Correct. If the decision-maker prefers more money to less (which most people I know do), then the higher the dollar amount associated with the prospect, the more preferred is the prospect.

S: And utility is also needed in some cases?

I: Yes. When we are dealing with decisions involving uncertainty about the prospect that we will get, we need the utility of a prospect. We shall revisit this concept in future chapters.

S: Now isn't the utility of a prospect a measure of happiness?

I: Absolutely not! Utility of a prospect has a clear meaning in terms of indifference probability assessments. The higher the utility, the more preferred is a prospect, but it has no reflection on how happy or sad you are.

S: What about placing a value on a decision alternative, or a value for receiving a number of prospects, each with a given probability? Can we place a dollar value on an alternative that has uncertainty?

I: Yes, we shall use both the utility and the value of each prospect, as well as the corresponding probability of each prospect to determine a quantity called the certain equivalent. The certain equivalent places a dollar value on the alternative you face. We shall discuss the certain equivalent in detail in Chapter 13. We shall also discuss numerous methods to assign preference, value, and utility to prospects.

S: Great.

I: But you have to remember one thing as we go through this book and present several models.

S: What is that?

I: It is people, not a model, that make decisions. Models are useful constructs to help us think about our decision and generate insights into what we should do. The model does not make the decision; we do.

S: Very insightful. I shall remember that.

Figure 1.5 summarizes the preference, value, and utility interpretations for three prospects A, B, and C.

1.5 USING VALUE STATEMENTS TO HELP WITH THE UTILITY OF A DETERMINISTIC PROSPECT

Having a value statement is not required to assign the utility of a deterministic prospect. However, as we shall see in Chapter 13, using a value measure has several benefits for calculating the certain equivalent (value) of an uncertain deal you face. Using a value measure can also be used to calculate the value of information on an uncertainty. As we shall also see later in this book, a value measure can be very useful in determining the utility of a deterministic consequence particularly when it is characterized by multiple attributes that contribute directly to value. For example, suppose that House *A* was characterized by multiple attributes of location, size, cost, and view. Once we have determined that the value of House *A* is $1 million, we can think about the utility of $1 million in terms of the best and worst prospects ($3 million and $100,000) instead of thinking about the utility of House *A* that was characterized by these multiple attributes. We can then write

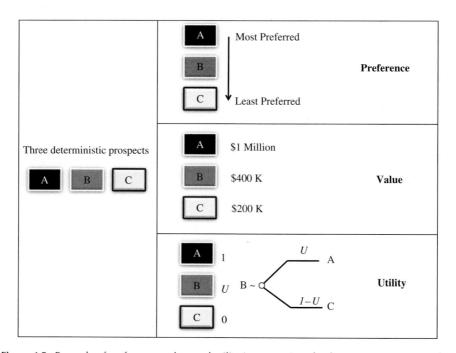

Figure 1.5. Example of preference, value, and utility interpretations for three prospects A, B, and C.

Figure 1.6. Two equivalent ways of thinking about the utility of a prospect: the first in terms of the prospect itself, and the second in terms of its value.

$$U(\$1 Million)\,|_{\$0.1 Million}^{\$3 Million} = 0.7.$$

In fact, we do not think about House A at all when making this utility assignment once we have determined its value. Figure 1.6 shows this equivalence and the two ways of thinking about the utility value of House A in terms of $3 million and $100,000.

As we shall see, this simple observation about characterizing a prospect in terms of its value will play an important role in assigning utility values when prospects are characterized by multiple attributes: we assign utilities to the value of the prospect instead of the characterization of the prospect.

Throughout this book, we shall help you develop ways to think about your preference, value, and utility for deterministic prospects. We will show how this characterization can also help you determine the best decision even if there is uncertainty about the consequence you will get.

1.6 SUMMARY

A prospect (consequence) of a decision is one of possible states of the world that may follow one or more chosen alternatives and the revelation of any uncertainties.

Preference is a statement about deterministic prospects. If the decision situation is deterministic, the preference statement is sufficient to help you determine the best decision alternative.

The value of a prospect is needed if we wish to express our indifference price for a prospect. It is also important if we wish to determine the amount we would be indifferent to paying (or receiving) to receive a more preferred (less preferred) prospect to another.

Preference order of prospects can be deduced from the value of the prospects, but the value of the prospects cannot necessarily be deduced from the preference order.

Utility of a prospect is an expression of an indifference probability in terms of receiving a more preferred prospect and a less preferred prospect.

Utility is not needed when there is no uncertainty about the prospects. Preference and/or value statements are sufficient to help you determine the best alternative when there is no uncertainty.

Preference order of prospects can be deduced from the utility values of prospects, but the utility values of prospects cannot in general be deduced from the preference order.

Except for particular situations (such as stochastic dominance, which is out of the scope of this book), the utility value of prospects is needed when there is uncertainty.

If a prospect is assigned a value, the utility of the prospect is also equal to the utility of the value of that prospect. We can replace the prospect characterization with its value and assign a utility to that value.

Preference is an ordering relation. You do not say "My preference for this house is "this amount" on some scale of units". Rather, this would be a value statement about the house.

You can say, "My value for this house is $ 1 million," because value is an absolute scale and it is your indifference selling price for the house.

You cannot say, "My utility for this house is 0.7" without further clarifying the assessment that was made, because utility is a statement that expresses the prospect you have in terms of an indifference probability of getting a better and a worse prospect.

Our focus in this book is on the preference, value, and utility of deterministic prospects. This is important for several reasons:

1. When there is no uncertainty in the decision, the choice you make determines the prospect you will get. Therefore, the choice between two decision alternatives is determined by expressing a preference between the prospects resulting from the decision alternatives. For example, if you had to choose between receiving $10 or $5, the situation can be characterized by two alternatives (the choice of $10 or $5) and the corresponding prospects you will finally receive ($10 or $5 depending on your choice). There is no uncertainty about what you will get, and, presumably, for this range of consequences, there is no uncertainty about how each consequence will affect your future life. The choice of the decision is determined by your preference for the two prospects as characterized by their monetary amounts.

2. Even if you incorporate uncertainty into the analysis of the decision, all you need to do to determine the best decision alternative(s) is to think about your preferences, values, and utilities for the deterministic prospects, as well as their likelihood. This is the good news. So while there is always uncertainty following our decisions, the basic building blocks of making a decision under uncertainty will be the preference, value, and utility of deterministic prospects.

REFERENCES

Howard, R. A. and A. E. Abbas. 2015. *Foundations of Decision Analysis*. Prentice Hall, New York.
von Neumann, J. and O. Morgenstern. 1944. *Theory of Games and Economic Behavior*. Princeton University Press, Princeton, NJ.

ADDITIONAL READINGS

Abbas, A. E. 2010. Constructing multiattribute utility functions for decision analysis. In J. J. Hasenbein, ed. *TutORials in Operations Research*, Vol. 7. INFORMS, Hanover, MD, pp. 62–98.
Howard, R. A. 1992. In praise of the old time religion. In W. Edwards, ed. *Utility Theories: Measurements and Applications*. Kluwer, Boston, MA, pp. 27–55.

Keeney, R. and H. Raiffa. 1976. *Decisions with Multiple Objectives: Preferences and Value Tradeoffs.* John Wiley, New York. Reprinted, Cambridge University Press, New York (1993).

Matheson, J. E. and A. E. Abbas. 2005. Utility transversality: A value-based approach. *Journal of Multi-Criteria Decision Analysis* 13(5–6): 229–238.

Matheson, J. E. and R. A. Howard. 1968. An introduction to decision analysis. In R. A. Howard and J. E. Matheson, eds. *The Principles and Applications of Decision Analysis*, Vol. I. Strategic Decisions Group, Menlo Park, CA.

APPENDIX: HOW PEOPLE GET IT WRONG

Many chapters in this book will include a Section titled "How People Get it Wrong." The purpose of this section is to highlight some common misconceptions or misuses of preference, value, and utility, or the misuse of decision analysis in general. Many of the discussions in these sections have been inspired by practical situations that have occurred in personal life, academic research, as well as professional business and government settings. Some of the topics will be disguised for confidentiality, but the main message will be conveyed to illustrate some common pitfalls to the readers. While reading these sections, think of how you would have reacted if you were present in these situations, and what you might do to caution the decision-makers about the possible flaws. This section is one of the main motivations for writing this book.

1) The Terms "Preference," "Value," and "Utility" Are Widely Misused

The terms "preference," "value," and "utility" that we have described in this chapter are widely misused. In economics, the word "utility" is widely used to mean value in use of an item, and it is incorporated when there is no uncertainty in the decision. You might hear statements like

> "*My utility for this jacket is $50.*"

This is not our definition of utility. What is really meant here is that the value I get by owning and using this jacket is $50. This is also the least amount I would be willing to accept to part with it. It is a value statement, not a utility statement. This point needs to be highlighted, because people often use arbitrary scales for utility values when analyzing decisions, and this can lead to errors.

You will also hear quite frequently that

> "*Utility is a measure of happiness.*"

We did not think about any happiness or sadness measure when talking about utility. In fact, we do not even need utility when there is no uncertainty. Throughout this book, we shall stick to the meaning of utility as an indifference probability that makes you indifferent between receiving a prospect for certain or receiving a binary deal having the best and worst possible prospects with that same indifference probability of receiving the best.

2) Thinking Only in Terms of Monetary Values Related to the Prospect

We have illustrated how to characterize prospect of a decision. Future chapters will discuss thinking about the valuation of the prospects. Quite often people anchor on some monetary values associated with the prospect, and they forget about other aspects of the prospect that might contribute to its preference, value, or utility. For example, think of the lottery decision associated with winning $ 1 million. People might use the $ 1 million immediately for characterizing the prospect instead of thinking about other aspects of the prospect such as family/friend relations or lifestyle, among others. The prospect of choosing a job might be (incorrectly) characterized by salary alone without taking into account the geographic location, the nature of the job, the working environment, opportunities for advancement, and the colleagues you will be working with every day.

Similarly, gambling in a casino is often characterized by the actual winnings or losses instead of the whole prospect of entertainment associated with being at a casino and some of the entertainment value associated with the thrill and excitement you might experience. Some friends often say that gambling is a bad decision, motivated by the idea that there is only a small chance you will get your money back without thinking about the entertainment value. Regardless of whether gambling is a good or bad decision, the response to that would be: going to a movie theatre is a bad decision, because you have no chance of getting your money back.

3) Using Constructed Scales to Characterize Prospects

Quite often people use constructed scales to characterize the prospects of a decision instead of using its value or any other meaningful measure. For example, you might hear people say

"On a scale of 0 to 10, how much is the pain that you feel?"

They would then characterize the prospect of pain with the answer you provide. These types of questions are usually ambiguous for a decision-maker to answer, and the corresponding scale is not only arbitrary but also lacks the precision needed to accurately characterize the prospect.

For example, they might specify that 0 is no pain and 10 is extreme pain. But what does extreme pain really mean? And what is a 9 or an 8 on this scale? Does this scale allow for fractions or decimals such as 8.7? If not, then there might be some loss in resolution due to this construct. Furthermore, different people might have different interpretations for what an 8 or a 9 really is.

It is better to keep the original characterization of a prospect in mind when assigning preferences rather than use arbitrary constructed scales for its characterization. If a prospect is expressed in its original units such as temperature or humidity or money, it is much more meaningful than a constructed scale that allows for ambiguity among different people.

Other examples of this type of constructed scale include:

"On a scale of 0 to 5, how would you rate our service to you today?"

This a statement whose interpretation might change from one person to another. Other examples include:

> *"On a scale of 0 to 10, how was the dinner buffet?"*

Think about other situations where these types of constructed scales occur and think about how you would characterize the prospects differently.

CHAPTER 2

Foundations of Expected Utility

Chapter Concepts

- Why is the alternative with the highest expected utility the best decision?
- The Five Rules of Actional Thought
- Preferences for alternatives (uncertain deals with multiple prospects)
- Utility for uncertain deals having multiple prospects
- An attribute of a prospect.
- The Same Rules Apply for Prospects Characterized by Multiple Attributes
- Decision-Making with Uncertainty and Multiple Attributes

INTRODUCTION

Having defined in Chapter 1 what we mean by a deterministic prospect, as well as the preference, value, and utility of a prospect, we shall now demonstrate why thinking about prospects is essential in helping you determine the best decision alternative even when you are uncertain about the prospect that you will get.

The approach for making the best decision in the face of uncertainty relies on five rules. The first four rules reduce any uncertain deal into an equivalent deal having only two prospects; the best and the worst. The fifth rule helps you choose among binary deals having the exact same prospects but with different probabilities of achieving them. These five rules are all you need to make decisions under uncertainty.

2.1 THE FIVE RULES OF ACTIONAL THOUGHT: THE DECISION CALCULATOR

Dialogue between Instructor (I) and Student (S)

I: In this chapter, we are going to discuss the choice criterion for helping us decide. We will further understand the motivation for caring about the preference, value, and utility of a prospect.

Before we begin, let me first ask: suppose I told you to choose between the two deals A and B in Figure 2.1. Which would you prefer?

S: It is difficult to answer that right away.

Deal A gives a 40% chance of $30,000; a 10% chance of $5,000, and a 50% chance of $10.

Deal B gives a 20% chance of $60,000; a 30% chance of $1,000, and a 50% chance of $0.

For simplicity, if I ignore monetary amounts less than, say, $100, I can compare the deals by observing that Deal A has a 40% chance of getting $30,000 – a reasonably good chance – but only a 10% chance of $5,000. Deal B gives only a 20% chance of getting $60,000 (a large sum of money) and a 30% chance of a lower amount of $1,000. It is a trade-off between getting more money with a lower probability or less with a higher probability. There are also multiple different prospects, so I need to think about many factors in making this decision.

I: Yes, it is difficult. We have not even discussed the $10 and $0 prospects. Now suppose that, instead, I asked you to choose between the two deals C and D of Figure 2.2. Can you do that?

S: Yes. They are binary deals, having the exact same prospects $60,000 and $0, but deal D has a higher chance of getting the $60,000 than deal C.

I: Correct. How do you feel about making that choice?

S: Yes, I can do that. There is no trade-off in this case. One deal has a higher probability of the same (better) prospect. I simply choose the alternative with the higher probability of getting the better prospect. Therefore, I choose Deal D.

I: A word of caution, this only works if we have binary deals having the exact same prospects. In this case, we can definitely choose the deal with the highest probability of getting the best prospect. It is not a general rule for choosing between deals, however.

S: I cannot just choose based on the higher probability, even for binary deals that have different prospects. This is an important insight.

I: Correct. Only if the binary deals have the exact same prospects can I do that. Now suppose that I told you that I am indifferent between Deal A and Deal C, and I am also indifferent between Deal B and Deal D. Would you be able to choose the best deal for me of Figure 2.1?

S: Yes. Because you say you are indifferent between Deals A and C and between Deals B and D, you must prefer Deal B to Deal A.

I: Correct. Suppose further that I have a way of converting general deals, like the deals of Figure 2.1, into the simpler binary deals of Figure 2.2. That means I constructed the deals of Figure 2.2 in a structured way that makes me indifferent, and that I can do that for any deals. Would making decisions under uncertainty be easier?

S: Yes. I first choose between the two binary deals having the same prospects. Then I infer your preference for the more complicated deals.

I: This is precisely the right way of thinking about it.

S: But how did you come up with the binary deals in Figure 2.2 that have the same prospects and are indifferent to the deals in Figure 2.1?

I: This is the theory behind how we will choose. The theory will allow us to convert all complicated deals into simpler binary deals with the exact same prospects in such a way that we are indifferent between the complicated deals and the simple deals. Then we will make the choice based on the simple deals to determine our choice for the complicated deals.

S: This sounds very clever.

I: It is. And once we develop this criterion, we will discuss the general way of choosing between deals.

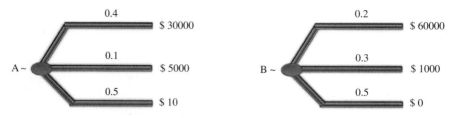

Figure 2.1. Choice between two deals A and B.

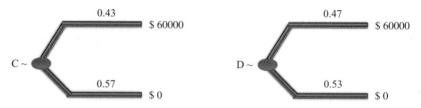

Figure 2.2. Choice between two binary deals having the same outcomes.

To understand the method of conversion from complicated deals into simple binary deals, we use the axioms (rules) used in Howard (1992) and Howard and Abbas (2015). These rules are as follows:

The Five Rules of Actional Thought for Making Any Decision

1. **The Probability Rule:**

   This rule states that for each decision alternative I face,

   (i)   *I can characterize a set of deterministic prospects that may follow; and*
   (ii) *I can assign a probability of receiving each prospect for every alternative that I face.*

   Put very generally, this rule states that we can draw a decision tree with all of its nodes and numbers. The second part of the probability rule requires the assignment of probabilities to indicate the likelihood of the prospects.

2. **The Order Rule:**

   The order rule is a preference statement about the prospects of the decision. It states that I can rank order the deterministic prospects that I have characterized in a list from the most preferred to the least preferred. Ties are allowed but preferences must be transitive, i.e., if I prefer prospect A to prospect B, and if I prefer prospect B to prospect C, then I must prefer prospect A to prospect C.

   In future chapters, we shall discuss numerous methods to help us make preference statements to order prospects of a decision.

3. **The Equivalence Rule:**

   The equivalence rule is a utility statement about the utility of prospect B in terms of the more preferred prospect A and the less preferred prospect C. It states that given three ordered (deterministic) prospects with strict preference, $A \succ B \succ C$, I can assign *a preference probability, p_B, that would make me indifferent between receiving prospect B for certain, or a binary deal having a probability p_B of receiving A and $1 - p_B$ of receiving C.*

Using the notation of Section 1.4, this rule allows us to write:

$$U(B)|_C^A = p_B.$$

The idea of assigning an indifference probability (utility) to a prospect will be discussed in numerous settings in this book.

4. **The Substitution Rule:**

This rule states that whenever I face a prospect B, for which I have stated a preference probability p_B of receiving A and $1 - p_B$ of receiving C in the Equivalence Rule, I would be indifferent between receiving this binary deal or receiving prospect B.

This rule allows us to make various substitutions of a binary deal (with prospects A and C) and the deterministic prospect, B, whenever either occurs in the decision tree.

5. **The Choice Rule:**

This rule states that if I face two binary decision alternatives, "L1" and "L2", both yielding the exact same prospects A and B (prospect A is preferred to B), and if "L1" has a higher probability of getting A, then I should choose "L1" over "L2."

The Choice rule does not make the general statement that you should choose between deals based on the alternative that has the highest probability of getting the best outcome; it requests this choice criterion only when comparing binary deals having the exact same prospects.

The five rules allow us to determine our preference order for uncertain deals by first converting the deals into binary deals having the exact same outcomes but different probabilities of achieving them. Figure 2.3 illustrates this idea.

EXAMPLE 2.1: How the Five Rules Reduce a Complex Deal into a Binary Deal

---

Figure 2.3 (a) shows an uncertain deal (decision alternative) characterized by a set of prospects, $R_1, ..., R_n$, and their corresponding probabilities, $p_1, ..., p_n$, as determined by the Probability Rule. The figure also shows the Order Rule ranking (assume that R_1 is the most preferred prospect and R_n is the least preferred).

Figure 2.3 (b) shows an example of the Equivalence Rule in which the indifference preference probability prospect R_j is given in terms of the best and worst prospects, R_1 and R_n. The utility, U_j, in the figure represents the indifference assessment $U(R_j)|_{R_n}^{R_1} = U_j$.

Figure 2.3 (c) shows the substitutions made for the prospects in terms of their utilities (preference probabilities). Following the Substitution Rule allows you to make the various substitutions and requires you to be indifferent between Figures 2.3(a) and 2.3(c).

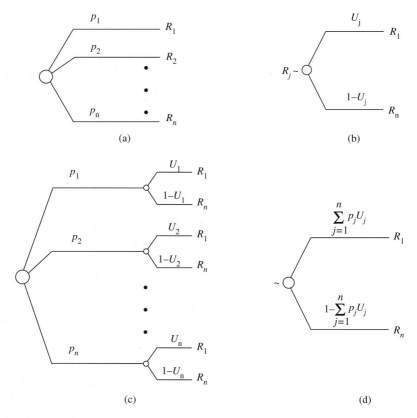

Figure 2.3. Interpretation of expected utility of a lottery as the probability of the best outcome of an equivalent lottery containing only two outcomes: the best and the worst.

Figure 2.3 (d) shows the equivalent lottery after multiplying the probabilities in Figure 2.3 (a) by the indifference probabilities (utility values). The probability of the best prospect in Figure 2.3 (d) is equal to the expected utility of the original alternative, $\sum_{j=1}^{n} p_j U_j$. A decision-maker who follows the five rules will be indifferent between receiving the lotteries of Figures 2.3(a) and 2.3 (d).

Thus *any decision alternative can be reduced into an equivalent binary deal using the Equivalence Rule utilities (preference probabilities) and the five rules.* Furthermore, *the expected utility of the original alternative is equal to the probability of getting the best prospect in the equivalent binary alternative.*

Because any other alternative can also be reduced into a binary alternative with the same prospects, R_1 and R_n (but with different probabilities of achieving them), the Choice Rule determines the best decision alternative: we choose the decision alternative (lottery) with the highest expected utility.

Definition
The Expected Utility Criterion: Following the Five Rules of Actional Thought requires us to choose the alternative with the highest expected utility of its prospects.

The expected utility calculation for any alternative in a decision requires identifying the set of possible prospects of that alternative; determining the utility of each prospect U_i of the alternative; determining the probability of getting each prospect, p_i; and then calculating the quantity

$$\text{Expected Utility} = \sum_{i=1}^{n} p_i U_i$$

for each alternative. The best decision alternative is the one with the highest expected utility.

Preference Statements for Alternatives (Uncertain Deals)

Besides determining the best decision alternative, the Five Rules also enable us to make preference statements and a complete preference ranking of the decision alternatives even if they involve multiple possible prospects. We simply determine the expected utility of each alternative. Any alternative with a higher expected utility than another must also be more preferred.

Utility of a Decision Alternative (Uncertain Deal)

The Five Rules of Actional Thought also enable us to make utility statements for alternatives (or uncertain deals having multiple prospects). We simply determine the expected utility of each deal. The utility of an alternative (uncertain deal) is its expected utility.

If the utility is normalized to range from zero to one, then the expected utility of any deal also has an interpretation in terms of indifference probability assessments. The expected utility of a deal is the indifference probability that makes us just indifferent to keeping the deal, or exchanging it for a binary deal with two prospects, the best (with a probability equal to the expected utility) and the worst (with a probability equal to one minus the expected utility).

Note: Utility values are not always needed: In some cases (such as _deterministic deals_) where we have no uncertainty in the decision, preferences statements are sufficient to help us determine what to do. The Order Rule is in fact the preference statement that is sufficient in this case. In other situations (known as deterministic dominance, where the worst you can get with one alternative is better than the best you can get with another), you still do not need utility values, because the choice is clear.

Dialogue between Instructor (I) and Student (S)

I: Now we have discussed the choice criterion for making decisions. Is it clear?

S: Yes. You choose the alternative with the highest expected utility.

I: Correct, **expected utility maximization is a direct result of the rules**. The approach requires the probability of getting each prospect, as well as the utility of each prospect (defined using the indifference probability assessment of the prospect in terms of the best and worst prospects).

S: And I know why the alternative with the highest expected utility is the choice criterion: because the expected utility of each alternative is in fact the probability of getting the best prospect in the equivalent binary deal.

I: Correct. The conversion of a decision altnernative having multiple possible prospects into an equally preferred binary deal also implies that the probability of getting the best outcome in the equivalent binary deal is the expected utility of the decision alternative. Therefore, we choose the deal with the highest expected utility. Not any arbitrary numeric measure representing some form of preferences can be used (and maximized) in this formulation. The measure whose expected value is to be maximized over the set of alternatives must be the utility values obtained by the Equivalence Rule.

S: In Chapter 1 we discussed preference, value, and utility, but you did not mention a rule about value statements for the prospects.

I: The rules do not require a value statement for each prospect. The rules do not require you to assign a value for each prospect, because the indifference probability can be expressed using the prospect itself. As we shall see, a value statement can be used to determine the value of an investment and the value of information, but it is not required by the rules. A value statement can also facilitate the construction of a multiattribute utility function significantly.

S: Are there any specific constraints on the form of the utility or preferences?

I: No major restrictions on the form or magnitude of preference or utility: The rules do not pose major restrictions on the form or magnitude of the preference or the utility of a prospect except that consistency requires that a more preferred prospect be higher on the list than a less preferred prospect and that is also have a higher utility value.

S: And now we know how to determine preference statements and utility values for uncertain deals and not just prospects. The utility of a decision alternative is its expected utility, and the higher the expected utility, the more preferred is the decision alternative.

I: Correct. In Chapter 13, we shall also discuss how to determine the value of a decision alternative.

S: Are we always going to make this conversion into binary deals whenever we make a choice?

I: No! This is an illustration that demonstrated why calculating the expected utility is needed to make the conversion. All we have to do now is to calculate the expected utility of each alternative and choose the one with the highest expected utility. You can forget about the binary deals now. It was the explanation of why we choose this way.

2.2 USING THE FIVE RULES WITH MULTIATTRIBUTE PROSPECTS: THE SAME RULES APPLY

In Chapter 1, we discussed how a prospect may be characterized by a value measure. It is often the case that prospects of a decision can be characterized by more than a single measure. A medical decision, for example, might involve health state and wealth, each presented on its own units. We might then assign a value measure that incorporates both health and wealth and assigns a value to the prospect, or we might characterize each prospect by the health and wealth levels associated with it. We refer to health and wealth in this case as two attributes characterizing the prospects of this decision.

Similarly, the purchase of a new home might involve attributes such as price, size, and distance to work, among other things. The same rules apply: we still need to characterize the prospects (now in terms of the multiple attributes) and assign a probability to them; we need to order the prospects from best to worst; we need to assign a utility value to the prospects; we need to follow the Substitution Rule; and we need to follow the Choice Rule to make the best decision.

Definition
Attribute: An attribute is a factor (or an aspect of the decision) that characterizes the preference for deterministic prospects. Multiple attributes may be used to characterize the preference of a prospect.

Attributes often have a measurable scale, such as dollars or units of time, and will be denoted by capital letters. For example, Y might be an attribute representing total wealth. An instantiation of an attribute will be represented in small letters. For example, y_1 might represent a particular wealth level.

Attributes do not need to have a numeric scale. For example, an attribute in a car purchase decision could be its color or the type of transmission involved. Quite often people confuse attributes with other factors that merely reveal information about the decision. We shall refer to this distinction in detail in Chapter 5 in our discussion of direct and indirect values.

Notation: Characterizing a Prospect with Two or More Attributes
If prospects are characterized by some levels, x and y, of two attributes, X and Y (respectively), we shall represent a prospect using the bracket notation (x, y). Similarly, if prospects are characterized by three attributes, we use the notation (x, y, z), and for more than three, we use the notation $(x_1, x_2, ..., x_n)$ to characterize a prospect with n attributes.

For simplicity of expression, we may also use the notation $(x_i, \bar{x}_i)$ to represent a prospect characterized by an instantiation x_i of attribute X_i, and an instantiation $\bar{x}_i$ of the remaining attributes (the set of complement attributes to attribute X_i). We shall refer to the complement attributes to X_i as $\bar{X}_i$.

We use the superscript "*" to denote an attribute at its most preferred (often maximum) value and the superscript "o" to denote an instantiation at its least preferred (often minimum) value. For example, we may write (x, y^*) to represent a prospect with an instantiation x of attribute X at the maximum value of attribute Y.

Note: The Five Rules do not dictate how to characterize a prospect in terms of multiple attributes. This characterization merely helps you think about your preference for the prospects when ordering them. The Five Rules are about making the best decision using a given order of the prospects, as well as the utility and probability of achieving the prospects.

2.2.1 Applying the Rules to Multiattribute Prospects without a Value Measure

The attributes in a job decision might represent salary ($100k) and vacation time (days). In principle, we can still order the prospects in accordance with the Order Rule without much difficulty. For example, if the decision-maker prefers more vacation and more salary to less, then the prospect (4,6) is preferred to either (2,6) or (4,3). But if we face two prospects, one having a higher salary but less vacation, such as (4,6) and (6,5), then the Order Rule ranking requires a bit more thought. If we only face a few multiattribute prospects, then the ordering is still relatively easy. For example, when comparing these two jobs in terms of salary and vacation (with all else constant), you think about how much salary you are willing to forgo to get more vacation. If you face a large number of prospects, however, you might want to use some tools to help you order them. Chapters 7 and 8 discuss the use of preference functions for ordering multiattribute prospects. The preference function captures the preference ordering of the prospects and maps each prospect (x, y) to some numeric value. The higher the magnitude of this preference function, the more preferred is the prospect. When there is no uncertainty, this preference function determines the best decision alternative.

The Equivalence Rule is also straightforward when multiple attributes are present. For example, consider two attributes, X and Y. Let (x^*, y^*) be the most preferred prospect that we face and let (x^o, y^o) be the least preferred. For each prospect (x, y), the Equivalence Rule implies that we can state a preference probability, p_{xy}, of receiving the best prospect (x^*, y^*) and a probability $1 - p_{xy}$ of receiving the worst prospect (x^o, y^o), that makes us indifferent to receiving (x, y) for certain (Figure 2.4).

If there is a relatively small number of prospects, then we can state a preference probability, p_{xy}, for each prospect individually by direct assessment, and the problem is solved. For salary and vacation time, as an example, we would need to assign an indifference probability of getting (2, 3) or a binary deal giving (4,6) or (0,0). Once these utility values are assigned, the best decision alternative is the one with the highest expected utility. However, if the alternatives we face have numerous prospects (such as a continuum of prospects on the domain of the multiple attributes), then this utility elicitation task might be tedious, and we might need some tools to help us assign utility values in these situations. Furthermore, this assessment requires thinking about variations in two attributes. In future chapters, we shall see how to decompose this assessment into multiple assessments that require thinking about the variations in a single attribute at each stage.

2.2.2 Applying the Rules to Multiattribute Prospects using a Value Measure

If we have made a value statement about the prospects, then we can replace the indifference assessment of Figure 2.4 with another assessment of Figure 2.5 using a deal where the prospects are replaced by their value. A value function, $V(x, y)$ provides a value measure for each prospect. The units of this measure need not be money, but money is a simple measure to assign a utility function to in comparison to other scales. It also enables the calculation of the certain equivalent and the value of information on the uncertainties of interest.

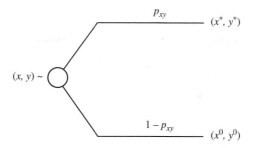

Figure 2.4. Equivalence Rule is the same with multiple attributes.

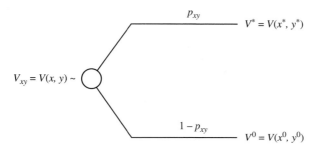

Figure 2.5. Equivalence Rule using a value function for multiple measures.

Let $V_{xy} = V(x, y)$ be the magnitude of the value function for any prospect (x, y), and let $V^* = V(x^*, y^*)$, $V^0 = V(x^0, y^0)$, be the magnitude of the value function for the most and least preferred prospects (respectively).

The preference probability for any prospect (x, y) can now be determined by simply asking the decision-maker for his indifference between the two deals in Figure 2.5. The decision-maker need not even be concerned about the actual values of X and Y. This assessment is simply a utility function (preference probability) assessment over a single measure, V in terms of the best value, V^*, and the worst, V^0.

2.3 DEALING WITH UNCERTAINTY: PROBABILITY ASSIGNMENT

The Five Rules require us to choose the alternative with the highest expected utility. The expected utility calculation for each alternative, $\sum_{i=1}^{n} p_i U(x_i)$, requires us to assign:

1. A probability to each prospect of the decision alternative, and
2. The indifference probability (utility) of the prospects in terms of the best and the worst prospects of the decision.

The distinction between the probability of getting a prospect and its utility (indifference probability in terms of the best and worst prospects) should now be clear. Our focus in this book will be on assigning utility values to prospects (as well as preference and value statements about prospects of the decision). We shall not discuss methods for assigning probabilities to the prospects, but refer the reader to Howard and Abbas (2015) for more information on probability assignment.

2.4 DEALING WITH BOTH UNCERTAINTY AND TRADE-OFFS

The choice between two medical treatments could involve two attributes such as money and health expressed in terms of equivalent healthy life years, where the decision-maker needs to think about the trade-offs between life and money. But the decision is not always that simple. There is always uncertainty about the prospects of a decision, such as prospects of medical treatments, and so each alternative might result in a number of prospects each involving different wealth levels and remaining lifetimes.

Thinking about uncertainty and trade-offs among multiple objectives without a coherent procedure requires too much cognitive effort from the decision-maker. Fortunately, we can think about each of these aspects in isolation. We can first ignore the uncertainty and think about our deterministic trade-offs between money and remaining lifetime (as will be discussed in many of the future chapters). This will entail making preference and value statements about prospects with different wealth and remaining life time levels. Next, we think about the uncertainties in the decision and assign probability and utility values to the prospects.

The privilege of having to consider these two aspects separately (uncertainty and deterministic trade-offs) is one of the main benefits of using decision analysis. You can think about deterministic trade-offs and then think about utility values without having to consider both simultaneously.

The idea of separating the deterministic analysis that includes preferences and value statements from the uncertainty phase of the analysis goes back at least to the early stages of decision analysis (Matheson and Howard 1968, Howard and Abbas 2015). To emphasize this separation, we shall organize some future chapters as follows:

Chapters 6, 7, 8, and 9 discuss preference and value statements for deterministic decisions. The idea is to think about deterministic trade-offs and values without considering and uncertainty about the prospects.

Chapter 10 discusses utility values for prospects characterized by univariate value measures to reason about decisions with uncertainty for a single attribute.

Chapters 11 and 12 illustrate how to construct multiattribute utility functions, thereby allowing for decisions with both uncertainty and multiple attributes, using a univariate utility function over value.

2.5 SUMMARY

The expected utility choice criterion is based on the Five Rules of Actional Thought.

The rules do not require a value statement, but as we shall see, a value measure will help when constructing multiattribute utility functions and will help with the valuation of uncertain deals.

An attribute is one of several possible factors that determine the preference order for the prospects of a decision.

In decisions with multiple attributes, it is possible to replace the multiattribute characterization of a prospect with its value when making indifference probability assessments.

In decisions with uncertainty and multiple attributes, it is possible to first think about the preferences and values deterministically over the multiple attributes and then think about the uncertainty using utility values over the value measure.

REFERENCES AND ADDITIONAL READINGS

Abbas, A. E. 2010. Constructing multiattribute utility functions for decision analysis. In J. J. Hasenbein, ed. *TutORials in Operations Research,* Vol. 7. INFORMS, Hanover, MD, pp. 62–98.

Howard, R. A. 1992. In praise of the old time religion. In W. Edwards, ed. *Utility Theories: Measurements and Applications*. Kluwer, Boston, pp. 27–55.

Howard, R. A. and A. E. Abbas. 2015. *Foundations of Decision Analysis*. Prentice Hall, New York.

Matheson, J. E and A. E. Abbas. 2005. Utility transversality: A value-based approach. *Journal of Multi-Criteria Decision Analysis* 13(5–6): 229–238.

Matheson, J. E. and R. A. Howard. 1968. An introduction to decision analysis. In R. A. Howard and J. E. Matheson, eds. *The Principles and Applications of Decision Analysis*, Vol. I. Strategic Decisions Group, Menlo Park, CA.

von Neumann, J. and O. Morgenstern. 1944. *Theory of Games and Economic Behavior*. Princeton University Press, Princeton, NJ.

APPENDIX 2A MISCONCEPTIONS, MISINTERPRETATIONS, AND MISUSES OF THE FIVE RULES

Many people have provided various opinions about why the Five Rules (or the expected utility criterion) should not be used as a basis for making decisions. Most of these opinions are based on either misinterpretations about of the motivation for the rules or misunderstandings of what the rules imply.

We discuss some of these opinions below:

Recurring Argument # 1:
People do not follow the Five Rules in their decision-making, so why are we using them as a basis for decision-making in our models?

Expected utility theory is a calculator for making decisions. It is not meant to mimic or predict how people make decisions. It is meant to help them make decisions, and this is why we need it.

The Calculator Analogy

Take the analogy of a calculator. If you give an average person a problem in long division, say you ask them to divide 25.67/13.22 without using a calculator, most people will not get the answer right to the full range of decimals involved. People may not give the same answer to this question as does the calculator, and different individuals will probably give different answers. This does not mean that we should change the laws of long division to mimic how people respond to this question. Calculators do not predict or mimic how people answer arithmetical problems. We are okay with that, and we observe that this is precisely why we need calculators when we face long-division problems. Calculators provide the logic needed to determine an answer. Similarly, because people deviate from the recommendations of the Five Rules in their daily lives, we need to rely on a decision calculator (The Five Rules) to make important decisions. The fact that people do not follow the rules is the motivation for why we need this theory of decision-making. The role of the decision analyst is to help decision-makers provide the inputs to the calculator and to help with the calculations, not to propose arbitrary methods of decision-making that are motivated by simplicity.

Recurring Argument # 2:
Expected utility theory assumes that all people are rational.

The theory does not assume that people are rational. The theory provides a means to help you make a rational decision. The theory helps you break down the problem into a simpler problem that you can reason about. For example, the theory enables the decision-maker to think about beliefs about the likelihood of the prospects independently from the preference/value/utility of the prospects, and provides a mechanism to combine the different elements to help you make the best decision.

Recurring Argument # 3:
Because expected utility theory is not a good predictor of how people make decisions, it should be augmented to predict how people behave.

The response to this misconception is clear: expected utility is not meant to predict how people behave. Expected utility is the calculator that helps you make decisions when faced with uncertainty.

Indeed, other theories are needed if you would like to mimic what people do and how people actually make choices. For example, observing that people are more likely to purchase chewing gum if it is placed near the counter as you leave the store can be used with normative decision analysis. The expected utility criterion would use this information to update the probability of gum purchase and would recommend that you place the gum near the counter. This recommendation incorporates knowledge of human behavior into the expected utility theory. It uses the expected utility criterion, not a modified version of it. As long as you have this concept clear, there is no problem with having other theories that predict behavior. We do not augment the rules of arithmetic because they are bad predictors of how people might respond.

Indeed, it has been proven that people are prone to many cognitive biases that might make the inputs to the decision calculator difficult. For example, when reasoning about small probabilities, people to tend to underestimate small probabilities and overestimate large probabilities. But the role of behavioral research is to identify these cognitive biases and to conduct the assessments in a way that minimizes their effects. The theory does not need to be modified. Rather, behavioral research should help provide inputs that better match beliefs and preferences of the decision-maker. Multiple researchers have attempted to modify the rules themselves to better mimic how people behave and have criticized the expected utility criterion because it is not a good predictor. The resulting modified theory would no longer be a calculator of choice, but a method to predict behavior. We shall see many examples in the following chapter of how this might lead to errors in decision-making (such as the min-max regret criterion).

Recurring Argument # 4:
Expected utility is complicated and it requires probability calculations

True, expected utility theory is not simple, and this is why it should be learned and taught in academic and professional settings. Instead, people often use arbitrary and incorrect methods of decision-making that are simpler. But *simplicity is not an excuse for using a wrong approach*. The next chapter presents some examples of simple, but flawed, methods of decision-making to further illustrate this point.

Recurring Argument # 5:
If I am not guaranteed a good outcome using expected utility theory, how do I know that I have made a good analysis? How do I know I entered the correct probabilities or utility values? This is why I should rely on other (simpler) methods of decision-making even if they are arbitrary.

Let us make one thing clear: unless you have a crystal ball, no method of decision-making can guarantee the best outcome when uncertainty is present. Everybody wants a good outcome, but using a bad decision-making method is not the way to get there. There is no such thing as a true probability. The probability that you use in the analysis is the one that reflects your beliefs. Using expected utility, you at least have a rigorous method that can help you justify why you made the decision. Sensitivity analyses can also help you test the robustness of the parameters and validate the model. The role of the analyst is to help provide meaningful inputs into the model and not to use arbitrary methods for simplicity. As we shall see in the next chapter, there are many simple (and arbitrary) methods of decision-making that can lead to bad decisions (and that also do not guarantee a good outcome).

APPENDIX 2B ORIGINAL VERSION OF THE AXIOMS OF EXPECTED UTILITY

The expected utility criterion described above was first proposed by von Neumann and Morgenstern (1944), who laid out a set of axioms pertaining to lotteries: completeness, transitivity, continuity, and independence using indifference between lotteries. We describe these in more detail here, but focus our discussion in this book on the Five Rules presented in this chapter.

Completeness asserts a complete ranking of preferences for alternatives: for two decision alternatives, either one is preferred to the other or the decision-maker is indifferent between them.

Transitivity means that if alternative A is preferred to alternative B, and if alternative B is preferred to alternative C, then alternative A must be preferred to alternative C.

Continuity states that given three ordered lotteries with strict preference, $A \succ B \succ C$, I can assign a probability, p, that would make me indifferent between receiving lottery B for certain, or a deal having a probability p of receiving lottery A and 1-p of receiving lottery C.

Independence refers to the independence of preferences with respect to added alternatives, i.e., if A, B, C are three lotteries with $A \succ B$, and let $t \in [0,1]$, then $tA + (1-t)C \succ tB + (1-t)C$, and likewise if $A \sim B$, then $tA + (1-t)C \sim tB + (1-t)C$.

CHAPTER 3

Some Flawed Methods of Decision-Making

Chapter Concepts

- Approximate vs. fundamentally flawed methods of decision-making
- Simplicity is not an excuse for making a bad decision
- The concept of rank reversal
- Flaws in using pairwise comparisons and elimination to provide a total order
- Flaws in ranking the desirability of attributes independently
- Flaws in the Min-Max Regret method of decision-making

3.1 INTRODUCTION: APPROXIMATE VS. FUNDAMENTALLY FLAWED METHODS OF DECISION-MAKING

To better appreciate the expected utility criterion and the analysis that will follow in future chapters, it is important to examine other methods of decision-making that are widely used. Many people motivate the use of arbitrary methods of decision-making by many factors, some of which are:

1. Simplicity;
2. The reluctance to using probability to describe uncertainty; and
3. The reluctance to making explicit preferences and trade-offs in decisions.

An often cited argument is that making preference, value, or utility statements, and assigning probabilities to the prospects, is not an easy thing to do. You will often hear statements motivating arbitrary methods of decision-making along the lines of

"Yes, this is not a perfect method but it is approximate"

Or even statements like

"Expected utility is an academic exercise but this is the real world"

We need to make a distinction between an approximate method of decision-making and a fundamentally flawed method.

Approximate vs. Fundamentally Flawed Methods
An *approximate method* is one that is built on some rational logic – one that gives you a better answer when you put more effort into it. Examples of sound approximate methods include using Taylor expansions within regions of convergence: the more terms you include in the expansion, the closer you get to the correct answer.
A *fundamentally flawed method* of decision-making is one that does not give the right answer no matter how much effort you put into it. The method violates some reasonable logical requirements in a decision-making system.

As we shall see in the examples in this chapter, s*implicity is not an excuse for using a wrong decision method* because the consequences can be disastrous. Failing to incorporate uncertainty can result in a bad decision, and the reluctance to assert preferences, and then using an arbitrary choice criterion results not only in poor decision-making but also in preferences and trade-offs that do not match those of the decision-maker. The methods presented in this chapter and the corresponding numerical examples are disguised, but they are all based on real-life situations, and these methods are still widely used. The purpose of this chapter is to demonstrate that the mere existence of a decision-making method does not mean that it is necessarily a good decision-making method. We shall refer to other flawed methods of decision-making in various stages throughout the book. We introduce these methods at this early stage to motivate the need for a rigorous decision-making criterion: the expected utility criterion.

3.2 USING PAIRWISE COMPARISONS OF PROSPECTS AND THE METHOD OF ELIMINATION

A common method for ranking deterministic decision alternatives (or prospects of a decision) is to use pairwise comparisons between each two alternatives and then use sequential elimination to determine the best alternative.

Pairwise comparisons, if conducted appropriately, are not necessarily a problem by themselves because of transitivity: if we face prospects A, B, and C, and if we think about our preference for A vs. B alone, then we think about our preference for B vs. C, we can finally think about our preference for A vs. C. If we have transitive preferences, we may be able to arrive at a consistent order. If used incorrectly, however, pairwise comparisons can lead to what is known as *rank reversal*.

Definition
Rank reversal is the change in preference order of prospects by the removal or inclusion of new (uninformative) prospects.

Rank reversal has been one of the fundamental criticisms of the Analytic Hierarchy Process (AHP) – see, for example, Saaty (1990) and Dyer (1990). Besides rank reversal, several variations of pairwise comparisons can also result in a preference order that is counterintuitive. The following example demonstrates a flawed method of decision-making using pairwise comparisons. It is based on an example by Saari (2001).

EXAMPLE 3.1: Choosing a CEO by Pairwise Comparisons and Elimination

---

Suppose your company is deciding on who will be the next CEO. It has six candidates in mind: A, B, C, D, E, F. Figure 3.1 illustrates the six candidates being considered.

To help with the selection process, the company conducts a survey and asks employees about their preferences. The response was coherent in each of three different groups. Each group included 10 people, and the preferences were as follows.

$$Group\ 1:\ A \succ B \succ C \succ D \succ E \succ F$$
$$Group\ 2:\ B \succ C \succ D \succ E \succ F \succ A$$
$$Group\ 3:\ C \succ D \succ E \succ F \succ A \succ B$$

It is clear from the preference statements that everybody prefers Candidate C to D to E to F. This might suggest that the company would exclude candidates D, E, F immediately. But the company chooses an alternative route. They decide to do pairwise rankings and sequential elimination.

That is, they will ask for votes on pairwise rankings. They start by asking for the pairwise ranking:

Who do you prefer: Candidate D or E?

The result of this vote would be Candidate D because everybody prefers candidate D to E. So candidate E is eliminated. Next, the company asks:

Who do you prefer: Candidate C or D?

The result of this vote would be Candidate C because everybody prefers candidate C to D. So candidate D is eliminated. Next, the company asks:

Who do you prefer: Candidate B or C?

From the survey results, two groups (20 people) prefer candidate B and one group (10 people) will prefer candidate C. The result of this vote would be Candidate B so candidate C is eliminated. Next, the company asks:

Who do you prefer: Candidate A or B?

Candidate A   Candidate B   Candidate C   Candidate D   Candidate E   Candidate F

Figure 3.1. Potential candidates for the CEO job.

From the survey results, two groups (20 people) prefer candidate A and one group (10 people) prefers candidate B. The result of this vote would be Candidate A so candidate B is eliminated. Next, the company asks:

Who do you prefer: Candidate E or A?

From the survey results, two groups (20 people) prefer candidate E and one group (10 people) will prefers candidate A. The result of this vote would be Candidate E so candidate A is eliminated. Next, the company asks:

Who do you prefer: Candidate F or E?

From the survey results, two groups (20 people) prefer candidate F and one group (10 people) prefers candidate E. The result of this vote would be Candidate F so candidate E is eliminated. They are left with Candidate F.

> *Using this selection criterion, candidate F is selected to be the next CEO, even though everybody prefers Candidate C to D to E to F.*

A situation like this can occur quite often. The three groups can represent different departments in a company or groups of stakeholders having the same preferences. But why is it arbitrary? It is possible that a group might choose something that every individual does not prefer. This happens with insurance and re-insurance companies that are willing to take more risk than an individual would. The problem with Example 3.1, however, is that if the group had started by asking:

Who do you prefer: Candidate E or F?

Candidate *F* would have been eliminated at the start, and so the decision depends on the order of the questions asked. This example illustrates some of the issues that might result with arbitrary rankings when using pairwise comparisons and elimination to order prospects. It also shows the potential for tactical manipulation using an arbitrary method.

3.3 MAJORITY VOTES OVER THE INDIVIDUAL ATTRIBUTES

Sometimes prospects of a decision are described by multiple factors or attributes. For example, in a medical decision, the prospects might be described by attributes of health state and wealth. The attributes may be represented numerically on a continuous scale or categorically such as attributes of color of a vehicle or its transmission type. It is tempting, when multiple attributes are present, to ask for preferences over each attribute independently and then combine the preferences to determine the order of the prospects and the best decision. This method can result in poor decision-making.

EXAMPLE 3.2: Designing an Automobile Using Customer Feedback over the Individual Attributes

An automobile manufacturer is choosing among design alternatives for its new smart automobile (Figure 3.2). The different alternatives include engine size (expressed in

Figure 3.2. Design of a smart vehicle. © Tom Wood / Alamy Stock Photo. Results of the Market Survey.

terms of the number of cylinders; 4 or 6), vehicle size (family sedan or compact), and transmission type (manual or automatic transmission). The firm would like to conduct a market survey before making the design selection decision, and so it decides to ask customers about their preferences for each of the design attributes. It interviews three segments: the big city folk, the family sedan folk, and the sports vehicle folk. The company asks each segment to circle their preference using the following questionnaire:

> **Q1: Vehicle Size**: Which do you prefer: Family Sedan or Compact?
> **Q2: Engine Size**: Which do you prefer: 4 Cylinders or 6 Cylinders?
> **Q3: Transmission Type**: Which do you prefer: Automatic or Manual Transmission?

Group 1: The "Family Sedan" Folk

The "family sedan" folk express their preferences as follows:

- They prefer a family sedan to a compact vehicle (they will not purchase a compact car).
- They prefer six cylinders to four (but might purchase a four-cylinder engine to reduce fuel consumption).
- They prefer automatic transmission to manual transmission (but they might purchase a vehicle with manual transmission).

Group 2: The Big City "Commuter" Folk

The "big city" folk express their preferences as follows:

- They prefer a compact vehicle to a family sedan because of ease of parking (but might buy a sedan).
- They prefer a four-cylinder engine to a six-cylinder one (they will not purchase a six-cylinder engine because of fuel consumption in traffic congestion).
- They prefer automatic transmission to manual transmission (but might purchase manual transmission).

Table 3.1. Results of the Market Survey.

Vehicle Size	Engine Size	Transmission Type
1 Group Prefers Family Sedan	1 Group Prefers 4 Cylinder	2 Groups Prefer Automatic
2 Groups Prefer a Compact car	2 Groups prefer 6 Cylinder	1 Group Prefers Manual

Group 3: The Sports Folk

The sports folk express their preferences as follows:

- They prefer a compact vehicle to a family-size sedan.
- They prefer a six-cylinder engine to a four-cylinder engine.
- They prefer manual transmission to automatic transmission (they will not purchase automatic transmission because they prefer the feeling of manual control).

Given the survey results, the firm then constructs a matrix of preferences as shown in Table 3.1.

From the responses, the company observes that

- Two groups prefer a compact car and only one group prefers a family sedan, so they choose a compact vehicle design.
- Two groups prefer a six-cylinder engine and only one group prefers a four-cylinder engine, so they choose a six-cylinder engine for the design.
- Two groups prefer automatic transmission and only one group prefers manual transmission, so they go with automatic transmission.

The company decides to use majority customer vote over each attribute independently. Using this choice criterion, the company designs:

A compact vehicle with a six-cylinder engine and automatic transmission,

a vehicle that none of its surveyed customers will purchase: the "family sedan" folk will not buy it because it is compact, the "big city" folk will not buy it because it has six cylinders, and the sports folk will not buy it because it has automatic transmission.

Example 3.2 is based on a real story of how an automobile manufacturer incorporated survey data and used a choice criterion for making the final design. The product was deemed a failure for the manufacturing firm and named the worst vehicle of the year.

Note: The main idea here, as we shall see repeatedly throughout the book, is that thinking in terms of the individual attributes independently can lead to poor decisions.

To determine the best design alternatives, the firm should have determined the expected utility of each design alternative. If the firm is interested in profit, then it would calculate

$$\text{Profit} = \text{Revenue} - \text{Cost}$$

The decisions in Example 3.2 would be the different design alternatives as well as the product price. The uncertainties involved would be the demand for the product as a function of its design and price, as well as the uncertainty about the costs (manufacturing, marketing, etc). The best design decision is the one with the highest expected utility of profit.

The firm may use the market survey to update is demand distribution for each design alternative using Bayesian analysis (see, for example, Traverso and Abbas 2009, Sun and Abbas 2013). The value of the market survey in this decision can also be calculated using value of information analysis (Howard and Abbas 2015) and Bayesian updating.

3.4 THE MIN-MAX REGRET CRITERION

The min-max criterion is based on the idea of incorporating regret into the analysis of a decision and choosing the alternative that provides the lowest regret. Common forms of incorporating regret include subtracting a certain amount of money from the valuation of "bad" prospects due to the experienced regret, i.e., a prospect would be valued less because you could have obtained a better outcome had you chosen a different alternative.

The min-max criterion was discussed in Savage (1951, 1972) and it suggests choosing the decision alternative that provides the least amount of maximum regret experienced. Savage (1951) motivates this choice criterion by the fact that it does not require the explicit assignment of probability to represent uncertainty, and no explicit statements about trade-offs or risk aversion are required. Savage (1951) mentions: "In fact, it [the min-max rule] is the only rule of comparable generality proposed since Bayes' was published."

This type of modeling the experienced regret might provide descriptive representations that mimic how people actually behave in the world, and they might even be used as predictors of how some (unwise) people might behave. But, as we shall see, incorporating regret into the valuation of outcomes and then using this "min-max" criterion can lead to errors in decision-making.

The following (disguised) example uses the famous Party Problem (Howard and Abbas 2015) to illustrate how using the "min-max regret" criterion may lead to what is known as rank reversal. The example extends directly to any other types of decisions.

EXAMPLE 3.3: Rank Reversal with Min-Max Regret

Joan is interested in having a party. She has three alternatives from which to choose: Indoors, Outdoors, and the Porch. She is uncertain about the weather, which could be Sun or Rain. Joan faces six prospects and assigns a dollar value for each prospect as follows:

- Outdoors – Sun has a value of $100
- Outdoors – Rain has a value of $0

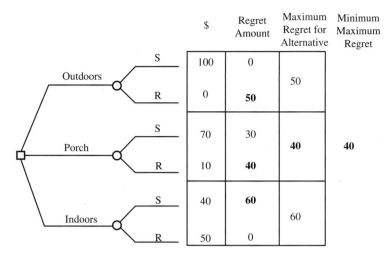

	$	Regret Amount	Maximum Regret for Alternative	Minimum Maximum Regret
Outdoors S	100	0		
			50	
Outdoors R	0	**50**		
Porch S	70	30		
			40	**40**
Porch R	10	**40**		
Indoors S	40	**60**		
			60	
Indoors R	50	0		

Porch > Outdoors > Indoors

Figure 3.3. Using the Min-Max Regret Criterion with Three Alternatives.

- Porch – Sun has a value of $70
- Porch – Rain has a value of $10
- Indoors – Sun has a value of $40
- Indoors – Rain has a value of $50

Figure 3.3 shows the decision tree for these prospects. The tree does not include any probabilities or utility values for the monetary amounts.

The regret amount for each prospect would then consider the difference between:

a) The value of any prospect (for a chosen alternative and a given outcome for the uncertainty) and

b) The best prospect that could have been materialized (with the same outcome of the uncertainty but with any other alternative).

For example, the Outdoor-Rain prospect has an outcome of the weather uncertainty being "Rain" and a chosen alternative "Outdoors." Its value is $0. This prospect would then be compared to the best prospect that can be materialized when the weather is Rain regardless of the chosen alternative. This would correspond to the prospect "Indoors-Rain," which has a value of $50, because this is the best prospect that can be materialized when the weather is Rain. The regret amount for the prospect Outdoor-Rain would then be $50 – $0 = $50.

Similarly, Indoors-Sun (whose value is $40) would be compared to Outdoor-Sun (whose value is $100), because going Outdoors provides the highest value with the outcome Sun. The difference in value between the two prospects (the regret amount) is $100 – $60 = $40. Likewise, Porch-Sun would be compared to Outdoor-Sun to give a regret amount of $30, and Porch-Rain would be compared to Indoors-Rain to give a regret amount of $40. The regret associated with Indoors-Rain is zero because going Indoors is the best you could have done if the weather is Rain, and Outdoors-Sun also has zero regret because of the same argument: going Outdoors is the best you can do when the weather is Sun.

The amount of regret for each prospect is the difference between the highest prospect value that can be materialized with that outcome of the uncertainty less the value of that prospect.

The next step determines the maximum regret that can obtained for any alternative:

- The Outdoors alternative has a maximum regret of $50.
- The Porch alternative has a maximum regret of $40.
- The Indoors alternative has a maximum regret of $60.

The final step chooses the alternative with the lowest maximum regret (hence the name min-max regret). In this example, the alternative with lowest maximum regret is the alternative Porch, which has a maximum regret amount of $40. The second-best alternative according to this criterion would be Outdoors, with a maximum regret amount of $50. The third-best alternative would be Indoors, which has the highest maximum regret, $60.

The preference order for the alternatives is

$$\text{Porch} \succ \text{Outdoors} \succ \text{Indoors}$$

So far this sounds simple, and indeed many people, including Savage (1951), advocate the use of the min-max principle. Let us now illustrate why this min-max regret principle causes a problem.

EXAMPLE 3.4 The Problem with the Min-Max Regret Principle

Suppose that Joan lost the Indoors alternative, and as a result has the decision tree shown in Figure 3.4. One would expect that her preferences would then be to have the party on the Porch and then the Outdoors alternative, because there is no reason why Porch is not the best alternative anymore if she loses the Indoors alternative (a less-preferred alternative).

But let us carry out the analysis one more time given that we have lost the Indoors alternative. Figure 3.4 illustrates the tree, the regret amounts, and the maximum regret for each alternative.

As we can see now, the alternative with the lowest maximum regret is Outdoors, which has a maximum regret of $10. The Porch alternative has a maximum regret of $30. Therefore the min-max choice criterion would recommend the alternative Outdoors, and so the ranking of the alternatives would be

$$\text{Outdoors} \succ \text{Porch}$$

Preferences have reversed as a result of removing an alternative that was not even at the top of the list of available alternatives. This is phenomenon is known as ***rank reversal***. Rank reversal is a common feature when pairwise comparisons are made (here we compared a prospect to the best prospect that could have occurred with the materialization of the uncertainty). The problem with incorporating regret is that you can no longer rank (or value) a prospect independently without considering other prospects. This is why removing an alternative may lead to a change in rank.

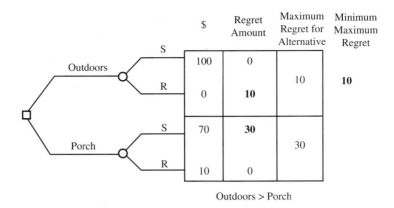

Figure 3.4. Decision tree without the Indoors alternative.

Would you want to use a choice criterion that is prone to this type of rank reversal behavior? Would you not worry about whether including additional alternatives would have changed the rank? How many alternatives do you include? What is the limit?

The Min-Max regret criterion has gained appeal because of its simplicity and because (descriptively) people may experience regret it in their lives. While this is a descriptive phenomenon, it is by no means a recommendation you would make to your beloved ones: "to experience regret" and to live life with that feeling. Rather, you would ask them to learn from their actions and to move on to make better decisions in the future. In fact, the quality of the outcome should not determine the quality of the decision that is made. Therefore, incorporating some form of regret once you have made a good decision is meaningless. To best summarize this learning, let us propose the phrase:

Do not value your life based on what was or what could have been.

Of course, the example presented here can be scaled to any type of decision with uncertainties. For example, a project manager may be considering investing in design alternatives: low upgrades, moderate upgrades, and major upgrades, each with different costs and an uncertainty, may be whether the product will be obsolete in a few years. The same type of rank reversal phenomenon may occur.

3.5 SUMMARY

This chapter discussed the difference between an approximate and a fundamentally flawed method of decision-making.

Simplicity is not an excuse for using an arbitrary method of decision-making.

The job of the analyst is not to propose an arbitrary method (because of its simplicity) but to help the decision-maker provide the appropriate inputs to the expected utility model. The focus of the remaining chapters is on how an analyst can help the decision-maker provide inputs on preference, value, and utility.

REFERENCES

Abbas, A. E. 2015. Perspectives on some widely used methods of multi-objective decision making in systems engineering. *Proceedings of the 2015 Industrial and Systems Engineering Research Conference*, S. Cetinkaya and J. K. Ryan, eds.

Dyer, J. 1990. Remarks on the analytic hierarchy process. *Management Science* 36(3): 249–258.

Howard, R. A. and A. E. Abbas. 2015. *Foundations of Decision Analysis*. Prentice Hall, New York.

Saari, D. 2001. *Decisions and Elections: Explaining the Unexpected*, Cambridge University Press, Cambridge.

Saaty, T. 1990. How to make a decision: The analytic hierarchy process. *European Journal of Operational Research* 48: 9–26.

Savage, L. 1951. The theory of statistical decision. *Journal of the American Statistical Association* 46: 55–67.

Savage, L. 1972. *The Foundations of Statistics*. Dover Publications, New York.

Sun, Z. and A. E. Abbas. 2013. Bayesian updating on price elasticity of uncertain demand. *IEEE International Systems Conference (SysCon)*, 222–228.

Traverso, M. G. and A. E. Abbas. 2009. Demand curve prediction via Bayesian probability assignment over a functional space. *Winter Simulation Conference*, 2971–2976.

ADDITIONAL READINGS

Howard, R. A. 1992. In praise of the old time religion. In W. Edwards, ed. *Utility Theories: Measurements and Applications*. Kluwer, Boston, pp. 27–55.

von Neumann, J. and O. Morgenstern. 1944. *Theory of Games and Economic Behavior*. Princeton University Press, Princeton, NJ.

PART II

MAKING DECISIONS WHEN THERE IS NO UNCERTAINTY

The purpose of this part is to discuss methods for ordering deterministic prospects of a decision as required by the Order Rule. This part presents several techniques for determining the order including visualization, as well as more technical methods of constructing preferences and value functions. It also introduces a fundamental distinction between direct and indirect values for characterizing the prospects of a decision.

CHAPTER 4   The Order Rule: Preference for Deterministic Prospects

The great road is not difficult for he who has no preferences.

Chapter Concepts

- The preference order of prospects is sufficient to help you make the best decision when there is no uncertainty
- Ordering deterministic prospects of a decision by spending sufficient time visualizing them

4.1 INTRODUCTION

In this chapter, we shall discuss methods to help you make preference statements for the prospects of a decision. We shall not discuss the value of a prospect or the utility of a prospect in this chapter because you do not need a value measure or utility to rank order deterministic prospects.

Note: Because we are talking about ordering deterministic prospects, these prospects are not characterized by any probability of receiving them. The preference order of the prospects is independent of this probability. Many formulations inappropriately include probability in the characterization of a prospect.

4.2 ORDERING PROSPECTS BY VISUALIZATION

Visualization is a particularly useful method of ordering the prospects when they are few. You need to think clearly about your life with each of the prospects that you might get. It is quite helpful to visualize the prospects of a decision before ranking them.

What makes ordering prospects difficult is that we often do not put enough effort into thinking about all the aspects involved, and we do not spend enough time thinking about our future lives with these prospects.

To visualize important prospects, it is helpful to "live your life" for a few days as if you have attained one prospect and then live your life another set of days as if you

Figure 4.1. Visualizing your current Life.

have another prospect. This visualization can help you order the prospects. Sometimes, visualizing the prospects alone is sufficient to help you decide.

Before we consider visualizing the prospects of a particular decision, take a moment to reflect on (and visualize) your own life. Think about your home, family, meals, financial resources, and work environment. Figure 4.1 highlights some aspects of a person's life. You can tailor those to your own. Can you think of other aspects of your personal life and visualize them? Think of the commute time for example. Are you happy with your life? Are there any aspects that you feel can be improved? Take your time thinking about this exercise. Disconnect from what you are doing and think from the perspective of someone who is viewing a video of your life. Are there things that you can do to improve your current life? Why aren't you doing them? Remember: decisions are the only means you have to change your life.

Now, let us consider a decision you might face and visualize its prospects.

EXAMPLE 4.1: Visualizing Prospects of Receiving a Bribe

Suppose you were approached by an agent to help him facilitate some business transactions in exchange for some monetary amount, "a bribe" for your services. What prospects can you think of? Two main prospects stand out: (1) "getting away with the bribe" and receiving the money and (2) "getting caught" and going to jail. Can you visualize them? Figure 4.2 illustrates these two prospects.

What makes visualizing the prospects powerful (besides enumerating them) is that we often do not realize what a prospect really entails. For example, the prospect of not getting caught and receiving the money might sound appealing until you think

Figure 4.2. Visualizing two prospects of accepting a bribe.

of the guilt that you might encounter for providing an unequal opportunity for one person over others. Then you might think about the ethical implications this would entail. Suddenly, the prospect might not seem as appealing as it did initially. The prospect of going to jail is important to think about too. Think of your life and the lives of those around you, including your family. Suddenly the "getting caught" prospect becomes more severe. The list of prospects in this decision is numerous: the agent who is approaching you might be an FBI agent who is testing you; you do not get caught but you do not receive the money, you get caught and confess and tell on the agent so you get a minimal penalty, etc. The point is that besides enumerating the prospects, visualizing the prospects provides a deeper perspective about them, and helps with the preference order.

EXAMPLE 4.2: Visualizing the Prospects of a New Job

This particular situation arose with a friend at Stanford who received a job offer from another institution. To give some perspective on this situation, suppose you lived in Palo Alto and had a family that resides there. You also have a good job at a reputable school. You are then offered a very high-ranking job at a university on the East Coast. The offer provides a much higher salary. It would take you at least ten to fifteen years in your current job to get to the salary level being offered (if at all). You also know that it will be difficult for your whole family to move with you because of school allocations, but the East Coast school is open to the idea of your spending three days a week on the East Coast and the rest of the week on the West Coast with your family. You are clear about the job prospects because you have obtained sufficient information to resolve the uncertainties about both jobs. Should you accept the job offer?

To think clearly about both prospects, it is helpful to visualize them. To do that, you might live your life for a few days as if you had accepted the offer. Think of the travel

Figure 4.3(a) Visualizing the prospect of accepting a new job with higher salary, more work, and family relations do not suffer.

Figure 4.3(b) Visualizing the prospect of accepting a new job with higher salary, more work, while family relations do suffer.

that would entail each week, the time and effort that would take; the time spent away from family a few days a week, and also the type of responsibilities that the new job would entail. Then live your life as if you had turned down the job offer. Think about how you would feel when you are back at your current job with fewer responsibilities and lower pay but spending more time with family and less time on the road. Visualize these prospects. After living your life both ways, you should have a clearer picture about which prospect you like better.

If there is any uncertainty about the prospects (such as whether family relations will suffer, or whether you will like your colleagues in the new place), you just create this new distinction and now you have three (or more) prospects: your current job, the new job where you will like your colleagues, and the new job where you will not like your colleagues.

Figure 4.3 illustrates three prospects for this decision. Figure 4.3(a) illustrates wealth from the new job, some occupation at work, and family relations do not suffer. Figure 4.3(b) considers the situation where family relations do suffer in the new job and you are left with wealth and hours of work. Figure 4.3(c) depicts a person staying at

(c)

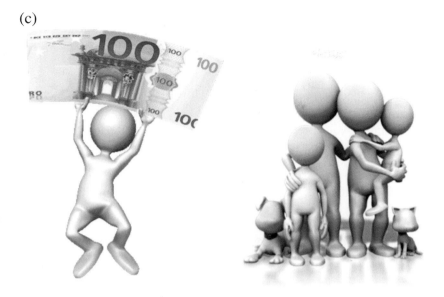

Figure 4.3(c) Visualizing the prospect of staying in current job and family relations do not suffer.

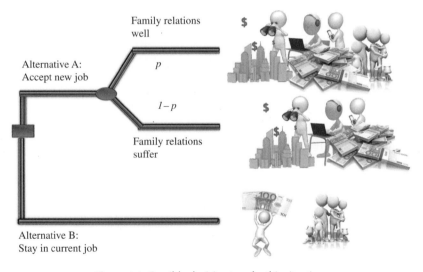

Figure 4.4. Possible decision tree for this situation.

their current job with less wealth but family relations remaining the same. Once again, you can visualize the prospects yourself in your own way to determine the order of your preferences and there are many other prospects for this decision. For example, you might stay in your current job and family relations still suffer.

Once you have visualized the prospects, you can place them in a decision tree. Figure 4.4 depicts the decision tree for this situation, where there is a probability, p, that family relations will not suffer with the new job.

Visualizing the prospects has not yet helped you decide, but it has helped you with the Order Rule.

We can envision many situations:

1) **New Job, Family Relations Well ≻ New Job, Family Relations Suffer ≻ Current Job**

You prefer the prospect of the new job with healthy family relations to that of the new job with poor family relations to the prospect of keeping your current job. In this case, the decision is simple; you accept the new job because you would have a preferred prospect regardless of whether or not family relations suffer.

2) **Current Job ≻ New Job, Family Relations Well ≻ New Job, Family Relations Suffer**

You prefer your current job (with less stress and less money) to your new job whether or not family relations suffer. In this case, you should stay in your current job because you are always better off.

3) **New Job, Family Relations Well ≻ Current Job ≻ New Job, Family Relations Suffer**

You prefer the prospect of the new job where family relations are well to your current job to the prospect of the new job where family elations suffer. This is where most people find decision-making difficult. On one hand, you would like to go to the new job and get a better prospect than the one you currently have, but on the other hand, you are afraid to get the worse prospect of family relations suffering. This best decision in this situation can be determined by the Five Rules. Figure 4.5 shows the Equivalence Rule indifference probability assessment. Figure 4.6 shows the decision situation after the Substitution Rule has been invoked.

From the tree in Figure 4.6, the choice criterion is clear: if $p > q$, then you should accept the new job, and if $q > p$, then you should stay in your current job.

EXAMPLE 4.3: Visualizing the Prospect of a Red Car, Yellow Car, or Black Car

Suppose you are considering your preference for the prospects of buying a yellow car, a red car, or a black car with everything else the same for all cars (same brand, model, type of car [sedan], transmission type [automatic], price, mileage, etc). There is little uncertainty about what you will get. You know exactly what each car color looks like because you have seen it at the dealer and you have a choice.

To visualize the prospects, it is helpful to think about how you will use the car; where you will drive it, and what you will value when driving it. Think about the prospect of driving a yellow sedan to work. Do you care about the impression you are giving to other people? Will color be an important factor in this impression? Then think about the red sedan and the black one. Continue to do this until you have a clear ordering of these prospects after this visualization.

If the make of the car is not yet decided, you might say: "If it is a sports car, I would prefer the yellow car or the red car, but if it is a sedan, then I would prefer the black car." Here your preference for the color of the car might depend on its type and make.

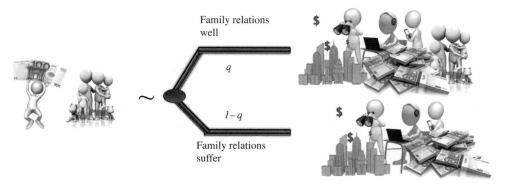

Figure 4.5. Indifference Probability Assessment for current job vs. prospects of new job.

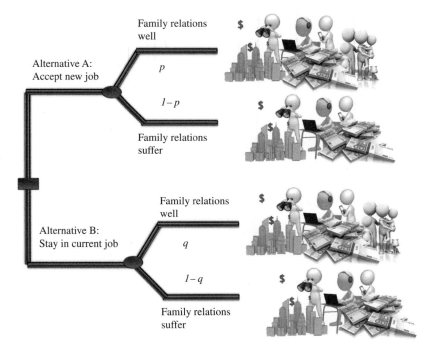

Figure 4.6. Equivalent Decision Tree using the Substitution Rule.

If the price of the car is not fixed, you can also visualize the prospects. A Mercedes SLK, for example, is more expensive than the CLK model. It has two seats, though, and a larger engine. In thinking about your preference, several factors might now come into play. The first is your usage of the car. Will you be driving it mostly alone? Will you be driving it mostly with friends or family? This distinction alone can help you think about your preference. But suppose that you will be driving with another passenger most of the time; you might want to consider other factors such as your enjoyment of driving a smaller car with a bigger engine and the convenience it may provide in parking lots.

If you will have other passengers with you most of the time, then you will have a trade-off that can also be clarified by visualizing the prospect. How often will you have passengers? Would you be willing to reduce your enjoyment of driving a smaller sports car instead of a larger vehicle because of the amount of time you will have more than one passenger? By thinking about these prospects you can get clarity on this order.

EXAMPLE 4.4: Visualizing the Prospect of Sitting "Outside" or "Inside" at a Restaurant

---

Have you ever been to a restaurant and the greeter asked you if you prefer to sit inside or outside? What comes to your mind when you think about this question? You might want to visualize a dinner outside. If it is lunchtime, you might consider factors such as the temperature or the brightness (sun or shade), a preference for air conditioning, as well as the opportunity to breathe fresh air. You might also consider the possibility of rain and the prospect of having to move if it does rain. Then you can visualize the prospect of sitting inside, breathing less fresh air, a possibly darker environment, but perhaps with more controlled temperature. Visualizing these prospects while at the restaurant can often be sufficient to help you determine your preferred seating option. Figure 4.7 illustrates two situations of sitting at a restaurant.

As we have seen, visualizing the prospects can help get clarity about the best decision. But visualization is not enough to guarantee that a good decision is made, particularly when uncertainty is present.

Dialogue between Instructor (I) and Student (S)

S: This idea of visualizing the prospects and also living your life with the prospects is really helpful. We can see how it can be used to order the prospects. But what if we are still unclear about the preference order?

I: You can create more scenarios. As we have seen in Chapter 1, for example, the prospect of winning the lottery can be further described by a distinction of whether relations with friends would suffer. You now have two prospects and you can think about your future life in each case.

S: Are ties allowed when we order the prospects?

I: Yes, it is okay to be indifferent between two prospects. But you cannot have cyclic preferences.

S: What does that mean?

I: For example, you cannot say, I prefer House A to House B because of the view, and House B to House C because of the size, and House C to House A because of the location.

S: Why not?

I: You need to state your preferences for the houses taking view, size, and location into account together. Suppose you owned House A. If indeed you preferred House C to House A, then you would be willing to give up House A and pay money to get House C. Correct?

S: Yes.

I: You also prefer House B to House C, so you would be willing to pay money to move from House C to House B. Correct?

(a)

(b)

Figure 4.7. Prospects of sitting inside and outside at a restaurant. © Paul Airs / Alamy Stock Photo;
© Mechika / Alamy Stock Photo

S: Yes.
I: But you prefer House A to House B, so you would be willing to pay money to give up
House B and get House A. You are back to House A after spending money.
This argument is called a money pump argument. You can be made a money pump
because of these cyclic preferences.

S: But what if we are uncertain about our preferences? For example, I am thinking about two jobs but I do not know if I will like my colleagues in one job and the possibility of advancement in the other.

I: Then you need to create distinctions. For example, you think about each job when you like or do not like your colleagues, and when you advance sufficiently or you do not. You end up with a tree of prospects for each job, but each prospect is deterministic, and you should be able to state your preference for each prospect.

S: Aha.

I: The main philosophy of this book is to help you think about your preference for decision alternatives by thinking about deterministic prospects and their corresponding likelihood. When you have a set of prospects each occurring with a given probability (i.e., you have a decision alternative with uncertainty), you use utility values for the prospects and then calculate the expected utility of each alternative to determine the preference order.

S: I have a question. We sometimes hear statements like "this person is a good decision-maker and makes decisions quickly." Can you be a good and fast decision-maker without analysis?

I: In repeat situations, maybe! For less important decisions, we could potentially go ahead and decide without analysis. But as we shall see, once uncertainty is present, our intuition fails and we can make errors unless we rely on our tools. A good decision-maker, in this case, is one who is clear about their preferences for the prospects, and can express likelihoods of events. Some decision-makers might be clearer than others about their preferences and beliefs, while others might need more time to gain such clarity. The rest is a computation that uses these inputs to determine the best decision.

4.3 SUMMARY

Visualizing the prospects is a very useful method of ordering the prospects when they are few. Think of your life with the prospect. For important decisions, it is useful to live your life for a few days and imagine you have obtained that prospect. Repeat this for a few other important prospects. This alone might be sufficient to gain clarity.

Preference is a statement about the order of deterministic prospects. The actual magnitude of the preference function is not important. The higher the magnitude of the preference function, the more preferred is the prospect.

APPENDIX: RANKING TWO PROSPECTS USING PROS/CONS

Because of its historical perspective, it is useful to discuss the pros/cons method of ordering prospects (or deterministic decision alternatives). This method goes back to Benjamin Franklin. The following historic letter from Ben Franklin replying to Joseph Priestly who asked him for advice about making a decision is self-explanatory. The letter presents pros/cons in terms of choosing between alternatives, but as we have seen, if you follow the five rules, then the expected utility criterion is the method for doing that. Pros/cons analysis could be useful in thinking about preferences for deterministic prospects.

To Joseph Priestley

London, September 19, 1772

Dear Sir,

In the Affair of so much Importance to you, wherein you ask my Advice, I cannot for want of sufficient Premises, advise you what to determine, but if you please I will tell you how.

Benjamin Franklin © iStock.com/Joe Cicak

When these difficult Cases occur, they are difficult chiefly because while we have them under Consideration all the Reasons pro and con are not present to the Mind at the same time; but sometimes one Set present themselves, and at other times another, the first being out of Sight. Hence the various Purposes or Inclinations that alternately prevail, and the Uncertainty that perplexes us.

To get over this, my Way is, to divide half a Sheet of Paper by a Line into two Columns, writing over the one Pro, and over the other Con. Then during three or four Days Consideration I put down under the different Heads short Hints of the different Motives that at different Times occur to me for or against the Measure. When I have thus got them all together in one View, I endeavour to estimate their respective Weights; and where I find two, one on each side, that seem equal, I strike them both out: If I find a Reason pro equal to some two Reasons con, I strike out the three. If I judge some two Reasons con equal to some three Reasons pro, I strike out the five; and thus proceeding I find at length where the Balance lies; and if after a Day or two of farther Consideration nothing new that is of Importance occurs on either side, I come to a Determination accordingly.

And tho' the Weight of Reasons cannot be taken with the Precision of Algebraic Quantities, yet when each is thus considered separately and comparatively, and the whole lies before me, I think I can judge better, and am less likely to take a rash Step; and in fact I have found great Advantage from this kind of Equation, in what may be called Moral or Prudential Algebra.

Wishing sincerely that you may determine for the best, I am ever, my dear Friend,

Yours most affectionately
B. Franklin

Source: Mr. Franklin: A Selection from His Personal Letters. Contributors: Whitfield J. Bell Jr., editor, Franklin, author, Leonard W. Labaree, editor. Publisher: Yale University Press: New Haven, CT 1956.

CHAPTER 5 Getting the Attributes Right

Chapter Concepts

- Direct vs. indirect values
- Do not rush into assuming arbitrary preference structures because of their simplicity
- Using measurable (and meaningful) attributes
- Using a structural model of preferences
- Reducing the number of direct value attributes using monetary equivalents

5.1 INTRODUCTION

In Chapter 3, we discussed the errors that might occur when using an arbitrary method of decision-making. In Chapter 4, we discussed preference statements made by visualizing the prospects of a decision. In many cases, the prospects may be characterized by one or more attributes. For example, prospects of a medical decision might be characterized by measures such as health state (which may be interpreted in terms of remaining life time) and consumption. An oil company might be interested in expanding its drilling operations, and so the prospects could be characterized by multiple measures such as profit ($); safety of its operation (expressed in terms of the number of lives saved); and environmental effects of its operation (which may be characterized by a carbon footprint or in some cases converted into monetary equivalents).

When a decision involves multiple factors of interest, we need to think carefully about which of those factors we should use as attributes to characterize preferences for deterministic prospects. One of the common mistakes that people make in the selection of attributes is that they select attributes that do not really characterize a preference for the prospect itself, but that are mere contributors to achieving that prospect or they shed information about the likelihood of achieving it. Consider, for example, a business decision where a company is thinking about enhancements in a product or about promotions in the next quarter. It is tempting to rush to characterize preferences for prospects of this decision by market share, increased demand, and profit, among other things. But why does a profit-maximizing firm care about market share? Ultimately it is because market share increases the likelihood of getting more profit in the long term. Therefore, increasing market share is a means to achieving profit – the attribute the company really cares about. To include both market share and profit in the character-ization of the prospect would be double counting and can result in inconsistencies. To further distinguish the attributes that should characterize the preference for the pros-pect in this scenario, think about the following question: If you know how much profit you will make, regardless of market share or what you can do to enhance it, would you still care about the market share? The answer is, most probably, no! You might not want to spend more money to increase market share if it will not lead to more profit. Therefore, market share is not an attribute of direct value.

We refer to the factors that contribute directly to the preference order of determin-istic prospects as *direct value attributes* and to those that convey information about the likelihood of getting the prospects as *indirect values*. Prospects should be charac-terized by direct value attributes. This chapter discusses the distinction between direct and indirect values and the appropriate choice of measures to characterize prospects with multiple attributes. We shall also discuss decomposing a direct value attribute into other components to help with the construction of preferences.

In many decisions, it might also be important to capture any domain knowledge – using existing physical, engineering, economic, accounting, medical, or other principles that exist – to determine the order of the prospects in a particular field and ot maintain consistency. For example, if you are interested in profit obtained through energy gener-ation, it would be important to include the relation between energy, mass, and velocity of a turbine pump if we wish to order prospects based on their characterization by mass and velocity. Failing to capture these domain knowledge relationships might result in inconsistencies or double counting. It would be difficult to rank prospects in this case based on mass and velocity of the pump alone without making the conversion to energy and profit, the attributes we care about. We refer to this domain-knowledge relation as a *structural model*. We shall discuss examples of structural models and their use in constructing preferences in this and many future chapters.

5.2 DIRECT VS. INDIRECT VALUES

To begin, it is important to make a distinction between direct and indirect values (Howard 1992, Howard and Abbas 2015).

> **Definition**
> **Direct value attributes** are factors that contribute to the Order Rule preference for deterministic prospects. Changing the level of a direct value attribute might change the preference for a prospect.
>
> **Indirect values** are factors that do not contribute to a change in the Order Rule ranking of deterministic prospects. They merely serve as indicators about the like-lihood of achieving a level of one or more prospects characterized by direct value attributes.

To further clarify the distinction between direct and indirect values, consider the following example from Howard (1992) and Howard and Abbas (2015).

EXAMPLE 5.1: The African Resort Hotel Owner

An African resort hotel owner is vitally interested in tourism, because of the revenues associated with tourists, but is indifferent to the presence of African wildlife. However, the owner knows that the wild animals increase the chance of attracting more tourists, and, therefore, the owner would prefer having more wildlife to less. We say that the owner places a ***direct value*** on tourism and an ***indirect value*** on wildlife.

Figure 5.1 shows a diagram with two ovals: "Tourism" corresponding to profit from tourists, and "Wildlife" corresponding to the abundance of the wildlife habitat. In the diagram, there is an arrow from Tourism to value. This notation implies that tourism is a direct value attribute. The diagram shows no arrow from Wildlife to Value but an arrow from Wildlife to Tourism. Because both Tourism and Wildlife are uncertain a priori, the absence of this arrow implies that Wildlife is an indirect value. The prospects in this decision are characterized by "Tourism" profit. The hotel owner is interested in wildlife only because it contributes to more tourism profit. If this hotel owner was guaranteed a fixed amount of tourism profit regardless of the level of wildlife, he would no longer be concerned about the wildlife.

The tree representation of this situation is on the right side of Figure 5.1. The figure shows that the decision-maker is indifferent between the two prospects "Wildlife – High Tourism" and "No Wildlife – High Tourism," as they product the same value, V_1. He is also indifferent between "Wildlife– Low Tourism" and "No Wildlife–Low Tourism," as he values them both at V_2. Consequently, wildlife does not play a role in the valuation of the prospects. It is an indirect value. Therefore, this decision-maker is indifferent between any two prospects having identical values of the direct value attributes, even if they have different indirect values. The characterization of prospects by the second attribute, Wildlife, in this example, does not affect the preference ordering of the prospects.

In contrast, an Environmentalist cares only about the welfare of African wildlife, and is indifferent to the presence of tourists. Figure 5.2 shows the assignments of value in a diagram. The environmentalist places a direct value on Wildlife but not on Tourism.

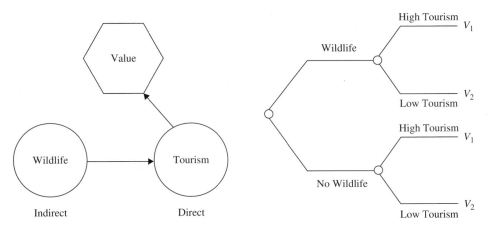

Figure 5.1. Direct vs. Indirect Values for Hotel Owner.

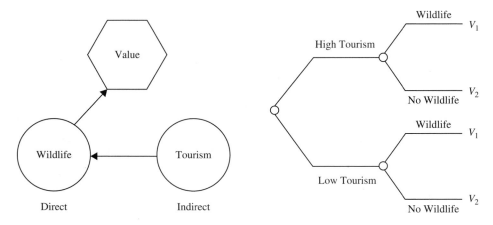

Figure 5.2. Direct vs. Indirect Values for Environmentalist.

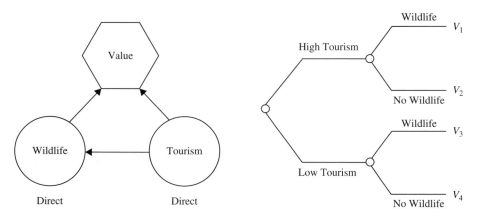

Figure 5.3. Direct Values for Both Tourism and Wildlife.

However, this environmentalist believes that tourism would provide funding to preserve animal habitat and defend animals against poachers, and therefore places an indirect value on tourism. Once again, if new information showed that there was no correlation between Tourism and Wildlife, then tourism would no longer be an indirect value. As before, if the information showed that tourism was deleterious to wildlife, then the indirect value of tourism would be negative. Once again, the tree representation shows that prospects having the same level of direct value attributes are valued equally, regardless of the level of the indirect value.

Someone who placed a direct value on both Tourism and Wildlife would have arrow connections like those in Figure 5.3. The Value node would show how overall value depends on combinations of Wildlife and Tourism. In the tree, we see that both Wildlife and Tourism contribute to the valuation of the prospect. If he believed that there was no relevance between the two attributes of tourism and wildlife, the hotel owner would still care about both attributes and place a direct value upon them.

> ***Indirect values will not contribute to a change in the Order Rule ranking of the prospects.***

5.3 DO NOT MIX THE DIRECT AND INDIRECT VALUES IN A DETERMINISTIC OBJECTIVE FUNCTION

Note: The concepts of direct and indirect values can considerably simplify the formulation and solution of decision problems. While there are many features of a decision that a decision-maker might care about, we often identify only a few direct value attributes and recognize other factors as indirect. To order deterministic prospects, we need only consider the levels of their direct value attributes. Two prospects having different indirect values, but the same direct values would have the same preference. This implies that the ordering of deterministic prospects should depend only on the direct values. Unfortunately, there is major confusion about this point, particularly when direct and indirect values are mixed into some additive objective function.

The following example illustrates a classical misuse of objective functions that include both direct and indirect values.

EXAMPLE 5.2: Is Quality a Direct Value?

Consider the Dilbert Cartoon of Figure 5.4, in which a manager is advocating the importance of quality to his employees. He starts "Quality is our top priority," and the employees remind him that there are also other important priorities and factors such as safety, obeying the law, maximizing shareholder return, as well as ethical aspects.

Here are the factors that we see in the cartoon:

a) Quality
b) Safety
c) Obeying the Law (Legal)
d) Maximizing Shareholder Return ($)
e) Lie (Ethics)

What is funny about this cartoon is that it highlights a situation that happens quite a lot in business. A manager advocates the importance of something that need not be a direct value, but it is repeated throughout the organizational hierarchy so that the manager feels the need to continuously advocate it. It also highlights the emphasis that people put on quality. When the quality movement started, it was heard everywhere: in seminars, research papers, web sites – you could even get a black belt in some quality initiatives. But the quantification of the value of achieving quality in comparison to the expense of achieving it was seldom provided.

There are several problems with the method of reasoning that the manager advocates in Figure 5.4 as we illustrate here.

First of all, for a profit-maximizing firm, quality might not even be a direct value attribute in the first place. Quality is a means for achieving higher profit by selling more products at a given price. The higher the quality, the more likely it is that the demand

Figure 5.4. Is Quality a Direct or Indirect Value? DILBERT © 2004 Scott Adams. Used By permission of UNIVERSAL UCLICK. All rights reserved.

for the product, and the sales, will be higher, but the direct value attributes are money (total shareholder return) and possibly safety (lives/injuries saved in the process).

Methods for Identifying Direct and Indirect Values in a Decision

Many ways can be used to identify whether an attribute is a direct value attribute or an indirect value in a conversation with decision-makers. **One way** is to ask:

> *"Why do you care about this?"*

And continue to ask this question until you reach the direct values. For example, *why do we care about quality*? People will reply, *"because it leads to higher demand"* and then you ask *"but why do you care about higher demand?"* and the response would ultimately be *"to make more money."* These types of questions identify whether a factor of concern is a direct or an indirect value immediately.

Another method to identify direct value attributes is to ask:

> *"If you knew that the levels of other attributes were fixed regardless of what you do, would you still care about this factor?"*

For example, if the organization were guaranteed a fixed level of profit no matter what the quality of the product were, would they still care about improving quality? The answer would most likely be "No," and so quality would be an indirect value.

In the cartoon, we also see a comment made by one of the employees:

> "If we could maximize shareholder value by selling lower quality products, wouldn't we have a fiduciary responsibility to do it?"

The answer is positive. Quite often organizations initiate intervention schemes that work on improving the indirect values in the hopes of achieving higher direct values. For example, they might set up a department whose sole responsibility is to improve quality. This could be useful in some occasions, but the departments often forget that their mission is to improve quality for the sake of improving the direct value, profit.

Therefore, the costs needed to improve quality should be weighed against actual profit. And if spending nothing on quality and selling at a cheaper price would indeed lead to higher overall long-term return, then a profit-maximizing firm should not spend more money on quality.

Lying and disobeying the law are ethical and legal actions that the firm might wish to exclude from its set of alternatives. The justification could be that the firm does not want to engage in anything illegal, and that it will not engage in activities it regards as unethical. These items might not be traded off against profit; they limit the set of alternatives that the company would consider.

Therefore, the actual decision situation in the cartoon could be the choice among the different alternatives of how to design and price a product. The company would exclude alternatives that involve any illegal or unethical aspects. The uncertainties might be demand for the product (whose distribution depends on its quality, as well as its price and design), the cost of the product, and the revenue generated. The direct values for this decision would be money and safety, only two direct-value attributes. The indirect values would be demand of the product and its quality.

Caution with Advocating "Top Priorities" When Characterizing Prospects

Besides the issues encountered with direct and indirect values while framing this decision, another problem that usually arises with these types of statements in decisions with multiple factors to think about is the meaning of "priority" or which attribute is more important than another. The declaration of one of the attributes as a *top priority* without considering the levels of other attributes can lead to errors in decision-making. First of all, the *"importance"* or *"priority"* of an attribute has no clear meaning. Different people would interpret the level of importance differently even if they referred to the same thing. Furthermore, the "priority" of an attribute might well depend on the levels of other attributes. For example, when buying a house in the city, cost might be a "priority" and you are willing to spend more time commuting to the suburbs if it would save you money. You might think of a formula for money saved per commute time or commute miles within a certain distance from work, and so living in the suburbs could be appealing. But as this distance you drive gets larger, and you realize that you are spending a lot of your time on the road, cost might not be as important. For example, you might not want to continue to drive extra hours if your commute time is five hours each way, even if it will save you the same amount of money. In this case, we say your trade-off for money saved and commute time depends on the actual commute time, and so cost might not be a priority after a certain commute time from work. Stating that one attribute has more importance than the other is not only unclear in terms of quantification; it also ignores the level of the other attributes. Furthermore, it usually implies some form of "importance weights" and a "weight and rate" scoring system, which is widely misused, as we discuss in further detail below.

Caution with the Weight-and-Rate Scoring System

Another common error with decisions involving multiple factors of concern is that decision-makers usually combine direct and indirect values using some weighted combination that is usually additive or multiplicative in a method known as the "weight-and-rate" scoring system. To illustrate, below we demonstrate a common misuse of the weight-and-rate system that could be used with this quality situation.

Do Not Model This Situation by Immediately Imposing a Form Like This:

$$P(Quality, Safety, Legal, \$, Ethics) = w_1 Quality + w_2 Safety + w_3 Legal + w_4 Shareholder\ Return + w_5 Ethics$$

Or like this:

$$P(Quality, Safety, Legal, \$, Ethics) = Quality^{w_1} \times Safety^{w_2} \times Legal^{w_3} \times Shareholder\ Return^{w_4} \times Ethics^{w_5}$$

While these functional forms might be applicable in some settings, you need to think clearly about the attributes and the actual structure of the problem. The appendix of Chapter 8 discusses some of the main concerns with this widely used (and widely misused) weight and rate scoring approach. As we have discussed, quality is not really a direct value attribute for a profit maximizing firm, and presenting it as some formula requires a clear scale and meaningful weights. It is hard to quantify quality on a meaningful scale unless there are some specific dimensions that can be specified and are measurable, such as smoothness of a surface in a machining job. But even then it should not be placed into an arbitrary formula. Rather, the contribution to quality to the direct value attribute of profit would need to be determined, possibly by assessing the demand for a given "smoothness of surface" at a given price.

Safety also needs to be clarified and not just assigned some arbitrary scale. When formulating the problem, try to think of something meaningful. Is it the number of lives saved per year? What do we mean by "saved"? Is it the decrease in the number of accidents? This latter definition would also require clarity such as: "What counts as an accident and what does not?"

Legal also needs to be specified. It would not be appropriate as an attribute in this setting. Does the company want to avoid anything illegal? If so, then it is a matter of removing any illegal alternatives from consideration. The company might also feel that an alternative is borderline and could be interpreted either way, and so they might want to consider the costs associated with legal time that would explain any legal issues if a problem arose. The company would then need to consider the uncertainties associated with any legal pursuits. Ultimately it might translate into cost. Presenting "Legal" alone and asking people to assign some arbitrary scale would be an arbitrary way of handling this problem.

Ethics is also an interesting factor. Once again, if the company has some ethical dilemmas with some of the alternatives it is considering, and if it chooses to avoid such alternatives based on ethical issues, then it should simply remove the alternatives from consideration instead of including what is ethical as a direct value attribute and trying to assign some scale to it.

Shareholder return needs further clarification. Is it the return on the investment measured as a percentage or is it the net present shareholder value? Ratios (such as return) are not appropriate measures to include as direct value attributes.

A first step in the direction of improving the additive and multiplicative functions above is to remove the indirect value attributes and present them on meaningful scales. When you are set on the attributes, and are clear about the direct and indirect values, you still need to think about the actual formula for determining the preferences more clearly: What are the weights? Why do we have weights in the first place? What do they mean? How do we assign them? What trade-offs are implied by assuming an additive or a multiplicative functional form? Is there another form that should be used? Unless we are clear about the interpretations of what we are asking others, we will be on a wrong track from the start before we even conduct any analysis. We shall revisit these questions in Chapters 7, 8, and 9, and many times throughout the book.

EXAMPLE 5.3: Is "Probability of Success" a Direct Value Attribute?

A space mission is concerned with capturing an asteroid and bringing it to Earth to conduct experiments on it. Some of the decisions involved include the choice of the space vehicle, the choice of the asteroid to bring back, and the capturing mechanism for "grabbing" the asteroid.

Preferences for prospects of this mission could include the cost of the mission, the safety of the crew, as well as the value derived from scientific discovery from experiments on the asteroid.

What about the probability of mission success (the probability of being able to bring an asteroid back and to bring the crew back safely)? Is this a direct value? Should this be used in the preference function as a direct value attribute? The answer is negative.

We care about the success of the mission because it will lead to scientific discovery and lives saved. We would also like a mission with high probability of success, but this is not entered into the characterization of the deterministic prospects. What enters are the cost, safety, and scientific value conducted from the experiments.

If the mission were successful, we can think about our preference for various deterministic prospects with different degrees of cost and scientific discovery. If the mission was unsuccessful in a sense that an asteroid was not delivered, then this would also lead to a deterministic prospect with low value for scientific discovery. The probability of success itself does not enter into the characterization of the deterministic prospects. Therefore, the preference, value, and utility of the prospects are not affected by this probability. Probability of success is not a direct value attribute.

5.4 USING MEANINGFUL AND MEASURABLE ATTRIBUTES

When thinking about the attributes that comprise your preference for a prospect, it is good practice to use attributes that have a meaningful and measurable scale and ones that have no ambiguity about them. Otherwise, different people will have different interpretations for the magnitude of each attribute. For example, an attribute such as "complexity" in the design of an engineering system might not be easily interpreted, not to mention it is not even a direct value attribute in the first place. For example, it might be that the real issue here is whether a company will be able to achieve technical success for launch, and so there is a need to capture the probability of success at a given plant for expected utility calculations, but the prospect itself is characterized by whether or not technical success has been achieved. This is a big difference in thinking from including an arbitrary attribute like "complexity" – giving it an arbitrary scale, and then combining it into some arbitrary weighted formula to determine preferences. When an attribute has not been clearly defined, it is usually a result of a lack of clear thinking or, as we shall see in the next section, the absence of some clear structural model that defines the attribute in terms of some other components. Even if some physical structural model does not exist, we might still be able to define preference for attributes in terms of clear components, as we illustrate in the next four examples.

EXAMPLE 5.4: Measurable Attributes for a Seat in the Movie Theater or a Concert

---

Think about the attributes that contribute to your preference for a seat in the theater. What would make you prefer one seat over the other? It is difficult to answer this question without a decision context. For example, it might be that the person (or group of people) who are at the movie theater are not actually interested in the movie in the first place. But with all else constant, suppose you were going alone to the theater; what factors would lead to your preference for one seat over another for a given movie?

One factor that could affect your experience is the quality of the view, but simply stating "view" is by itself not a clear specification, and assigning an arbitrary score from 0 to 10 (0 being worst and 10 being best) is arbitrary and could be a recipe for disaster.

We can further decompose "view" into other components that are meaningful and express our preference in terms of these components. One component, for example, could be the radial distance from the screen (or stage). The distance is clear regardless of the units you use. There is no ambiguity about a radial distance of 12 feet, for example. In thinking about your preference for the radial distance, you might not want a seat that is too close to the screen or one that is too far, and so you might have a peak in your preferences at some radial distance: there might be an optimal radial distance below which and above which your preference would decrease.

Another component that could contribute to the "view" is the symmetry by which you view the screen (or stage). You might prefer to be centered. This can be modeled by the angle from the line perpendicular to the screen to your location. Figure 5.5 presents a schematic view of a layout of a theater.

Another consideration could be the "inconvenience" incurred if you have to pass many people in the aisle to get to your seat. This attribute is not immediately measurable, and so you might interpret it in terms of the number of seats you have to pass, or even as a cost for every seat you would have to pass to get to your seat.

Other attributes might also be involved in this decision and pertain to a particular theater, such as sound. For example, if you knew the location of the speakers, you might prefer a seat that is within some radial distance from the speakers, or one that is not too close and not too far.

You can continue to model certain aspects of the prospect of a seat until the characterization will contribute very little to your deterministic preference, or that the costs are so small compared to the modeling effort.

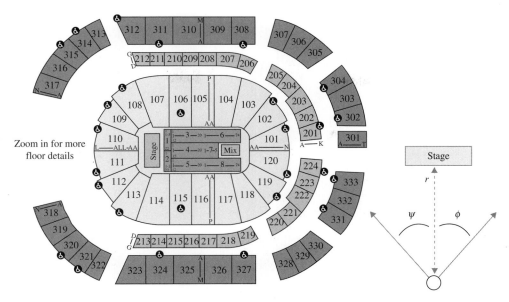

Figure 5.5. Layout for seats in a theater.

We have identified several factors contributing to our preference for the seat, and have formulated them in a meaningful way. We have not yet discussed how to combine them to determine your preference for a given seat. In Chapter 7, we shall discuss how to combine these factors to think about our preference for view using a preference function.

EXAMPLE 5.5: Preference for a Peanut Butter and Jelly Sandwich

---

Think of the factors that might affect your preference for the prospect of eating your peanut butter and jelly sandwich (Figure 5.6).

Consider, for example, two sandwiches; what factors might contribute to your preference for one sandwich over another? You might say "taste," but that is not a clear attribute. Some of the factors that might contribute to taste include:

- The type of bread used (wheat, white bread, etc)
- The thickness of the slices of bread
- Whether the bread is toasted or untoasted
- The thickness (and type) of the peanut butter
- The thickness (and type) of jelly
- Other ingredients of the sandwich.

This list can go on. What important, though, is to include those factors in a given decision that have an effect on the alternative you will choose. For example, if you are set on eating a peanut butter and jelly sandwich, and you have a choice among different sandwiches, but the restaurant only uses sourdough, and if sourdough will not change your preferences for the other attributes in the sandwich, then you might exclude the characterization of bread with sourdough from your prospects. If, however, having only

Figure 5.6. Visualizing a peanut butter and jelly sandwich. © Bernard Prost/StockFood Collective/ Getty Images.

sourdough will make you reconsider your alternative of eating a peanut butter and jelly sandwich at that (or another) restaurant, or if it will change your preferences for thickness of the sandwich ingredients, then it should be included because it affects your preferences for the prospects and the alternative you will choose.

EXAMPLE 5.6: Measurable Attributes for a Car

Many attributes characterize preferences for a car. They need not all be on a continuous domain. For example, the make of the car could be an attribute, as well as the transmission type, its size (family sedan or compact), and whether or not it is a convertible. Other attributes that are continuous may include cost of the car, miles per gallon, leg space, trunk space, and a collection of other attributes that could be included under total cost of ownership. Color of the car is also an important attribute, which can be modeled as a discrete attribute (blue, red, etc), even though color is actually a continuum on an electromagnetic spectrum. All these attributes can be explained clearly and determined such that there is no ambiguity about their meaning.

EXAMPLE 5.7: Measurable Attributes for a Job

Direct value attributes for a job may include salary, geographic location, commute time, and weather, all of which are measurable. There are of course other factors associated with the prospect of a new job that are often stated without clarity such as lifestyle. When used as an attribute, it could be further clarified in terms of measurable attributes such as the amount of hours spent at work, the amount of time available to pursue hobbies of interest, personal reflection time, and the proximity of certain locations to the place of work or residence (such as hiking trails).

Sometimes attributes that are widely stated when considering a job need a range of prospects because of the uncertainty associated with them. An example of such attributes is "advancement," which may include several prospects characterized by possible positions within the company and time needed to achieve them. This group of prospects would then require a probability distribution representing our belief about the likelihood of achieving such positions within a time frame.

5.5 USING STRUCTURAL MODELS TO ORDER PROSPECTS (AND TO CHARACTERIZE DIRECT VALUE ATTRIBUTES)

In Example 5.4 we discussed how to think about your preference for a seat in a movie theater using clear and meaningful factors without having a specified domain-knowledge relationship. In this section we illustrate how *structural models* can also help us think about our preferences by incorporating domain knowledge to ensure consistency. By themselves, these models can help with the preference order, but in addition they can translate the direct value attribute into physical/engineering/ or accounting terms that could be relevant to the problem. To illustrate, suppose you were deciding on the type of heater for a basement, and you have several alternatives including oil, gas or electricity. Cost of the heater and cost of operating the heater might be one factor, but you might also think about temperature you would like to have, and how long it would take to heat the given space. These direct value attributes can be related to the design features of the heater, such as the power of the heater, the energy or fuel source, as well as the efficiency of the heater, using structural models. Ignoring these structural models in the choice of the heater and simply applying a weight-and-rate scoring system could lead to a bad decision.

EXAMPLE 5.8: Using a Structural Model for Height of a Projectile

---

To further illustrate the use of structural models, suppose you are interested in the height of a projectile (such as when viewing fireworks from a distance). The projectile is projected vertically at an initial vertical speed u under the force of gravity, g. The height of the projectile at any time, t, depends on the initial velocity and gravity. Can you express your preference for the view of the projectile as a function of time just by thinking about the initial velocity and gravity? It would be a lot easier (and would lead to more consistency) if we can think about our preference for the height of the projectile and then relate that to the initial velocity and gravity as a function of time. This is the role of the structural model.

Ignoring friction, the vertical elevation of this projectile at any time t is given by the formula

$$y = ut - \frac{1}{2}gt^2$$

This logical relationship has decomposed the attribute of height into two other factors, launch speed and time in orbit, and so we may write for some function, f, the vertical distance as

$$y = f(u,t).$$

Now you can think in terms of preferences in terms of speed of launch and time in orbit to compare different launch speeds. You can also use this relation to design a better height for the projectile at a given time, and what the optimal view would be at that time. It is clear that preferences as a function of time are not monotonic, and that they increase and then decrease. It would be double-counting to think about those attributes as three independent attributes y, u, t using some arbitrary preference function if what you really care about is height.

Another example of a structural model is one for energy generation (or dollars from energy generation). There might exist a structural model that decomposes the direct value attribute of energy into sub-factors such as

$$Energy = \frac{1}{2} \times mass \times (velocity)^2$$

In some cases, the power of velocity or some additional scalars might also be used with the formula and determined by experimentation. It would lead to inconsistent preferences if you stated your preference for energy but then expressed it differently in terms of mass and velocity if a structural relation exists.

EXAMPLE 5.9: Preference for Mass and Velocity of a Turbine Generator Using A Structural Relation

---

Suppose we are comparing prospects characterized by two numeric measures: mass of a turbine generator and its velocity (speed of rotation) for use in energy generation. Both of these measures contribute to energy production. The higher the value of each measure, the higher is the energy generated, and so "more of each attribute is preferred to less" in this example.

If we compare two generators, where one has both higher mass and higher angular velocity, then this statement that "more of each attribute is preferred to less" is sufficient to determine the better prospect.

But suppose we compare two generators where one has a higher mass but rotates at a lower velocity, and the other has a lower mass but rotates at a higher velocity. The two prospects are represented in Figure 5.7. Which generator should we prefer? This statement is insufficient to determine the better prospect. How much increase in velocity do we need to compensate for the decrease in mass?

If our preference is the energy generated, then this should be the criterion for ranking the prospects. We choose the prospect that provides higher energy. Figure 5.8 plots the energy generated as function of the mass and velocity using the structural relation

$$Energy = \frac{1}{2} mv^2 = \frac{1}{2} \times mass \times (velocity)^2.$$

With this observation, let us refer back to our earlier question. Do we prefer prospect (m_1, v_1) or prospect (m_2, v_2), if $m_1 \geq m_2, v_1 \leq v_2$?

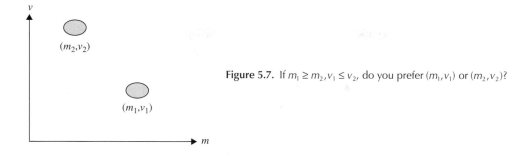

Figure 5.7. If $m_1 \geq m_2, v_1 \leq v_2$, do you prefer (m_1, v_1) or (m_2, v_2)?

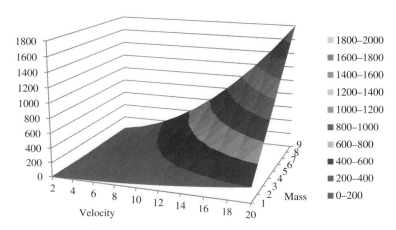

Figure 5.8. Energy as a function of mass and velocity.

If we are clear about our direct value attribute (energy generation), then we can determine the preference order for these prospects right away. We simply have to calculate the energy generated by the mass and velocity levels of each prospect. The most preferred prospect is the one that provides the higher energy. The ordering is simple once we are clear about what it is we really care about, and if there is a functional relationship that relates it to the different attributes.

We can also ask what combinations of mass and velocity would produce the same energy. If prospect (m_a, v_a) provides the same energy as prospect (m_b, v_b), then it must be that

$$\frac{1}{2}m_a v_a^2 = \frac{1}{2}m_b v_b^2$$

These two prospects (m_a, v_a) and (m_b, v_b) are said to lie on the same level set of energy production. Because our preference is for energy production, the two prospects are said to lie on the same isopreference contour. Figure 5.9 plots what is known as a contour plot of Figure 5.8. The figure provides a color for prospects providing energy within a specified range. As the range gets narrower, the contour plot traces the isopreference contours. In fact, the boundaries of the regions in Figure 5.9 are themselves isopreference contours representing energy generated in amounts of 500, 1,000, and 1,500 units. As we shall see in the next section, identifying the isopreference contours is sufficient to determine the ordering of the prospects.

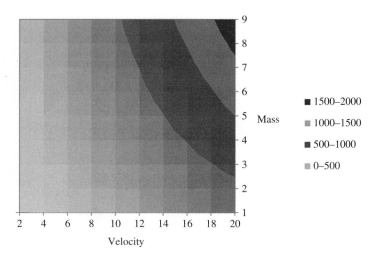

Figure 5.9. Contour plot for preference function of energy generation.

The decomposition of the preference for energy into mass and velocity requires knowledge of the relationship

$$Energy = \frac{1}{2}mv^2 = \frac{1}{2} \times mass \times (velocity)^2.$$

Such domain knowledge about a particular problem is often needed to conduct an appropriate decomposition. In this case, it has converted the preference function into that of a single attribute, energy, in terms of its components through a physical relationship.

If we had not known this physical relationship, we could have expressed preferences for mass and velocity combinations in a way that is inconsistent with our preferences for energy. This is why it is important to understand the logical relationships and the domain knowledge that is present.

5.6 CONVERTING MEANINGFUL AND MEASURABLE ATTRIBUTES INTO DOLLAR EQUIVALENTS

It is useful to think about reducing the number of attributes used in its characterization to reduce the dimensionality of the problem. One method of reducing the number of attributes is by converting them into monetary equivalents. For example, commute time in a job is by itself an attribute but may be taken into account by converting this time into a dollar amount. Once this is done, it can be added to the salary attribute and it no longer needs to be treated as a separate attribute of value. By converting direct value attributes into dollar equivalents we can reduce the burden of thinking about multiple attributes in our preference, value, and utility statements.

EXAMPLE 5.10: Monetary Equivalents for Attributes When Buying a House

---

When buying a house, we might place a direct value on the location, because it affects desirability. If it is for our personal use, location can be broken down into other factors such as distance to work, school, and/or some of our favorite attractions. We might then place a monetary equivalent on time spent commuting to work and to the attractions. In this case, we can convert location into some monetary equivalent.

We might also place a direct value on the total cost associated with the house. We might think about factors that contribute to this total cost, such as cost of maintenance, insurance, and utilities.

Once the monetary equivalents have been made, we can think of this decision in terms of money. We do not distinguish between money from a location attribute and money from a cost attribute. It is all money at this stage, and the total monetary amount is what matters.

EXAMPLE 5.11: Monetary Equivalents for some of the attributes of a seat in the movie theater

---

In Example 5.4, we discussed three attributes – view, convenience of getting to a seat, and quality of sound – and clarified them in terms of meaningful and measurable attributes. In some cases, we can reduce the number of attributes by converting some of the attributes into dollar equivalents.

If we can place a value on every seat needed to be crossed to get to our seat, and if this dollar amount does not depend on the view or the equality of sound, and if we can also express a dollar amount for quality of sound independent of the other attributes, then we can simple add up dollar amounts from the direct values. We do not treat monetary amounts differently.

Money Is Money: A study about machining once considered three attributes: profit, environment, and safety. The study converted environmental impacts into an independent monetary equivalent, but instead of treating money from profit and money from monetizing environmental impacts the same, it treated each separately. The result was three attributes: money from profit, money from environment, and safety. The study went on to combine the three attributes into a product form to determine preferences under uncertainty.

EXAMPLE 5.12: Converting the Whole Prospect into a Dollar Equivalent

---

It might also be possible to convert a whole prospect into a monetary equivalent. For example, the prospect of energy generation might be converted into dollars by multiplying by the unit price of energy. Now we can treat the whole prospect as a dollar value. We can now talk about the value of a prospect. In some cases, as we shall see in Chapter 9, a structural model that uses domain knowledge and experimentation might be used to place a monetary amount over the prospects.

5.7 SUMMARY

Before rushing to select the attributes of a decision, pause and ask yourself the following questions:

1. Are they direct or indirect values?

   Direct value attributes are what should be used to characterize preference, value, and utility. Indirect value attributes play a role in characterizing the uncertainty about the value obtained by direct value attributes. To help identify the direct value attributes, ask yourself why you care about a particular value. Usually, this process continues until the direct values are identified. Another method is to ask if the levels of other attributes are guaranteed to be fixed, would you still care about this value. If the answer is negative, then this is not a direct value.

2. Are the attributes meaningful and measurable?

   Do not use attributes that have arbitrary interpretations such as "comfort" and "taste" without clarifying these factors using meaningful measures.

3. Do the chosen attributes have arbitrary constructed scales?

   Do not use arbitrary scores, like a scale of 0 to 5, as measures for the attributes if the scale is not meaningful.

4. Can you convert any of the direct value attributes into monetary terms?

   If you can convert some of the attributes into monetary equivalents, then go ahead and do it. This will reduce the dimensionality of the problem significantly.

5. Are you using any probabilities, such as probability of success, as direct values?

   Do not use probabilities as attributes. Instead, define success, and consider a prospect with success and a prospect without success.

6. Can you further break down the direct value attributes into other factors that help you think about your preferences?

   Do not rush into assuming arbitrary preference structures because of their simplicity. If you get the objective function wrong, you can very easily end up with an inappropriate solution. Refer to the movie theater example.

7. Does there exist a structural model than can help you think about your preferences?

   If a structural model exists, it is important to incorporate to make sure you are being consistent with your preferences and are avoiding double counting.

REFERENCES AND SUGGESTED READINGS

Howard, R. A. 1992. In praise of the old time religion. In W. Edwards, ed. *Utility Theories: Measurements and Applications*. Kluwer, Boston, pp. 27–55

Howard, R. A. and A. E. Abbas. 2015. *Foundations of Decision Analysis*. Prentice Hall, New York.

Keeney, R. L. 1992. *Value-Focused Thinking: A Path to Creative Decision-Making*. Harvard University Press, Cambridge, MA.

APPENDIX: HOW PEOPLE GET IT WRONG

The literature is (unfortunately) full of examples of how people get the attributes that characterize a prospect wrong. Consequently, they end up with inappropriate formulations for the problem they are solving and inappropriate solutions.

Examples of some of the bad practices that are widely used with preference functions:

1) Using Attributes That Are Not Clear or Measurable or Direct

More often than not, attributes such as "Taste," "Complexity," "Pain," "Public Perception," or "Service Quality" are used to characterize a prospect. These types of attributes are by themselves not very clear. Different people will have different interpretations for what complexity really means, and it will be difficult to assign any preference, value, or utility statement if the definition of the attributes characterizing the prospects is not clear. In most of these situations, people have not spent enough time characterizing what they really want. Furthermore, people do not take the time to think about whether the attributes they are using are direct or indirect values.

2) Using Arbitrary Constructed Scores and Importance Weights

Another widely misused method of representing the attributes besides using attributes that are not clear is the use of some constructed scale (like 0 to 10). For example, you might hear statements like "On a scale of 0 to 10, how do you rate the service you received today?" or "On a scale of 1 to 5, how do you rate the pain level?" We now have two problems: first, the meaning of pain will be interpreted differently; and second, the scale itself will have different perceptions by different people because different people may interpret differently what a score of 8 out of 10 means in Taste or Complexity. These types of constructed scales also do not provide interpretations for what fractions of these integers really mean, and so they might not provide sufficient resolution for characterizing a prospect. Further, assigning arbitrary weights and interpreting them in terms of some unclear importance factor provides room for unclear thinking and bad decisions. In future chapters, we shall provide meaningful interpretations for any parameters used in preference, value, and utility functions.

3) Using Scores Whose Scales Are Relative to the Levels of Another Prospect

This occurs quite often when people compare two prospects and use the reference base of an attribute in one prospect to characterize the second. An example of where this occurs quite frequently (and incorrectly) is when you have a product design decision and are considering a current design. Then you compare the current to other designs and the levels of their attributes. For example, you say, "complexity" of this product is higher so I give it a "+2" in comparison to the existing design while complexity of another design is lower so it gets "-1" relative to the current design. This type of relative ranking is not desirable, because you lose the ability to characterize a prospect independently of other prospects. They can also lead to rank reversal, which we discussed in Chapter 3, where the removal of an alternative changes the rank order of the remaining alternatives.

These relative assignments of scores to a prospect in comparison to another also appear when people attempt to incorporate "regret" into the characterization of a prospect. You do not value a prospect in comparison to another prospect. You value each prospect separately.

While regret is a descriptive phenomenon that people may exhibit, incorporating regret into a normative process makes you lose the ability to value a prospect or an alternative without considering the remaining alternatives. We have already discussed some of the normative issues with incorporating regret into the characterization of a prospect in Chapter 3.

4) **Including "Probability of Getting a Prospect" or an "Expectation of an Attribute" into the Characterization of the Prospect**

Recall from Chapter 2 that the Order Rule is a ranking of deterministic prospects that does not depend on the probability of getting the prospects. There is no uncertainty about a prospect. Therefore, there can be no probability of getting the prospect in the characterization of the prospect. The preference, value, and utility of a prospect should be independent of the probability of getting it.

Many formulations of engineering design include attributes such as "probability of success" in the characterization of the prospect. Probability of success is not a direct value. Instead of incorporating probability into the characterization of a prospect, the way to deal with this situation is to include different prospects with varying degrees of success and then assign a probability to achieving these prospects.

Other common examples of incorporating some form of probability into the characterization is to include attributes like "expected profit" or "mean room temperature" into the valuation. Once again, it is the profit that is the direct value, not the expected profit. You do not carry expected profit in your pocket. Instead of using an attribute such as expected profit, you model each profit level as a separate prospect and then assign a probability to receiving each prospect.

5) **Ratios and Figures of Merit: You Do Not Carry Ratios in Your Pocket**

Ratios are another widely used (and misused) method of characterizing a prospect. For example, an attribute may be the return on an investment or the percentage of marijuana caught at the border relative to other illegal drugs. Ratios are often blind to scaling. For example, a 20% return could represent a $2 return on a $10 investment, while a 5% return could represent a $5 million return on a $100 million investment. Most people would prefer the latter even though it has a lower percentage return. It is a lot better to characterize the prospect using actual monetary amounts of the investment and to use actual units of the attributes, not ratios. Similar types of arbitrary methods include the internal rate of return and benefit-cost ratios.

6) **Money Is Money: They Are All Pictures of George**

It is often convenient to reduce the number of direct value attributes used in the characterization of a prospect by finding a monetary equivalent for some of the attributes. Once you have converted the attributes into monetary equivalents, however, you treat money the same regardless of the attribute from which it was converted. The purpose of highlighting this is to explain that money is money. There is no need to treat different sources of money differently. A colleague once mentioned: "They are all pictures of George."

7) **Using Arbitrary Metrics as Direct Value Attributes and Goals**

People love to measure things, and they love to measure performance. Quite often you will hear people say that they need a metric to see how well a project is doing. These metrics are often picked arbitrarily and are then used in decision-making. For example, if a metric goes down, it would be an indication that more resources need to be spent to bring it up again. What ends up happening is that these metrics often become objectives or goals that need to be met. This is where the problems arise because the metric is often picked arbitrarily, and maximizing might not lead to good decisions. If a metric is to be used in decision-making, such as resource allocation decisions, then why not analyze the decision of how much resources (if any) to spend in the first place? Why construct an arbitrary metric and then use that in the decision-making process? If an arbitrary metric decreases, it does not mean that the situation is necessarily worse.

When metrics become objectives or goals, they can lead to poor decision-making. Consider a metric whose purpose is to improve the waiting times for planes before given permission to land in an airport. One such metric could be the total waiting time for all passengers of airplanes before the plane is given permission to land as it gets closer to the airport within a certain radius. A plane with a large number of passengers and a given wait time would contribute more to the metric than a plane with fewer passengers for the same waiting time because the metric would sum the waiting times for all people on each plane.

A study once suggested using this metric as a measure of efficiency of the airport. An immediate implication of having this metric as an objective to reduce would be to give landing priority to planes with a large number of passengers and have planes with less passengers circle the airport until there are less passengers in other planes. This might seem reasonable at first, except when a plane runs out of fuel waiting for its turn and then crashes. Clearly this is not a desirable situation or an intended consequence, yet it has improved the metric. While extreme, this example is a mere illustration of the consequences of setting arbitrary metrics as objectives or goals. We end up with these types of unintended consequences when we use arbitrary metrics as objectives in decision-making. Similar examples abide in financial markets where arbitrary metrics are used as measures of performance.

CHAPTER 6

Preferences for Prospects with a Single Direct Value Attribute

Chapter Concepts

- Motivation for a preference function
- Characterizing prospects by a numeric measure
- Preference functions for ordering prospects by
    - Capturing qualitative features of the problem
    - Thinking about changes in preferences
    - Structural models using functional or logical relationships
- Invariance of the preference function to a strictly increasing transformation

6.1 INTRODUCTION

In Chapter 4 we discussed how to order prospects of a decision by visualizing our life with each of them. In Chapter 5, we discussed the difference between direct and indirect values. We also showed how a direct value attribute can be decomposed into other factors of interest using a structural model, and how structural models can provide consistency in decision-making by drawing on domain knowledge. This chapter discusses how to order prospects characterized by a single direct value attribute. The focus will be on attributes that can be described by a numeric measure. Chapters 7 and 8 discuss methods for ordering prospects characterized by multiple direct value attributes.

Representing a prospect by a measurable direct value attribute is particularly useful when the number of prospects is large. In this case, we can think of some function of this measure to help us order the prospects according to our preference. We call this function the ***preference function***. The higher the magnitude of the preference function, the more preferred is the prospect. Two prospects that have the same Order Rule preference will have the same magnitude on the preference function.

This chapter illustrates how to construct a preference function for prospects that can be characterized by a single numerical direct value attribute. The main concepts introduced in this chapter will also be used in future chapters to reason about preferences for prospects characterized by multiple direct values.

6.2 PROSPECTS CHARACTERIZED BY A SINGLE NUMERIC MEASURE

A measure is a number assigned to a prospect or to a direct value attribute that is used to characterize that prospect. An important criterion when assigning measures to prospets or to attributes is that they be clear and meaningful.

Note: Do Not Lose Perspective of the Prospect When You Assign a Measure to It: When assigning a measure to a prospect (or an attribute), it is important to remember that this numeric assignment is merely a characterization of the prospect in terms of some number and that it is not the prospect itself. You should still visualize your life with this prospect when thinking about your preferences using this measure, and you should not lose perspective of the prospect you are actually experiencing just because of the number you have assigned to it. For example, if the temperature in a certain location of the room (measured in Fahrenheit) is associated with a prospect, it can entail different preferences if you are sitting in a sauna or if you are studying for an exam. While the number assigned to the prospect is the same in both cases, preferences for the prospect characterized by this number will change depending on the context that you are in and the decision that you face.

EXAMPLE 6.1: Avoiding Arbitrary Attributes and Arbitrary Measures

Consider a decision about the design of a seat. If an attribute such as "comfort" is used as a direct value, it could cause confusion. It is important to define what "comfort" really means and in what context it is being used. Too often the process of clarifying the direct value attributes is skipped, resulting in lack of clarity and confusion. What adds to the problem is that arbitrary scales are often used as measures for unclear direct value attributes. For example, a scale such as

    0 = Uncomfortable,
    5 = Medium comfort,
    10 = Very comfortable … etc.

might be used. These scales are quite difficult to reason about and have different interpretations for different people. They also offer less precision. For example, it is difficult to interpret what a measure of 6 or 7.5 would imply for different people, especially when the attribute "comfort" is not clearly defined in the first place.

Similar types of attributes appear in many settings, such as "public perception" of some legislation when devising policy. This attribute is not clear. How do you know it when you see it? Is it the result of a poll? Which poll? And if so, then that poll result would be a better descriptor of that attribute. Attempts to quantify "public perception" on an arbitrary scale, such as 0 to 10, are arbitrary.

Compare the lack of clarity of an attribute like "public perception" to the clarity of an attribute like temperature of a particular location in a room whose units are clearly defined. It is always helpful to spend the time defining the direct value attributes and using clear and meaningful measures for their representation.

In most professional situations where unclear attributes are used and are associated with unclear scales, it is because the analyst has not taken the time to clarify what the attribute of value really is, or when an analyst wishes to bypass some inherent structural model or some more rigorous analysis that converts an attribute such as "public perception" into meaningful direct values.

6.3 PREFERENCE FUNCTIONS OBTAINED BY THINKING QUALITATIVELY ABOUT YOUR PREFERENCES

Having identified the direct value attributes, and spent sufficient time clarifying them and representing them with meaningful measures, the next step is to think about a preference function to order the prospects that are characterized by these measures. Our focus in this chapter will be on prospects characterized by a single direct value attribute.

Definition

Preference Function for a Single Attribute: A preference function for a prospect characterized by a single direct value attribute, Y, is a function whose argument is a measure representing Y and returns a higher magnitude for a more preferred prospect and returns equal magnitudes if two prospects are equally preferred, i.e., for two prospects characterized by measures y_1 and y_2,

If a prospect characterized by y_1 is more preferred to the prospect characterized by y_2, then $P(y_1) > P(y_2)$, and
if they are equally preferred, then $P(y_1) = P(y_2)$.

EXAMPLE 6.2: Preference Function When More of an Attribute Is Preferred to Less

The simplest form of deterministic preferences over a direct value attribute is when the decision maker states:

"I prefer more of this attribute to less."

If the attribute is represented by a numeric measure – that has a higher magnitude for a higher level of the attribute – then this statement means that the higher the numeric measure assigned to the prospect, the more preferred is the prospect. A good example of such an attribute is money. The higher the dollar value, the more preferred is the prospect.

Any monotone preference function can represent this preference that more is preferred to less. For example, if the direct value attribute is represented by a measure, y, then a preference function such as $P(y) = y^3$ or $P(y) = e^y$ could also represent this preference.

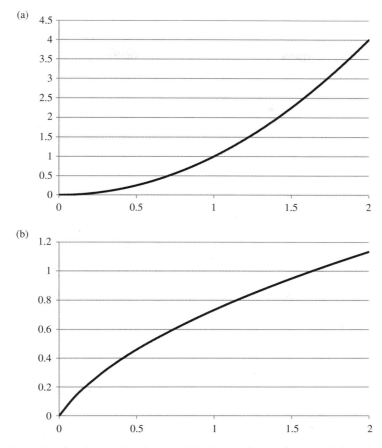

Figure 6.1. Examples of preference functions satisfying "more of an attribute is strictly preferred to less."

Note: Preferred vs. Strictly Preferred – Monotone vs. Strictly Monotone: We need to be careful when we use the phrase "more is preferred to less." If more is ***strictly preferred to less***, then the preference function needs to be ***strictly increasing***. Figure 6.1 shows examples of strictly monotonic preference functions. On the other hand, if the statement is merely "more is preferred to less," then a preference function such as that shown in Figure 6.2 would also satisfy this preference statement.

EXAMPLE 6.3: Preference Function for the Temperature of a Room

Imagine the temperature of a particular location in a room where you are studying for an important exam. It is clear that you might not want to study in a room at 0°F. At this temperature you would probably prefer more temperature to less. If you were given a choice between 0°F and 32°F, you would probably choose the latter. On the other extreme, you might not want to study in a room in which the temperature is 100°F. At that temperature level you would prefer less temperature to more. If given a choice between 100°F and 80°F, you might prefer 80°F.

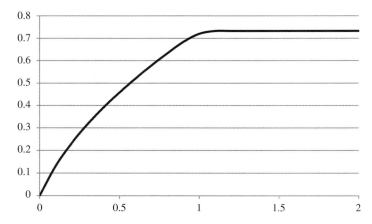

Figure 6.2. Example of a preference function satisfying "more of an attribute is preferred to less."

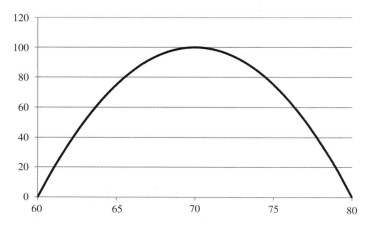

Figure 6.3. Example of a preference function for the temperature of a room.

Somewhere in between 0°F and 100°F is the temperature you would prefer most. There is no logical formula for this behavior; it is just a matter of your personal preference. Every person might have a different preferred temperature, above which they prefer less temperature to more, and below which they prefer more to less.

Having thought about these qualitative features, we now have a feel for the preference for room temperature. This is where the preference function can help us order the prospects, particularly when the number of prospects is large, such as when various temperatures are under consideration.

Figures 6.3 shows an examples of a function that may characterize preference for temperature in a room while studying. The figure shows a parabolic function of the form

$$P(t) = 100 - (t - 70)^2.$$

The most preferred temperature in this example is 70°F, and preferences drop in either direction above and below 70°F. Note that the vertical scale of this preference function is irrelevant. What matters is the relative magnitude of any two prospects (or temperature values). In the figure, the temperatures 65°F and 75°F have the same preference, and temperatures 60°F and 80°F also have the same preference.

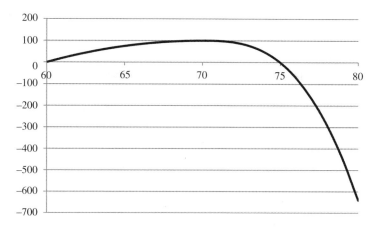

Figure 6.4. Preference function that is not symmetric around the preferred temperature.

How Has This Preference Function Helped Us?

Instead of thinking about preferences for numerous prospects, you can simplify the task by constructing a preference function. For example, you can start thinking about the preferred room temperature for studying, say t^*. You might say something like $t^* = 70$. You can then think about how preferences change as you deviate from t^*. For example is your preference for a temperature of $t = 65$ the same as your preference for a temperature $t = 75$? This could imply some symmetric preferences around $t^* = 70$. Then you can ask yourself a few more questions to verify whether indeed this symmetry holds or whether it holds up to a certain temperature after which there could be steep asymmetry.

Indeed, the preference function need not be symmetric around the most preferred value. An increase in temperature might lead to a steeper decrease in preferences than does a corresponding decrease in temperature. For example, you might be indifferent between a temperature of 60°F and a temperature of 75°F and you would like your preference function to reflect that. Figure 6.4 shows an example of such a preference function that is not symmetric around the peak. This enables us to model asymmetries in references around the preferred temperature.

$$P(t) = \begin{cases} 100 - (t - 70)^2, t \le 70 \\ 100 - (t - 70)^{2.87}, t \ge 70 \end{cases}$$

By reasoning about your preferred temperatures this way, you now have a preference function that can determine your preference when multiple prospects are present. You can also offer your preference function to agent who can now state deterministic preferences on your behalf.

You can continue to modify your preference function for asymmetries or particular preferences of your choice. For example, you might be indifferent to the range of temperatures between 65°F and 75°F. Figure 6.3 shows an example of a preference function that satisfies these preferences.

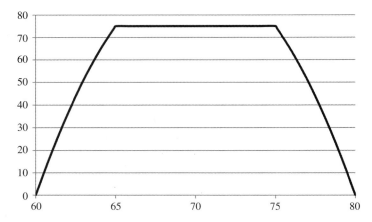

Figure 6.5. Indifference to a range of room temperatures.

EXAMPLE 6.4: Preference for the Thickness of Bread in a Sandwich

---

Suppose you ordered a deli chicken sandwich. Everything in the sandwich is fixed (the thickness of slices of chicken, the amount of lettuce, mustard, and tomato) except for the thickness of the bread. The person preparing the sandwich asks you about the thickness of bread you prefer for this sandwich given this ingredient configuration.

Most people would prefer some bread thickness to no bread at all, but they would not prefer a slice of bread that has a thickness of one foot. That would be rather hard to eat.

The preference function for the thickness of bread for the sandwich might also have a turning point like that for the room temperature. The preferred thickness of bread might depend on the given thickness of the slices of chicken, lettuce, mustard, and tomato. Figure 6.6 shows an example of such a function, with a preferred thickness of one inch.

On the other hand, if you were on a very low-carb diet, then you might prefer zero thickness of bread to any thickness at all and would rather have a sandwich wrapped in lettuce than a sandwich with bread. In this case, your preference function for the thickness of bread might be a decreasing function. The preference function might look like Figure 6.7.

6.4 THE VERTICAL SCALE OF THE PREFERENCE FUNCTION

The vertical scale that is used in the plot of the preference function is irrelevant to the ordering of the prospects. The higher the magnitude, the more preferred is the prospect. Therefore, any strictly increasing transformation applied to this preference function will preserve the order. For example, if we raise a preference function $P(t) = (t - t^*)^2$ as an argument of the exponential function, we would get

$$e^{P(t)} = e^{(t-t^*)^2}$$

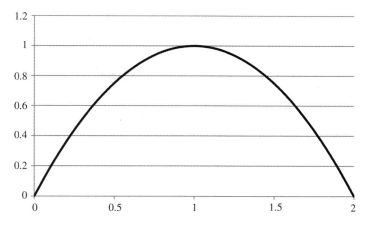

Figure 6.6. Example of a Preference Function for thickness of bread.

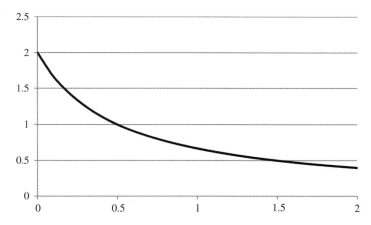

Figure 6.7. Example of a Preference Function for thickness of bread when you are on a low-carb diet.

which still has the same preference order. If one temperature is preferred to another with the preference function $P(t) = (t - t^*)^2$, for example, then it will still be preferred with the preference function $e^{P(t)} = e^{(t-t^*)^2}$.

On the other hand, if the transformation is not strictly increasing – for example, it might be constant over an interval – then some preferences might become equally preferred over a range of prospects due to this transformation. Therefore, preferences will not necessarily be preserved for all prospects.

Note: If a strictly decreasing transformation is applied to the preference function, then the lower the magnitude of the resulting preference function, the more preferred is the prospect. A strictly decreasing transformation, therefore, will still preserve the preference relationships for the prospects if we take into account the reversed magnitude relationship of the new preference function. Taking this inverse relationship into account, we can now assert that applying a strictly monotone transformation (either strictly increasing or strictly decreasing) will preserve the preference relations for the prospects.

We summarize this observation below.

> *The preference function is invariant to a strictly monotone transformation.*

6.5 EFFECTS OF A PREFERENCE FUNCTION APPLIED TO A DIRECT VALUE ATTRIBUTE ON ITS COMPONENTS IN THE STRUCTURAL MODEL

If a preference function is applied to a direct value attribute, and if that attribute is expressed deterministically in terms of other subcomponents of value using a deterministic (or structural model) relation, then the preference function will also imply preferences for these subcomponents.

To illustrate, let us refer back to the decomposition of the vertical height of a projectile into velocity of launch and time from Section 5.5. The vertical elevation of this projectile, y, at any time t is given by

$$y = ut - \frac{1}{2} gt^2,$$

where the projectile is projected vertically at a speed u under gravity, g. This logical relationship has decomposed the direct value attribute of height into two other factors, velocity and time, and so we may write for some function, f,

$$y = f(u,t)$$

Now comes the role of the preference function, where you think about your preference for the vertical height. If you prefer a higher elevation to less then your preference function $P(y)$ is monotone. The preference function can also be expressed in terms of the components as

$$P(f(u,t)).$$

If you prefer more elevation to less, for example, then your preference for time of flight is parabolic as described above, and you have a precise relation that depends on the speed at which the projectile was projected as well as gravity. Your preference for the elevation of the projectile is monotone increasing with velocity u and has a turning point with t. Just like the height of the projectile depends on both t and u, in general, preferences for t and u are specified by the preference function, but it is plausible that the preferences for t alone might depend on the magnitude of u. We refer to this as a conditional preference for t.

On the other hand, if you were in a situation where the view were blocked for given heights, then your preferences might be minimum beyond some vertical height, then increasing with the vertical height, and minimum again after another threshold height. This translates into preferences for various combinations of u and t.

> *If a direct value attribute can be decomposed into several components using a structural model, then once the preference function for the direct value attribute is specified, the preferences for all deterministic components of the structural model are also specified.*

6.6 INTERPRETING PREFERENCE FUNCTIONS BY THINKING ABOUT CHANGES IN PREFERENCES

Another way to think about the preference function for prospects characterized by a direct value attribute, say X, is to consider how preferences for increments of X change as we get more of X.

Thinking about Changes in Preferences for a Single Attribute

If we face two prospects characterized by two values of a direct value attribute, X, say x_1 and x_2, and if you prefer level x_1 to level x_2, how does this preference change if an amount w is added to the base levels of x_1 and x_2?

The idea of thinking about changes in preferences will be used in the development of preference and utility functions in future chapters. We illustrate this idea in its simplest form below.

6.6.1 Zero-Switch Preferences for Prospects as an Attribute Increases (or Decreases)

Consider your preference for two prospects characterized by a monetary amount. Suppose, for example, that you will receive either $5,000 or $10,000 (once again, this monetary amount is the characterization of the prospect, and you should visualize your life with these monetary amounts to better characterize your preferences).

If you prefer more money to less, then your preference for the prospects is clear: you prefer $10,000 to $5,000.

Now suppose that you will receive each prospect with an additional monetary amount added to it. For example, suppose that you will receive an extra amount of $1,000 added to each prospect:

If you pick the $5,000 prospect, you will receive $6,000, and if you pick the $10,000 prospect, you will receive $11,000. Will your preferences for the prospects change by adding this monetary amount?

It is clear that if you prefer more money to less, then you will still choose the $10,000 prospect over the $5,000 regardless of the additional monetary amount that will be added (or even subtracted).

We say that you have "zero-switch" ordinal preferences for money if your preferences for prospects characterized by money do not change when additional monetary amounts are added to (or subtracted from) the prospects.

Definition: Zero-Switch Ordinal Preferences
You have "zero-switch" ordinal preferences for prospects characterized by a direct value attribute if your preference for the prospects does not change as additional amounts of that attribute are added to (or subtracted from) the base values of the prospects.

To formalize this idea, suppose you face two prospects characterized by different levels of an attribute, X, say x_1 and x_2, which could represent monetary values. From the properties of preference functions,

If you prefer x_1 to x_2, then $P(x_1) > P(x_2)$.

Now suppose that this preference relationship remains the same as the level of both attributes either increases or decreases by any amounts, i.e,

You prefer $(x_1 + w)$ to $(x_2 + w)$ for all levels of w (positive or negative) for all $x_1 \neq x_2$

This implies that

$P(x_1 + w) > P(x_2 + w)$ *for all levels of w (positive or negative) for all $x_1 \neq x_2$*

Also, if you are indifferent between two prospects, x_1 and x_2, then you remain indifferent between the prospects for all levels of w.

If $x_1 \sim x_2$, *then $(x_1 + w) \sim (x_2 + w)$ for all levels of w (positive or negative) for all $x_1 \neq x_2$*

A preference function that satisfies these conditions is said to satisfy the ***zero-switch condition for a single attribute*** and it must be either strictly monotone or flat.

Definition
Zero-Switch (Deterministic) **Preferences for a prospect characterized by a single attribute**
   A preference function $P(x)$ satisfies the zero-switch preference condition if the difference $P(x_1 + w) - P(x_2 + w)$ does not change sign with w, for all $x_1 \neq x_2$

A function does not change sign over an interval if it is strictly positive or strictly negative or zero over that interval.

A preference function $P(x)$ satisfies the zero switch condition if and only if it is either a strictly monotone or a flat function.

Note: This monotonicity result is useful in understanding the properties of changes in preferences. For example, if we have a preference function, $P(x) = x^3$, which is strictly monotone, and if we prefer x_1 to x_2, which implies that $x_1 > x_2$ in this case, then we will always prefer to $(x_2 + w)$, regardless of the level of w, and indeed $(x_1 + w)^3$ will always be larger than $(x_2 + w)^3$.

6.6.2 One-Switch Preferences as an Attribute Increases (or Decreases)

Suppose that your preferences between two prospects characterized by a single attribute might change, but only once, if an additional amount is added to each attribute. What must be shape of the preference function?

Consider for example the temperature of a room we discussed earlier. If given a choice between 0°F and 32°F you might prefer 32°F. But what if we add a certain amount of temperature to each prospect, would your preference change? If we add 100 degrees to each prospect you choose, so that you will end up with either 100°F or 132°F, then you might now choose the prospect 0°F so as not to end up studying in a room whose temperature is 132°F. Your preference for the two temperatures, 0°F and 32°F, changes as the temperature you get increases. It is quite plausible that further temperature increases will not change your preference.

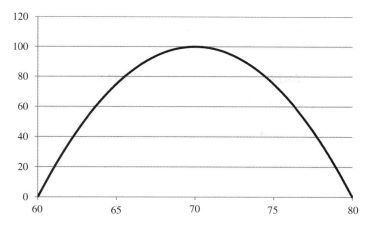

Figure 6.8. Preference function with at most one turning point.

> *A preference function satisfies ordinal one-switch preferences for a single attribute if whenever $x_1 \succ x_2$, then either $(x_1 + w) \succ (x_2 + w)$ as w increases or there is a unique value of w, say w_0, below which $(x_1 + w) \succ (x_2 + w)$ and above which $(x_1 + w) \prec (x_2 + w)$. If $(x_1 + w) \sim (x_2 + w)$ for all w, then the preference function also satisfies the ordinal one-switch condition because preferences do not switch.*

A preference function $P(x+w)$ is ***ordinal one switch*** if for any pair of consequences either one is preferred to the other for all w, or they are indifferent for all w, or there exists a unique level of w above which one consequence is preferred, below which the other is preferred.

> *A preference function satisfies the ordinal one-switch preference condition for a single attribute if and only if the function has at most one turning point.*

If the function has a single minimum, e.g., $P(x) = x^2$, which has a minimum at zero, then it is clear that preferences can change only once as we increment any two prospects. Similarly, if the function has a single maximum, as shown in Figure 6.3, which we repeat for convenience in Figure 6.8, then preferences can still switch at most once.

We shall revisit the idea of changes in preferences numerous times throughout the book. We presented it here in its simplest form to familiarize the reader with this concept.

6.7 A NOTE ON THE TYPES OF PREFERENCE FUNCTIONS THAT WE HAVE PRESENTED

The preference functions we have discussed so far have arguments comprised of numerical measures of direct value attributes on a continuous domain. These preference functions apply to a wide variety of situations, e.g., money, temperature, humidity, acidity, and remaining life time, among many others.

We have not considered other types of preference functions that involve categorical attributes. Think about preference for a car where the attributes are transmission type (Automatic, Manual), the make of car, convertible or non-convertible, and its color, beyond other numerical attributes such as cost, leg space, and mileage. In this case, you might want to consider logical relations for the preference function. For example,

> IF the car is a Convertible, Then I prefer Color = Red to Color = Blue.
> ELSE IF the car is not a Convertible, then I prefer Color =Blue to Color=Red.

This type of preference statement also characterizes preferences over attributes color and make of the car, and it is sufficient for an agent or a computer to choose the car color on your behalf depending on whether it is a convertible or not. There is a lot of literature in the artificial intelligence community that deals with these types of preference structures. Our focus in this book is on functions whose attributes are numerical and those that span a continuous domain.

Note: For sophisticated readers, color can also be represented as a measure on a continuous scale in terms of its wavelength or frequency. But it might be difficult to explain that to a decision-maker facing a car purchase decision.

6.8 SUMMARY

A measure is a number assigned to a prospect or to an attribute characterizing the prospect. The measure is not the prospect itself; it is a characterization of the prospect. You still need to visualize the prospect (and your life) with this numerical measure to help you think about your preferences. A measure assigned to a direct value attribute should be meaningful.

A preference function is a function that returns a higher magnitude for a more preferred prospect and a lower magnitude for a less preferred prospect. It is particularly helpful when a large number of prospects is present.

Sometimes merely thinking about the shape of your preference function may provide sufficient clarity about your preferences for a particular decision.

The preference function is invariant to a monotone transformation because a monotone transformation is order-preserving and the magnitude of the preference function is not important.

A monotone preference function preserves the ordering of two prospects even as the base level of the attribute changes. A preference function with a single turning point allows for only one change in preferences.

When constructing a preference function for a single attribute, we can think of

- Capturing qualitative features of the problem, and
- Thinking about preference changes as we increase the value of the attribute

  1. Zero-switch preference functions
  2. One-switch preference functions

ADDITIONAL READING

Abbas, A. E. and D. E. Bell. 2011. One-switch independence for multiattribute utility functions. *Operations Research* 59(3): 764–771.

Abbas, A. E. and D. E. Bell. 2012. One-switch conditions for multiattribute utility functions. *Operations Research* 60(5): 1199–1212.

Abbas, A. E. and D. E. Bell. 2015. Ordinal one-switch utility functions. *Operations Research* 63(6): 1411–1419.

Aczél, J. 1966. *Lectures on Functional Equations and Their Applications.* Academic Press, New York.

Bell, D. E. 1988. One-switch utility functions and a measure of risk. *Management Science* 34(12): 1416–1424.

Howard, R. A. and A. E. Abbas. 2015. *Foundations of Decision Analysis.* Prentice Hall, New York.

Matheson, J. E. and R. A. Howard. 1968. An introduction to decision analysis. In R. A. Howard and J. E. Matheson, eds. *The Principles and Applications of Decision Analysis,* Vol. I. Strategic Decisions Group, Menlo Park, CA, 1968.

Keeney, R. and H. Raiffa. 1976. *Decisions with Multiple Objectives: Preferences and Value Tradeoffs.* John Wiley, New York. Reprinted, Cambridge University Press, New York (1993).

APPENDIX: HOW PEOPLE GET IT WRONG

Do Not Rush to Build a Preference Function If You Do Not Need to

Remember

The goal is not to build a preference function; the goal is to make a good decision

Many analysts often rush to build a preference function when called upon to help with a decision even if the decision can be resolved by visualizing the prospects or by having a conversation with the customer to achieve clear action. They do that because this is what they are comfortable doing.

In many classes on decisions with multiple attributes, the semester is spent focusing on forms of preference functions and the mathematical content of building a preference function. Arbitrary constructed scales for the attributes are often used and the validation check for whether the preference function actually matches preferences of the decision-maker or whether the assumptions made are meaningful is often overlooked.

When You *Do Not* Need to Build a Preference Function

When the prospects are few: Comparing **two** jobs: Salary, Vacation, Commute Time. They are just a few prospects. You might not want to specify a preference on the whole domain. Instead, try to monetize what the difference in each attribute is worth to you. How much are the extra vacation days worth to you? And is this compensated by the higher salary you are getting at another job?

When Visualizing the Prospect Is Sufficient and Getting More Information Is More Important

If you are helping a student with a college decision, you probably would not want to build a preference function for temperature. Rather, you would have a conversation about the different attributes involved in a college decision to help them think about their preferences.

For example, you can think about the college ranking, the geographical location, the student life, the job prospects, and the social network they might build. You can gather information about all these aspects as well as the faculty who teach there, and all other aspects of the school. In this particular situation, gathering more information about the schools and being clear about the factors affecting the preferences might be more useful than building a preference function.

When the Attribute Is Deterministically Immaterial

Remember that a preference function is built for direct value attributes. If the attribute is indirect, then it will not change the deterministic rankings of the prospects.

When You *Do* Need to Build a Preference Function

1. When you have a large number of prospects: Remember a preference function requires the characterization of a prospect in terms of attributes. Those attributes are the direct value attributes. The function is particularly helpful when you have a continuum (or a large number of prospects) such as possibly health outcomes and costs of medical surgery because it would be difficult to reason about the continuum of preferences, particularly in important decisions such as remaining lifetime.
2. When you have repeat situations and you would like to have consistent preferences.
3. When you would like to have transparency in how you ranked your prospects.
4. When you would like an agent or a computer to make a decision on your behalf.
5. When the process of thinking about the preference function helps you get clarity on your preferences.

CHAPTER 7

Preferences for Prospects with Multiple
Direct Value Attributes

Chapter Concepts

- The concept of an isopreference contour when a prospect has multiple direct value attributes.
- The concept of trade-offs among the attributes using the isopreference contours.
- Methods for characterizing trade-offs when multiple direct value attributes are present:
  - Laws of physics, engineering, or accounting or logical relationships
  - Direct assessment of trade-offs
  - Assuming a functional form of isopreference contours and then assessing its parameters
    - Linear isopreference contours (with constant slope)
    - Nonlinear isopreference contours
  - Capturing important features of the problem
  - Zero-switch and one-switch ordinal preferences.

7.1 INTRODUCTION

We have discussed how to characterize prospects of a decision in terms of direct value attributes and the importance of using clear and meaningful measures for their representation. In Chapter 6, we also discussed methods for constructing preference functions for prospects characterized by a single direct value. This chapter discusses how to construct preference functions for prospects characterized by multiple measures representing multiple direct value attributes.

In some cases, especially when the prospects are few, we can (again) go through a mental exercise of thinking about our future life with each prospect, as we did in Chapter 4. By so doing we can rank the prospects without much calculation. In other cases, we might have multiple prospects, but ordering them might still be straightforward. For example, if prospects are characterized by a single attribute and if we prefer more of the attribute to less, then the ordering can be achieved by any monotone function. If we prefer more of the attribute to less, up to a certain point, after which we prefer less to more, then we may use a function with a single turning point and adjust its parameters to suit the decision-maker's preferences.

When prospects are characterized by multiple attributes, it might still be possible to order the attributes easily using qualitative statements, but in most cases, there will be an added complexity when ordering them.

To illustrate, suppose that a prospect is characterized by two attributes, X and Y, such as health and wealth, and suppose that "you strictly prefer more of each attribute to less." Suppose also that one prospect, A, has more of attribute X and less of attribute Y than does prospect B. Which prospect do you prefer, A or B? The one with the higher value of X and lower value of Y, or the one with the lower value of X and higher value of Y?

Note: Unlike the case of a single attribute, the statement "more of an attribute is strictly preferred to less" is not sufficient to provide a total order for the prospects in this case. You need to think about how much less of one attribute would compensate for an increase in the other. This is the idea of ***making trade-offs***. Of course, if prospect A has more of each attribute than prospect B, then the order is simple, as A would dominate prospect B on all attributes, and it would therefore be more preferred. But this type of dominance relation for the prospects is not always present. Quite often one prospect will indeed have more of one desirable attribute and less of another. The idea of making trade-offs between the attributes leads naturally to concept of an *isopreference contour*, which is the set of points having the same preference.

7.2 BASIC NOTATION AND DEFINITIONS

To order prospects characterized by multiple attributes, it is still important to use meaningful measures for each attribute and preferably those that can be represented by a clear numerical scale. A preference function can still be used to rank the prospects in this case. The magnitudes of the direct value attributes comprise the arguments of the preference function.

For direct value attributes such as Money and Health, with values x and y, we use the notation,

$$P(x, y) \text{ or sometimes write } P(\text{Money}, \text{Health})$$

for the preference function. For more than two attributes, we use the general notation

$$P(x_1, x_2, ..., x_n)$$

for levels of direct value attributes, $X_1, X_2, ..., X_n$.

In general there is no requirement for the preference function to have a particular form. If a structural model that relates the attributes exists, then the preference function takes the form of the structural model. In some cases, however, there might not always be a clear structural relationship among the attributes, and so a preference function that provides the order is needed.

It is tempting at first to think of an individual preference function for each attribute and then to combine those functions into some form to get the overall preference of the prospect, i.e., we might think of some preference functions $P_1(x_1), P_2(x_2), ..., P_n(x_n)$, and some aggregation of the form

$$P(x_1, x_2, ..., x_n) = f\big(P_1(x_1), P_2(x_2), ..., P_n(x_n)\big).$$

A special case of this might even use the additive form

$$P(x_1, x_2, ..., x_n) = P_1(x_1) + P_2(x_2) + ... + P_n(x_n),$$

where f is an additive function.

The bulk of the literature has used multiattribute preference functions using individual preference functions, and especially those with the additive form. While the additive form captures a variety of trade-offs, it is not always the case that preferences can be expressed this way, particularly when there are interactions among the attributes (when preferences for an attribute changes with the level of the other attributes), or when the shape of the preference function for one attribute changes as the levels of the other attributes vary.

This chapter describes methods for constructing preference functions for prospects characterized by multiple attributes. Chapter 8 discusses the additive preference function in more detail to better understand its properties and implications.

7.3 ISOPREFERENCE CONTOURS: THE ADDED COMPLEXITY WITH MULTIPLE ATTRIBUTES

One big difference between preference functions for single and multiple direct value attributes is the sufficiency of the order relation that is specified when we say:

more of any attribute is strictly preferred to less.

For a single attribute, such as money, this statement is sufficient to order the prospects completely. Any strictly monotone function would be a suitable preference function for the prospects. When prospects are characterized by more than one direct value attribute, the statement **"*more of any attribute is strictly preferred to less*" is not sufficient to order the prospects completely.** This is why we need to construct a preference function for the multiple attributes, and to think about trade-offs, even in this simple case where more is preferred to less.

> **Definition: Multiattribute Preference Function**
> A multiattribute preference function is a function of two or more direct value attributes that returns a higher value for the more preferred prospect and equal values if two prospects that are equally preferred. For example, for two attributes,
>
> If a prospect (x_1, y_1) is more preferred to (x_2, y_2), then $P(x_1, y_1) > P(x_2, y_2)$, and if (x_1, y_1) is equally preferred to (x_2, y_2), then $P(x_1, y_1) = P(x_2, y_2)$.

A second difference between preference functions for single and multiple attributes when more is strictly preferred to less is that for a single attribute, the equation $P(x) = a$ defines a unique value of attribute X. For multiple attributes, however, even when more of an attribute is strictly preferred to less, the relation $P(x, y) = a$ defines a set of points,

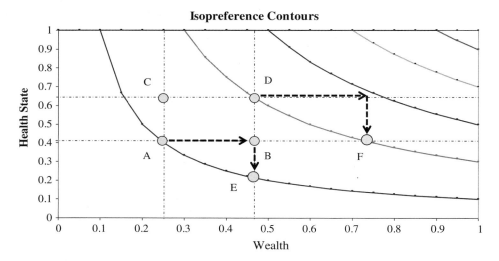

Figure 7.1. Examples of isopreference contours of a preference function.

known as the **level set**. The level set, a, of a preference function is determined by the set of prospects that satisfy $P(x,y) = a$, i.e.,

$$\{(x,y) : P(x,y) = a\}$$

Because prospects on the level set have the same preference, they are also said to lie on the same *isopreference contour*.

Definition

An *isopreference contour* is a level set of the preference function. All points on the same isopreference contour have the same preference.

Note: Given a set of prospects on specified isopreference contours, we can determine if a prospect is more preferred or less preferred to another by identifying the contour on which each prospect lies.

When more of an attribute is strictly preferred to less, the isopreference contours look like those in Figure 7.1.

Figure 7.1 represents isopreference contours of a preference function of two attributes such as health state (as determined by the equivalent remaining lifetime) and wealth. On the domain shown in the figure, more of these attributes is strictly preferred to less. Prospects A and E lie on the same isopreference contour and are, therefore, equally preferred. The same applies to prospects D and F. In both of these cases, each of the prospects that lies on the same isopreference contour has more of one attribute and less of another to be indifferent. Simple stated, this means that we need to

"win some – lose some"

to be indifferent.

Prospects A and B do not lie on the same contour. Both A and B have the same health level, but B has a higher wealth level. Since we always prefer more money to less,

B is preferred to A. Similarly, prospects C and A have the same wealth level, but C has a better health level. Since we prefer a better health level to a worse one, C is preferred to A. We also know that D is preferred to A, B, or C, since it provides a better health level and more wealth.

The difficulty arises when we compare prospects B and C, because B has a higher wealth and C has a better health level. Here, a trade-off between having better health and more wealth needs to be determined before we can order the prospects.

Shape of the Isopreference Contours When More of an Attribute Is Preferred to Less

To understand the intuition behind the shape of the isopreference contours if you strictly prefer more of an attribute to less, suppose you currently have a fixed level of attribute X and a fixed level of attribute Y, i.e., you have a prospect characterized by levels (x, y).

You cannot be indifferent between your current prospect and a prospect that has more of X and the same value of Y, i.e., $(x+w, y)$, $w > 0$. This would be a contradiction to the statement "more is strictly preferred to less."

If more of an attribute is preferred to less, you have to "*win some and lose some*" to be indifferent.

Similarly, you cannot be indifferent between the prospect (x, y) and a prospect that has more of Y and the same value of X, i.e., $(x, y+b)$, $b > 0$.

If you are offered a prospect that has a higher value of X, and you are indifferent between that prospect and the prospect (x, y), then that new prospect must have a lower value of Y, i.e., $(x+w, y-b)$, $w, b > 0$.

The more you get of one attribute, the less you need of another for you to be indifferent. This relationship explains the shape of the contours in Figure 7.4. The challenge is to be able to specify the values of w, b that make you indifferent.

We can also verify the shape of the isopreference contours in this case by calculus as we illustrate below (and if you do not know calculus, do not worry about it, the intuition provided above is sufficient).

Shape of the Isopreference Contours When More of an Attribute Is Preferred to Less – Using Calculus

If we are indifferent between two prospects, they must lie on the same isopreference contour. Let the magnitude of the preference function at the contour be a constant, a. Therefore, the isopreference contour on which the points lie is defined by the equation

$$P(x, y) = a.$$

Because the preference is constant across this contour, the total change in preference (total derivative of the preference function) must be zero and so

$$dP(x, y) = 0 \Rightarrow \frac{\partial}{\partial x} P(x, y)dx + \frac{\partial}{\partial y} P(x, y)dy = 0. \tag{1}$$

If more of an attribute is *strictly* preferred to less, then the partial derivatives are positive, and so

$$\frac{\partial}{\partial x} P(x,y) > 0 \text{ and } \frac{\partial}{\partial y} P(x,y) > 0.$$

Rearranging (1) gives

$$\frac{dy}{dx} = -\frac{\frac{\partial}{\partial x} P(x,y)}{\frac{\partial}{\partial y} P(x,y)} < 0.$$

This equation implies that the slope of the isopreference contours must be negative, and so they must have the shape of the contours of Figure 7.1.

Note: Isopreference contours cannot cross. If they did, then this would be an inconsistency, as we are indifferent between a set of prospects that lie on one contour and a set of prospects that lie on the other, yet we have a preference for one contour over the other. This would be a contradiction.

Once the isopreference contours have been determined, we have a complete ordering of the preferences for the prospects. The actual magnitude of the preference function is not important. What matters with the preference function is that the higher its magnitude, the more preferred is the prospect.

7.4 CONSTRUCTING PREFERENCE FUNCTIONS FOR MULTIPLE ATTRIBUTES

7.4.1 Structural Models Using Laws of Physics or Engineering or Economic Principles

Engineering or accounting principles and physics may (quite often) determine the preference function (or the trade-offs) using a structural model. Being clear on the direct value attributes and the structural relations that may be present simplifies the determination of the preference function significantly. We have seen, for example, the decomposition of energy into mass and velocity using a structural model. Preference for mass and velocity is determined completely by identifying the direct value attribute of energy and the structural domain knowledge decomposition of energy into its components. The preference function is in fact the structural model itself in this case.

One common error when constructing preference functions is to ignore the structural preferences that might be present. For example, while it is possible to assess trade-offs between mass and velocity of a generator using isopreference contours, it is highly unlikely that the assessed contours will match those produced by the structural model formula. Therefore, if preference is for energy and if we ignore the structural relationships, then we will likely end up with a preference function that is inconsistent with our preference for higher energy.

Likewise, if we are considering a bridge with two pillars, and we would like to decrease the diameter of one due to space limitations and increase the diameter of another, we do not assess the isopreference contours. The bridge will collapse. Rather, we rely on the structural model that determines requires trade-offs.

7.4.2 Direct Assessment of Trade-Offs Using Isopreference Contours

If an immediate functional relationship between the direct value attributes is not clear, either by an engineering quantity or by an accounting principle or some other structural model or logical means, then one method to determine the isopreference contours of the preference function is by direct elicitation. We discuss this idea in more detail below.

Consider the trade-offs between consumption and remaining life years. There is no explicit accounting formula for the trade-offs between these attributes. It is a matter of personal preference. But we do know that more life years and more wealth are preferred to less (for most scenarios). Therefore, as we have discussed, we might expect the contours to look like those of Figure 7.1. To determine those contours by direct elicitation, we start with a prospect, say (c,l), that provides c units of consumption ($) and l units of remaining life years. We know that the contours have the shape of Figure 7.1 because more is preferred to less. We would then ask the decision-maker:

> *"How much increase in consumption Δc makes you just indifferent to a decrease in remaining life time Δl?"*

Figure 7.2 illustrates this question graphically. Of course, the value of Δc required to achieve indifference for a given Δl might depend on the actual levels of (c,l) being considered, and they might vary for each person. It might very well be the case that you would trade more money for life years if you had fewer years to live than if you had more. Therefore, the slope of the isopreference contours, while negative, need not be constant.

The approach of direct assessments of the contours is feasible for a few contours or points, but it can be tedious if conducted on the entire domain of the attributes. It

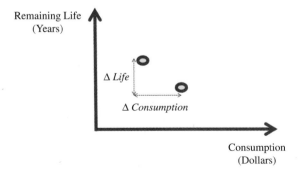

Figure 7.2. Direct assessment of trade-offs.

is, however, usually a good first step toward fitting the assessments obtained to a given functional form for the isopreference contours or for verifying a chosen form for the preference function.

Direct trade-off assessments are also more practical when more of an attribute is preferred to less because it is relatively easier to think about the increase in one attribute needed to compensate for a decrease in another. The next section describes some possible functional forms of isopreference contours to which some assessments might be fitted and used to order the prospects.

Note: Some people might at first be hesitant to answer these types of explicit trade-off questions that involve health and consumption. In my experience, it is a matter of proper coaching, and most people can eventually state a number, particularly for small changes in health, after some training and familiarity with the questions. Government officials might also be reluctant to put an explicit price on lives saved, and companies might be reluctant to put a price on safety or injuries avoided. Yet, they continue to make decisions that involve these quantities and they are implicitly making that trade-off anyway. It is better to think about the trade-offs directly and understand the implications of the decisions than to make decisions and then later discover inconsistencies and unintended implications. By explaining to companies and government officials that are they making explicit trade-offs about safety anyway by the decisions they make, people become more willing to answer these types of trade-off questions.

Dialogue between Instructor (I) and Class (C)

I: Is everyone clear on the idea of trade-offs for prospects characterized by multiple direct value attributes?

C: Yes, when more of an attribute is preferred to less, it characterizes the amount of increase in one attribute needed to make you indifferent to a decrease in another.

I: Correct, and this amount need not be constant, it might very well depend on the level you have of each attribute.

C: Does an increase in one attribute always need to correspond to a decrease in the other?

I: If more of every attribute is preferred to less, then yes. "win some – lose some." These are the types of contours that we shall focus on for direct assessment. But of course, it is plausible that you encounter situations where the problem is formulated such that more of an attribute is not preferred to less, and contours have different shapes. You might even need an increase in two attributes to be indifferent in this case.

C: And so, by repeating these assessments, we can envision contours of constant preference. Each point on the contour has the same preference.

I: Correct, and we call those contours isopreference contours.

C: Can we order prospects using the isopreference contours?

I: Yes, recall that an isopreference contour is just a level set of a preference function: the higher the magnitude associate with the isopreference contour, the more preferred is the prospect.

C: What if there are many prospects? Like prospects of medical surgery? It would be quite tedious to make all these assessments.

I: In this case, we can think in terms of some functional forms for the contours and then assess their parameters. Changing the parameters changes the shapes of the isopreference contours.

C: How do we assess the parameters of those functional forms?

I: By making a few trade-off assessments and then fitting the parameters of the functional forms to match these assessments, so that prospects that have the same preference lie on the same isopreference contours. But if you do that, you need to use functional forms that allow for general shapes.

C: How do we choose the functional forms?

I: First, you need to understand the implications of the functional forms that you use. In the next section, we shall discuss a few forms of contours characterizing the preference relation – more of an attribute is preferred to less – and illustrate their implications.

7.5 ASSESSMENT OF TRADE-OFFS USING A FUNCTIONAL FORM FOR THE ISOPREFERENCE CONTOURS

When there are a large number of prospects in the decision, direct assessment of trade-offs for each prospect could be tedious. One possible remedy for this problem is to assume a functional form for the shape of the isopreference contours and then assess its parameters. This is a good approach if that chosen functional form allows for a wide range of shapes. It should also meet certain qualitative features about the decision-maker's preferences. Assuming an arbitrary form, however, especially when there is already a physical or engineering relation between the attributes is simply asking for trouble.

Our focus in this session is on functional forms of isopreference contours that characterize the relation "more of any attribute is preferred to less." As we have seen, these contours must have a negative (non-positive) slope. Contours that do not characterize this relationship are better assessed using other means, as we shall see in Section 7.4.

7.5.1 Linear Trade-Offs (Linear Isopreference Contours)

LINEAR EQUATION FOR ISOPREFERENCE CONTOURS The simplest form of isopreference contours satisfying the relation "more of any attribute is preferred to less" are linear contours having a negative slope.

Linear isopreference contours can be described by the linear relation,

$$y = -cx + b, \quad c > 0, \ b \text{ arbitrary}$$

An isopreference contour following this relation asserts that the decrease in y corresponding to a unit increase in x is a constant equal to a, regardless of the values of x and y.

Figure 7.3 shows an example of such linear contours. This type of trade-off could be useful for unit conversion (for example if x is units of energy, the constant c might

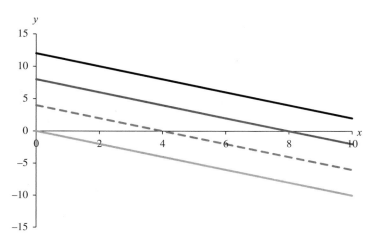

Figure 7.3. Slope of isopreference contours.

represent the unit conversion [price per unit energy], and y and b represent monetary values).

By direct calculus, the slope of the isopreference contours is constant, i.e.,

$$\frac{dy}{dx}\Big|_{P(x,y)=a} = -c = \text{ constant.}$$

CORRESPONDING LINEAR PREFERENCE FUNCTION A preference function that satisfies this linear isopreference contour condition (and is popular because of its simplicity) is the additive linear form,

$$P(x,y) = w_x x + w_y y, \tag{2}$$

for two attributes, where w_x, w_y are weight parameters.

Note that $P(x,y) = a$ in the above form implies

$$w_x x + w_y y = a,$$

or

$$y = \frac{a}{w_y} - \frac{w_x}{w_y} x,$$

which is the linear equation for the isopreference contour.

This preference function implies that (i) there are some weights that need to be assessed, and (ii) that the higher the weight, the "higher the priority" of its corresponding attribute.

To illustrate that the isopreference contours of this function do indeed have constant slope, note that if $P(x,y) = w_x x + w_y y$, then along any isopreference contour of magnitude, a, we have

$$dP = 0 \Rightarrow \frac{\partial P(x,y)}{\partial x} dx + \frac{\partial P(x,y)}{\partial y} dy = 0$$

This implies that the slope is indeed constant because

$$w_x dx + w_y dy = 0 \Rightarrow \frac{dy}{dx} = -\frac{w_x}{w_y} = \text{constant.} \qquad (3)$$

EXAMPLE 7.1: Assessing the Weights of the Linear Function

To assess, the weights, we need to interpret them in terms of the implied trade-offs. To provide an intuitive interpretation for the weights, consider two prospects, $(x_1, y_1), (x_2, y_2)$. On the same isopreference contour, we must have

$$P(x_1, y_1) = P(x_2, y_2).$$

This implies that

$$w_x x_1 + w_y y_1 = w_x x_2 + w_y y_2.$$

Many times, the weights are non-negative and sum to one, i.e., $w_x + w_y = 1; \quad w_x, w_y \geq 0$
Re-arranging gives

$$\frac{\Delta y}{\Delta x} = \frac{y_2 - y_1}{x_2 - x_1} = -\frac{w_x}{w_y} = \text{constant.}$$

This implies that the trade-offs between the attributes are constant. In other words, the increase in levels of attribute X needed to compensate for a decrease in the level of attribute Y is constant and does not depend on the levels of either X or Y. The ratio of the weights would be assessed by asking:

"*How much increase in* X *is needed to compensate for a unit decrease in* Y*?*"

Suppose the decision-maker answers, c. If the weights are non-negative and sum to 1, then this question is sufficient to determine the weights because we have two equations:

$$\frac{w_x}{w_y} = c, \qquad w_x + w_y = 1$$

This implies that $w_y = \dfrac{1}{1+c}, \quad w_x = \dfrac{c}{1+c}$

Changing the values of the weights of the preference function changes the slope of the isopreference contours and the corresponding trade-offs.

Note: Caution with the Assessments Using "Importance" Questions: Example 7.1 demonstrated how the weights could be assessed with this linear preference function in terms of trade-offs. Unfortunately, they are often assessed arbitrarily using questions like:

"*How many times more important is this attribute than that?*"

It is difficult to reason about (and even understand) the meaning of this question, yet this type of modeling is widely used. The weights in this linear form represent trade-offs between the attributes. It is also useful to think of them as unit conversion factors. For example, if preference is for money and energy, then the weight corresponding to energy represents the conversion from energy to money, or the unit price of energy. It should not be interpreted in terms of how much more important energy is to money.

IMPLICATIONS OF LINEAR ISOPREFERENCE CONTOURS This constant trade-off implied by linear isopreference contours poses a major restriction on the decisions where this preference function can be used. It requires situations where you always have a constant trade-off across the domain of the attributes.

One such situation might be trade-offs for investment cash flows (a net present value function), where the increase (or decrease) in one attribute needed to compensate for the decrease (or increase) in another may be determined by the discount factor.

For consumption cash flows, however, this might not be an appropriate function over the entire domain of the attributes. You cannot constantly trade your entire consumption for this year in return for next year's consumption, because you will not live to receive it.

Other examples where this preference function might not be suitable include the two attributes of remaining life years and consumption, which need not have constant trade-offs. If we really want to show the limits of this function, think of two attributes: "wealth" and "amount of air we breathe." It is obvious that the amount of "wealth" increase needed to compensate for a decrease in "air supply" will depend on the available air we have. The additive function would not be suitable for this situation.

7.5.2 Nonlinear Isopreference Contours

We have seen that linear preference functions have a constant (negative) slope, and that this, while simple, need not be a desirable property on the entire domain because it implies that trade-offs are constant regardless of how much you have of each attribute. The contours had the form

$$\frac{dx}{dy} = \text{constant.}$$

A more general property of the slope of the isopreference contours is that it could be some function of the attributes themselves. This would imply contours of the form

$$\frac{dx}{dy} = f(x, y) \tag{4}$$

While useful, $\frac{dx}{dy} = f(x, y)$ is a very general formulation that does not provide much specificity into the shape of the isopreference contours. Let us provide some more specific families of contours that might be used.

ISOPREFERENCE CONTOURS BY THE RECIPROCAL EQUATION A simple form of an isopreference contour having a negative nonlinear slope could be

$$\boxed{\frac{dx}{dy} = -\frac{x}{y}, \quad x, y > 0}.$$

(5)

Rearranging gives

$$\boxed{\frac{dx}{x} = -\frac{dy}{y}, \quad x, y > 0}$$

or

$$\frac{\left(\dfrac{dx}{x}\right)}{\left(\dfrac{dy}{y}\right)} = -1$$

This implies that the (small) fractional increase in x needed to compensate for a (small) fractional decrease in y is a constant and is equal to 1. The difference between this formulation and the linear contour formulation is that for the linear contours we asked whether the change in X needed to compensate for a change in Y is constant. Here we ask the decision-maker to verify the following question.

Assessment Questions

"Is the fractional increase in X needed to compensate for a small fractional decrease in Y constant?"

If the answer is yes, then we ask:

"Is the fractional change in X equal to the fractional change in Y needed for indifference?"

Direct integration shows that the isopreference contours that satisfy this property are determined by the reciprocal equation

$$\boxed{y = \frac{a}{x}, \quad x, y > 0}$$

(6)

for some positive constant a.

Figure 7.4 shows the isopreference contours. The figure shows a clear difference between the shape of these contours and those of the linear form. It is clear that the contours are no longer linear and that the amount of x needed to compensate for a decrease in y increases as the level of y decreases: the less amount of attribute y, the more of attribute x is needed for a unit reduction. This is a deviation from the linear contours. Note that each value of the parameter a corresponds to a specific isopreference contour.

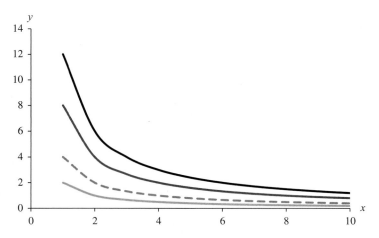

Figure 7.4. Isopreference contours corresponding to the "equal small fractional change" equation.

EQUATION OF THE CORRESPONDING PREFERENCE FUNCTION Rearranging shows
the preference function corresponding to these contours has the form

$$\boxed{P(x,y) = xy, \ \ x,y > 0}. \tag{7}$$

ISOPREFERENCE CONTOURS BY THE SCALED RECIPROCAL EQUATION Suppose we
add another degree of freedom to vary the slope of the isopreference contours to get
the form

$$\boxed{\frac{dx}{dy} = -\eta\frac{x}{y}, \ \ x,y > 0} \tag{8}$$

where η is a constant that needs to be assessed. Rearranging gives

$$\frac{\left(\dfrac{dx}{x}\right)}{\left(\dfrac{dy}{y}\right)} = -\eta,$$

which implies that the fractional change is a constant, η, and not necessarily equal to
one as it was in the previous reciprocal case. Direct integration gives an equation for
the isopreference contours as

$$\boxed{x = \frac{a}{y^{\eta}}}.$$

Figures 7.5 and 7.6 show the shapes of these contours corresponding to different values
of η. The parameter η allows for further flexibility in the shape of the contours than that
provided by the reciprocal equation. Now we have more flexibility in the trade-offs that
we can model. It is also clear that more of x is needed to compensate for a decrease in
y as the level of y decreases.

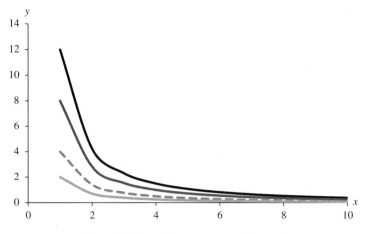

Figure 7.5 Isopreference contours for $\eta = 2$.

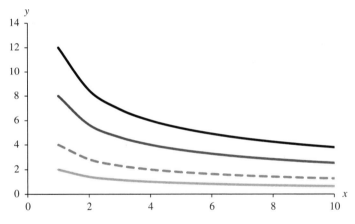

Figure 7.6. Isopreference contours for $\eta = 0.5$.

Corresponding Preference Function

These contours corresponding to the equation $x = \dfrac{a}{y^{\eta}}$ correspond to a preference function of the form

$$P(x, y) = xy^{\eta}, \quad x, y > 0, \tag{9}$$

which is known as the ***Cobb-Douglas constant elasticity of substitution preference function.*** Of course, any monotone transformation of this preference function has the same contours.

Assessment Questions

To elicit the parameter η, we ask the decision-maker the following two questions:

> *"Is the fractional increase in X needed to compensate for a small fractional decrease in Y constant?"*

If the answer is yes, then we ask,

> *"How much fractional increase in X is needed to compensate for a small fractional decrease in Y?"*

ISOPREFERENCE CONTOURS WHOSE SLOPE SATISFIES A POWER EQUATION
We can further generalize the equation of the slope of the isopreference contours
to get

$$\frac{dx}{dy} = -\eta \left(\frac{x}{y}\right)^{\mu}, \eta \neq 0 \qquad (10)$$

Note: The case of $\mu = 0$ corresponds to the linear isopreference contours, and $\mu = 1$ corresponds to the reciprocal Equation (8). Therefore, we shall focus on the case where $\mu \neq 0, 1$.

Rearranging shows that this equation can be written as

$$\frac{\left(\frac{dx}{x}\right)}{\left(\frac{dy}{y}\right)} = -\eta \left(\frac{x}{y}\right)^{\mu-1}.$$

This time the fractional increase in x needed to compensate for a fractional decrease in y is not constant, and it depends on the values of x and y. By direct integration, the isopreference contour satisfy the equation

$$x^b - \frac{\eta}{y^b} = c \quad \text{or} \quad x = (c + \frac{\eta}{y^b})^{\frac{1}{b}}, \qquad (11)$$

for some constants c, η, and $b = 1 - \mu$..

By making trade-off assessments for an isopreference contour, we can find the best values of these parameters that match the assessed values. Figure 7.7 shows examples of these contours for values of $b = 0.5, c = 2$, and $\eta = 12, 8, 4, 2$ (from top curve to bottom curve).

CORRESPONDING PREFERENCE FUNCTION   The simplest preference function whose level sets correspond to the contours $x^b - \frac{\eta}{y^b} = c$ is

$$P(x, y) = x^b - \frac{\eta}{y^b} \qquad (12)$$

7.6 CONSTRUCTING PREFERENCE FUNCTIONS BY CAPTURING IMPORTANT FEATURES OF THE PROBLEM

An appropriate preference function is one that captures the important qualitative and quantitative features of the problem in as much detail as is needed to determine the decision-maker's preferences (ranking) of the prospects. In many cases, ***more of an attribute is not necessarily preferred to less***. In this case, it might be better to construct the preference function using other methods besides direct assessment.

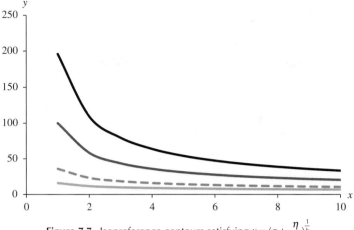

Figure 7.7. Isopreference contours satisfying $x = (c + \frac{\eta}{y^b})^{\frac{1}{b}}$.

Choosing an appropriate preference function relies on an art of mathematical modeling that captures these important features. In Chapter 6, we discussed preference functions that captured important qualitative and quantitative features for prospects characterized by a single numeric measure. We showed, for example, that if we prefer more of an attribute to less, then the preference function must be monotone increasing. On the other hand, if more is preferred to less up to a certain level, after which less is preferred to more, then the preference function must have at most one peak. As a first step, we can extend this method of reasoning to multiple attributes by first thinking about our preferences for each attribute at reference values of the remaining attributes and then combining the individual preferences and thinking about the interactions that result from the combination, even if there is no clear physical relation. We illustrate this idea in more detail below.

EXAMPLE 7.2: Preference Function for Seat in a Movie Theater

Consider your preference for a seat in the movie theater. First, it is helpful to visualize the prospects and to think about the important attributes.

One attribute that comes to mind is the **view**. This might be characterized by two components. The first component is **radial distance** from the screen. Most people would not want to sit in the first row, and so there is a preference for seats farther away from the screen up to a certain point, after which the view might be difficult, and so there is a preference for closer seats beyond that point. A first step in modeling this type of behavior is to capture some single-peaked preference with radial distance. For example, the function

$$R - (r - r_0)^2,$$

where R is a constant, has a peak at $r = r_0$. The value of r_0 is determined by your most preferred distance. There is nothing magical about the power of two in the previous function. It can be further refined. Figure 7.9 plots this preference function for values $R = 400$, $r_0 = 20$ feet.

Figure 7.8. Preference for a seat in the movie theater. © Artem_Furman/iStock/Getty Images Plus.

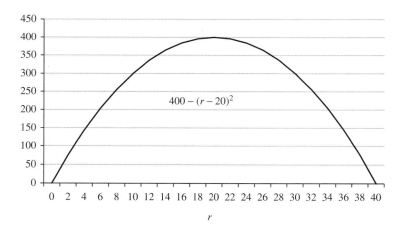

Figure 7.9. Preference as a function of the distance from the screen.

Now that we have thought about the distance from the screen, we should note that this statement of thinking about an individual preference is only meaningful at a specific reference value of the other attributes considered.

The second attribute that could affect your preference for the view in a seat in a movie theater is its ***angular view***. Preference might be higher for a seat located in front of the center of the screen (perpendicular to the screen making a 90 degree angle) than that for a seat at the end of the row where the view is not comfortable. We can specify some angle ψ that is included in the arc between your position and that of the perpendicular to the screen. Figure 7.10 depicts this relation graphically.

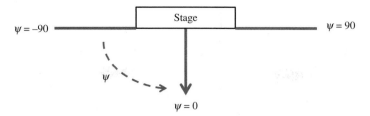

Figure 7.10. Angular view from screen.

Preference is for $\psi = 0$ if you prefer seats facing the center and it decreases as the absolute value of ψ increases. But now you want to see the nature of the decrease. Is it $\dfrac{1}{\psi}$, for example? Or is it $a^2 - \psi^2$? If you use a function like $\dfrac{1}{\psi}$, you get an infinite value at $\psi = 0$, which might not be realistic, and might also cause problems because of the unboundedness. Furthermore, you need to account for negative values of ψ. Another way to model this peaked preference is to think about $\dfrac{1}{\psi^2 + a}$ or $\dfrac{1}{|\psi| + a}$. Note that the latter function $\dfrac{1}{|\psi| + a}$ is not continuously differentiable on the entire domain, and so it might be easier to model the behavior with $\dfrac{1}{\psi^2 + a}$ or a different power of ψ if the preference function were to be differentiable (this would simplify some optimization procedures). Once again, this assessment is to be conducted at a specific reference value of the remaining attributes.

Figure 7.11 shows examples of univariate preference functions that could be used to model the decrease in preference as you move away from the center.

Other factors that might also be involved in the preference for the seat might pertain to the view in a particular theater. For example in an amphitheater setting, other considerations might also be important, such as vertical angle from the horizontal plane to the person.

There might be other direct value attributes considered in this situation: for example, convenience of getting to your seat as determined by the number of people you need to cross to get to that seat. As we shall see in Chapter 9, it is often best to convert this type of effort into money to minimize the number of attributes. Other factors pertaining to a particular theater also include the effects of the sound amplifiers in relation to the seat in that particular move theater.

As a first step, you might want to consider the view in this setting, combining only the effects of distance and angle from the screen. If these individual preferences remain the same regardless of the levels of the other attributes, you might consider a product (or a sum) of these individual preferences; for example, Figure 7.12 plots the preference surface as a function of both r and ψ for the functions $\dfrac{1000}{\psi^2 + 20^2}[400 - (r - 20)^2]$ and $(90^2 - \psi^2)[400 - (r - 20)^2]$.

Figure 7.13 plots the isopreference contours corresponding to these two surfaces.

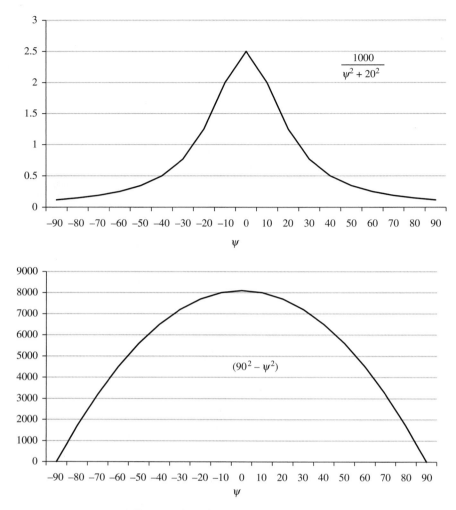

Figure 7.11. Different preference functions for the seat as a function of ψ.

Validating the Model: Testing for Interactions among the Attributes

At this stage, you might want to think about how preferences interact with each other for these two attributes. Does your optimal radial distance depend on the angular view where you are sitting? If not, then a product or additive form might be sufficient. If there is dependency, then additional terms could be added to the preference function.

The following example illustrates a case where preferences do indeed interact and do depend on the level of the remaining attributes.

EXAMPLE 7.3: Preference Function for a Peanut Butter and Jelly Sandwich

---

Think about your preference for a peanut butter and jelly sandwich. Start by thinking about what you like in the sandwich. Three attributes come naturally to mind: thickness of slices of bread, b, thickness of peanut butter, p, and thickness of jelly, j. Now think

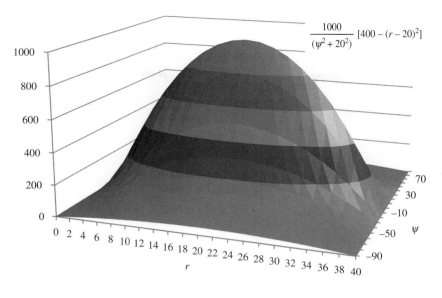

$$\frac{1000}{(\psi^2 + 20^2)}[400 - (r - 20)^2]$$

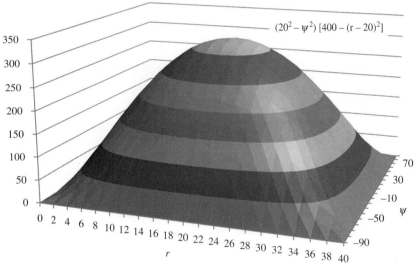

$$(20^2 - \psi^2)[400 - (r - 20)^2]$$

Figure 7.12. Preference functions for two attributes: radial and angular distances.

about your preference for the behavior of each individual attribute in the sandwich. It is clear that you do not want zero thickness of bread (unless you are allergic to it or on a very low-carb diet). At the same time, you do not want a thickness of one foot of bread. That would be hard to eat. You prefer more bread to less up to a certain point after which you prefer less bread to more. A first shot at capturing this behavior could be

$$P(p, j, b, f) = pjb(2p^* - p)(2j^* - j)(2b^* - b).$$

This preference function has a maximum value when $p=p^*$, $b=b^*$, $j=j^*$, your best sandwich.

The problem with this preference function is that the optimal values of jelly, for example, do not depend on the thickness of peanut butter. But in thinking about that,

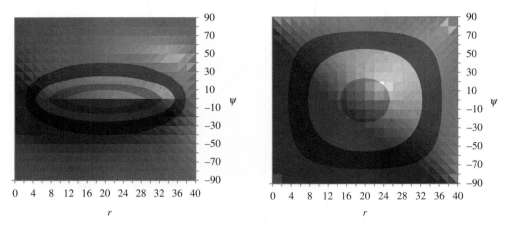

Figure 7.13. Contour plots for two attributes: radial and angular distances.

Figure 7.14. Peanut butter and jelly sandwich and its components. © Bernard Prost/StockFood Collective/ Getty.

your optimal thickness of jelly might in fact change as you want a certain ratio of thicknesses. For example, you might want the two thicknesses to be equal, i.e, if the thickness of peanut butter increases, you might prefer more thickness of jelly. It would be convenient to capture this preference dependence into the preference function.

One way to do this is to add the ratio, f, into the preference function following an example from Howard and Abbas (2015). The attributes involved are thickness of slices of bread, b, thickness of peanut butter, p, thickness of jelly, j, and fraction of thickness of peanut butter to jelly, $f = p / j$,

$$P(p, j, b, f) = pjbf(2p^* - p)(2j^* - j)(2b^* - b)(2f^* - f),$$

where p^*, j^*, b^*, f^* are the optimal values of p, b, j, f (respectively).

Note that the preference function is not necessarily increasing with any of the attributes on the entire domain; it is not additive either, and there is dependence in preferences among the attributes. For example, the preference for the thickness of peanut butter

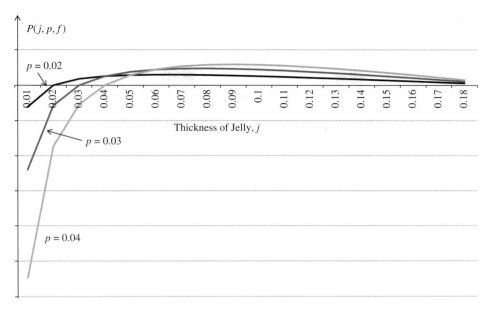

Figure 7.15. Preference for thickness of jelly as the thickness of peanut butter changes.

will depend on the thickness of jelly. Note further that this function does not require any form of "weights," which are widely accepted as the immediate common first step people think of when they attempt to construct preference functions.

EXAMPLE 7.4: Preference for Jelly as the Level of Peanut Butter Changes

---

Consider the peanut butter and jelly sandwich preference function

$$P(p, j, b, f) = pjbf(2p^* - p)(2j^* - j)(2b^* - b)(2f^* - f)$$

where $b^* = 1, j^* = 0.1$, $p^* = .05$, and $f^* = 0.5$. Figure 7.16 plots the preference function for peanut butter given different levels of jelly. Note that thickness of bread merely adds a scaling factor to the preference function for peanut butter, but it does not change its shape. Therefore, we merely plot the equation.

Figure 7.15 plots the preference for the thickness of jelly for three different thicknesses of peanut butter: 0.02, 0.03, and 0.04 inches. As shown in the figure, the peak in preference for the thickness of jelly changes depending on the thickness of peanut butter. This is the idea of preference dependence. We cannot specify our preference for the thickness of jelly unless we specify the thickness of peanut butter.

On the other hand, the preference for either the peanut butter or the jelly would not change with thickness of bread in this example, and likewise, the preference for thickness of bread would not depend on either the thickness of peanut butter or jelly. We can therefore specify our preference of thickness of bread without specifying the thickness of peanut butter or jelly. Similarly, as we have seen, we can specify our preference for peanut butter and jelly without specifying the thickness of bread.

7.7 CONSTRUCTING PREFERENCE FUNCTIONS BY THINKING ABOUT CHANGES IN PREFERENCES

In Chapter 6, we discussed the shape of preference functions over an attribute when preference change once or do not change as the level of that attributes changes. The corresponding preference functions were unimodal and monotone, respectively. Can we use the same ideas to construct preference functions for multiple attributes? That is, if we have two prospects (x_1, y) and (x_2, y), and if our preferences for the prospects do not change as we increase y, what does this imply about the preference function? We refer to this condition as zero-switch ordinal preferences and discuss its implications in the next chapter. Furthermore, if preferences can change, but only once as we increase y, what can we infer about the shape of the preference function? We refer to this condition as ordinal one-switch preferences and refer back to it in Chapter 32.

7.8 SUMMARY

1. Methods for constructing preference functions for multiple attributes
2. Using laws of Physics, Engineering, or Accounting to construct the preference function
3. When more of an attribute is preferred to less:
   a. Making trade-offs among the attributes by direct assessment
   b. Using functional forms of isopreference contours and then assessing their parameters
4. Constructing preference functions by capturing important features of the problem: This approach is particularly useful when more of an attribute is not necessarily preferred to less.
5. The concept of preference dependence
6. The concept of change in preferences between two prospects as the level of an attribute increases.

REFERENCES AND ADDITIONAL READINGS

Abbas, A. E. and J. Matheson. 2010. Normative decision making with multiattribute performance targets. *Journal of Multicriteria Decision Analysis* 16 (3–4): 67–78.
Howard, R. A. and A. E. Abbas. 2015. *Foundations of Decision Analysis*. Prentice Hall, New York.
Keeney, R. and H. Raiffa. 1976. *Decisions with Multiple Objectives: Preferences and Value Tradeoffs*. John Wiley, New York. Reprinted, Cambridge University Press, New York (1993).
MacCrimmon, K. R. and M. Toda. 1969. The experimental determination of indifference curves. *Review of Economic Studies* 36(4): 433–451.

APPENDIX 7 HOW PEOPLE GET IT WRONG

1) **Validation Is Essential! Put Some Time and Effort into Validating the Model**

   *"All **models are wrong**, some are useful."*

The ultimate test for a preference function is whether or not it can provide the same deterministic rank order of prospects as those that the decision-maker would provide. Therefore, the process does not end with merely constructing a preference function.

After constructing the preference function, there is an important step that must not be forgotten: *validating the function that you have constructed.*

Validation can be conducted by asking the decision-maker to rank certain prospects characterized by the inputs to the preference function and then seeing whether or not the preference function will provide an order that matches these preferences.

In the peanut butter and jelly sandwich example, you can test whether or not you are truly indifferent between several sandwiches having the same preferences, or if you have a preference for various sandwiches. If you do have a preference, then the preferences function should return a higher value for those sandwiches that are ranked higher.

It is important to remember that it is the decision-maker, not the model, that makes a decision. The purpose of the model is to generate insights to help the decision-maker achieve clarity of action. Too often analysts rush to pick a particular form of preference function (because of its simplicity and analytical tractability) and then forget to verify at the end whether this function does indeed match the preferences of the decision-maker. A recommendation is then made based on the output of this arbitrary function.

2) Focusing on Requirements without Specifying the Trade-offs

This chapter discussed how to determine isopreference contours and preference functions for multiple attributes. The choice of a better prospect given the preference function is a relatively easy task.

It is often the case that when multiple attributes are present in a decision, a threshold value is specified for each attribute, such that exceeding these threshold values is regarded as satisfactory. This threshold value often represents a "target" or a performance metric, and the requirement is that this threshold values are met for each attribute.

We encounter such situations on both personal and professional levels. For example, an enterprise might reward employees using a "management-by-objectives" scheme where a manager is rewarded upon meeting threshold values for each target or goal.

Design organizations often set design requirements where each subcomponent of the design must meet certain threshold values. These design requirements for a product or an engineering system are handed to the design engineers who are asked to meet (or exceed) the requirements in their design. This often helps translate preferences into requirements and distributes decision-making. But what are the implications of setting threshold targets independently for each attribute in a multiattribute decision?

The first problem with this is why we set requirements in the first place. Some might argue that they help provide focus and also help decentralize decision-making in design. Of course, it is also plausible that some requirements are due to physical constraints, but quite often requirements are set arbitrarily to distribute decision-making without taking note of the implications of these requirements.

Another problem is that trade-offs among the design requirements are seldom expressed. As we shall see, this method of specifying independent requirements can lead to suboptimal designs. Figure 7.16 shows an example of the issues that can result

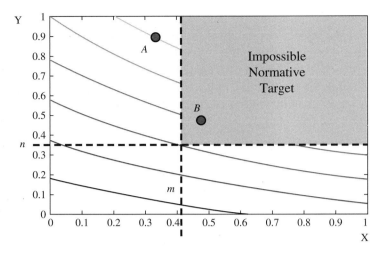

Figure 7.16. Target region formed by setting independent requirements.

from setting arbitrary design requirements in engineering design and how they might lead to design choices that are inconsistent with preferences for the different designs.

To illustrate, suppose you have two design components, X and Y, where more is preferred to less as shown in Figure 7.16. We have seen that the slope of the isopreference contours must be negative in this case. If requirements are set for each attribute independently, and the design engineer is required to meet or exceed the specified requirement, then the acceptable designs are limited to those in the rectangular (shaded) region in the figure.

Now consider two designs, A and B. Design A lies on a higher isopreference contour than design B and is therefore more preferred. However, *design A does not meet the requirement for attribute X. On the* other hand, *design B meets all requirements and lies within the target region but it has a lower value than design A.*

> *The design that would be induced by the specified requirement region has a lower value than one that does not satisfy the requirement.*
> *The requirement has led to a suboptimal decision and the difference in choice comprises a value gap.*

To avoid this type of behavior, a simple axiom must be introduced within the design process.

Axiom

> *There shall be no design outside the requirements region that is preferred to a design within the requirements region.*

Abbas and Matheson (2010) show that this axiom implies that the target region must be bounded by an isopreference contour and not the rectangular region shown in Figure 7.16. This means that trade-offs among the requirements must be specified.

The issue we have just discussed appears regularly in many instances of target setting and in "Management by Objectives." Multiple requirements are set without specifying the trade-offs among them. If more of each requirement is preferred to less, then achieving more of one attribute should be considered if you achieve less of another.

We often say in our daily lives, "I would like to make at least X hours of work time for salary, Y hours of travel in vacation time, Z of study time, and W hours at the gym for fitness." Trade-offs are never specified, such as, for example, that for every extra hour I spend at the gym, I am willing to spend less time making money or traveling. Failing to specify trade-offs results in the "rectangular target region" shown in the figure, and in suboptimal decisions.

The following interaction between Steve Austin, from the famous TV series, *The Six Million Dollar Man*, and Johnny, a NASA employee, exhibits the same phenomenon. Check it out.

Steve Austin: A Bionic Christmas Carol
In the Steve Austin TV series, *A Bionic Christmas Carol*, Horton Budge, a contractor, is developing a life-support system for a NASA Mars flight. However, the deliverable is only just within compliance. Steve Austin is sent to examine the situation over Christmas. In the following conversation, he is on site where he has just used his bionic skills to open a door to save a man's life. He is in a conversation with Johnny, one of the scientists on site.

© Photos 12 / Alamy Stock Photo

Johnny: I can't even figure out how you got the door open.
Steve Austin: Probably substandard metal. It seemed to me pretty easy.
Johnny: And that's what I thought, but this material is within the tolerance set of this job contract.
Steve Austin: Well within?
Johnny: Just barely.
Steve Austin: Johnny, I want a complete copy of the specifications laid of the materials that went into this project and I want a sample of each one of these materials.
Johnny: Sure, Colonel Austin. Anything else?
Steve Austin: Yeah, the most accurate mechanical and electronic testing equipment that you have and a place to work.
Johnny: You got it.

[Steve Austin examines the samples and calculates various parameters.]
Johnny: Find the trouble?
[Steve Austin nods.]
Johnny: Substandard material?
Steve Austin: Nope. *All the materials meet the standards, just barely.* If you want to even settle the component's been built to the lower level of specifications, there wouldn't have been a problem. Budge *used the lowest allowable standards on every single component.*
Johnny: Yeah, well that's all Budge's operation. He never tries to get away with anything below contractual specs. He's honest, but greedy.

What this interaction tells us is that the contractor was given requirements thresholds "specs" and that he built the product at the lowest level of each spec, but met the design requirements. He was given requirements and met each one of them (barely), and so the product was designed at the lowest isopreference contour of the acceptable region in Figure 7.17.

Think about that for a second. The contractor was given specs that he was asked to meet, and he met them, and yet the product was deficient. This by itself shows that the specifications that were given to him were not sufficient to yield a suitable product if each threshold was met. The people who gave him the specs did not anticipate that everyone would be met barely. But this means that the requirements were set independently and arbitrarily in a way that the resulting target does not result in a satisfactory product.

Point A in Figure 7.17 shows an example of independent requirements when each spec is met barely. The narrative implies that the isopreference contour that passes by point A is actually not satisfactory, even though it lies within the specified region. If instead they had thought of preferences and isopreference contours, they would not have set specs independently and would have selected an isopreference contour that would have resulted in an acceptable product. This would require identifying trade-offs among the specs or requirements.

Recall the issue we had with thinking about the attributes independently in Chapter 3. *This is another demonstration of how thinking about each attribute independently might result in poor decision-making.*

The contractor was given specs independently and he was asked to meet them, and so he did. Why do they blame the contractor?

Moral of the Story: Even Requirements Need to Have Trade-Offs Specified among The Attributes

Quite often you hear expressions about meeting some requirements that need not be physical constraints. For example, "we need to exceed a 20% increase in sales and a 10% reduction in cost," or "we need to exceed this much level in screening safety and we need to reduce waiting times at the airports." These types of statements do not take into account the trade-offs among the different attributes, and so you cannot even compare two prospects that satisfy these requirements. Whenever you set multiple requirements,

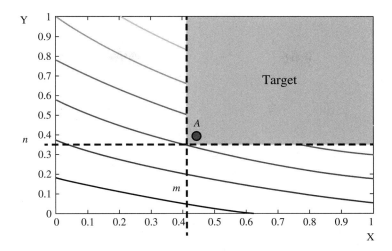

Figure 7.17. A project design that meets the requirements but has the lowest value in the target region.

you need to specify the trade-offs among them. In this example, there is some economic structural model that could determine these multiple requirements in terms of a direct value – say, profit. The requirements should then be related to profit, and the trade-offs would be automatically determined. A higher-level conversation would be why set a requirement on profit in the first place, and just make the best decisions you can.

CHAPTER 8

Additive Preference Functions

Chapter Concepts

- Additive preference functions
  - Preferential independence
  - Shapes of preference functions
    - Grounded (attribute dominance) preference functions
    - Non-grounded preference functions
  - Archimedean representation of additive preference functions
  - Constant elasticity of substitution preference functions

8.1 INTRODUCTION

In the previous chapter, we discussed methods to determine the trade-offs between direct value attributes by direct assessment or by using functional forms for the isopreference contours. We also discussed how to construct preference functions by thinking about the qualitative features of the problem. In this chapter, we discuss preference functions of the additive form, and several of their implications. In particular, we illustrate how they represent situations where there are no preference interactions among the attributes – the preference for any attribute (or any subset of the attributes) does not depend on the levels of the remaining attributes. The chapter elaborates on this concept and introduces the notion of *preferential independence*. This chapter also includes an equivalent representation of additive preference functions, known as an ***Archimedean combination*** of preferences, which will play an important role in multiattribute utility functions. Finally, the chapter introduces a qualitative feature of some preference functions known as the ***grounding*** or ***attribute dominance*** property where the function has a fixed (minimum) value if any other attribute is minimum.

8.2 THE ADDITIVE PREFERENCE FUNCTION

The additive preference function is particularly useful when you can think about your preference for each of the direct value attributes separately – without considering the levels of the other attributes. Because preferences for each attribute are separable in this case, you can think about the individual preference function for each attribute individually and then combine those functions using an additive form.

For two attributes, the additive preference function has the form

$$P(x, y) = f_1(x) + f_2(y)$$

where the functions f_1 and f_2 are arbitrary functions of the attributes. For multiple attributes, the additive preference function has the form

$$P(x_1, ..., x_n) = \sum_{i=1}^{n} f_i(x_i),$$

where the functions f_i, $i = 1, ..., n$ are arbitrary functions.

Note that the additive preference function is more general than the linear preference function,

$$P(x_1, ..., x_n) = \sum_{i=1}^{n} w_i x_i,$$

that we discussed in the previous chapter. Here we have a summation of individual functions of the attributes and not a weighted combination of the attributes themselves. As we shall see, this formulation allows for a richer set of isopreference contours that need not be linear. Furthermore, we will show that all the functional forms of isopreference contours that we discussed in Section 7.5 can be modeled using the additive form.

Because deterministic preferences remain the same when applying a strictly monotone transformation, the shape of the isopreference contours will not change if we apply a strictly monotone transformation to the preference function.

The Additive Preference Function
The general form of the additive preference function is

$$P(x_1, ..., x_n) = g\left(\sum_{i=1}^{n} f_i(x_i) \right),$$

where g is a strictly monotone function, and f_i, $i = 1, ..., n$ are arbitrary functions.

This definition implies that any strictly monotone transformation of an additive preference function has the same preference structure as an additive preference function. An example of a monotone transformation is the exponential transformation, which, when applied to an additive preference function, converts it into the product form,

$$e^{P(x,y)} = e^{f_1(x) + f_2(y)} = e^{f_1(x)} e^{f_2(y)},$$

and for multiple attributes,

$$e^{P(x_1,\ldots,x_n)} = e^{\sum_{i=1}^{n} f_i(x_i)} = \prod_{i=1}^{n} e^{f_i(x_i)},$$

which is a product form. This product form has the same ordinal preferences as the additive form.

8.2.1 Slope of the Isopreference Contours for Additive Preference Functions

The additive function allows for a rich set of isopreference contours. In fact, the functional forms of isopreference contours that we discussed in Sections 7.5 can all be modeled by additive preference functions.

To illustrate, we now derive the slope of the isopreference contours for a two-attribute preference function:

$$P(x,y) = f_1(x) + f_2(y).$$

Along any contour of constant slope, the preference is constant, and so the change in preference along the contour is zero. If we take the derivative of the preference function and set it to zero along an isopreference contour, we get

$$dP(x,y) = 0 \quad \Rightarrow \frac{\partial f_1(x)}{\partial x} dx + \frac{\partial f_2(y)}{\partial y} dy = 0$$

Rearranging gives the slope of the isopreference contour by an additive preference function as as

$$\frac{dy}{dx} = -\frac{\left(\partial f_1(x)\middle/\partial x\right)}{\left(\partial f_2(y)\middle/\partial y\right)}, \tag{1}$$

which is not constant on the domain of the attributes, but is a separable function in each variable.

The following example demonstrates the generality of the additive preference function.

8.2.1.1 SLOPE OF THE ISOPREFERENCE CONTOURS OF ADDITIVE PREFERENCE FUNCTIONS The contours of the additive preference function generalize the functional forms of isopreference contours that we discussed in Section 7.5 as we illustrate below.

EXAMPLE 8.1: Additive Preference Functions with Contours Having Constant Slope

---

If we desire a constant slope for the isopreference contours, then we can use an additive preference function and observe from (1) that if you set

$$\frac{\partial f_1(x)}{\partial x} = \text{constant} = w_x \text{ and } \frac{\partial f_2(y)}{\partial y} = \text{constant} = w_y,$$

then you get by direct substitution into (1), contours with a slope of

$$\boxed{\frac{dy}{dx} = -\frac{w_x}{w_y} = \text{constant}}.$$

The functions $f_1(x)$ and $f_2(y)$ for the additive preference function are obtained by direct integration of their constant derivatives. This implies that

$$f_1(x) = w_x x + c \text{ and } f_2(y) = w_y y + d,$$

which also implies the linear preference function,

$$P(x, y) = w_x x + w_y y + \text{constant}.$$

EXAMPLE 8.2: Additive Preference Functions with Constant (Small) Fractional Trade-Offs

---

Additive preference functions can also be used to model contours that have the property that small fractional changes in the attributes have constant trade-offs, i.e., that the small percentage increase in attribute X, $\frac{\Delta x}{x}$, needed to maintain indifference for a small percentage decrease in attribute Y, $\frac{\Delta y}{y}$, is a constant along the same isopreference contour, and so

$$\frac{dx/x}{dy/y} = -\eta, \quad x, y > 0, \tag{2}$$

is a constant. To see how this can be achieved from (1), we can rearrange (2) to get

$$\frac{dx}{dy} = -\eta \frac{x}{y}, \quad x, y > 0.$$

If we set

$$\frac{\partial f_1(x)}{\partial x} = \frac{1}{x} \text{ and } \frac{\partial f_2(y)}{\partial y} = \frac{\eta}{y}$$

in the additive preference function, we can obtain this slope.

By direct integration, this implies that

$$f_1(x) = \ln(x) + \text{constant and } f_2(y) = \eta\ln(y) + \text{constant}$$

giving the additive preference function

$$P(x,y) = \ln x + \eta\ln y + \text{constant.}$$

Because the preference function is invariant to a monotone transformation, this preference function has the same contours as Cobb-Douglas function

$$P(x,y) = xy^{\eta}$$

as can be seen by taking the logarithm of this Cobb-Douglas preference function.

EXAMPLE 8.3: Additive Preference Functions Whose Contours Have a Power Function Slope

---

Finally, if we wish to match the isopreference contours defined by the slope

$$\frac{dx}{dy} = -\eta\left(\frac{x}{y}\right)^{\mu}$$

then we can set

$$\frac{\partial f_1(x)}{\partial x} = \frac{1}{x^{\mu}} \text{ and } \frac{\partial f_2(y)}{\partial y} = \frac{\eta}{y^{\mu}}$$

to get the power functions

$$f_1(x) = x^{1-\mu} \text{ and } f_2(y) = \eta y^{1-\mu}$$

and the additive preference function

$$P(x,y) = x^{1-\mu} + \eta y^{1-\mu}.$$

8.2.2 When More of an Attribute Is Not Necessarily Preferred to Less

The additive preference function also allows for preferences where more of an attribute is not necessarily preferred to less, and so the slope of the isopreference contours need not be negative. We have seen examples of such additive preference functions in the movie theater preference function of Chapter 7, with

$$\frac{1000}{\psi^2 + 20^2}[400 - (r-20)^2] \text{ and } (90^2 - \psi^2)[400 - (r-20)^2]$$

where r is the radial distance from the screen and ψ is the angular view.

Note that these preference functions can be converted into an additive functions of r and ψ by a simple monotone logarithmic transformation. Therefore, they are both additive preference functions.

8.3 VERIFYING THE ADDITIVE FORM BY PREFERENCE SWITCHES: PREFERENTIAL INDEPENDENCE

Two important considerations must be made when using the additive form. The first is whether it is indeed applicable to the decision – does it really represent the decision-maker's preferences? The second is how to assess the functions $f_i(x_i)$. Both of these questions are essential to the appropriate use of the additive form.

The additive preference function asserts that you can think about your preferences for each of the attributes individually and that the remaining attributes do not change those individual preference functions. This also means that when ranking prospects with multiple attributes, you can rank the preference for the level of any attribute by focusing only on that attribute itself, without considering the other attributes, and then combining the individual attribute rankings to obtain the ranking of the prospect.

Peanut Butter and Jelly Sandwich Using an Additive Preference Function

We have seen in the previous chapter two examples of the peanut butter and jelly sandwich. One is where there are no interactions, and so the additive form is applicable, because the product

$$P(p, j, b) = pjb(2p^* - p)(2j^* - j)(2b^* - b)$$

can be converted into an additive form through a monotone logarithmic transformation to get

$$\ln\left(p(2p^* - p)\right) + \ln\left(b(2b^* - b)\right) + \ln\left(j(2j^* - j)\right).$$

This form also asserts that the preference for any level of the attributes of the peanut butter and jelly sandwich does not depend on the levels of the remaining attributes. The optimal thickness of peanut butter will not depend on the levels of bread or jelly.

We can therefore write that if thickness of peanut butter p_1 is more preferred to p_2 at any combinations of j, b, then it will be preferred for all other combinations.

Peanut Butter and Jelly Sandwich That Cannot Be Modeled Using an Additive Preference Function

Another example is where the interactions and the preferences for the amount of peanut butter depend on the level of jelly (and likewise preference for jelly depended on peanut butter) using the formula,

$$P(p, j, b, f) = pjbf(2p^* - p)(2j^* - j)(2b^* - b)(2f^* - f)$$

where $f = \dfrac{p}{j}$ is the ratio of thickness of peanut butter to jelly.

This function cannot be converted into an additive form of all attributes, but we can separate the thickness of bread by taking the logarithm to get

$$\ln\left(pjf(2p^* - p)(2j^* - j)(2f^* - f)\right) + \ln\left(b(2b^* - b)\right).$$

Therefore, we cannot think about our preference for peanut butter alone without thinking about the level of jelly, and vice versa. We say that the thickness of peanut butter and the thickness of jelly exhibit preference dependence. However, we can indeed think about our preference for the thickness of bread without considering the other attributes in this example, and we can also think about our preference for peanut butter and jelly combinations without considering the thickness of bread.

8.3.1 Two Attributes: Zero-Switch Preferences

This section discusses the idea of thinking about the preference function as the level of each individual attribute varies. We have already discussed a similar idea for the case of a single attribute in Chapter 6, where preference for prospects characterized by a single attribute changed as the base level of that attribute varied. We showed, for example, that if we prefer $x_1 \succ x_2$, and if we assert that $(x_1 + w) \succ (x_2 + w) \; \forall w$, then the preference function satisfies

$$P(x_1 + w) - P(x_2 + w) > 0 \; \forall w,$$

and, furthermore, that it must be a monotone function. In this section, we extend this method of reasoning to multiple attributes and illustrate how this idea helps us think about the preference function.

Consider a decision whose prospects are characterized by two attributes, X and Y. Suppose you prefer level $x_1 \succ x_2$ of attribute X, and suppose that this preference relation all values of Y and for all x_1, x_2, i.e., preferences for x_1 over x_2 do not change as Y increases. Suppose also that this is true for all x_1, x_2, then this would imply that it makes sense to discuss an individual preference function for attribute X. Mathematically, it also implies that

Preferential Independence: *Attribute X is preferentially independent of attribute Y if*

$$P(x_1, y) - P(x_2, y) \text{ does not change sign for all } y \text{ and this relation is true for all } x_1, x_2.$$

By "does not change sign" we mean that the sign is either always positive or always negative or zero for all y.

We have also seen this situation in the movie theater example in the previous chapter where we assumed that the preference function for radial distance did not depend on the angular view from the screen, but it was not the case for the second version of the peanut butter and jelly sandwich where the optimal thicknesses depended on the levels of the other attributes.

Proposition: *X is preferentially independent of Y if and only if the preference function can be written as*

$$P(x, y) = \phi(v(x), y),$$

where ϕ is either strictly monotone or a constant function of its first argument for all y.

EXAMPLE 8.4: A Preference Function Satisfying X Is Preferentially Independent of Y

Consider the function,

$$\phi(v, y) = y^2 e^{yv}, y > 0,$$

which is strictly increasing in v for all positive y. This implies that the preference function

$$P(x, y) = y^2 e^{yv(x)},$$

satisfies X preferential independent of Y for any choice of a function $v(x)$.

To illustrate, note that for any fixed value of y, say y_0, the function $P(x, y) = y_0^2 e^{y_0 v(x)}$, is strictly monotone in $v(x)$, and so preferences for levels of attribute X will not change with Y – i.e., given any two levels of attribute X, say x_1, x_2, if $P(x_1, y) > P(x_2, y)$ for any value of Y, then $P(x_1, y) > P(x_2, y)$ for all values of Y.

Of course, if X is preferentially independent of Y, it does not mean that Y is necessarily preferentially independent of X.

Mutual Preferential Independence: If X and Y are both preferentially independent of each other, then we say that they are mutually preferentially independent, and we can write

$$P(x, y) = \phi(v_1(x), v_2(y)),$$

where ϕ is either strictly increasing or strictly decreasing in each argument.

Note that the concept of preferential independence does not specify the functional form of the preference function for two attributes. The idea of preferential independence, however, does provide more specificity when it applies to three or more attributes, as we illustrate below.

8.3.2 Zero-Switch Preferences: Preferential Independence for Three (or More) Attributes

One of the earliest conditions that specified the form of the preference function for three or more attributes was discussed in Debreu (1960). This condition, known as mutual preferential independence, requires that

Preferences for any subset of the attributes do not depend on the levels of the remaining attributes.

In other words, the preference function for any subset of the attributes does not vary as the levels of the remaining attributes vary. Equivalently, we can say that *trade-offs among any subset of the attributes do not depend on the remaining attributes*. This condition is stronger than saying that preference for any single attribute does not depend on the remaining attributes: it also requires the joint preference for any subset of the attributes not to change.

Consider, for example, the three attributes: Money, Health (as determined by the equivalent remaining life time), and Vacation days per year. Mutual Preferential Independence would imply the following:

Suppose we plot the isopreference contours for Money vs. Health at a given level of Vacation days per year. Trade-off independence means that the shape of the isopreference contours of Money vs. Health will not change if we plotted it (or assessed it) for any other level of Vacation days per year. This means that the subset (Money and Health) is preferentially independent of Vacation days. Likewise, we say that the trade-off between Money and Vacations days is independent of Health, and also that the trade-off between Health and Vacation days is independent of Money.

Furthermore, mutual preferential independence requires that the shape of the preference function for Money alone does not depend on either Health or Wealth; that the preference for Health alone does not depend on either Wealth or Vacation days; and finally that the preference function for Vacation days does not depend on either Money or Health.

> **Definition**
> **Mutual preferential independence** means that the trade-offs among any subset of the attributes do not depend on the remaining subset.

The other interpretation of preferential independence is in terms of changes in preference (or lack thereof) instead of trade-off independence. This implies that if we have two prospects, A and B, comprising different Money and Health levels but the same level of Vacation days per year, and if we prefer prospect A to B at one level of Vacation days, then we should always prefer A to B regardless of the level of Vacation days. The difference,

$$P(Money_1, Health_1, Vacation\ Days) - P(Money_2, Health_2, Vacation\ Days)$$

does not change sign with the number of Vacation days. If this is true for all subsets of the attributes and for all values of Money, Health, and Vacation days, then mutual preferential independence is satisfied.

Debreu (1960) showed that this condition on preferences corresponds to a simple form of preference function when the number of attributes, $n \geq 3$, which by now should be quite familiar:

> **Proposition:** A preference function with three or more attributes satisfies mutual preferential if and only if it is a strictly monotone transformation of an additive function of the attributes, i.e.,
>
> $$P(x_1,...,x_n) = g\left(\sum_{i=1}^{n} f_i(x_i) \right),\qquad (3)$$
>
> where g is a strictly monotone function and f_i, $i = 1,...,n \geq 3$ are arbitrary univariate functions.

This condition provides a consistency check that may be used to verify that the additive preference function holds. As we have seen with the peanut butter and jelly sandwich example, sometimes dependencies do exist, in which case it would not be appropriate to use an additive preference function.

A Note about the Construction of Additive Preference Functions

We now have a clear test that can be used to verify the additive form for three or more attributes. While the additive preference function is simple, and allows for a variety of forms and trade-offs, the most difficult part is identifying the functions f_i in (3) that are used for its construction.

To remedy this problem, we may think about additional assumptions regarding trade-offs among the attributes. These additional assumptions usually result in a special class of additive preference functions that provide further specificity to the functional form. Another approach, as we have seen with the movie theater example, is to think qualitatively about your preferences. For example, the term $(90^2 - \psi^2)[400 - (r-20)^2]$ represented preferences for a seat in terms of radial distance and angular distance from the screen. This function is in fact an additive preference function, as can be seen by taking its logarithm.

Another approach for reasoning about the functions f_i is to think in terms of an absolute scale. As we shall see, the additive preference function results in a particular form of value function (on an absolute scale) and utility function (on an absolute that can be assessed using lottery assessments). We shall discuss this approach in more detail in Chapters 9, 11, and 12.

We considered the case of two attributes where

$P(x_1, y) - P(x_2, y)$ does not change sign for all y and this relation is true for all x_1, x_2.

What if preferences can switch, but only once, i.e.,

$P(x_1, y) - P(x_2, y)$ can change sign only once as we increase y and this relation is true for all x_1, x_2 ?

We shall revisit this property in Chapter 32 and refer to it as ordinal one-switch independence.

8.4 ALTERNATE REPRESENTATION: ARCHIMEDEAN COMBINATIONS OF INDIVIDUAL FUNCTIONS

As we discussed, one of the challenges of constructing additive preference functions

$$P(x_1,...,x_n) = \sum_{i=1}^{n} f_i(x_i)$$

is the identification of the individual functions $f_i(x_i)$, $i = 1,...,n$. We have seen that making additional assumptions about the trade-offs among the attributes or the elasticity of substation among them can provide additional specificity to the functional form.

Another approach for identifying the individual functions $f_i(x_i)$, as we shall see in Chapter 30, is interpreting them in terms of utility assessments and then eliciting them using lottery assessments. This approach constructs multiattribute utility functions that satisfy additive ordinal preferences.

This section provides an equivalent representation of additive preference functions that will provide several insights when constructing multiattribute utility functions satisfying additive ordinal preferences. The purpose of this section is to familiarize the reader with this "Archimedean" form. Appendix A at the end of the book provides further details on the operations and the notation used in this section.

Suppose that the functions $f_i(x_i)$, $i = 1,...,n$ are continuous, strictly increasing, and bounded. For any continuous and strictly monotone function η, whose range contains the range of the functions $f_i(x_i)$, define

$$f_i(x_i) \triangleq \eta\big(P_i(x_i)\big), i = 1,...,n.$$

Rearranging gives

$$P_i(x_i) \triangleq \eta^{-1}\big(f_i(x_i)\big) \ \forall i,$$

where η^{-1} is the inverse function of η (Appendix A at the end of the book provides more details on inverse functions).

We can now write the additive form as

$$P(x_1,...,x_n) = \sum_{i=1}^{n} \eta\big(P_i(x_i)\big)$$

As we have seen, a preference function is invariant to a monotone transformation. If the term $\sum_{i=1}^{n} \eta\big(P_i(x_i)\big)$ lies in the range of η, we can define an equivalent ordinal preference function as

$$\eta^{-1}\big(P(x_1,...,x_n)\big) = \eta^{-1}\left(\sum_{i=1}^{n} \eta\big(P_i(x_i)\big)\right) \tag{4}$$

which we refer to as an ***additive Archimedean combination*** of the individual functions P_i. Because η^{-1} is a strictly increasing function, both $\eta^{-1}\big(P(x_1,...,x_n)\big)$ and $P(x_1,...,x_n)$ have the same isopreference contours and have identical preferences. Therefore, we can simply write

$$\boxed{P(x_1,...,x_n) = \eta^{-1}\left(\sum_{i=1}^{n} \eta\big(P_i(x_i)\big)\right),}$$

with the understanding that equivalence is due to invariance under a strictly increasing transformation.

We can further define $\eta(t) = \ln(\phi(t))$. Direct substitution gives

$$P(x_1,...,x_n) = \eta^{-1}\left(\sum_{i=1}^{n}\ln\big(\phi(P_i(x_i))\big)\right) = \eta^{-1}\left(\ln\prod_{i=1}^{n}\phi(P_i(x_i))\right)$$

By definition of the inverse of composite functions, $\eta^{-1}(t) = \phi^{-1}(e^t)$. This implies that

$$P(x_1,...,x_n) = \phi^{-1}\left(e^{\ln\prod_{i=1}^{n}\phi(P_i(x_i))}\right) = \phi^{-1}\left(\prod_{i=1}^{n}\phi(P_i(x_i))\right)$$

We refer to the form

$$P(x_1,...,x_n) = \phi^{-1}\left(\prod_{i=1}^{n}\phi(P_i(x_i))\right) \tag{5}$$

as the ***multiplicative Archimedean combination*** of functions.

We shall revisit the functional forms (4) and (5) numerous times throughout this book in many different contexts, and particularly in Chapters 9, 12, 29, and 30, while placing various conditions on the functions η and ϕ. Our purpose here was merely to familiarize the reader with these Archimedean forms and to show that there exists both an additive and multiplicative Archimedean representation for additive preference functions. From here on, we refer to the functions η and ϕ as the generating functions of the Archimedean form.

Because a preference function is invariant to a monotone transformation, it is clear that applying a monotone transformation to an additive preference function results in an equivalent preference function. It should not come as a surprise, therefore, that applying a monotone transformation to an Archimedean preference function also results in an Archimedean preference function. We prove this observation in the next chapter.

Note: Just like the additive preference function allows for a variety of shapes and forms of trade-offs, the equivalent Archimedean combination of functions also allows for such a variety because it is related by a monotone transformation that does not change the shape of the isopreference contours.

Dialogue between Instructor (I) and Student (S)

On the Use of Additive Preference Functions

S: This additive preference function seems to be simple and that would encourage people to use it.

I: Indeed it is simple. But simplicity does not justify using it when it is not appropriate.

S: True. When can we use this type of additive preference function?

I: If you have three or more attributes, and if your preference for any of the attributes (or any group of the attributes) does not depend on the values of the remaining attributes, then you have an additive preference function. This condition is referred to as preference independence. The additive form was discussed by Debreu (1960). For example, if you have three attributes, X, Y, and Z, and if I ask you to make trade-offs between levels of attributes X and Y, then this trade-off should not depend on the level of attribute Z.

S: What if we have two attributes?

I: Then preferential independence takes a more general form. For example, if preferences for attribute X do not depend on the level of Y, then the preference function takes the form $\phi(v(x), y)$, where phi is monotone in $v(x)$ for all y. Note that $v(x)$ itself need not be monotone. For example, if $v(x) = \sin x$, then the function $e^{y^2 \sin x} - y^3$ is monotone in $\sin x$ for all values of y. Therefore, it satisfies the condition that X is preferentially independent of Y.

S: OK, we get the idea of preferential independence. However, in previous classes, we were told that we have some scoring function that assigns a measure (or a rate) to each attribute and a weight that reflects the importance of that attribute. Then we multiply the weights by the rates and sum up. But where are the weights in the expression $P(x_1, ..., x_n) = \sum_{i=1}^{n} f_i(x_i)$?

I: What you are saying is referred to as the "weight and rate" preference function. It further assumes that the functions $f_i(x_i)$ are normalized on the same scale, and assigns weights that are non-negative and that sum to one. This preference function is simple, but is quite often used arbitrarily and this arbitrariness leads to errors, as we shall see in Appendix A of this chapter. The question of how important is one attribute relative to another has no meaning. Therefore, the results can actually be manipulated to suit whichever answer you want. The functions $f_i(x_i)$ need not even be monotone and certainly do not need to have the same scale. What is even worse is that the rates are often chosen arbitrary as some measures of performance without a clear interpretation.

C: Wow, but this "weight and rate" is widely used.

I: I know. In fact, by using an arbitrary score and having arbitrary scales for the attributes, I can make any alternative you wish be the best alternative. This type of analysis also often ignores the structural models and attempts to use attributes that are not clearly defined. It also confuses direct and indirect values quite often.

C: So now we know that an additive preference function is not necessarily a weight and rate preference function. How do we determine the functions $f_i(x_i)$ for the additive form?

I: You have asked a very important question. In principle you will see that when you make statements over deterministic consequences, you will be left with the problem of having to identify many unknown functions. We shall see in future chapters how to assess those functions using utility assessments. But recall that we have already discussed many methods. For example, we might specify the shapes of the isopreference contours. We can also think about our preferences qualitatively to build the function, as we did in the movie theater example. Or specify a property of the functional form (such as constant elasticity of substitution that we discuss in the next section) and then assess its parameters. We must of course also assess a few points on the surface or assess some points on a contour to verify the functional form.

C: Are there other specifications that could help with the functional form?

I: Yes, for example, the grounding (or attribute dominance) property that we discuss later in this chapter, is a qualitative feature that can also help with the choice of the function. Also, before we leave this topic, recall the Archimedean representation of preference functions. This will be used in many future chapters to determine utility functions that assess the functions and determine the trade-offs.

8.5 ELASTICITY OF SUBSTITUTION

Let us recall the way we introduced the functional forms of isopreference contours in the previous chapter. We considered an equation of the form $\dfrac{dx/x}{dy/y}$ (a small fractional change in the levels of the attributes) as being either constant or equal to some function of x and y. The elasticity of substitution compares this expression to the fractional change in the slope of the isopreference contours, $\Delta\left(\dfrac{dy}{dx}|_c\right)/\dfrac{dy}{dx}|_c$.

For two attributes, and a preference function, $P(x, y)$, the elasticity of substitution between x and y, $\eta(x, y)$, is given by the ratio

$$\eta(x, y) = \frac{\left(\dfrac{\Delta(y/x)}{(y/x)}\right)}{\Delta\left(\dfrac{dy}{dx}|_c\right)/\dfrac{dy}{dx}|_c}.$$

If the elasticity of substitution, $\eta(x, y)$, is constant across the domain, then the function is said to have constant elasticity of substitution. Below we present two classes of preference functions that satisfy constant elasticity of substitution. They are both additive functions, with more specificity in the functional form, and each provides desirable properties that could be applicable in decision analysis. We then relate these functions to the functional forms of isopreference contours discussed in the previous chapter.

8.5.1 Grounded Constant Elasticity of Substitution Preference Functions

One preference function that is monotonically increasing with each of the attributes and may be of practical interest on a domain where both attributes are positive is the function

$$\boxed{P(x, y) = xy^\eta, x, y \geq 0}, \tag{6}$$

where η is a trade-off parameter. This preference function is also known as the Cobb-Douglas preference function. It is also a special case of the additive preference function as can be seen by taking the logarithm of this function to get $\ln(x) + \eta \ln(y)$. In Appendix C we show that the preference function (6) does indeed satisfy constant elasticity of substitution.

As we shall see below, the trade-off parameter η in (6) also has an intuitive interpretation. To interpret trade-off parameter η, note that the change in preference across an isopreference contour must be zero. Hence, as we discussed in the previous chapter,

$$dP(x, y) = 0 \Rightarrow \frac{\partial}{\partial x} P(x, y)dx + \frac{\partial}{\partial y} P(x, y)dy = 0.$$

This implies that

$$y^\eta dx + \eta x y^{\eta-1} dy = 0.$$

Rearranging gives

$$\boxed{\eta = -\frac{dx/x}{dy/y}}.$$ (7)

This also means that for small percentage changes, the percentage increase in attribute X needed to compensate for a percentage decrease in attribute Y is constant. Rearranging shows that this preference function provides isopreference contours whose slope is equal to

$$\frac{dx}{dy} = -\eta \frac{x}{y}, \quad x, y > 0$$

Therefore, another way to determine the value of η, and also to verify that this preference function is indeed appropriate in the first place, is to assess several points on an isopreference contour and determine the best fit and the goodness of fit. If the points on the isopreference contour do not match contour equation for any values of η, then a different preference function should be used.

Let us now explore a qualitative property of this preference function.

Attribute Dominance (Grounding) Condition

The preference function $P(x, y) = xy^\eta$ has an important property on the non-negative quadrant $x, y \geq 0$: if any attribute is minimum (equal to zero in this case), then the whole preference function is minimum (and is equal to zero).

We refer to this property as the "*attribute dominance condition*," because a single attribute set at its minimum value dominates the remaining attributes and sets the preference of the prospect to a minimum.

This property of the preference function is also referred to as the "*grounding*" property. To illustrate the rationale behind this name, note that if any attribute falls below a minimum threshold value, the whole preference function achieves its minimum value. Therefore, the function will be a minimum on certain axes and it satisfies the relation:

$$P(x_{min}, y) = 0.y^\eta = 0 = P_{min}, \quad P(x, y_{min}) = x.0 = 0 = P_{min} \quad \forall x, y.$$

Figures 8.1, 8.2, and 8.3 show the Cobb-Douglas preference function and the shape of the isopreference contours for different values of η. The preference functions have the property that they are minimum (in this case zero) at both axes: there is some level of each attribute that sets preference function to a minimum value regardless of the level of the other attribute. Examples of attributes that would satisfy this preference function include remaining life time and consumption.

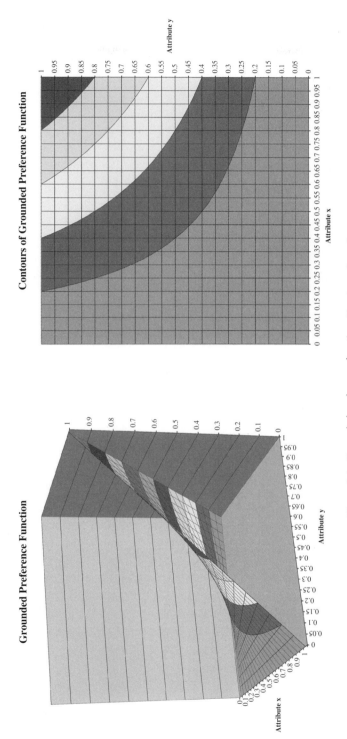

Figure 8.1. Grounded preference function, $P(x,y) = xy^\eta$, $\eta = 1$.

141

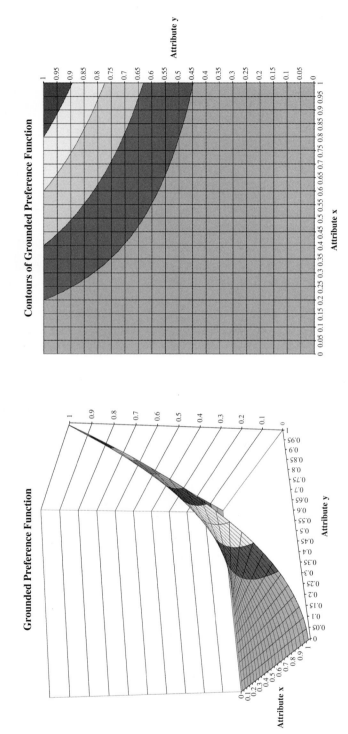

Figure 8.2. Grounded preference function, $P(x, y) = xy^{\eta}$, $\eta = 2$.

142

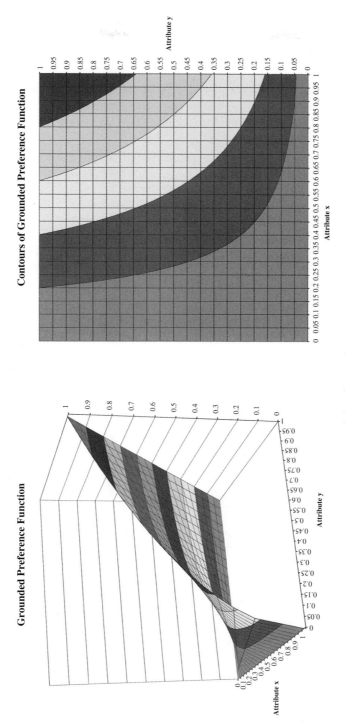

Figure 8.3. Grounded preference function, $P(x, y) = xy^\eta$, $\eta = 0.5$.

143

> **To the Reader:** Think of other situations where the attribute dominance grounding condition holds and some situations where it does not hold. This qualitative feature is helpful in recognizing the shape of the preference function.

Note that, in Figures 8.1, 8.2, and 8.3, the isopreference contours all intersect the top horizontal axis and the right vertical axis. This implies that any point in the domain of the attributes has an equally preferred point on these axes. This is an important property of grounded (attribute dominance) functions that will play an important role in constructing multiattribute utility functions.

Several variations of this type of preference function exist, for example one that is shifted with respect to reference vales of x and y, and that is grounded at $x = x_0, y = y_0$ is

$$P(x,y) = (x - x_0)(y - y_0)^\eta, \quad x \geq x_0, y \geq y_0$$

If the region of interest is $x > x_0, y > y_0$, then the preference function will not have the grounding property on this domain. Therefore, this function can also be used for decisions where the grounding property need not hold given the corresponding domain.

Multiattribute extensions of this form also exist. For example,

$$P(x_1, \ldots, x_n) = \prod_{i=1}^{n} (x_i - x_i^0)^{\lambda_i}. \tag{8}$$

8.5.2 Ungrounded Constant Elasticity of Substitution Preference Functions

Another class of constant elasticity of substitution preference functions that is monotonically increasing with each of the attributes (and may be of practical interest in situations where the attribute dominance grounding conditions do not exist) is the function

$$\boxed{P(x,y) = \left(\lambda_x x^\beta + \lambda_y y^\beta\right)^{\frac{1}{\beta}}, \quad \beta \neq 0, \ \lambda_x, \lambda_y > 0, \ \lambda_x + \lambda_y = 1, \ x, y \geq 0}. \tag{9}$$

To illustrate that this preference function does not satisfy the grounding condition on the domain of definition of (9) on the unit square, we get by direct substitution,

$$P(x_{\min}, y) = P(0, y) = \left(\lambda_y y^\beta\right)^{\frac{1}{\beta}} = y\left(\lambda_y\right)^{\frac{1}{\beta}},$$

which increases as y increases, and similarly,

$$P(x, y_{\min}) = P(x, 0) = \left(\lambda_x x^\beta\right)^{\frac{1}{\beta}} = x\left(\lambda_x\right)^{\frac{1}{\beta}},$$

which increases as x increases.

Figures 8.4 and 8.5 show examples of this preference function for different values of the parameters.

Note that this preference function is also a special case of the additive preference function as can be seen by raising it to a monotone power transformation t^β to

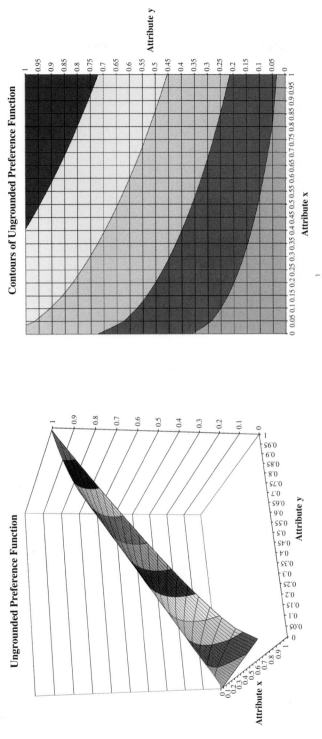

Figure 8.4. Ungrounded preference function, $P(x,y) = (\lambda_x x^\beta + \lambda_y y^\beta)^{\frac{1}{\beta}}$, $\lambda_x = 0.3, \lambda_y = 0.7, \beta = 0.6$.

145

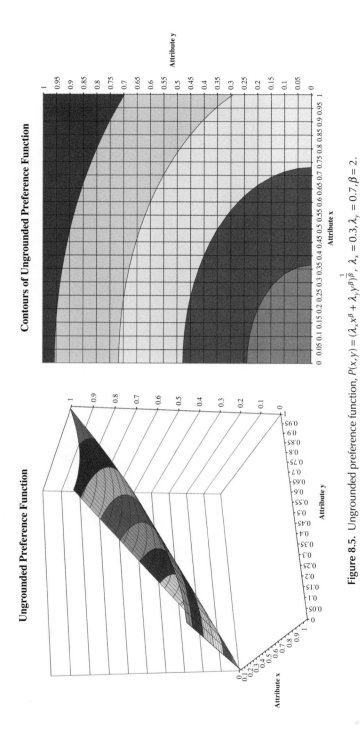

Figure 8.5. Ungrounded preference function, $P(x,y) = (\lambda_x x^\beta + \lambda_y y^\beta)^{\frac{1}{\beta}}$, $\lambda_x = 0.3, \lambda_y = 0.7, \beta = 2$.

get $\lambda_x x^\beta + \lambda_y y^\beta$. On a domain $[0,1]\times[0,1]$, note the we need three parameters $\lambda_x, \lambda_y, \beta$. Appendix C also shows that the preference function (9) does indeed satisfy constant elasticity of substitution.

Note that the preference function $\left(\lambda_x x^\beta + \lambda_y y^\beta\right)^{\frac{1}{\beta}}$ has the same shape of isopreference contours as the preference function $\left(\lambda_x x^\beta + \lambda_y y^\beta\right)$ because they are related by a monotone transformation. The slope of the isopreference contours of this function can be obtained by setting the derivative to zero to get

$$dP(x,y) = 0 \Rightarrow \frac{\partial}{\partial x} P(x,y) dx + \frac{\partial}{\partial y} P(x,y) dy = 0.$$

Work on the simpler form $\left(\lambda_x x^\beta + \lambda_y y^\beta\right)$, this implies that

$$\frac{\partial}{\partial x} P(x,y) = \beta \lambda_x x^{\beta-1}, \quad \frac{\partial}{\partial y} P(x,y) = \beta \lambda_y y^{\beta-1}$$

and so

$$\lambda_x x^{\beta-1} dx + \lambda_y y^{\beta-1} dy = 0.$$

Rearranging gives the slope as

$$\frac{dy}{dx} = -\frac{\lambda_x}{\lambda_y}\left(\frac{x}{y}\right)^{\beta-1}.$$

We have also seen this preference function in Chapter 7. Furthermore, by setting the preference function to a constant, the equation for the isopreference contours is of the form $x = (c + \frac{b}{y^\alpha})^{\frac{1}{\alpha}}$.

The following example illustrates the generality of this preference function.

Special Cases
Note that if $\beta = 1$, direct substitution into (9) shows that it is a linear preference function,

$$P(x,y) = \lambda_x x + \lambda_y y,$$

which has infinite elasticity of substitution.

As we now show, as $\beta \to 0$, we get closer to the Cobb-Douglas form of (6) described above. To illustrate, note again that the change in preference along an isopreference contour must be zero and so

$$dP(x,y) = 0 \Rightarrow \frac{\partial}{\partial x} P(x,y) dx + \frac{\partial}{\partial y} P(x,y) dy = 0.$$

We have

$$\frac{\partial}{\partial x} P(x,y) = \lambda_x x^{\beta-1}\left(\lambda_x x^\beta + \lambda_y y^\beta\right)^{\frac{1}{\beta}-1}, \quad \frac{\partial}{\partial y} P(x,y) = \lambda_y y^{\beta-1}\left(\lambda_x x^\beta + \lambda_y y^\beta\right)^{\frac{1}{\beta}-1}$$

This implies that

$$\lambda_x x^{\beta-1}\left(\lambda_x x^\beta + \lambda_y y^\beta\right)^{\frac{1}{\beta}-1} dx + \lambda_y y^{\beta-1}\left(\lambda_x x^\beta + \lambda_y y^\beta\right)^{\frac{1}{\beta}-1} dy = 0$$

For $\beta \neq 0$, $x, y > 0$,

$$\lambda_x x^{\beta-1} dx + \lambda_y y^{\beta-1} dy = 0$$

Equivalently,

$$\lambda_x x^\beta \frac{dx}{x} + \lambda_y y^\beta \frac{dy}{y} = 0$$

Rearranging gives

$$\frac{\left(\dfrac{dx}{x}\right)}{\left(\dfrac{dy}{y}\right)} = -\frac{\lambda_y}{\lambda_x}\left(\frac{y}{x}\right)^\beta \tag{10}$$

Comparing (10) and (7) shows the role of β in determining trade-offs for this preference function. Whereas (7) showed that the Cobb-Douglas preference function had a ratio $\dfrac{dx/x}{dy/y}$ that was constant, (10) shows that the general form (9) enables this quantity to vary. It is equal to a constant $\dfrac{\lambda_y}{\lambda_x}$, multiplied by the ratio $\left(\dfrac{y}{x}\right)^\beta$.

Equation (10) also shows that as $\beta \to 0$, we get closer to the constant fractional change described by (7), since

$$\lim_{\beta \to 0} \frac{\left(\dfrac{dx}{x}\right)}{\left(\dfrac{dy}{y}\right)} = \lim_{\beta \to 0} -\frac{\lambda_y}{\lambda_x}\left(\frac{y}{x}\right)^\beta = -\frac{\lambda_y}{\lambda_x}$$

Moreover, the preference function gets closer to satisfying the **attribute dominance** condition of the Cobb-Douglas form (6) because

$$\lim_{\beta \to 0}\left(\lambda_x x^\beta + \lambda_y y^\beta\right)^{\frac{1}{\beta}} = x^{\lambda_x} y^{\lambda_y}$$

EXAMPLE 8.5: Multiattribute Extensions

Multiattribute extensions of this constant elasticity of substitution preference function also exist and take the form

$$P(x_1,...,x_n) = \left(\sum_{i=1}^{n} \lambda_i x_i^\beta\right)^{\frac{1}{\beta}}. \tag{11}$$

We can therefore think of the constant elasticity of substitution preference function as some weighted mean of some order β. It is also important to get a feel for the bounds provided by this preference function. Fortunately, such a bound exists in terms of the minimum and maximum values of the attributes for given prospect for $\lambda_i \geq 0, \sum_{i=1}^{n} \lambda_i = 1$, i.e.,

$$\min(x_1,...,x_n) \leq \left(\sum_{i=1}^{n} \lambda_i x_i^{\beta} \right)^{\frac{1}{\beta}} \leq \max(x_1,...,x_n)$$

As $\beta \to 0$, we get the multivariate Cobb-Douglas form

$$\lim_{\beta \to 0} \left(\sum_{i=1}^{n} \lambda_i x_i^{\beta} \right)^{\frac{1}{\beta}} = \prod_{i=1}^{n} x_i^{\lambda_i}.$$

As $\beta \to 1$, we get the linear form,

$$\lim_{\beta \to 1} \left(\sum_{i=1}^{n} \lambda_i x_i^{\beta} \right)^{\frac{1}{\beta}} = \sum_{i=1}^{n} \lambda_i x_i.$$

As $\beta \to \infty$, we get

$$\lim_{\beta \to \infty} \left(\sum_{i=1}^{n} \lambda_i x_i^{\beta} \right)^{\frac{1}{\beta}} = \max(x_1,...,x_n)$$

and finally, as $\beta \to -\infty$, we get

$$\lim_{\beta \to -\infty} \left(\sum_{i=1}^{n} \lambda_i x_i^{\beta} \right)^{\frac{1}{\beta}} = \min(x_1,...,x_n)$$

8.6 SUMMARY

- Additive preference functions correspond to mutual preferential independence for three or more attributes. $P(x_1,...,x_n) = g\left(\sum_{i=1}^{n} f_i(x_i) \right)$, where g is a monotone function.

- If you ask the decision-maker about their preference for levels of each attribute and they are comfortable stating that preference without the other attributes, then the additive form applies ($n \geq 3$).

- Preferential Independence for two attributes: Attribute X is preferentially independent of attribute Y if
  - $P(x_1,y) - P(x_2,y)$ does not change sign for all y and this relation is true for all x_1, x_2

- Think about the implications of this relation, which we shall revisit in Chapter 32:
  - $P(x_1,y) - P(x_2,y)$ can change sign only once as we increase y and this relation is true for all x_1, x_2

ADDITIONAL READING

Abbas, A. E. and D. E. Bell. 2011. One-switch independence for multiattribute utility functions. *Operations Research* 59(3): 764–771.

Abbas, A. E. and D. E. Bell. 2015. Ordinal one-switch utility functions. *Operations Research* 63(6): 1411–1419.

Abbas, A. E. and R. A. Howard. 2005. Attribute dominance utility. *Decision Analysis* 2(4): 185–206.

Aczél, J. 1966. *Lectures on Functional Equations and Their Applications.* Academic Press, New York.

Debreu, G. 1960. Topological methods in cardinal utility theory. In K. Arrow, S. Karlin, and P. Suppes, eds. *Mathematical Methods in the Social Sciences.* Stanford University Press, Stanford, CA, pp. 16–26.

Howard R. A. 1984. On fates comparable to death. *Management Science* 30(4): 407–422.

Keeney, R. and H. Raiffa. 1976. *Decisions with Multiple Objectives: Preferences and Value Tradeoffs.* John Wiley, New York. Reprinted, Cambridge University Press, New York (1993).

APPENDIX 8A HOW PEOPLE GET IT WRONG

Caution with the "Weight and Rate" Preference Function

As we discussed, the ultimate test for a preference function is whether or not it can provide the same preferences as those of the decision-maker. Quite often, people use a particular functional form for the preference function because of its simplicity and also use arbitrary methods for its construction without validating the preference function at the end. In so doing they ignore the assumptions on which the form should be used and they often end up with implications that they had not anticipated. For example, using an additive function, as we have seen, has many implications for preferential independence. This condition (or other equivalent conditions) would need to be verified before using this additive form in the first place.

Many applications rush to use such additive functions even if there are in fact interactions among the attributes. By doing this, they ignore the fact that individual preferences over each attribute cannot be determined separately, but depend on the levels of the remaining attributes. As we shall see throughout the book, this is a recurring problem and one that is widely used in practice.

A special case of additive preference functions that is widely used (and widely misused) is often referred to as the "***weight and rate***" preference function or the "weight and rate" approach. With this special function, more of an attribute is preferred to less (which need not be the case for additive preference functions), and the individual functions f_i of the additive preference function are not only monotone but are also assumed to have the same range (for example, to range from zero to one, or any other scale, where zero represents the worst value and one represents the best). The resulting preference function is usually normalized using some weights w_i that are greater than or equal to zero and that sum to one. The preference function is written as

$$P(x_1,...,x_n) = \sum_{i=1}^{n} w_i f_i(x_i)$$

where $w_i \geq 0, \sum_{i=1}^{n} w_i = 1$ and where the functions f_i have the same range with $0 \leq f_i \leq 1$.

This function then returns a score of 0 to 1.

By itself, the use of this preference function is not the main point of concern here, although it must be noted that the structure of the function is ***much more restrictive*** now than the general forms of the additive preference function because of the normalization of the individual functions and because of the weighting. The main concerns will be demonstrated by examples below.

1) **Arbitrary Scores used for the Additive Functions**

Because the functions $f_i(x_i)$ are difficult to assess, a common approach is to replace each function $f_i(x_i)$ with some arbitrary score, $S_i(x_i)$. For example, in a house purchase decision, the attributes could be cost, view, size, and location. A decision-maker considering the choice of several houses would then assign a score to the attributes of each house independently. For example, he might say:

> *House A gets a score 8 out of 10 on cost, 4 out of 10 on view, 6 out of 10 on size, and 3 out of 10 on location.*

Herein start the problems, because we have suddenly converted the attributes into some arbitrary constructed scale. Constructed scales quite often lead to arbitrary solutions and assessments. A constructed scale for view could go something like this:

Unobstructed Ocean View, High Floor Elevation: 10/10
Unobstructed Ocean View, Low Floor Elevation: 9/10
Partially Obstructed Ocean View, High Elevation: 8/10
Partially Obstructed Ocean View, Low Elevation: 7/10
City View, High Elevation: 6/10
City View, Low Elevation: 5/10

It is clear that the scale is set arbitrarily. What does 6 really mean? What about 6.5?

2) **Arbitrary Weights**

Besides the constructed scales, we now have to weigh the constructed scales of the attributes instead of thinking about trade-offs for real and meaningful attributes. The weights in the "Weight and Rate" formulation are referred to as "importance" weights. They are (incorrectly) assessed using questions like

> *"How important is the attribute View in comparison to the attribute Cost?"*

The answer to this question is arbitrary because the question itself does not have a clear meaning. First, "importance" is not well defined and is subject to various interpretations; and second, these weights relate to the score assigned to the attribute, and not to the attribute itself. Because the score itself is arbitrary, we are assigning weights to arbitrary scores.

3) **Arbitrary Functional Form**

By using an arbitrary score and arbitrary weights, the preference function now has the form

$$P(x_1,...,x_n) = \sum_{i=1}^{n} w_i S_i(x_i).$$

The isopreference contours for this function represent trade-offs between constructed scales and are often difficult to reason about.

> By using "weight and rate" with arbitrary scores, we have not only introduced an arbitrary formula, but we are also making trade-offs using arbitrary constructed scales and arbitrary weights.

4) Direct vs. Indirect Values

Another issue with weight and rate analysis (even if applicable and even if the attributes are set correctly) is that people do not pay attention to having *direct value* attributes in the arguments. For example, in a space mission, attributes like "complexity" or "capability" or "public perception" or "probability of success" or some "performance ratio" are often used and assigned a score from 0 to 10, or 0 to 100. Complexity or Capability are seldom direct value attributes. As we have discussed, prospects are characterized by their direct value attributes for the Order Rule. Indirect values merely serve to update information about the probability distributions of the direct values using Bayes' Rule. They should not be weighted into some additive formula.

Quite often you will hear in a space mission attributes like "public perception" that are being used in a weight and rate formulation. But is that a direct value attribute? Sometimes probing further with questions like "Why do you care about public perception?" could lead to answers like "continuation of the space mission" or "to encourage more people to get involved with space research," and these explanations ultimately lead to the real direct values. In many government organizations, you might hear statements like "we cannot trace the public perception back to the direct values, and given the authority we have, we just need to make sure that there is good public perception about our program" or they might insist that "public perception" is indeed a direct value for their purpose. The problem that now arises is how to define it, and how you know that there is strong public perception? Can it be defined by the results of a poll asking a select group of people whether they believe that the government should continue to fund the space program? Or whether researchers would be willing to get involved in space research? If you cannot define it properly, then the rest of the analysis will become a waste of time, because you will later assign an arbitrary score and arbitrary weight to it.

5) Using Metrics as Direct Values

Quite often *metrics* are specified as inputs to the weight and rate preference function and are then assigned importance weights. Decision-makers might insist that these metrics are direct value attributes. The role of the analyst in this case is to make sure that these metrics are really attributes of direct value. Consider, for example, metrics like Debt-Equity Ratio or Internal Rate of Return that might (inappropriately) be used as direct value attributes in a weight an rate function, and be given a constructed scale and weights using the weight and rate approach. The direct value attributes we care about are money. Nobody carries an internal rate of return in their pocket. You can have a project with a higher internal rate of return but will provide much less dollar values in return than a project with a lower internal rate of return. Assigning a score of 0 to 5 (for example) to how well an investment does in terms of these quantities is arbitrary and making trade-offs between the score is even more arbitrary. An investment

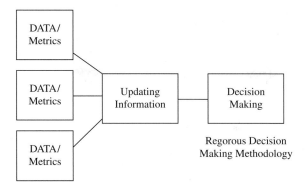

Figure 8.6. Decision-making with metrics: They are used to update information.

manager might insist that these are indeed the metrics on which his performance will be evaluated and therefore they are direct values for him. The analyst might then explain to management the implications of the incentive system they are imposing, because they will result in lower value for the organization.

Quite often metrics are used as a shortcut to bypass the need to determine the structural model that relates them to the direct value attributes. In the case of Debt-Equity Ratio and Internal Rate of Return, for example, a structural model should be used to map these metrics to the direct values. Figure 8.6 depicts a schematic diagram for how metrics should be used in this process. This would entail first identifying the direct value attributes. Then identifying the decision alternatives to enhance these direct value attributes, and also identifying the main uncertainties present. The metrics are then used to update the probability distributions for the uncertainties, and a decision should be made to maximize the utility of the direct value attributes given the updated information.

Instead, what usually happens is that the different metrics are themselves normalized and combined into some arbitrary weight and rate formula to come up with some score to determine the best alternative. If after several conversations the organization insists that these metrics are themselves direct values that they care about, then the analyst should conduct a *validation* by providing the enterprise with several prospects (characterized by these metrics) and then asking them to rank the prospects and see whether or not a weight and rate preference will indeed match these preferences. The preference function might actually be more involved than "weight and rate," and validation will be an important part of illustrating the issue with its use.

6) **Interpreting the Results of the Weight and Rate Method**

Weight and Rate is often used as a score to justify the need for further funds for projects when comparing two situations. If a score is lower, it is used to request funds to improve a situation. This is not good practice, because the "arbitrary" score and arbitrary weight using "importance" questions do not necessarily imply that a situation is better or worse. Weight and Rate scores can be *manipulated* because of their arbitrariness.

Consider the following hypothetical example that further illustrates some of these points.

Example: Using Weight and Rate as a Security Measure for a Military Base

Consider a hypothetical situation of a need for some score to determine the security of a military base at the border of a country. The base might decide on two high-level categories to determine its security:

Threat and **Situational Awareness**.

The military base then decides that both threat and situational awareness have equal "importance" and hence equal weights, and this usually happens because everybody is reluctant to state a preference for one over the other, and so Threat gets a weight of ½ and Situational Awareness gets a weight of ½,

$$\text{Score} = \frac{1}{2}(\text{Threat}) + \frac{1}{2}(\text{Situational Awareness})$$

The agency then identifies three "metrics" of indirect value that belong to these three categories:

1. Assaults on soldiers at the border
2. Ratio of illegal marijuana caught at the border (lbs) to total illegal drugs caught (lbs)
3. Percentage of people caught at the border who are in the terrorist-screening database

These metrics are available from data that is collected daily, and so the agency feels that they are well defined. They are of course indirect values, however. Why should an agency care about percentage of people or ratio of Marijuana caught not even actual amounts?

The agency then assigns these three metrics to the two high-level categories of interest as follows.

Threat: (Two Metrics are Used to Define It)

1. Assaults on soldiers at the border, and
2. Ratio of marijuana caught at the border (lbs) to total illegal drugs caught (lbs)

Situational Awareness: (One Metric is Used to Define It)

1. Percentage of people caught at the border who are in the terrorist-screening database

The military base decides to use these metrics but is reluctant to assign weights to the metrics themselves within each category, so they divide the category weight by the number of metrics in each category. Therefore, Assaults on Soldiers and Ratio of Marijuana get equal weight of ½ within the Threat category.

A weight and rate preference function is now assigned as follows:

$$\text{Score} = \frac{1}{2}\left[\frac{1}{2}(\#\text{ Assaults on Soldiers}) + \frac{1}{2}(\%\text{Marijuana caught})\right]$$
$$+ \frac{1}{2}(\%\text{Caught on Terrorist DataBase})$$

Simplifying gives

$$\text{Score} = \frac{1}{4}(\# \text{ Assaults on Soldiers}) + \frac{1}{4}(\% \text{Marijuana caught})$$
$$+ \frac{1}{2}(\% \text{Caught on Terrorist DataBase})$$

The agency then uses this score to compare the situation from year to year as an indication of the improvement (or not) of the border situation and a justification to Congress for why more funds are needed.

What Are the Issues with This Approach?
The issues are numerous.

First, why do we need a score in the first place?
   Most people would answer: to see how well we are doing with a particular program. But, let us dig further: Why do we need to know how well we are doing? Ultimately, it is about making a decision, and so the situation should be formulated as the decision:

> *Should we continue with this program or should we terminate it, or should we do something else with border security?*

Instead, the approach of focusing on an arbitrary score and then using the score to make an arbitrary decision can lead to poor decision-making.

Second, the metrics themselves are not the direct value attributes that the military base should care about. The direct value attributes could be things like prosperity (measured in terms of measurable attributes like GDP, etc.) and security (which could be measured using numerous metrics such as crime rate). The metrics chosen are indirect measures that contribute probabilistically to the expected utility formulation.

Third, the indicators should be used to update information and then used for rigorous decision-making for maximizing the utility of prospects characterized by direct value attributes. They should not just be combined arbitrarily into a formula. The resulting score in this hypothetical example is arbitrary and should not be used as an indication of the security of the border. Do not "weight and rate" the indirect values.

Forth, the reluctance to assign weights usually leads decision-makers to assign equal weights, which poses a major problem by itself because the assignment of equal weights is an assertion of some preferences and trade-offs among the arbitrary metrics. Furthermore, the weight assigned to each metric using this approach depends on the number of metrics grouped together into each category. Had the military base grouped more factors, say 6, into the threat category (as an example), then each metric in this category would have obtained a weight of 1/6, while the weight assigned to the (% caught on terrorist database) metric would have remained the same and equal to ½.

Fifth, preferences for some of these metrics are not necessarily monotone, so it is not the case that the higher the score, the better. For example, consider the metric: number of assaults on agents. If this number is higher, does this necessarily mean that the border situation is worse? In fact, if you were successful at smuggling drugs or people through various positions of the border, would you attempt to assault an agent or would you continue to smuggle silently? Attacking an agent can actually be considered an act of despair that all other avenues are not working for the smugglers.

Sixth, a metric like the percentage of marijuana caught is also problematic and is an indication of the poor choice of the metrics. Suppose that the government decides to legalize marijuana (or some forms of it), and as a result, the price of illegal marijuana would drop significantly. Smugglers would then find it much more lucrative to ship other illegal substance such as cocaine to make more profit. This would lead to a drop in percentage of marijuana caught, and hence a drop in the score, but it is not a reflection of the status of the border. It is a consequence of changes in the situation.

All these issues can be solved by incorporating the metrics to update information using Bayesian analysis and then using the updated information for decision-making using an expected utility approach.

Conclusion: The type of preference function reasoning, often referred to as the *"weight and rate"* approach, is widely used (and widely misused). It is an additive construct that uses arbitrary attributes, arbitrary weights, and arbitrary scores for each metric to come up with some arbitrary score. It often uses indirect value attributes and arbitrary scales for each indirect value. The score is compiled and used to justify performance improvement instead of thinking about the actual decision. Many people who use this approach in practice say they are using "decision analysis." In fact, they are not.

APPENDIX 8B HOW PEOPLE GET IT WRONG

Using Preference Functions to Incorporate Uncertainty

The word "Probability" is often intimidating when mentioned in a decision analysis practice. It might even remind people of the struggles they may have experienced in probability classes. This has led many people to (incorrectly) avoid using probability correctly using decision trees in the analysis of decision problems, and instead use arbitrary methods of incorporating uncertainty.

Using Probability as an Argument (Attribute) of the Preference Function

Using the probability of getting a prospect as a direct value attribute in the preference is a very common error. If there is an uncertainty, such as "Technical Success," the correct way of implementing it would be to consider various prospects of different levels of "Technical Success" and build a decision tree with a probability assigned to each prospect. Instead, a common error is the tendency to incorporate "Probability of Technical Success" as a direct value attribute within the preference itself. As we have discussed, a prospect is deterministic, and the Order Rule does not consider the probability of getting the prospect when determining the preferences. When uncertainty is present, the expected utility criterion requires us to calculate the expected value of the utility of the prospects and not some preference that includes the probability of getting the prospect. By doing that, we would be in violation of the expected utility criterion.

Failing to Incorporate New Information Due to Inappropriate Formulation of the Problem

When using arbitrary metrics as direct value attributes, a common error is the failure to recognize the arbitrariness of the score when new information arises or when the situation changes. Consider, for example, a situation where a score being assigned to answer the question of "how safe is the border?" First, this question is not clearly defined. The real question should be "what do we do (if anything) to make the border more secure?" and then consider the different alternatives and choose the alternative with the highest expected utility.

But often there is an attempt to answer this question (incorrectly) as a score resulting from some preference function as a weighted combination of some attributes. Quite often metrics or performance ratios are used. For example, the percentage of marijuana caught at the border in relation to the total amount of drugs caught could be a metric. The value of the preference function resulting from this and other metrics is then used as the score. Herein lies the problem. First, the percentage of marijuana caught is not a direct value attribute. Suppose (for some reason) that the percentage of marijuana is indeed incorporated into some preference function that is non-decreasing with its arguments. Then the higher the percentage, the higher is the score of the border security.

Now suppose that due to legislation, such as legalization of some forms of marijuana in some states, drug lords smuggling marijuana decide to shift their business to another drug (because there is less profit in legal marijuana) and instead ship other types of illegal drugs. Because of this shift, the percentage of marijuana caught at the border will drop, and so the score will also drop. But does that mean that the border is less secure? No. The problem was ill-formulated, and the preference function as-is should not have been used to represent the score in the first place. Had the decision been framed correctly, the event of legalizing marijuana would have had led to a Bayesian updating of probability and a choice of a different decision alternative to make the border more secure. Very often people use arbitrary methods for scores, and they might even work up until a problem or a change of situation occurs, and then there is panic. The change in situation quite often reveals the arbitrariness of the formulation that was used, but often it is too late.

APPENDIX 8C PREFERENCE FUNCTIONS SATISFYING CONSTANT ELASTICITY OF SUBSTITUTION

Constant Elasticity of Substitution: Single Attribute

For a single variable x and a function $f(x)$, the elasticity of x, $\eta(x)$, is defined as

$$\eta(x) \triangleq \frac{\left(\dfrac{df(x)}{f(x)}\right)}{\left(\dfrac{dx}{x}\right)} = \frac{d\ln x}{d\ln f(x)},$$

which is the ratio of the (small) fractional change in the function corresponding to a (small) fractional change in the variable.

This implies that $\eta(x)$ is also the slope of the curve $\ln x$ when plotted vs $\ln f(x)$. Using the chain rule for derivatives, gives additional interpretations for $\eta(x)$ as

$$\eta(x) \triangleq \frac{d \ln x}{dx} \frac{dx}{d \ln f(x)} = \frac{1}{x} \frac{f(x)}{f'(x)} = \frac{\left(\dfrac{df}{f}\right)}{\left(\dfrac{dx}{x}\right)}. \tag{12}$$

Equation (12) shows that $\eta(x)$ is the ratio of the slope of the line from the origin connecting $(x, f(x))$ to the derivative of $f(x)$.

Equation (12) also shows that $\eta(x)$ is equal to the ratio of a (small) fractional change in some function $f(x)$ divided by the (small) fractional change in X. As such, it measures the sensitivity of $f(x)$ to x on a fractional level. When $\eta(x)$ is large, a small fractional change $\frac{dx}{x}$ corresponds to a large fractional change $\frac{df}{f}$, and vice versa. It is possible that $\eta(x)$ be increasing with x, in which case the fractional change $\frac{df}{f}$ corresponding to a fraction $\frac{dx}{x}$ increases with x. A converse effect occurs if $\eta(x)$ is decreasing with x.

It is natural to consider the case where this quantity $\eta(x)$ is a constant (constant elasticity of substitution), and so the fraction $\frac{df}{f}$ corresponding to a fraction $\frac{dx}{x}$ does not depend on x. Rearranging (12) gives

$$\eta \frac{dx}{x} = \frac{df}{f}.$$

Integrating gives

$$\ln f(x) = \eta \ln x + b.$$

Or equivalently,

$$f(x) = ax^{\eta}.$$

The power function has constant elasticity of substitution, that is the ratio of the fractional change in x to the fractional change in $f(x)$ is constant.

Constant Elasticity of Substitution: Two Attributes

We can extend this reasoning to two attributes X and Y by thinking about the elasticity of their ratio $\frac{y}{x}$ to the trade-offs between them (as measured by the slope on an isopreference contour of constant value, c, as $\frac{dy}{dx}|_c = -\frac{\partial P(x,y)/\partial x}{\partial P(x,y)/\partial y}$. The elasticity of substitution of X and Y is given by

$$\text{Elasticity of substitution between } X \text{ and } Y = \frac{d\ln\left(\frac{y}{x}\right)}{d\ln\left|\frac{\partial P(x,y)/\partial x}{\partial P(x,y)/\partial y}\right|} \tag{13}$$

The elasticity of substitution can also be expressed as the (small) fractional change in the ratio $\frac{y}{x}$ divided by the (small) fractional change in the trade-off $\frac{dy}{dx}\big|_c$

$$\text{Elasticity of substitution between } X \text{ and } Y = \frac{\left(\frac{\Delta(y/x)}{(y/x)}\right)}{\Delta\left(\frac{dy}{dx}\big|_c\right)/\frac{dy}{dx}\big|_c}$$

Once again, the elasticity of substitution can be increasing with x and y, decreasing or constant. If the elasticity of substitution is constant, the preference function for X and Y is said to be a constant elasticity of substitution preference function.

Let us first now verify that the Cobb-Douglas preference function of (6) does indeed satisfy constant elasticity of substitution as per (13). First, we calculate the slope of the Isopreference contours. We have

$$\frac{\partial}{\partial x}P(x,y) = y^\eta, \frac{\partial}{\partial y}P(x,y) = \eta x y^{\eta-1} \quad \Rightarrow \frac{dy}{dx}\big|_c = -\frac{\left(\frac{\partial P(x,y)/\partial x}{\partial P(x,y)/\partial y}\right)}{} = -\frac{y}{\eta x}$$

Define $\theta(x,y) = \frac{y}{\eta x} \quad \Rightarrow \frac{y}{x} = \eta\theta(x,y)$. From (13),

$$\text{Elasticity of substitution between } X \text{ and } Y = \frac{d\ln(\eta\theta(x,y))}{d\ln(\theta(x,y))}$$

$$= \frac{d\ln(\eta\theta)}{d\theta}\cdot\frac{d\theta}{d\ln(\theta)} = \frac{1}{\theta}\cdot\theta = 1.$$

Therefore, the preference function (6) does indeed satisfy constant elasticity of substitution.

We can also verify that the preference function of (9) does indeed satisfy constant elasticity of substitution. First, we calculate the slope of the Isopreference contours. We have

$$P(x,y) = \left(\lambda_x x^\beta + \lambda_y y^\beta\right)^{\frac{1}{\beta}}, \quad \beta \neq 0,1, \quad \lambda_x, \lambda_y > 0, \quad \lambda_x + \lambda_y = 1.$$

$$\frac{\partial}{\partial x}P(x,y) = \lambda_x x^{\beta-1}\left(\lambda_x x^\beta + \lambda_y y^\beta\right)^{\frac{1}{\beta}-1}, \quad \frac{\partial}{\partial y}P(x,y) = \lambda_y y^{\beta-1}\left(\lambda_x x^\beta + \lambda_y y^\beta\right)^{\frac{1}{\beta}-1}$$

This implies that

$$\frac{\left(\partial P(x,y)\big/\partial x\right)}{\left(\partial P(x,y)\big/\partial y\right)} = \frac{\lambda_x x^{\beta-1}\left(\lambda_x x^\beta + \lambda_y y^\beta\right)^{\frac{1}{\beta}-1}}{\lambda_y y^{\beta-1}\left(\lambda_x x^\beta + \lambda_y y^\beta\right)^{\frac{1}{\beta}-1}} = \frac{\lambda_x}{\lambda_y}\left(\frac{x}{y}\right)^{\beta-1}$$

Define $\theta(x,y) = \dfrac{\lambda_x}{\lambda_y}\left(\dfrac{x}{y}\right)^{\beta-1}$ $\Rightarrow \dfrac{y}{x} = \left(\dfrac{\lambda_y}{\lambda_x}\theta(x,y)\right)^{\frac{1}{1-\beta}}$

$$\text{Elasticity of substitution between } X \text{ and } Y = \frac{d\ln\left(\dfrac{\lambda_y}{\lambda_x}\theta\right)^{\frac{1}{1-\beta}}}{d\theta} \cdot \frac{d\theta}{d\ln(\theta)}$$

$$= \frac{\lambda_y}{\lambda_x}\frac{1}{1-\beta}\frac{1}{\dfrac{\lambda_y}{\lambda_x}\theta}\cdot\theta = \frac{1}{1-\beta}$$

Therefore, the preference function (9) does indeed satisfy constant elasticity of substitution. Note further that as $\beta \to 0$, the elasticity of substitution converges to that of the Cobb-Douglas function.

CHAPTER 9   Valuing Deterministic Prospects

Chapter Concepts

- Value is an absolute scale
- Constructing a value function for prospects of a decision
  - Using a structural model with engineering, accounting, or logical principles.
  - Using a monotone transformation of a preference function.
  - Monotone transformations of additive preference functions
  - Archimedean value functions
  - Incorporating Time Preference into the Valuation

9.1 INTRODUCTION: THE BENEFITS OF A VALUE MEASURE

We have discussed how to assign a preference function for single and multiple attributes. We also discussed a special case of preference functions known as the additive preference function. As we saw in Chapter 1, preference statements are not always sufficient to determine the best decision especially if monetary changes in the costs of obtaining the prospects are involved. This chapter illustrates how to assign a value to a set of prospects of a decision.

If the prospects of the decision are few, then we can think about our indifference values for the prospects by visualizing our life with that prospect, and our life with an extra monetary amount and a different prospect. For example, we can say that we are indifferent between the prospect of being at the movie theatre and the prospect of staying at home for the evening while watching TV and receiving $10. If we prefer more money to less, then this value statement also means that we prefer being at the movie theatre than staying at home and watching TV that particular evening. It also means that if you are at home watching TV that evening, you would be willing to pay $10 to be at the movie theatre. Note how the increments in value are now meaningful in contrast to the preference statement.

Value is an absolute scale. It is a special type of preference expression whose units are units of some value measure. Unlike general preference functions, these increments in value (and the corresponding value assigned to each prospect) are meaningful. Applying an arbitrary monotone transformation to all dollar values will preserve the rank order of the prospects but will not preserve the value assigned to each prospect or the increments in value.

When the decision involves a large number of prospects, assigning a value to each prospect can be a tedious task unless some logical relationship among the attributes is present. If the prospects are characterized by attributes that can be expressed as numeric measures, then a <u>value function</u> that assigns a value measure to each prospect can be incorporated. The output of the value function is the indifference selling price for each prospect.

Definition

Value Function: A value function is a function that assigns a value (indifference selling price) to each prospect that is characterized by one or more measures representing the direct value attributes. A more preferred prospect has a higher value and two equally preferred prospects have equal values. Value is an absolute scale. Differences in the value measure are meaningful.

Let's look at some examples of value functions.

9.2 VALUE FUNCTIONS FROM STRUCTURAL MODELS OF PHYSICS, ENGINEERING, OR ACCOUNTING PRINCIPLES

The tradeoff between revenue and cost of a project is a common choice in business settings. By thinking about the direct value attribute of interest, in this case profit, we identify a simple relation between them:

$$\text{Profit} = \text{Revenue} - \text{Cost}. \tag{9.1}$$

This observation enables us to assign value to many prospects that could involve additional structural phenomenon. We explore this idea further in the following examples and case study.

EXAMPLE 9.1: Value Functions from Basic Accounting Principles

You are choosing between two deterministic prospects related to Revenue (R) and Cost (C). You prefer more revenue to less and less cost to more. You face two prospects $(R_1, C_1), (R_2, C_2)$, where R_1 is a revenue level that is higher than R_2 (so it is more preferred), but C_1 is the associated cost that is higher than C_2, so it is less preferred. You have a trade-off; one prospect has higher revenue but it also has a higher cost.

Once we have established that the direct value is profit and that

$$\text{Profit} = \text{Revenue} - \text{Cost},$$

we simply calculate the difference $R_1 - C_1$ and the difference $R_2 - C_2$. The revenue-cost pair that has the highest profit will form the more preferred prospect. There is no need to think about trade-offs between revenue and cost once the profit formula has been established.

EXAMPLE 9.2: Energy Generation using a Structural Model

In example 5.8, we used the structural model Energy $=\dfrac{1}{2}mv^2$ as a preference function to order the prospects of mass and velocity. Any monotone transformation applied to this preference function preserves the ordinal preferences, but only one special transformation will lead to the value function for the unit of value we are interested in.

If our direct value is **profit** in dollars earned from energy generation less the cost, then the value function can take the form

$$\text{Value} = \text{Profit}\ (\$) = \textit{Price per unit energy generation}\ \times \frac{1}{2}mv^2\ -\ \textit{Cost}.$$

This expression tells us that we are indifferent between running the energy operation and receiving a certain dollar amount as determined by the value function. Having this value function also enables us to compare whether it is worth providing different generators that produce less energy at lower cost, or more energy at a higher cost.

Sometimes constructing the value function is not as simple as relying on basic accounting principles. However, a mixture of engineering and physical principles may be combined with accounting principles to determine the value function. The following case study demonstrates the idea of using a value function for high-speed milling in engineering applications.

9.2.1. Case Study: Value Function for Profit in High-Speed Machining Using a Structural Model

The analysis in this section was conducted in collaboration with Tony Schmitz through a National Science Foundation award to derive the value function for high-speed milling. Before we start, we need to understand the physics involved in a typical machining job and also to understand the direct value attributes of the machinist. We start with some basic principles.

In a typical milling job, a cutting tool cuts through a machining piece and removes metal as it passes over it. The machinist can adjust several factors during the job to remove more or less metal per tool pass over the metal by changing the angular spindle speed of the tool, the radial depth of cut the axial depth of cut, and the feed (progression made) per tooth of the cutting tool. Figure 9.1 highlights some of the settings during a typical machining job. Our focus (for simplicity) will be the angular spindle speed (machining speed) adjustment by which the tool rotates and the axial depth that determines how much thickness of metal is removed from the piece at each pass.

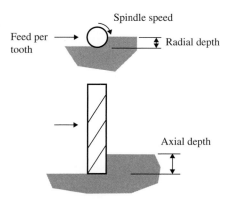

Figure 9.1. Schematic diagram representing settings in a machining job.

At first, it might seem that the best way to proceed is to increase the axial depth and the spindle speed to remove more metal. One problem, however, is that the higher the spindle speed, the lower is the cutting tool life time. The machinist would not want to keep changing cutting tools because of milling at high speeds. Another thing the machinist might want to avoid is "chatter". Chatter (instability in the machining) occurs for certain combinations of machining speeds and axial depths of cut among other things. Chatter may render the machining piece unusable, and it is more likely for higher machining speeds. The decision the manufacturer has is the levels of the angular speed and the axial depth during the machining job.

In a typical machining job, the manufacturer is ultimately interested in profit, a direct value attribute, that is given by the relation

$$\text{Profit} = \text{Revenue} - \text{Cost}.$$

Revenue is equal to the number of units of machined pieces sold multiplied by the price,

$$\text{Revenue} = \text{Units Sold x Price}$$

In its simplest form, the cost is determined by the sum of several components that include fixed costs and variable costs. Here we will use

$$\text{Cost} = C_{fixed} + C_{Machinist} + C_{Tool},$$

where

C_{fixed} is a fixed dollar quantity associated with a machining job, such as rent of the equipment and the facility…etc.

$C_{Machinist}$ is the amount paid to the machinist for the machining job. This is a variable cost that depends on the hourly rate and the time needed for the machining job,

$$C_{Machinist} = tr,$$

where t is the time in minutes and r is the machinist's fee per minute ($/minutes).

C_{Tool} is another variable cost determined by the wear on the cutting tool: it involves the cost of changing the cutting tool and time needed to change it. In its simplest form, this cost can be divided into two components:

(i) Cost due to the time needed to change a worn cutting tool, which is equal to the actual time needed to change the tool, t_{Ch} (minutes), multiplied by the machinist's rate, r ($/minutes), multiplied by the number of times this will happen during the machining job, $\frac{t}{T}$, where T is the tool life.

(ii) Cost to replace tool itself is equal to the cost of a cutting tool, C_t, multiplied by the number of tools needed for the machining job, $\frac{t}{T}$.

Therefore,

$$C_{Tool} = (t_{ch}r + C_t)\frac{t}{T}.$$

Combining these costs together gives

$$Cost = C_{fix} + tr + (t_{ch}r + C_t)\frac{t}{T}.$$

The Dilemma: The machinist would like to complete the machining job in less time to lower the cost and therefore to maximize the profit. But it is not that simple. What makes this problem particularly interesting is that the machining time cost calculation is not straightforward: it is determined by several machining parameters including axial depth of tool and its spindle speed. Each of these parameters affects the cost calculation. As we shall see, while increasing spindle speed and axial depth decreases machining time, increasing the spindle speed also reduces the tool life (leading to extra costs for tool replacement) and increasing the axial depth might lead to chatter. What should the machinist do?

This is a scenario where a value function can simplify the analysis, only it is clear that there are some physical and engineering dependencies that might not be immediately obvious. This is the role of engineering research and what is often referred to as the ***structural model*** or structural analysis. To determine the relation between a given spindle speed selection and the corresponding tool life, experiments were conducted with various speeds and the corresponding tool lives were measured (See references 1 through 4 at the end of this chapter and work done in collaboration with Tony Schmitz). The relation between the spindle speed and tool life is shown in Figure 9.2 and follows contours of the form

$$(\text{Spindle Speed})(\text{Tool Life})^n = \text{Constant}.$$

This is an inverse proportional relationship: the higher the spindle speed, the lower the tool life.

The level of modeling can be extended to include additional factors that may be more relevant to a given situation. For example, in high speed milling situations,

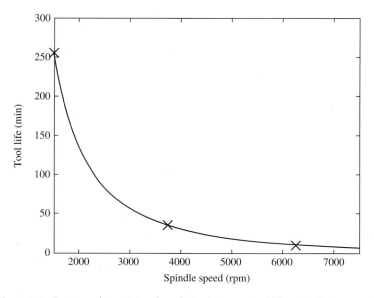

Figure 9.2. Contours determining the relation between tool life and spindle speed.

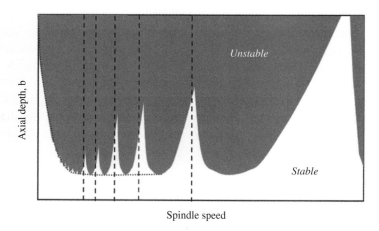

Figure 9.3. Regions of stable and unstable cuts.

increasing the axial depth at a given spindle speed may lead to chatter and instability in the machining piece and it may become waste. Figure 9.3 shows stability contours separating regions where combinations of spindle speed and axial depth lead to stable or unstable cuts. The regions were obtained experimentally.

The deterministic value function for a given setting, which is a prospect as a function of spindle speed and axial depth is now completely determined: For a given spindle speed and axial depth combination:

1) Using the spindle speed, we can calculate the tool life (from Figure 9.2).
2) For a given axial depth and spindle speed combination, we can determine whether there is chatter (Figure 9.3). If this situation is of concern, then there could be an additional

cost associated with reworking a piece with chatter or scarping it, and that portion is built into the value function.

We can now determine the cost parameters associated with the milling job given the fixed costs and the machinist rate, and the corresponding value function for profit.

Once the value function for profit is specified, we can determine many things including the best prospect (parameters of the machining job), and also how much we value any prospect and how much we would pay to be indifferent between one prospect and another.

How People Get It Wrong:

The most common way of getting this wrong is that people ignore the physics and try to use arbitrary functions to characterize the situation. For example, they might think of the attributes (such as spindle speed, tool wear) as direct value attributes, and then combine them into some additive or multiplicative form of value, without considering their physical properties. These simplistic (and arbitrary) attempts are mostly a result of trying to avoid understanding the domain knowledge and engineering principles behind a particular problem and relying instead on arbitrary forms. There is a lot of value that can be gleaned from understanding the domain knowledge and this should be incorporated.

9.3 VALUE FUNCTIONS OBTAINED BY APPLYING A MONOTONE TRANSFORMATION TO THE PREFERENCE FUNCTION

As we discussed in the introduction to this chapter, a value statement implies the same ordinal preferences as a preference statement. Therefore, if a prospect is characterized by multiple direct value attributes, then the value function and the preference function must have has the same isopreference contours. The difference is that the value function assigns a (dollar) value to each isopreference contour, and this dollar value is equal to the personal indifference selling price of the prospects that lie on that contour. The value function has units of an absolute scale.

Figure 9.4 shows an example of isopreference contours (or iso-value contours) for a value function and the corresponding dollar values (V_1, V_2, V_3, V_4) that are associated with them.

One method that might be useful in practice for constructing the value function is to start with a preference function and then apply a monotone transformation to convert the preference function into a value function. It is important to think about dollar values resulting from this transformation. It is often convenient to first normalize the preference function and then apply a monotone scaling function to these normalized preferences. If the monotone transformation is also normalized, [0,1] to [0,1], then the result is a normalized value scale that equals 1 for the most preferred prospect and 0 for the least preferred prospect. By applying a linear transformation to this normalized value scale, obtain a dollar value for each prospect. Furthermore, the increments in value are now meaningful. The following example illustrates this approach.

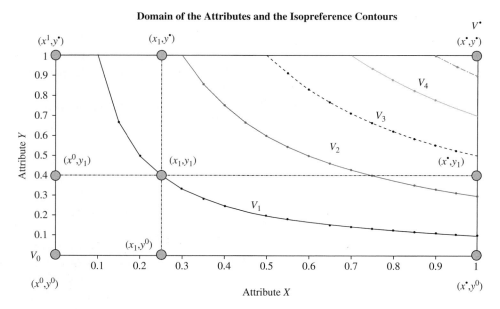

Figure 9.4. Example of a value function and values associated with isopreference contours.

EXAMPLE 9.3: Value Function for Peanut Butter and Jelly Sandwich

We can normalize the preference function for the peanut butter and jelly sandwich (equation 7.4) over a domain of the attributes to get the normalized preference function

$$P(p,j,b,f) = \frac{pjbf}{(p^*j^*b^*f^*)^2}(2p^* - p)(2j^* - j)(2b^* - b)(2f^* - f). \tag{1}$$

This preference function is equal to 1 when $p = p^*, j = j^*, b = b^*, f = f^*$ and is equal to zero when any of the attributes is zero or twice its optimal value. If the increments in preference are meaningful reflecting normalized values, then we can simply multiply by the indifferent selling price of the best sandwich, V_{max}, to get

$$V(p,j,b,f) = V_{max}\frac{pjbf}{(p^*j^*b^*f^*)^2}(2p^* - p)(2j^* - j)(2b^* - b)(2f^* - f).$$

To know whether this value function is an appropriate representation or not, we can test for a few sandwiches and see of this is indeed the value we would assign. If these values do not correspond to your indifference selling price, then you might want to scale the preference function before using the dollar multiplier using a normalized scaling function. The scaling function, S, scales the preference function to yield a fraction of the maximum value and makes the increments meaningful. The scaling function can also be applied to the normalized preference function to yield a normalized value function. For example, we may write

$$V(p,j,b,f) = V_{max}S\left(\frac{pjbf}{(p^*j^*b^*f^*)^2}(2p^* - p)(2j^* - j)(2b^* - b)(2f^* - f)\right).$$

The goal would be to find the scaling function, S.

To illustrate, consider the attributes representing the best sandwich:

$$p = p^*, j = j^*, b = b^*, f = f^*$$

How much is your personal indifferent selling price for this sandwich? If it is $10, then $V_{max} = 10$.

Now consider your value for a sandwich that has three quarters of the ingredients.

$$p = 0.3p^*, j = 0.4j^*, b = 0.42b^*, f = 0.75f^*$$

Direct substitution shows that the normalized preference function,

$$P(p, j, b, f) = 0.2$$

How much is this sandwich worth to you? Suppose it were worth $1, yet direct substitution results in $ $2. Therefore, the normalized preference 0.2 maps to $1, or a normalized value of 0.1.

Likewise, consider your preference for a sandwich that has $\frac{1}{2}$ of the optimal thicknesses.

$$p = \frac{1}{2}p^*, j = \frac{1}{2}j^*, b = \frac{1}{2}b^*, f = f^*.$$

How much is this sandwich worth to you? Direct substitution, however, would show that this sandwich gives $4.2. Suppose it were worth $3.8. Therefore, the normalized preference point 0.42 maps to $3.8 or a normalized value of 0.38.

Continuing along this path, we can consider the value of a sandwich having $\frac{1}{4}$ of the optimal thicknesses.

$$p = \frac{1}{4}p^*, j = \frac{1}{4}j^*, b = \frac{1}{4}b^*, f = \frac{1}{4}f^*.$$

Once again, we consider the value of this sandwich. Suppose it were worth $.18. Direct substitution, however, would show that this sandwich gives a normalized preference of 0.08. Therefore, the normalized preference point 0.08 maps to $0.18 or a normalized value of 0.018.

By considering additional points, we can now plot a transformation, $S(P)$, an example of which is shown in Figure 9.5.

The new value function has the form

$$V(p, j, b, f) = V_{max}S\left(\frac{pjbf}{(p^*j^*b^*f^*)^2}(2p^* - p)(2j^* - j)(2b^* - b)(2f^* - f)\right).$$

Exercise 9.1: A Seat in the Nashville Bridgestone Arena
Use a similar analysis to the examples in this chapter to determine your value function for a seat in the Bridgestone Arena in Nashville, TN. The lay-out is shown in Figure 9.6. Once you do that log on to the Arena site and look at the price of tickets. Determine which seats would be a good deal for you to purchase given your value for an event.

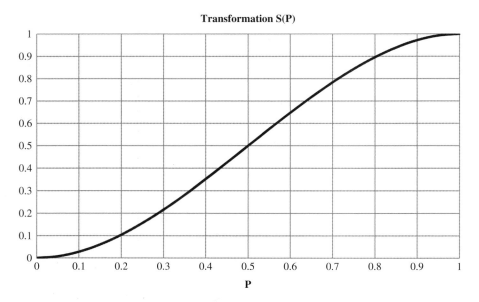

Figure 9.5. Plot of *S* from [0,1] to [0,1].

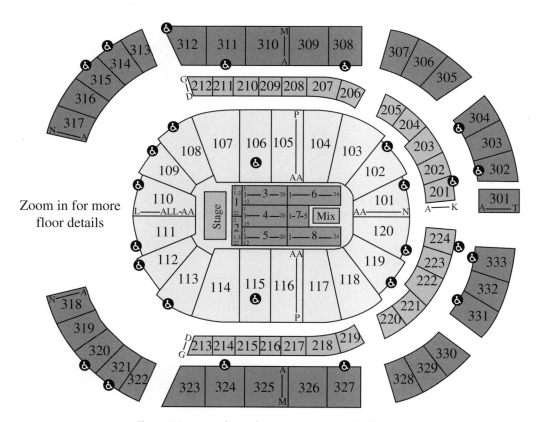

Figure 9.6. Layout for Bridgestone Arena in Nashville, Tn.

Hints: Start by thinking about the different contributors to value, and your indifference selling price for the best seat. You might start with a preference function and then apply a monotone transformation to get a value function.

Sometimes it is helpful to monetize some of the attributes to reduce their number. For example, if you wish to include the inconvenience to having to cross a certain number of seats before you get to your seat, then you can subtract a certain amount of money for each seat you cross.

Exercise 9.2: A Seat in a Boeing 737

Determine your value function for a seat in a Boeing 737 for a non-stop flight from Washington DC to Los Angeles, CA, whose layout is shown in Figure 9.7.

Thoughts for consideration: You might have a preference (and a premium) for an aisle or window seat, or a seat in the exit row, and you might deduct an amount for center seats. Ultimately, it is your preference.

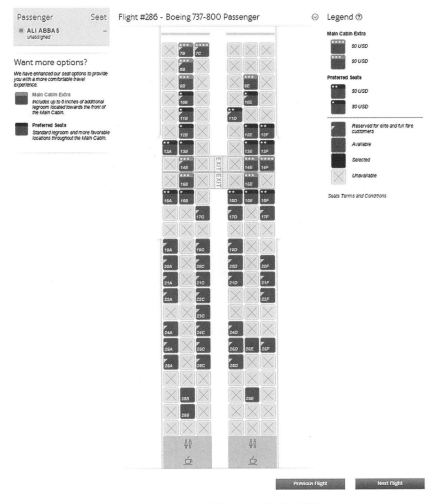

Figure 9.7. Layout of Seats in a Boeing 737.

9.4 MONOTONE TRANSFORMATIONS OF THE ARCHIMEDEAN PREFERENCE FUNCTION: ARCHIMEDEAN VALUE FUNCTION

As a special case of applying a monotone transformation to the preference function, consider the case of a monotone transformation applied to an additive (or Archimedean) preference function that we discussed in Chapter 8. We now show that applying a monotone transformation to an Archimedean preference function results in another Archimedean combination of functions, which we refer to as the Archimedean value function. We refer readers who are unfamiliar with composite functions and Archimedean combinations to Appendix 1 at the end of the book. For simplicity, we start with the case of two attributes.

9.4.1. Monotone Transformation of a Two-Attribute Archimedean Form

Suppose you have an Archimedean preference function of the form

$$P(x,y) = \phi^{-1}\left[\phi(P_1(x))\phi(P_2(y))\right].$$

Now suppose you apply a monotone transformation S to this Archimedean preference function to a value function. The value function has the form

$$V(x,y) = S\left(P(x,y)\right) = S\left(\phi^{-1}\left[\phi(P_1(x))\phi(P_2(y))\right]\right). \tag{2}$$

Define the composite transformation

$$g \triangleq \phi \circ S^{-1}.$$

The inverse of this composite transformation is by definition equal to

$$g^{-1} = S \circ \phi^{-1}$$

Substituting into (2) gives

$$V(x,y) = g^{-1}\left[\phi(P_1(x))\phi(P_2(y))\right] \tag{3}$$

Further, suppose that the domain of S enables the composite transformations

$$V_1 = S \circ P_1 \text{ and } V_2 = S \circ P_2$$

This implies that

$$P_1 = S^{-1} \circ V_1 \text{ and } P_2 = S^{-1} \circ V_2$$

Therefore

$$\phi \circ P_1 = \phi \circ S^{-1} \circ V_1 = g \circ V_1 \text{ and } \phi \circ P_2 = \phi \circ S^{-1} \circ V_2 = g \circ V_2$$

Direct substitution into (3) gives

$$V(x,y) = g^{-1}\left[g(V_1(x))g(V_2(y))\right] \tag{4}$$

Equation (4) implies that the value function that results from applying a monotone transformation to an Archimedean preference function also has an Archimedean form. Each of the individual value functions $V_1(x), V_2(y)$ in this new form is equal to the individual preference function scaled by the same transformation that is applied to the aggregate preference function, i.e.

$$V_1(x) = S(P_1(x)) \text{ and } V_2(y) = S(P_2(y)).$$

The generating function of the new form is $g(t) = \phi(S^{-1}(t))$.
We summarize this important result below.

Composite Transformation of an Archimedean Combination of Functions:
Applying a monotone transformation, S, to an Archimedean preference function with generating function, ϕ, and individual preference functions, P_i, with mild conditions on the domain and range of S, results in an Archimedean combination of functions with generating function $\phi \circ S^{-1}$ and individual value functions $S \circ P_i$.

9.4.2. Monotone Transformation of a Multi-Attribute Archimedean Form

Now suppose you have a multiattribute Archimedean preference function of the form

$$P(x_1,...,x_n) = \phi^{-1}\left(\prod_{i=1}^{n} \phi(P_i(x_i))\right).$$

Suppose you apply a monotone transformation S to this Archimedean preference function to get

$$V(x_1,...,x_n) = S\left(P(x_1,...,x_n)\right) = S\left(\phi^{-1}\left(\prod_{i=1}^{n} \phi(P_i(x_i))\right)\right).$$

Using similar analysis as the two-attribute case, define the composite transformation

$$g \triangleq \phi \circ S^{-1}.$$

The inverse of this composite transformation is

$$g^{-1} = S \circ \phi^{-1}$$

Further define the composite transformation

$$V_i = S \circ P_i$$

This implies that

$$P_i = S^{-1} \circ V_i$$

Therefore

$$\phi \circ P_i = \phi \circ S^{-1} \circ V_i = g \circ V_i$$

Direct substitution gives

$$V(x_1,...,x_n) = g^{-1}\left(\prod_{i=1}^{n} g(V_i(x_i))\right).$$

9.4.3. A Special Case of the Archimedean Form When $g(1) = 1$

If the individual value functions $V_i(x_i)$ are normalized (from zero to one) and if the generating function satisfies

$$g(1) = 1,$$

then we can interpret the individual functions in terms of points on the surface of the value function. Let x_i^* denote the maximum value of attribute X_i. Therefore $V_i(x_i^*) = 1$.
    Direct substitution gives

$$V(x_1, x_2^*,..., x_n^*) = g^{-1}\left(g(V_1(x_1)).1......1\right) = V_1(x_1).$$

Note: If the generating function of the Archimedean form passes by the point $(1,1)$, then value function has an important property: the magnitude of the value function when all, but one, of the attributes are at their maximum values is equal to the individual value function of the remaining attribute. This property is often referred to as the marginal property of an Archimedean form. This property also provides a clear method for assessing each of the individual value functions of an Archimedean form when $g(1) = 1$: we assess the value for each attribute when the remaining attributes are at their maximum values.

The Archimedean value function is not equivalent to an additive value function (even though it can be converted into an additive function by applying a monotone transformation). This is because the value function is an absolute scale that is not invariant to an arbitrary monotone transformation.

9.5 INCORPORATING TIME PREFERENCE INTO THE VALUATION OF PROSPECTS

What if the prospect will be received in the future? For example, you are offered a choice of receiving $10,000 today or receiving it in ten years. Should you be indifferent? The answer is in most cases negative because, intuitively speaking, if you received $10,000 today, you could save it for ten years, or keep it in a bank and earn interest, and you also have more options and investments to spend the money within these ten years if you receive it earlier. Furthermore, you are not guaranteed to live for ten years in the first place, you are only guaranteed the present. This is why there is a preference for receiving prospects of value today instead of in the future. Where does this valuation show up in the value model?

Recall that a prospect is deterministic and that the value of a prospect is your personal indifference selling price for the prospect. The way time preference is incorporated for a delayed prospect is by applying a discount factor to the value function itself to get the net present value. For example, if the value function for a prospect received in the future is $V(x_1, x_2, ..., x_n)$, then the equivalent value of receiving it today would be

$$\beta V(x_1, x_2, ..., x_n),$$

where β is the discount factor for that time period.

How people get it wrong: A common mistake that people make when they incorporate time preference is that they apply the discounting to the attributes themselves instead of the value function. Another common mistake is to apply the discounting to some preference function. Recall that the scale of the preference function is arbitrary, and discounting is much more meaningful when applied to a monetary quantity. In Chapter 19, we shall revisit the idea of discounting when uncertainty is present.

9.6 NO REGRETS

The approach we have used to construct the value function provides a value for each prospect, which is the personal indifference selling price of the prospect. This value is independent of the value assigned to other prospects. Each prospect is valued independently. Consequently the removal of some prospects from the decision will not change the value of any of the remaining prospects. In future chapters, this important feature will also allow us to value uncertain deals (or decision alternatives) independently.

If we allow the valuation of a prospect to depend on other prospects (whether existing in the current situation or non-existing), then we will not be able to rank (or value) prospects or decision alternatives independently from the remaining alternatives. As a result, the removal of an alternative might result in the change in ranking of the remaining alternatives. We have seen in Chapter 3 the issues that might result when using the Min-Max regret criterion that lead to rank reversal. The same situation would occur when incorporating regret into the valuation of the prospects.

Regret is a descriptive phenomenon that some people might experience and it can be modeled descriptively by modifying the valuation of the prospects, but it is best not to think in terms of such counterfactuals. Most situations of regret compare what you have with what could have been (that was better). There is no point in thinking about that counterfactual, except for the learning, especially if you cannot change it. Why live your life comparing it to some outcome that could have been better??,.

I like to summarize this by saying

"Do not value your life based on what was or what could have been."

There is no room for regret. You just learn from your past experiences and live the rest of your life. On that note, you are not even guaranteed the future: you are only guaranteed the present so live it.

9.7 SUMMARY

Value has an absolute scale that represents the personal indifference selling price of a prospect.

A value statement for the prospects provides more information than just a preference statement, and likewise a value function provides more information than a preference function.

The value function has an absolute scale. Its units are units of value described by some meaningful value measure.

Applying a monotone transformation to an Archimedean preference function results in an Archimedean combination of functions.

Value functions can be constructed using a variety of methods including logical or accounting principles, as well as engineering principles and the laws of physics using some structural models. Sometimes a mixture of both accounting and engineering principles can be combined using experimentation.

Time preference is applied to the output of the value function and not to the individual attributes.

Whenever your valuation of a prospect is a result of a comparison to a prospect that "could have been" you will end up not being able to value the prospect (or the alternative) independently. You may also be subject to rank reversal as indicated in the Min-Max regret example in Chapter 3.

READINGS ON VALUE FUNCTIONS FOR HIGH SPEED MILLING

Abbas, A., L. Yang, R. Zapata, and T, Schmitz. 2008. Application of decision analysis to milling profit maximization: An introduction. *Int. J. Materials and Product Technology*, Vol. 35 (1/2), 64–88.

Karandikar, J., Abbas, A., and Schmitz, T., 2014, Tool life prediction using bayesian updating, part 1: Milling tool life model using a discrete grid method, *Precision Engineering* 38(1), 9–17

Karandikar, J., Abbas, A., and Schmitz, T., 2014, Tool Life Prediction using Bayesian Updating, Part 2: Turning Tool Life using a Markov Chain Monte Carlo Approach, *Precision Engineering (Forthcoming)*

Karandikar, J, T. Schmitz, A. E. Abbas. 2012. Spindle speed selection for tool life testing using bayesian inference. *Journal of Manufacturing Systems*, 31, 403–411.

APPENDIX 1: HOW PEOPLE GET IT WRONG:

1) Ignoring the domain knowledge and structural relations among the attributes when constructing a value function.
2) Using arbitrary value functions that combine individual functions of the attributes into some arbitrary form.
3) Using arbitrary scales for the attributes e.g. pain on a scale of 0 to 10, and using that scale as an argument of the value function.

4) Incorporating time preference by discounting the attributes themselves instead of discounting the value function

5) Incorporating regret into the value function calculations and using that to make a normative decision. Regret is a descriptive phenomenon. It is prudent not to think about it except for learning to make future decisions. From a normative perspective, it prevents the valuation of prospects or alternatives independently without thinking about other prospects.

6) Many government organizations and many large enterprises do not like to state monetary values on some prospects, such as value of a human injury, and therefore do not provide explicit monetary values for prospects. Instead, they prefer to infer this trade-off indirectly using indifference utility assessments. In so doing, they often forget to validate the implied trade-offs. We shall refer back to this idea in Chapters 21 through 32.

PART III

DECISIONS WITH UNCERTAINTY: UTILITY FUNCTIONS USING PREFERENCE OR VALUE FUNCTIONS

This part illustrates how to assess utility functions over a value measure. Chapter 10 illustrates how to construct a utility curve over a direct value attribute using indifference probability assessments. Chapter 11 presents an important notion: that a single attribute utility function over the value function, or over an attribute of the preference function, is all that is needed to construct a multiattribute utility function. Chapter 12 presents the special case of multiattribute utility functions constructed by assigning a utility function over an additive preference or value function.

CHAPTER 10 Single-Attribute Utility Functions

Chapter Concepts

- The convenience of having a value measure when assigning utility values
- The meaning of a utility function
- The vertical scale of a utility function
- The exponential, logarithmic, and linear risk tolerance utility functions
- Assessing the parameters of a utility function
- Attitudes toward risk and time

10.1 INTRODUCTION

As we discussed in Chapter 2, a utility value is not needed if there is no uncertainty in the decision. When uncertainty is present, we need to assign a utility value for each prospect. When the prospects are few, the utility value of a prospect can be assessed (and interpreted) using an indifference probability assessment in terms of the best and worst prospects as per the Equivalence Rule of Chapter 2. The best decision alternative is the one with the highest expected utility.

Because the rank order of the expectation operator is invariant to any affine transformation of the functions for which it is calculated, the order of the alternatives (as determined by the expected utility criterion) remains unchanged if the utility values assigned to the prospects undergo an affine transformation. Therefore, the utility values need not be normalized to determine the alternative with the highest expected utility. When the utility values are not normalized, however, we lose the indifference probability interpretation for the expected utility of an alternative.

The indifference probability assignment using the Equivalence Rule does not require that the prospect be expressed in terms of a value measure. Recall the prospects of Figure 4.5, which we repeat in Figure 10.1 for convenience, where an indifference probability (utility) is expressed in terms of the best and worst prospects.

When the prospects of a decision are numerous, however, it would be tedious to conduct the indifference probability assignment. To simplify this task, we first assign a value measure to each prospect. Once a value measure is specified, we can envision

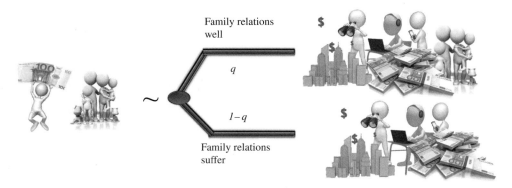

Figure 10.1. Utility assignment using indifference probability interpretations.

some function that operates on the value measure and returns the indifference probability assessment for each value in terms of the best and worst values considered. Because the prospect is expressed in terms of a value measure, in effect this function returns the indifference probability assessment for each prospect in terms of the best and worst prospects. We call this function the **utility function**. Our focus in this chapter is on utility functions over a single direct value measure.

Definition

Utility Function over a single value measure: A utility function U over a direct value measure, X, is a curve defined on the domain of the measure, such that each point on the curve, $U(x)$, is the indifference probability that makes the decision-maker indifferent between receiving the amount x for certain or receiving a binary deal that provides either the most preferred value, x^*, and the least preferred value, x^0.

If the decision-maker prefers more of the value to less, then the utility function should be increasing. If the utility function is normalized, then the best prospect will have a utility value of one and the worst will have a utility value of zero. The best and worst prospects represent the upper and lower bounds (respectively) of the domain of the utility function.

This chapter further discusses the meaning of a single-attribute utility function over a value measure, and presents some important utility functions including the exponential, logarithmic, and linear risk tolerance utility functions. Chapters 13 through 20 present further properties of utility functions and translate these properties into mathematical formulations that help characterize this functional form.

The utility function can be determined by several methods among which are (1) assessing a few indifference points and then fitting them to a given functional form. (2) Alternatively, the utility function can be constructed to match some desired properties stated by the decision-maker. Once the utility function is specified, the best decision alternative is the one with the highest expected utility. This chapter discusses examples of utility functions over value measures and methods for their direct assessment.

10.2 ASSESSING UTILITY FUNCTIONS OVER VALUE MEASURES

As we have discussed, by using a value measure to represent a prospect, we can facilitate the assignment of utility vales significantly. Furthermore, using a value measure provides more than just the best decision alternative, but it also determines the ***certain equivalent*** of any alternative (or uncertain deal).

> **Definition**
> The **certain equivalent** is the amount of money that makes you just indifferent to giving up the deal you are facing. We shall refer back to the certain equivalent in Chapter 13.

As we shall see in Chapter 13, the expected utility ranking is invariant to an affine transformation on the utility function. But ranking by the certain equivalent is invariant to a *linear* transformation on the utility function. This implies that we need not use a normalized utility function in our analysis, and that the certain equivalent remains the same even if the utility function is multiplied by a negative amount.

In most of the analysis in this book, however, we shall use a normalized utility function to maintain the indifference probability interpretation and provide numerous analogies between probability and utility.

Because more of a value measure is generally preferred to less, we shall focus on utility functions that are:

1. **Continuous** (you do not want any discontinuity in the utility function, otherwise you may value small increments in a deal by an abruptly large amount),
2. **Bounded** (this is not a major restriction: remember, utilities are just indifference probabilities so they can be normalized to range from zero to one),
3. **Nondecreasing** (because more money is preferred to less, the utility value increases as the value increases). If more of the value measure is strictly preferred to less, then the utility function is strictly increasing.
4. **Twice continuously differentiable**. This condition, which is widely used in many practical contexts, enables the calculation of what is known as the risk aversion function (Arrow 1965, Pratt 1964), which is discussed in Chapter 14.

The risk aversion function provides many insights into the valuation of uncertain deals. In Chapter 13, we shall present many analyses that rely only on the continuity and monotonicity of the utility function and that do not require it to be twice continuously differentiable.

10.2.1 Assessing a Utility Function by Direct Utility Assessments

One way to assess the utility curve is to empirically assess a few points on the curve (using indifference probability assessments) and then find the best fit that matches these points. The following example illustrates this point.

EXAMPLE 10.1: Constructing a Utility Function by Direct Utility Assessments

---

Suppose we wish to assess a decision-maker's utility function over the interval $0 to $1,000. We would ask the decision-maker a series of questions where he would state an indifference probability q that makes him indifferent between:

1. Receiving a monetary amount for certain, or
2. Receiving a binary deal that pays $1,000 with probability q and $0 with probability 1-$q$.

The monetary value can be $100, $200, $300, $400, $500, $600, $700, $800, and $900, for example.

For each of these values, the decision-maker would need to state the indifference probability, q.

Using our previous notation introduced in Chapter 1, these utility values can be written as

$$U(100)\,|_0^{1000} = 0.20$$
$$U(200)\,|_0^{1000} = 0.25$$
$$U(300)\,|_0^{1000} = 0.38$$
$$U(400)\,|_0^{1000} = 0.55$$
$$U(500)\,|_0^{1000} = 0.60$$
$$U(600)\,|_0^{1000} = 0.74$$
$$U(700)\,|_0^{1000} = 0.77$$
$$U(800)\,|_0^{1000} = 0.85$$
$$U(900)\,|_0^{1000} = 0.90.$$

Note that $U(0)\,|_0^{1000} = 0$ and $U(1000)\,|_0^{1000} = 1$.

Figure 10.2 shows an example of these utility assessments and plots these monetary values versus the corresponding values of q.

Several methods may then be used to curve-fit the assessments into a smooth (twice continuously differentiable) functional form. For example, Chapter 30 discusses a method to assign a smooth curve that fits indifference assessments.

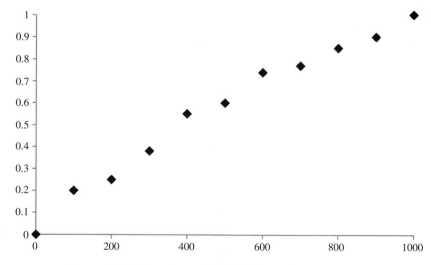

Figure 10.2. Utility values obtained by direct utility assessments.

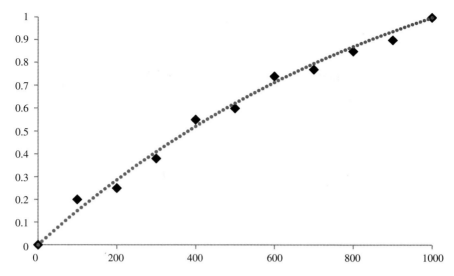

Figure 10.3. Curve-fitting the utility assessments.

Figure 10.3 shows a curve-fit of the assessed values onto the curve $U(x) = \dfrac{e^{-0.001x} - 1}{e^{-1} - 1}$ using a least-squares error estimation.

10.2.2 Extending the Domain of the Utility Assessments

Suppose that you have made the assessments of Figure 10.2. If later you would like to extend the interval of the assessments over a larger monetary domain such as $0 to $2,000, for example, then we would need to assess the indifference probabilities of more points. The previous assessments do not go to waste, however, as we illustrate below.

EXAMPLE 10.2: Extending the Utility Assessments on a Larger Domain

---

To extend the range of the utility function non a larger domain, we could ask the decision-maker a series of additional questions where he would state an indifference probability, q, for a particular monetary value for certain, say $1,000, $1,100, $1,200, $1,300, $1,400, $1,500, $1,600, $1,700, $1,800, and $1,900, or a binary deal that pays $2,000 with probability q and $0 with probability 1-$q$. For example, the decision-maker might provide the following assessments.

$$U(1000)\,|_0^{2000} = 0.70$$
$$U(1100)\,|_0^{2000} = 0.80$$
$$U(1200)\,|_0^{2000} = 0.85$$
$$U(1300)\,|_0^{2000} = 0.86$$
$$U(1400)\,|_0^{2000} = 0.88$$
$$U(1500)\,|_0^{2000} = 0.90$$
$$U(1600)\,|_0^{2000} = 0.93$$
$$U(1700)\,|_0^{2000} = 0.95$$
$$U(1800)\,|_0^{2000} = 0.97$$
$$U(1900)\,|_0^{2000} = 0.99$$

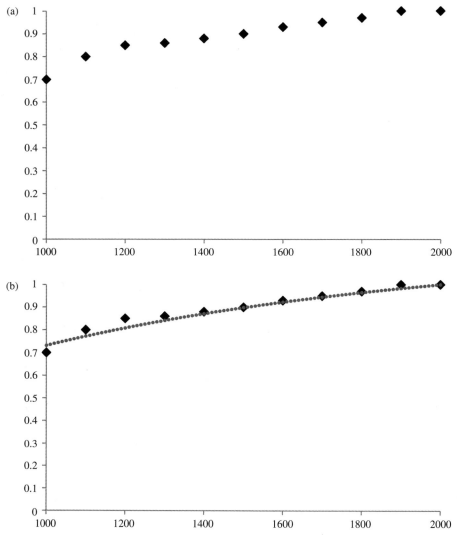

Figure 10.4. Utility assessments on an extended domain.

Figure 10.4 plots these assessments.

Of course the assessments can also be performed using indifference assessments for a binary lottery of $2,000 and $0.

We do not need to repeat the earlier assessments on the domain of $100, …, $900. We simply normalize the previous assessments using the Substitution Rule as follows:

If the indifference probability assessment of $900 were $q_1 = 0.9$ for a lottery with prospects $1,000 and $0, and if the indifference probability of $1,000 were $q_2 = 0.7$ for the lottery of $2,000 and $0, then we simply replace the prospect $1,000 with its own indifference probability assessment to get the new indifference probability of $900 as shown in Figure 10.5.

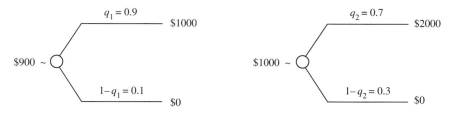

Figure 10.5. Assessing the utility of the highest prospect ($1,000) on an extended domain.

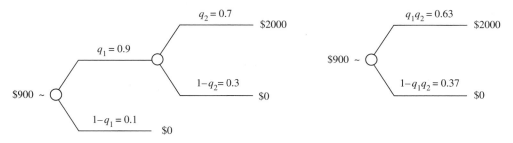

Figure 10.6. Combining assessment to obtain utility assessments on an extended domain.

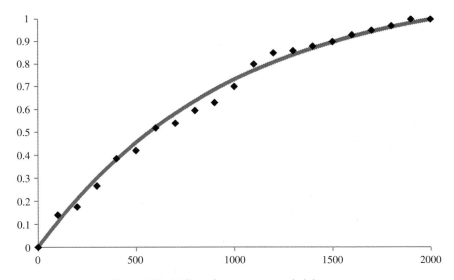

Figure 10.7. Utility values on an extended domain.

The indifference probability of $900 in terms of $2,000 and $0 is then $q_1 q_2 = 0.63$. The same applies for the other assessments on the domain $0 to $1,000. Figure 10.6 shows the normalized utility values on the whole scale of $0 to $2,000.

Figure 10.7 shows the utility assessments and the corresponding fitted curve on the whole domain.

10.2.3 Constructing a Utility Function by Assuming a Functional Form and Then Assessing Its Parameters

When the number of prospects is large, the construction of a utility function by direct assessment for all points might become a tedious task. One method for determining the utility function, in this case, is to assume a functional form for the utility function and then fit the parameters of this functional form to a few utility assessments.

EXAMPLE 10.3: Assessing the Parameter of an Exponential Utility Function

---

If the chosen functional form is the exponential form $U(x) = 1 - e^{-\gamma x}, \gamma \neq 0$, then there is only one parameter, γ, that needs to be determined. A single equation involving this parameter γ would be sufficient to determine its value. Equations relating to parameters of a utility function are usually in the form of indifference probability assessments. Equating the expected utility of both sides at indifference provides an equation that can be used to determine the value of the parameter.

For example, if the decision-maker is indifferent between receiving $\$b$ for certain and an α chance of receiving $\$a$ and $1 - \alpha$ chance of receiving $\$c$, then this implies that

$$U(b) = \alpha U(a) + (1 - \alpha)U(c).$$

Figure 10.8 depicts this relationship graphically.

Substituting for $U(x) = 1 - e^{-\gamma x}$ into this equation and simplifying gives

$$e^{-\gamma b} = \alpha e^{-\gamma a} + (1 - \alpha)e^{-\gamma c}.$$

Because α, a, b, c are known, the remaining unknown in this equation is γ, which is the parameter of the utility function. This equation can be solved numerically or by iteration.

For example, suppose that $a = \$1,000, c = \$0, b = \$500$ and the decision-maker states an indifference probability of $\alpha = 0.7$, then we have

$$e^{-500\gamma} = 0.7e^{-1000\gamma} + 0.3e^0$$

Rearranging and setting $x = e^{-500\gamma}$ gives

$$0.7x^2 - x + 0.3 = 0,$$

which has two solutions:

1. $x=1$ and so $\gamma = 0$. This is excluded because $\gamma \neq 0$ if the utility function is strictly increasing.

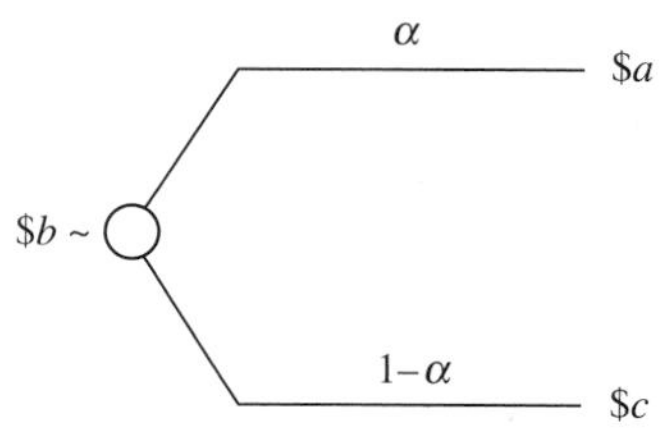

Figure 10.8. Utility assessment to determine the value of α.

2. $x=0.4285$ and so $\gamma = 0.001695$. This is the only feasible solution of γ for a strictly increasing utility function. By conducting a single assessment we were able to determine the value of the parameter of the utility function.

The valuation of the deal in this example – 0.7 probability of $1,000 and 0.3 probability of $0 – is $500. However, the expected value of the deal is $700. We shall discuss the difference between the valuation of a deal and its expected value in much more detail in future chapters.

Sometimes choosing particular values of a, b, c, however, can help simplify the computational effort for deducing γ as we illustrate in Example 10.4.

EXAMPLE 10.4: Simplifying the Assessments of Some Parameters Using Particular Lotteries

---

If we fix $\alpha = 0.5$, and set $c = -\dfrac{a}{2}, b = 0$, then the value of a that leads to indifference is approximately equal to the reciprocal of γ, i.e., $\gamma \approx 1/a$. Figure 10.9 shows the gamble for this type of question. The value of a that makes the decision-maker indifferent between receiving this deal or giving it up is equal to the reciprocal of γ. For example, if the decision-maker is indifferent between a 50/50 chance of receiving $1,000 or losing $500 versus not engaging in this deal, then $\gamma \approx \dfrac{1}{1000} = 0.001$.

There are many other functional forms of utility functions that can be used, some of which have one parameter and others have more – for example, the logarithmic utility function, $U(x) = \ln(x + \alpha)$, and the linear risk tolerance utility function, $U(x) = (\rho + \eta x)^{\frac{1-\frac{1}{\eta}}{\eta}}$, where $\eta \neq 0, 1$. We refer to these utility functions and others in more detail in Chapters 13 through 18.

10.3 DETERMINING THE FUNCTIONAL FORM OF A UTILITY FUNCTION

It is clear that choosing a functional form for the utility and then assessing its parameters is easier than assessing several utility values and then fitting them as we did in the previous section. The main problem, of course, is how to choose a functional form that appropriately represents the decision-maker's preferences and provides the same responses to the lottery questions using direct utility assessment. When choosing

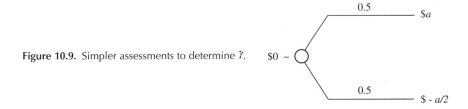

Figure 10.9. Simpler assessments to determine γ.

a particular functional form for the utility, the decision-maker might have some desirable properties that the utility function should satisfy. For example,

1. The decision-maker may wish to value all uncertain deals (or investments) by an amount that is less than the mean value of the investment. Alternatively,
2. The decision-maker may wish to value uncertain deals (or investments) by a higher amount as he gets richer. The rationale is that he might be willing to accept more risk. Alternatively,
3. The decision-maker may wish to provide qualitative (or quantitative) expressions about the way he values a change in a lottery or an investment when it gets modified. Alternatively,
4. The decision-maker might provide statements that preferences for any two investments might not change as he gets richer, or they might change but only once.

To determine the functional form that best matches the preferences of the decision-maker, it will be useful to understand the various desirable properties and their implications. As we shall see these properties translate into different functional forms of utility functions that the decision-maker might wish to use. Chapters 13 to 18 provide detailed analyses that help characterize the utility function under the various conditions discussed earlier.

10.4 UTILITY OVER VALUE VERSUS UTILITY OVER AN ATTRIBUTE

In principle, a utility function can be assigned directly over an attribute such as money or even leg space in a motor vehicle. But in some cases, it might be easier to reason about money (particularly when it involves the valuation or changes in valuation of a situation).

In some cases, a transformation might be applied to an attribute to make it more meaningful before the utility function is assessed. For example, for the preference function for peanut butter in Example 9.3, suppose that the levels of all other ingredients are held fixed. If we assign a utility function over the thickness of peanut butter directly, it might be difficult to reason about. However, if we convert the preference function into a value function for thickness of peanut butter, then we can assign a utility function over money without having to think about the thickness of peanut butter in the first place. This type of transformation will then allow us to use many of the tools from single-attribute utility functions to construct multiattribute utility functions. In the next chapter, we extend this approach to multiattribute utility functions constructed by assigning a utility function over a multiattribute preference or value function. This will enable us to think about our utility function for prospects (even when multiple attributes are involved) by reasoning about a one-dimensional utility function over value.

10.5 INCORPORATING TIME PREFERENCE

If prospects of a lottery are received in the future, or at various time intervals, then each prospect should be individually discounted to its present equivalent before assigning a utility value to it. A utility function is applied to the discounted net present value of a prospect. As we shall see in Chapter 19, this implies different utility functions (or different tastes for risk) for prospects received today and prospects received in the future. The expected utility is then calculated as usual.

A common mistake is to discount the utility of a future prospect instead of discounting its value and then assigning a utility to the discounted value.

Dialogue between Instructor (I) and Student (S)

I: Is everyone clear on the concept of a utility function?

S: Yes. It is a curve that represents the indifference probability assessments for a prospect in terms of the best and worst prospects.

I: Correct. And if more of the attribute is preferred to less, then the utility curve will be nondecreasing.

S: But, it seems unlikely that decision-makers will be consistent with the assessment of the utility curve. If we made the assessments in any arbitrary order, a decision-maker might give a higher indifference probability for a less preferred prospect.

I: Of course. But then you look at the assessments and you reconcile the differences. You use these inconsistencies as validity checks to help the decision-maker reason about the appropriate utility curve for them. For example, you can remind them that they prefer more to less and so the utility curve should not be decreasing. If they do not have a consistent utility curve, they will be inconsistent in their decision-making. It is worth spending the time to get this right.

The whole philosophy of the expected utility framework is to break down the reasoning of the uncertainties and the preferences into smaller building blocks where you reason about uncertainties separately, and you reason about preferences separately, and then you combine them together using the expected utility calculation. We shall discuss properties of the utility curve and its implications in Chapters 13 through 18.

S: We have seen that we cannot use a probability as an attribute in the preference or value function. Can we use probability as an attribute for the utility function?

I: Absolutely not.

S: But then how do we make trade-offs between a deal that has a higher probability of a lower gain and one with a lower probability of a higher gain?

I: We do that using the utility function and expected utility analysis. The appendix of this chapter illustrates how to do this.

10.6 SUMMARY

- This chapter showed how to construct a utility function over a value measure using

  1. indifference probability assessments
  2. a functional form and then assessing its parameters using indifference assessments.

- To incorporate time preference when making decisions about lotteries, we first determine the net present value of each prospect, and the utility function is then applied to the net present value of each prospect.

READINGS

Arrow, K. J. 1965. *The Theory of Risk Aversion. Lecture 2 in Aspects of the Theory of Risk-Bearing.* Yrjo Jahnssonin Saatio, Helsinki.

Howard, R. A. and A. E. Abbas. 2015. *Foundations of Decision Analysis.* Pearson, New York.

Pratt, J. 1964. Risk aversion in the small and in the large. *Econometrica* 32: 122–136.

von Neumann, J. and J. Morgenstern, 1947. *Theory of Games and Economic Behavior.* Princeton University Press, New York.

APPENDIX THE STATE OF THE BORDER IS *NOT* ORANGE!!

Recall the problem with the rectangular target regions defining rectangular isopreference contours that we discussed in the appendix of Chapter 7. This is a recurring problem, not just in engineering design but also in many risk analysis scenarios. Consider, for example, the common risk scenario plot on two axes: severity of a consequence and the probability of loss (or probability of success of an attack) as shown in Figure 10.10. The figure shows four rectangular regions: the bottom left region commonly colored green and referred to as the safe region; the top right region commonly colored red representing a danger zone; and the remaining two regions, usually colored orange, representing intermediate-risk regions.

What is the problem with this formulation?

There are several issues with this representation. First, representing an attack scenario by only one point, as is commonly done for comparative purposes, is misleading. An attack might well involve multiple levels of severity with different probabilities and not just one point. This implies that an attack should be a cloud of points on this space and not just one point representing only one possible loss scenario with a given probability.

Second, even if a successful attack represented only one loss point, the contours represented by this chart cannot be meaningful. The chart implies that we are indifferent between two scenarios having the same probability of success but one causing more damage. To illustrate, Figure 10.10. represents three possible attack scenarios, A, B, and C. The chart shows indifference between prospects B and C, as they both lie in the top right "Red" region. It is not possible to consider those scenarios equally because, although they have the same probability, one has a much higher severity of consequence. Therefore, C would be a much worse scenario than B, but they lie in the same region and have the same risk code. Using rectangular color code regions to represent risk is inappropriate and misleading.

Furthermore, it is difficult to provide a trade-off between probability of success and severity of a consequence without calculations using tools such as utility values. To illustrate, we will now show how to calculate indifference contours and demonstrate that the indifference regions defined by rectangular contours in Figure 10.10 can never occur with reasonable utility functions.

We can plot the relation between probability of success and severity using utility functions. Consider a scenario where there is 90% chance of a loss or damage worth $1 million and a 10% chance that the scenario would not occur resulting in $0 loss.

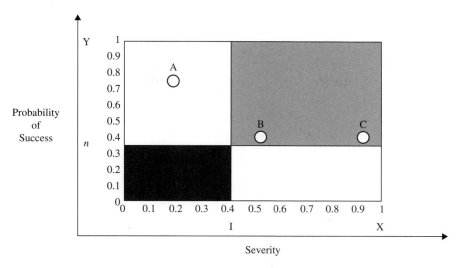

Figure 10.10. Contours presented by regions of border security.

Now suppose we wish to identify the amount of loss l that would make us indifferent to this scenario if the probability of loss were 80%. Using the principle of "win some – lose some," we should expect the loss value l to be more than $1 million for indifference because it has a lower probability of occurring. Figure 10.11 shows this scenario. Indifference between the two scenarios implies that the expected utility of each deal must be equal. Therefore, we need the utility values of the prospects involved.

Suppose that a decision-maker has an exponential utility function with risk aversion coefficient γ equal to 10^{-6}. For example, the utility function has the form $U(x) = 1 - e^{-10^{-6}x}$.

The expected utility of the deal on the left hand side of Figure 10.11 is

$$0.9\left(1 - e^{10^{-6} \cdot 10^{6}}\right) + 0.1\left(1 - e^{-10^{-6} \cdot 0}\right) = 0.9\left(1 - e\right).$$

The expected utility of the second lottery is

$$0.8\left(1 - e^{-10^{-6}l}\right) + 0.2\left(1 - e^{-10^{-6} \cdot 0}\right) = 0.8\left(1 - e^{-10^{-6}l}\right).$$

Equating the expected utilities of both deals gives

$$0.9\left(1 - e\right) = 0.8\left(1 - e^{-10^{-6}l}\right) \Rightarrow l = \frac{-1}{10^{-6}} \ln\left(1 - \frac{9}{8}(1 - e)\right) = -1.076 \text{ million.}$$

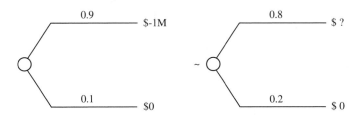

Figure 10.11. Indifference between two deals having different probabilities of loss.

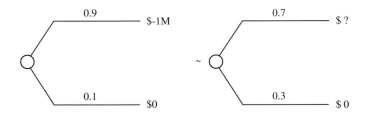

Figure 10.12. Indifference between two deals having different probabilities of loss.

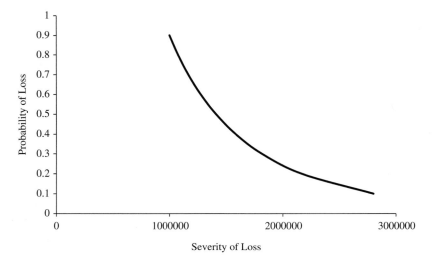

Figure 10.13. Trade-offs between probability of loss and severity of loss.

Indeed, the loss amount is more than $1 million as expected.

We can repeat the same analysis for different probabilities of loss. Figure 10.12 shows the deals for a loss probability of 0.7. Using similar analysis as above shows that the indifference loss amount for a probability of 0.7 is

$$l = \frac{-1}{10^{-6}} \ln\left(1 - \frac{9}{7}(1-e)\right) = -1.166 \text{ million.}$$

By repeating the same steps for various probabilities of loss, we can get the indifference contour for all values of probability of loss vs. loss. Figure 10.13 plots the indifference contour for losses and probabilities of loss that are equivalent to a 90% chance of losing $ 1 million. It is clear that the indifference region is not rectangular.

In principle, we can also repeat these steps starting with different probabilities of loss and determine the corresponding isopreference contours. Figure 10.14 shows the analysis conducted starting with probabilities of loss of $1 million corresponding to 0.1, 0.3, 0.5, 0.7, and 0.9 for comparison. As we can see these trade-offs were calculated using utility values and the shapes of these contours would vary based on the risk aversion coefficient that is used.

Another problem with the rectangular representation of Figure 10.10 is that we don't know why there is a *cutoff threshold* region in the first place. Why is it that if we

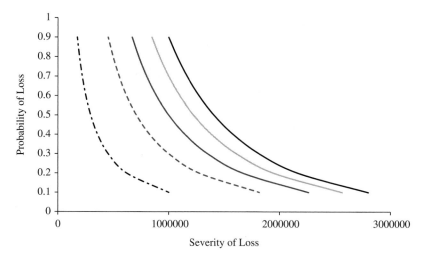

Figure 10.14. Trade-off contours between probability of loss and severity of loss.

are given a loss probability and a consequence such as B or C, then a small reduction in loss probability will result in a change of the risk situation from Red to Orange, and sudden reduction in readiness and alert, and a rationale for movement of resources? It is best to think of readiness and preparations as a decision about allocation of resources using full probability distributions instead of cutoff threshold values.

CHAPTER 11 Multiattribute Utility Functions Using Preference or Value Functions

Chapter Concepts

- Constructing multiattribute utility functions using a single-attribute utility function and a preference function or a value function or a structural model.
- Once the value function is specified, there is only one independent dimension of utility assignment in multiattribute decision problems.
- Historical comment about multiattribute utility from the early phases of a decision analysis.
- Change of variables in utility elicitation.

11.1 INTRODUCTION: THE GOOD NEWS

By now, you should have a good understanding of how to assess a single-attribute utility function over a value mearue. You should also be comfortable with the concept of a multiattribute preference function and the need to assign meaningful measures for the direct value attributes. In addition, you should be comfortable with the concept of a value function that assigns a value to each prospect.

If you have gone the through the effort of first constructing a preference or a value function, and if you understand how to assign a single-attribute utility function, then get ready for some good news.

The Good News

If you already have a value function, then a single-attribute utility function over value is all you need to construct the multiattribute utility function.

This observation enables many decision analysis applications involving utility functions over multiple attributes to be constructed using a one-dimensional utility assessment over value, instead of thinking about a utility function over the entire space of the attributes.

> **More Good News**
>
> *You can also construct a multiattribute utility function using a one-dimensional utility assessment over an attribute of the preference function, without having a value function.*

This chapter explains methods for constructing multiattribute utility functions using a one-dimensional utility assessment and a preference or a value function.

11.2 CONSTRUCTING MULTIATTRIBUTE UTILITY FUNCTIONS USING PREFERENCE FUNCTIONS

Suppose you have already determined the preference function for two attributes X and Y. Figure 11.1 shows an example of this preference function. Figure 11.1(a) shows the surface of the preference function and Figure 11.1(b) shows the corresponding isopreference contours. Note that all points on the same isopreference contour must have the same utility because they have the same Order Rule ranking. The utility of prospect A, for example, must be equal to the utility of prospect B. Therefore, if we assess the utility function along the curve $U(x_{max}, y)$ as shown by the solid arrows in Figure 11.1(b), we can determine the utility value of any point in this domain by tracing it back along its isopreference contour, $U(A) = U(B)$.

All that is needed to construct the multiattribute utility function is a one-dimensional utility assessment. It is helpful to assess this utility function for an attribute that is easy to reason about, such as money, and one that spans the set of isopreference contours. The one-dimensional assessment can be assigned using the concepts we discussed in Chapter 10, and also using the detailed analysis that will follow in Chapters 13 through 18.

11.2.1 Preference Functions Satisfying the Attribute Dominance Condition

In Figure 11.1, the preference function has the property that it is minimum at both axes, i.e., there is some level of each attribute that sets the preference function a minimum value regardless of the level of the other attribute. We referred to this property earlier in Chapter 8 as the grounding property or the attribute dominance property: any attribute that falls below a certain level dominates the remaining attributes and sets the preference function to its minimum value.

> **Note:** The isopreference contours of such grounded (attribute dominance) preference functions all intersect the lines (x, y^*) and (x^*, y) as shown in Figure 11.1(b).
>
> Given a preference function, $P(x, y)$, that satisfies the attribute dominance condition, you only need to assess a single-attribute utility function over any of the attributes at the maximum value of the other attributes. For example you may assess either the curve $U(x, y^*)$ or the curve $U(x^*, y)$, or assess them both and use one as a consistency check.

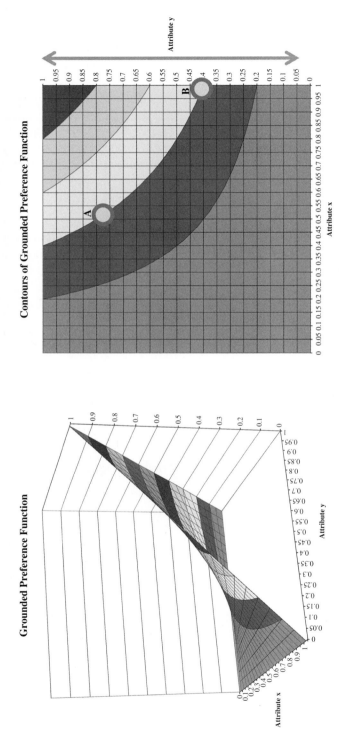

Figure 11.1. (a) Surface of Preference Function. (b) Two prospects on the same isopreference contour must have the same utility value.

After you have conducted the utility assessment, $U(x, y^*)$ for example, you can determine the utility values on the whole domain. For example, suppose you would like to determine the utility value of any prospect (x_1, y_1). All you need to do is find the corresponding point (x_2, y^*) that lies on the same isopreference contour, where

$$P(x_1, y_1) = P(x_2, y^*)$$

The value of x_2 can be determined analytically from this equation. It can also be determined graphically by tracing the isopreference contour that includes (x_1, y_1) until it intersects with the line $y = y^*$.

Once you have determined the value of x_2 that satisfies this relation, you simply refer to the assessed curve $U(x, y^*)$ and find the value of $U(x_2, y^*)$. This is the utility value needed because

$$U(x_1, y_1) = U(x_2, y^*).$$

As you see, the multiattribute utility problem is now reduced to a one-dimensional utility assessment over an attribute of value (as was discussed in Matheson and Abbas 2005).

EXAMPLE 11.1: Finding the Value of x_2 Analytically

Suppose you have a preference function of the Cobb-Douglas form

$$P(x, y) = xy^\eta, \ 0 \le x, y \le 1, \eta = 2.$$

Note that this preference function satisfies the attribute dominance condition. We are guaranteed that the isopreference contours converge at an axis. Suppose you have also assessed the utility function of attribute X at the maximum value of y, i.e., you have assessed

$$U(x, y^*) = U(x, 1) = \frac{1 - e^{-\gamma x}}{1 - e^{-\gamma}},$$

where $\gamma = .5$ is the risk aversion coefficient of attribute X when Y is maximum.

Now suppose you are interested in the utility value of the prospect $x = 0.5, y = 0.5$. We know that its utility value must be equal to that of a prospect that lies on its isopreference contour. In particular, for the prospect that lies at $y = y^* = 1$, we can get the corresponding value of X as

$$P(0.5, 0.5) = P(x_2, 1).$$

This implies that

$$(0.5)(0.5)^\eta = x_2(1)^\eta \Rightarrow x_2 = 0.125.$$

The utility value of the prospect $x = 0.5, y = 0.5$ is equal to the utility value of the prospect $x = 0.125, y = 1$. Hence,

$$U(0.5, 0.5) = U(0.125, 1) = \frac{1 - e^{-0.125\gamma}}{1 - e^{-\gamma}} = 0.5156.$$

To summarize, *if you have a grounded preference function, or a preference function with an attribute covering the range of isopreference contours, then a single-attribute utility function over this attribute is all you need to construct the multiattribute utility function.*

11.2.2 Preference Functions That Do Not Satisfy the Attribute Dominance Condition

If the isopreference contours do not converge into a single axis (as shown in Figure 11.2, for example, where the prospect characterized by point A meets the lower axis at point B, but the prospect characterized by point C does not intersect the lower axis on this domain, but rather connects to point D), you can still construct the multiattribute utility function using the preference function. All you need is enough utility assessments that span the set of isopreference contours. In this case, you simply need a utility assessment over the X- axis at the minimum value of Y and an assessment over the Y-axis at the maximum value of X. By doing this you will have covered the isopreference contours and you can determine the utility of any point in the domain in Figure 11.2.

Note that the utility assessment does not necessarily need to be conducted on a particular axis. A utility assessment across any path that crosses the isopreference curves would also suffice. Figure 11.3 shows an example of a path that covers the set of isopreference contours. We refer to such a path as a ***transverse***. In this example, the left vertical axis also happens to be a transverse.

Another way to conduct the utility assessment, particularly when the grounding condition is not satisfied, is to identify a value measure for each prospect using a value function and then assess a one-dimensional utility function over the units of value. We explain this approach and its implications in more detail in the next section.

11.3 CONSTRUCTING MULTIATTRIBUTE UTILITY FUNCTIONS USING VALUE FUNCTIONS

Recall that a value function can be determined from a preference function using a monotone transformation. If you have constructed a value function, $V(x,y)$, that has an absolute scale such as money, for example, then you can assess a one-dimensional assessment, U_V over value (money in this case). The utility function is then equal to the composite function,

$$U(x,y) = U_V(V(x,y)).$$

Geometrically, the value scale is the axis perpendicular to the page in Figures 11.1(b) and 11.2(b). By assessing a utility function over value, you are making a one-dimensional assessment over the axis of value. This axis for sure covers the isopreference contours.

Therefore, any point in the domain can now be assigned a utility function by considering its value as determined by the value function and then considering the utility of that value.

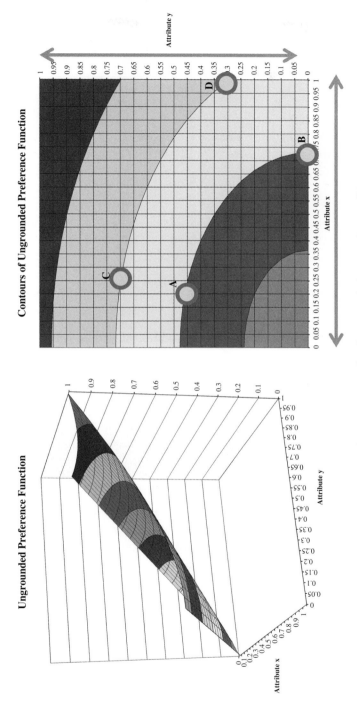

Figure 11.2. Isopreference contours that do not satisfy the grounding condition.

Contours of Ungrounded Preference Function

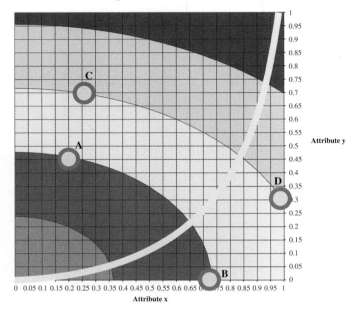

Figure 11.3. Transverse connecting isopreference contours that do not satisfy the grounding condition.

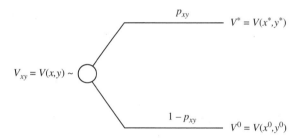

Figure 11.4. Characterizing a prospect by its value in the Equivalence Rule.

Refer back to Figure 2.3, which we repeat in Figure 11.4 for convenience. In essence, what we are doing with the Equivalence Rule is replacing the characterization of the prospect with its value. We then assign a utility function over value.

The indifference probability is the equivalent probability of getting either V^* or V^0 versus a sure deal of receiving the value of the prospect, V_{xy}. Note that this is a one-dimensional assessment over value.

The following examples illustrate the assignment of a utility function using a value function or a preference function.

EXAMPLE 11.2: Discounting Value, Not Utility

Consider a two-period cash flow, and a net present value function with a discount factor, β, of the form

$$V(x, y) = x + \beta y, \ x, y \ge 0,$$

where x is the amount received in the current year (therefore no discounting is applied to x) and y is the amount received in one year (with a discount factor β). Note that this value function does not satisfy the attribute dominance condition on a bounded domain.

Because we often encounter uncertainty in the formulation of multi-period cash flows, a decision-maker may wish to construct a utility function. He has already identified his value function as a net present value function, and so what remains is to assign a one-dimensional utility function over value.

Suppose the decision-maker has an exponential utility function over value, i.e., $U_V(V) = 1 - e^{-\gamma V}$. The term γ is known as the risk aversion coefficient and will be discussed in detail in Chapter 14. The reciprocal of γ is known as the risk tolerance.

The utility function for the two period cash flows is

$$U(x,y) = U_V\big(V(x,y)\big) = 1 - e^{-\gamma V(x,y)} = 1 - e^{-\gamma(x+\beta y)} = 1 - e^{-\gamma x}e^{-\gamma\beta y}.$$

This utility function is equal to some constant, 1, less the product of two utility functions, $e^{-\gamma x}e^{-\gamma\beta y}$. The first has a risk aversion coefficient, γ, and the second has a risk aversion coefficient $\gamma\beta$. Thus the risk aversion for the next year is equal to the risk aversion for the current year multiplied by the discount factor.

Equivalently, the risk tolerance for next year is equal to the risk tolerance for this year multiplied by the compounding factor, $\dfrac{1}{\beta}$.

Note that we did not discount the utility function; we discounted the value of the cash flow a year from now. In other words, we first thought of the deterministic trade-off between money received in today and money received a year from today through the net present value function. Following that, we thought about risk aversion over value. This was sufficient to construct the multiattribute utility function for money in the two years.

As we shall see, some people prefer to think about individual utility functions for each year and then try to aggregate them into some function. The problem is that they often forget to think about the deterministic time preference trade-off and assess individual utility functions for each year. As a result, they might end up with constructs that do not represent their time preference. What makes it even worse is when people assign a utility function over the cash flows for each year and then discount the utility function itself, not the cash flow. It is often easier to think about deterministic trade-offs to construct a preference or value function, and then think about one dimension of risk attitude.

EXAMPLE 11.3: Constructing a Multiattribute Utility Function Using a Cobb-Douglas Value Function

---

The following example is adapted from Howard (1984) on making trade-offs concerning situations where a decision-maker is exposed to fates comparable to death (such as outcomes of medical surgery). The decision-maker provides a value function over two attributes: consumption, X, and remaining life time, Y, using the form

$$V(x,y) = xy^{\eta}, \, 0 \le x, y \le 1 \qquad (1)$$

where x is expressed in dollars, y is the health state, and η determines the trade-off between x and y.

One method to determine the multiattribute utility function is to assess a utility function, $U_V(V)$, over the monetary attribute representing consumption. For example, if $U_V(V)$ were the utility function over consumption at the maximum value of $y = 1$, and if this utility function were exponential, then

$$U_V(V) = U_V(V(x,y)) = 1 - e^{-\gamma_V V(x,y)} = 1 - e^{-\gamma_V x y^\eta}, \tag{2}$$

where γ_V is the risk aversion coefficient over value. Note that the resulting two-attribute utility function does not have a product form (nothing in the Equivalence Rule dictates this).

EXAMPLE 11.4: Constructing a Multiattribute Utility Function Using a CES Value Function

---

Suppose the decision-maker provides a value function over two attributes: consumption, X, and remaining life time, Y, using the form

$$V(x,y) = \left(\lambda_x x^\beta + \lambda_y y^\beta\right)^{\frac{1}{\beta}}, \beta \neq 0, \quad \lambda_x, \lambda_y > 0, \quad \lambda_x + \lambda_y = 1, \; 0 \leq x, y \leq 1 \tag{3}$$

where x is expressed in dollars, y is the health state, and η determines the trade-off between x and y.

One method to determine the multiattribute utility function is to assess a utility function, $U_V(V)$, over the monetary attribute representing consumption: for example, if $U_V(V) = 1 - e^{-\gamma_V V}$ were the utility function over consumption

$$U_V(V) = U_V(V(x,y)) = 1 - e^{-\gamma_V V(x,y)} = 1 - e^{-\gamma_V \left(\lambda_x x^\beta + \lambda_y y^\beta\right)^{\frac{1}{\beta}}}, \tag{4}$$

where γ_V is the risk aversion coefficient over value. Note that the resulting two-attribute utility function does not have a product form.

On the other hand, if $U_V(V) = \ln V$, then

$$U_V(V) = U_V(V(x,y)) = \ln\left(\lambda_x x^\beta + \lambda_y y^\beta\right)^{\frac{1}{\beta}} = \frac{1}{\beta} \ln\left(\lambda_x x^\beta + \lambda_y y^\beta\right),$$

which is again neither additive nor a product form.

High-Speed Milling with Uncertainty about the Stability Contours

Refer back to the high-speed milling example of Chapter 9, where a machinist is deciding on the spindle speed and axial depth of the cutting tool. As we have seen, there is an inverse relationship between spindle speed and tool life. Furthermore, various combinations of spindle speed might lead to chatter.

Suppose that the machinist is uncertain about the exact location of the stability contours. He identifies three possible contours and their corresponding probabilities (Figure 11.5). For each axial depth and spindle speed combination, the machinist can

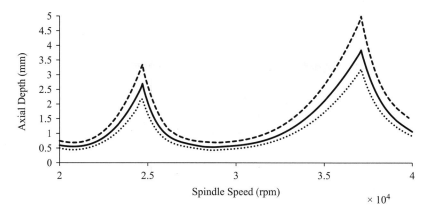

Figure 11.5. Example of three possible stability contours in high-speed milling.

determine the value of the machining job based on each of the three contours separately using the value function we discussed in Chapter 9.

Therefore, each spindle speed and axial depth combination corresponds to a
lottery with three possible prospects and their corresponding probabilities.

Next, the machinist would assign a utility function over value (profit in this
case). For each spindle speed and axial depth combination, he can now calculate the
expected utility of this combination. The best machining parameter decisions would
be those that maximize the expected utility.

Dialogue between Instructor (I) and Student (S):

I: So we have just discussed an important concept: the multiattribute utility
assessment can be constructed using a single-attribute utility assessment and the value
function.

S: Yes. But we can also construct it using a preference function, correct?

I: Correct. We can use the contours of the preference function and then assign a utility
function over any of the attributes. We assess the utility function at a particular curve
that crosses all contours. We refer to such a curve as a transverse.

S: If all the contours do not cross a particular axis, we can also assess two axes, right?

I: Yes, we just need any particular curve that all contours cross. Once you have done that,
you have a multiattribute utility function using the preference function.

S: And we can also construct a value function and then simply use utility over value.

I: Correct. Remember, this value function has an absolute scale and relates to the preference function using a monotone transformation.

11.4 HISTORICAL DOCUMENT FROM EARLY PHASES OF
DECISION ANALYSIS

Because of the historic significance, and various misconceptions about using the "utility
over value approach," this chapter summarizes a Stanford Research Institute (SRI)

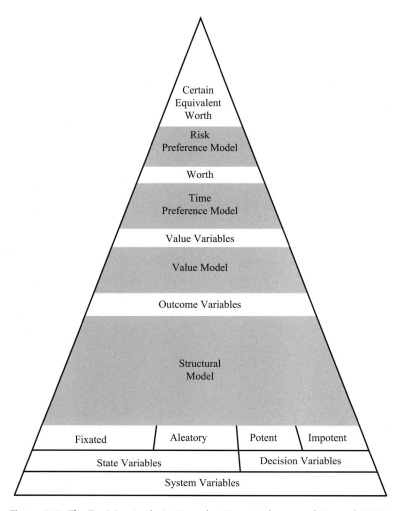

Figure 11.6. The Decision Analysis Hierarchy. (From Matheson and Howard, 1968).

Report written by Ronald Howard and James Matheson in 1968 for the European Long Range Planning Service.

Figure 11.6 is a figure from this report, and it was referred to as the Decision Analysis Hierarchy at the time. The figure shows the different aspects of incorporating the variables of a system into a decision analysis using a structural model, a value model, a time preference model, and a risk preference model. We have discussed each of these elements in the previous chapters. We shall review them within the context of this figure and comments on some different perspectives.

11.4.1 The Structural Model

The structural model incorporates the variables of a system into the analysis and defines the possible outcomes. It also converts such outcomes into meaningful direct value attributes. For example, in our high-speed machining example, the structural model would incorporate the spindle speed and axial depth to determine whether or not chatter and instability would occur in the machining piece given these settings. This would identify the value of the machining piece for given spindle speed and axial depth combinations. This type of structural model relies on domain knowledge and engineering or physical principles to incorporate the different aspects of the system in a meaningful way to determine the possible prospects or outcomes. The structural model is a deterministic model.

11.4.2 The Value Model

The value model assigns a value (which is the personal indifference selling price) to each prospect. In the high-speed milling example, this is the quantification of net profit for a given combination of spindle speed and axial depth taking into account the structural model and the tool wear, stability, and various other factors. The value model is a deterministic model.

11.4.3 The Time Preference Model

Because the value model might produce cash flows at various time frames, or there might be prospects that are materialized at a later date, there is a need for a model to convert these different cash flows into a net present value (NPV). This model takes into account the decision-maker's time preference. The time preference model is also a deterministic model applied to value.

11.4.4 The Risk Preference Model

After the time preference model is completed, the next step is to consider uncertainty about various aspects of the system. In the machining example, we considered uncertainty about stability given the axial depth and spindle speed combinations if we were uncertain about the precise location of the stability contours. A probabilistic analysis is now required to describe the uncertainty (probability of instability) as a given spindle speed and axial depth combination. In addition, a risk preference model is now required to determine the utility value of each prospect. This is achieved by assigning a one-dimensional utility function over the net present value model. The expected utility of each alternative (combinations of spindle speed and axial depth) can then be calculated. We can also determine the value of an alternative (or its certain equivalent). Chapter 13 will discuss the certain equivalent calculations in detail.

The main philosophy of this decision analysis hierarchy is that you can think about each of the following separately:

1. The reasoning about a structural model to provide direct value attributes or other insights into the possible outcomes;
2. The reasoning about deterministic trade-offs among the direct value attributes;
3. The reasoning the value of a prospect;
4. The time preference of the decision-maker; and
5. The risk attitude.

Now compare this hierarchy to the various modeling attempts that often assess utility functions over the system variables directly and attempt to include time preference within the utility assessment. It is a lot easier to reason about each of these components separately.

Comment: The approach for constructing multiattribute utility functions by first (1) thinking about the deterministic trade-offs among the attributes using a preference or a value function, and then (2) constructing a value function, then (3) incorporating time preferences, and finally (3) assessing a one-dimensional utility function over value (or over any of the attributes in a preference function) has unfortunately *not* been adopted by the bulk of the multiattribute utility literature.

There has been some confusion about the use of this hierarchy suggesting that it implies that all decision prospects should be converted into monetary equivalents before a one-dimensional utility can be assigned. Some might argue that making explicit trade-offs between attributes, such as money and safety for example, is difficult or in the least undesirable. Therefore, they attempt to construct the multiattribute utility function without assessing direct trade-offs, but by assessing individual utility functions over money and safety and then combining them using some forms (such as a product form or an additive form) to construct the multiattribute utility function. We shall refer to these methods in more detail in Chapters 23, 25, and 26.

Any multiattribute utility function, no matter how you assess it, will imply trade-offs among the attributes. If you make explicit assessments of trade-offs, between money and safety, for example, then at least you are thinking about this trade-off directly, and you can assign a one-dimensional utility function over money and use the contours of the preference function to determine the multiattribute utility function.

If a multiattribute utility function is constructed using individual utility assessments that are then combined into some functional form, we are also asserting trade-offs between the attributes, even though they were not assessed directly. It is better to reason about the trade-offs explicitly using a preference or a value function, and to understand the trade-off implications being asserted, than to end up with a multiattribute utility function whose trade-offs might not have been anticipated or validated because it was a result of some combination of two functions.

11.5 CHANGE OF VARIABLES IN UTILITY ASSIGNMENT

Let us discuss two questions to illustrate the idea of change of variables in utility elicitation. We shall focus our discussion in this section on structural models that provide the change of variables, but the same analysis applies of course to value functions that relate the attributes to value.

Consider a structural model of the form

$$\text{Force} = \text{Mass x Acceleration, or } F = ma.$$

Question 1

Suppose that you have a utility function over force, say $U_F(F)$. How can you determine your utility function over mass and utility function over acceleration given you know that $F = ma$?

To answer this question, we conduct a change of variables. Direct substitution for $F = ma$ gives

$$U_F(F) = U_F(ma).$$

For any fixed value of acceleration, say $a = a_0$, we can determine the utility over mass as $U_F(ma_0)$. Therefore, we can define a utility function over mass that will depend on the level of acceleration. We use the notation

$$U_m(m \mid a_0) = U_F(ma_0)$$

as the conditional utility function of mass for a given acceleration.

To illustrate, if $U_F(F) = 1 - e^{-\gamma F}$, then the conditional utility of mass for a given acceleration is

$$U_m(m \mid a_0) = 1 - e^{-\gamma m a_0}.$$

This observation has two important implications:

1. The Importance of Reference Values

    If you have assigned a utility function for Force, and if you are given a structural model, $F = ma$, then the conditional utility function of mass cannot be specified alone without specifying the level of acceleration for which it is to be determined. You cannot just say "utility over mass" in this case. The utility over mass depends on the magnitude of the acceleration, and it needs to be specified.

2. Limited Degrees of Freedom

    If you have the utility over force, and if you wish to be consistent with the structural model, then you do not have an additional degree of freedom to arbitrarily assign a utility over mass for any level of acceleration. The two utility functions over force and mass are functionally related. The same applies to the utility function over acceleration, and the same principles of relating utility functions apply when value functions relate the attributes.

We shall refer back to this idea of relating the utility functions using a structural model or a value function in Chapter 19 and many times throughout the book.

Question 2

Suppose that you have a utility function over mass at a given acceleration, and a utility function over acceleration at a given mass, can you determine the utility function over Force given you know the structural model that Force = mass x acceleration, $F = ma$?

This question is concerned with the utility function over some function (structural model) of the attributes given knowledge of the individual utility functions over each of the attributes. The analysis here is similar to that conducted in Section 11.2 for utility functions over an attribute in the preference function. The purpose of this example is to illustrate that it also applies when there are structural models.

First, we need to observe again that utility over mass alone is not a clear specification. At what level of acceleration was it assessed? And likewise the utility of acceleration requires the level of mass at which it was assessed. Quite often, people ignore the reference level of the remaining attributes for which the utility of an attribute is assessed. We shall see in many future chapters that this can lead to errors in decision-making. If we know that the utility over mass was assessed at the maximum value of acceleration on a given domain, then the answer to this problem is simple, as we illustrate below.

The first step is to plot the structural relation $F = ma$. The isopreference contours of Figure 11.7 represent the level sets of mass and acceleration yielding the same force.

Because the structural model $F = ma$ is grounded, it suffices to use either the utility over mass (at the maximum value of acceleration) or the utility over acceleration (at the maximum value of mass) to determine the utility value at any point in the domain. Any isopreference contour of force can project on the utility over mass or utility over acceleration as we illustrate below.

Let $U_{m|a^*}(m)$ be the utility value of mass at the maximum acceleration on the domain. The utility of any point (m_1, a_1) in the domain is determined by identifying the equally preferred value of force using mass m_2 at maximum acceleration a^* using the structural relation

$$m_1 a_1 = m_2 a^* \Rightarrow m_2 = \frac{m_1 a_1}{a^*}.$$

Therefore, the utility of any force $F_1 = m_1 a_1$ is equal to the utility of force using $F_1 = m_2 a^*$. We can also write

$$U_F(F_1) = U_F(m_1 a_1) = U_F(m_2 a^*).$$

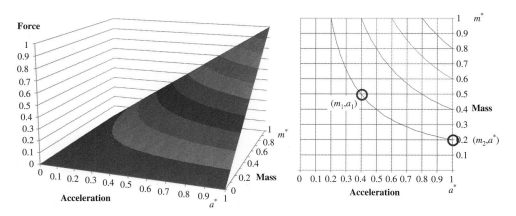

Figure 11.7. (Left) Structural model, $F = ma$. (Right) Isopreference contours for mass and acceleration.

Let $U_{m|a^*}(m)$ be the utility function of mass at the maximum acceleration. This curve is equal to the curve traced on the utility surface at the maximum value of acceleration. We can relate this to the point $m_1 a_1$ using the relation $U_{m|a^*}\left(\dfrac{m_1 a_1}{a^*}\right)$, and so

$$U_F(F_1) = U_{m|a^*}\left(\frac{m_1 a_1}{a^*}\right) = U_{m|a^*}\left(\frac{F_1}{a^*}\right).$$

To illustrate, let mass and acceleration of interest be defined on the unit square. Therefore, $m^* = a^* = 1$.

Suppose we have assessed the utility of mass at maximum acceleration as

$$U_{m|a^*}(m) = \frac{1 - e^{-\gamma m}}{1 - e^{-\gamma}},$$

then the utility function over force corresponding to any m, a is equal to the utility function of mass at m_2 and a^*, where $m_2 a^* = ma$, and so

$$U_F(F) = U_{m|a^*}(m_2) = U_{m|a^*}\left(\frac{ma}{a^*}\right) = \frac{1 - e^{-\gamma ma}}{1 - e^{-\gamma}} = \frac{1 - e^{-\gamma F}}{1 - e^{-\gamma}}.$$

Because we were given two variables in the structural model in this problem, we can also use the utility over acceleration at the maximum value of mass, $U_{a|m^*}(a)$. So we may define

$$m_1 a_1 = m^* a_2 \Rightarrow a_2 = \frac{m_1 a_1}{m^*}$$

and so the utility of any force $F_1 = m_1 a_1$ is determined by the utility over acceleration $U_{a|m^*}\left(\dfrac{m_1 a_1}{m^*}\right)$, and so

$$U_F(F_1) = U_{a|m^*}\left(\frac{m_1 a_1}{m^*}\right) = U_{a|m^*}\left(\frac{F_1}{m^*}\right)$$

Consistency also requires that

$$U_{m|a^*}\left(\frac{m_1 a_1}{a^*}\right) = U_{a|m^*}\left(\frac{m_1 a_1}{m^*}\right).$$

If $U_{m|a^*}(m) = \dfrac{1 - e^{-\gamma m}}{1 - e^{-\gamma}}$ in this example, then $U_{a|m^*}(a) = \dfrac{1 - e^{-\gamma a}}{1 - e^{-\gamma}}$. Therefore, we also need to observe that the assessments $U_{m|a^*}(m)$ and $U_{a|m^*}(a)$ are not independent of each other given the structural model.

If we have a structural model, we do not have the additional degree of freedom to assign $U_{m|a^*}(m)$ and $U_{a|m^*}(a)$ independently. They are related by the structural model.

This example emphasizes the need to incorporate structural models if we wish to be consistent in our utility assignment. We shall revisit this example in the next chapter and derive an analytical expression for the utility over force given these consistency conditions.

11.6 IMPORTANT REMARK

For many professionals, who are comfortable with using structural models or with constructing preference or value functions using the methods that we have discussed, you now have the tools needed to help you construct multiattribute utility functions in a wide variety of decisions. The rest of this book deals with implications of using certain functional forms, as well as properties single and multiattribute utility functions, and other methods for their construction that do not require preference or value functions.

The next chapter discusses the special case of utility functions over additive preference or value functions that results in an Archimedean combination of individual utility functions at specific values of the complement attributes. Part IV discusses the properties of single-attribute utility functions and the implications of using certain functional forms. Parts V and VI present methods for constructing multiattribute utility functions without using preference or value functions.

11.7 SUMMARY

Either a preference function or a value function is sufficient to determine the best decision alternative if there is no uncertainty about the outcomes.

When uncertainty is present, a utility function over the value function or over attributes of the preference function can determine the multiattribute utility function.

If either a preference or a value function or a structural model is specified, then the utility functions over the individual attributes are related. We do not have the additional degree of freedom to assign utility functions over the attributes independently in this case.

The main philosophy of using the utility over value approach is that you can separate the reasoning about utility (indifference probability assignments) from the reasoning about deterministic trade-offs among multiple attributes when constructing the multiattribute utility function.

If a preference or value function can be determined using the methods we discussed in Chapters 7, 8, and 9, then this would be a simple way to proceed. If, for any reason, these methods do not work, then the utility function can be constructed using the methods in Chapters 21–32 where the utility function is assessed directly without specifying a preference or a value function.

ADDITIONAL READINGS

Howard R. A. 1984. On fates comparable to death. *Management Science* 30(4): 407–422.
Howard, R. A. and A. E. Abbas. 2015. *Foundations of Decision Analysis*. Pearson, New York.

Matheson, J. E. and A. E. Abbas. 2005. Utility transversality: A value-based approach. *Journal of Multi-Criteria Decision Analysis* 13(5–6): 229–238.

Matheson, J. E. and R. A. Howard. 1968. An introduction to decision analysis. In R. A. Howard and J. E. Matheson, eds. *The Principles and Applications of Decision Analysis*, Vol. I. Strategic Decisions Group, Menlo Park, CA. Reprinted from Matheson, J. E. and R. A. Howard. 1968. A report by the European Long Range Planning Service, Stanford Research Institute Report 362.

CHAPTER 12

Utility Functions over Additive Preference or Value Functions

Chapter Concepts

- A class of multiattribute utility functions called Archimedean multiattribute utility functions (Abbas 2009).
- How to express a utility function over an additive preference function using an Archimedean combination of individual utility functions.
- A method for determining the preference functions of the attributes in the additive preference function.

Before proceeding with the main results of this chapter, I recommend reading Appendix A at the end of the book to gain some familiarity with composite functions, inverse functions, and Archimedean combinations if you are not already familiar with them. These basic concepts and mathematical operations will be used throughout the book.

12.1 INTRODUCTION: WHAT IS SO SPECIAL ABOUT THIS FORMULATION?

In the previous chapter, we discussed assigning utility functions over value functions or over individual attributes of the preference function. We dedicate this chapter to the special case of multiattribute utility functions constructed by assigning a one-dimensional utility function over an additive preference function or over any value function that can be converted into an additive function using a monotone transformation. The general form of the value functions that we shall discuss in this chapter has the form,

$$V(x_1,...,x_n) = m\left(\sum_{i=1}^{n} f_i(x_i)\right),$$

where m is a strictly monotone function. We will focus on the case where each of the functions f_i in the additive form is continuous, bounded, and strictly increasing.

> **Note:** Expressing a value function as a monotone transformation of an additive function still requires an absolute scale for V. This representation is merely an assertion that the functional form of the value function on this absolute scale can be converted into an additive function using a monotone transformation.

It will be important to recap the results of Chapter 8, where a monotone transformation of an additive function that is strictly increasing with each argument can also be represented as an additive or a multiplicative Archimedean combination of functions. Because of its importance to this chapter, we recap this formulation for convenience below.

Let

$$V(x_1,...,x_n) = m\left(\sum_{i=1}^{n} f_i(x_i)\right).$$

Defining $m^{-1}(t) = \eta(t)$ and $f_i(x_i) = \eta(V_i(x_i))$ gives the additive Archimedean form

$$V(x_1,...,x_n) = \eta^{-1}\left(\sum_{i=1}^{n} \eta(V_i(x_i))\right).$$

Further defining $\eta(t) = \ln(\phi(t))$ and $\eta(V_i(x_i)) = \ln(\phi(V_i(x_i)))$ gives the multiplicative Archimedean form

$$V(x_1,...,x_n) = \phi^{-1}\left(\prod_{i=1}^{n} \phi(V_i(x_i))\right). \tag{1}$$

Furthermore, we have shown that a monotone transformation of an Archimedean form results in an Archimedean form. Because a multiattribute utility function is simply a monotone transformation of a value function, and because a monotone transformation of an Archimedean form results in an Archimedean form, it is not surprising to note the main results of this chapter: *a multiattribute utility function that corresponds to additive ordinal preferences (or a value function that can be converted into an additive function using a monotone transformation) results in a multiattribute utility function that has an Archimedean form.*

To summarize the main results of this chapter, assigning a utility function over a value function that can be expressed as (1), results in a multiattribute utility function of the form,

$$U(x_1,...,x_n) = U_V\left(V(x_1,...,x_n)\right) = \varphi^{-1}\left[\prod_{i=1}^{n} \varphi(U_i(x_i))\right],$$

that is an Archimedean combination of individual utility assessments of the attributes, $U_i(x_i) = U(x_i, \overline{x}_i^*)$, for some generating function, φ.

Several reasons make this formulation particularly interesting:

1. The additive function, as we have seen, can take many shapes and forms. Therefore, it is important to understand the implications of using the widely used additive preferences when constructing multiattribute utility functions when uncertainty is present.

2. The structure of the corresponding multiattribute utility functions has some special features that will help with our general discussion of multiattribute utility functions in future chapters. You can think of this chapter as a link between the approach of constructing multiattribute utility functions using a utility function over value and other approaches in future chapters where individual utility assessments are made over the attributes and are incorporated into some functional form for the multiattribute utility function.

3. As we shall see, the formulation we present in this chapter will also form the basis of providing meaningful interpretations and assessments for the functions f_i in the additive preference or value function. In particular, the multiattribute utility function corresponding to such additive preference functions can be constructed by eliciting individual utility functions over each attribute (at the maximum value of the comple-ment attributes) as well as a monotone function known as the generating function. This type of structure is known as an Archimedean utility copula, first introduced in Abbas (2009). The individual utility function assessment can be conducted using indifference assessments, and the generating function can be assessed both empirically and using curve fitting methods (as we discuss in Chapter 30).

To illustrate the main idea of assigning utility functions over value functions that can be converted into an additive form using a monotone transformation, we start with the simple case of exponential utility functions over additive and multiplicative value functions (all of which are Archimedean) and show that this results in a multiattribute function that is a product of individual utility functions. We then generalize the results to multiattribute utility functions and show that a monotone generating function is needed.

12.2 EXPONENTIAL UTILITY FUNCTIONS OVER VALUE

12.2.1 Exponential Utility Functions over Additive Value Functions

Example 12.1

Consider a two-period cash flow whose net present value function is additive with a discount factor, β,

$$V(x,y) = x + \beta y,$$

where x is the amount received in the current year (no discounting) and y is the amount received in one year (with a discount factor β). It is straightforward to see that an additive value function can be expressed as a multiplicative Archimedean value function

$$V(x_1,...,x_n) = g^{-1}\left(\prod_{i=1}^{n} g(V_i(x_i)) \right),$$

with $g(t) = e^t$, where

$$V(x_1,...,x_n) = \ln\left(\prod_{i=1}^{n} e^{V_i(x_i)}\right) = \ln\left(e^{\sum_{i=1}^{n} V_i(x_i)}\right) = \sum_{i=1}^{n} V_i(x_i).$$

Now suppose that the decision-maker has an exponential utility function over net present value, i.e. $U_V(V) = e^{-\gamma V}$. The multiattribute utility function is equal to the composite transformation

$$U(x,y) = U_V(V(x,y)) = e^{-\gamma V(x,y)} = e^{-\gamma(x+\beta y)} = e^{-\gamma x}e^{-\gamma\beta y} = U_1(x)U_2(y),$$

where $U_1(x) = e^{-\gamma x}$ and $U_2(y) = e^{-\gamma\beta y}$.

Note that the two-attribute utility function is, in essence, a product of two univariate functions of each attribute. Furthermore, the product of two functions is a special case of the Archimedean combination with generating function equal to the identity function $\varphi(t) = t$.

This example highlights the idea of the possibility of constructing the two-attribute utility function directly by making utility assessments over each attribute and then combining them into some functional form to get the two-attribute utility function.

This product extends directly to multiple attributes, where the general form of the additive value function is

$$V(x_1,...,x_n) = \sum_{i=1}^{n} f_i(x_i),$$

where the functions $f_i(x_i)$ are strictly increasing. The multiattribute utility function resulting from assigning an exponential utility function over value is

$$U(x_1,...,x_n) = e^{-\gamma\sum_{i=1}^{n} f_i(x_i)} = \prod_{i=1}^{n} e^{-\gamma f_i(x_i)} = \prod_{i=1}^{n} U_i(x_i),$$

where $U_i(x_i) = e^{-\gamma f_i(x_i)}$. Again, the composition yields a function that is a product of univariate functions over the individual attributes.

12.2.2 Exponential Utility Functions over Multiplicative Value Functions

It is straightforward to observe that a multiplicative value function can be converted into an additive function by a logarithmic transformation. What if we assign an exponential utility function over a multiplicative value function? The following example illustrates the resulting functional form.

EXAMPLE 12.2: Two-Attribute Archimedean Form

Consider the ordinal preference function

$$V(x,y) = xy^\eta, 0 < x, y \le 1,$$

that has been used to represent trade-offs for health and consumption (Howard 1984). Note that this value function is strictly increasing with each argument at the maximum value of the complement attribute. Moreover, a logarithmic transformation converts this function into an additive preference function,

$$\log V(x, y) = \log x + \eta \log y.$$

Consider the two-attribute utility function, obtained by taking an exponential utility function that is normalized over this preference function, as used in Howard (1984),

$$U(x, y) = \frac{1 - e^{-\gamma x y^{\eta}}}{1 - e^{-\gamma}}, 0 < x, y \leq 1. \tag{2}$$

While the functional form in (2) is a bit more complicated than a simple product of two functions, we highlight, for now, that it can be expressed as

$$C(U(x, y^*), U(x^*, y)),$$

where the function C has the multiplicative Archimedean form

$$\boxed{C(u, v) = \varphi^{-1}\left[\varphi(u)\varphi(v)\right]}, \tag{3}$$

where the generating function,

$$\varphi(t) = -\frac{1}{\gamma} \log(1 - (1 - e^{-\gamma})t)$$

is a continuous and strictly monotone function with $\varphi(1) = 1$;

$$u \triangleq U(x, y^*) = U(x, 1) = \frac{1 - e^{-\gamma x}}{1 - e^{-\gamma}}$$

is the utility function of x assessed at the maximum value of y (the curve traced on the surface of the utility function in the x direction, when y is maximum); and

$$v \triangleq U(x^*, y) = U(1, y) = \frac{1 - e^{-\gamma y^{\eta}}}{1 - e^{-\gamma}}.$$

is the curve traced on the utility surface in the y direction when x is maximum.

Note that the Archimedean combination of this example used the univariate functions $U(x, y^*)$ and $U(x^*, y)$, both of which are defined at the maximum values of the remaining (complement) attributes.

The previous example illustrates that an exponential utility function over a multiplicative preference function results in an Archimedean form. Again, one might consider constructing the utility function by assessing individual functions over the attributes and combining them into some Archimedean functional form. We refer back to Archimedean multiattribute utility functions in Chapters 21, 29, 30.

12.3 LOGARITHMIC UTILITY FUNCTIONS OVER VALUE

12.3.1 Logarithmic Utility Functions over Additive Value Functions

---

EXAMPLE 12.3: Discounting Value, Not Utility

Consider again the additive value function

$$V(x,y) = x + \beta y,$$

over the domain $[1,2] \times [1,2]$.

For normalization, the minimum value of this value function is

$$V(1,1) = V^0 = 1 + \beta$$

and the maximum value is

$$V(2,2) = V^* = 2(1 + \beta).$$

Suppose that the decision-maker has a logarithmic utility function over value, and we use a normalized form such that

$$U_V(V) = \frac{\ln(V) - \ln(V^0)}{\ln(V^*) - \ln(V^0)}.$$

The resulting multiattribute utility function is

$$U(x,y) = U_V(V(x,y)) = \frac{\ln(x + \beta y) - \ln(1 + \beta)}{\ln(2(1 + \beta)) - \ln(1 + \beta)} = \frac{\ln(x + \beta y) - \ln(1 + \beta)}{\ln(2)}. \tag{4}$$

We highlight that the function (4) can also be expressed in the Archimedean form

$$\boxed{C(u,v) = \varphi^{-1}\left[\varphi(u)\varphi(v)\right],} \tag{5}$$

or an additive form $\theta^{-1}\left(\sum_{i=1}^{n} \theta_i(U_i(x_i))\right).$

We shall explain the details below, but for now note that this expression can be expressed when

$$\varphi(t) = e^{(1+\beta)(2^t - 2)}$$

$$u \triangleq U(x,y^*) = U(x,2) = \frac{\ln(x + 2\beta) - \ln(1 + \beta)}{\ln(2)}$$

is the utility of x at the maximum value of y and

$$v \triangleq U(x^*,y) = U(2,y) = \frac{\ln(2 + \beta y) - \ln(1 + \beta)}{\ln(2)}$$

is the utility of y at the maximum value of x.

12.3.2 Logarithmic Utility Functions over Multiplicative Value Functions

It is clear that assigning a logarithmic utility function over a multiplicative value function results in an additive utility function. For example, if $V(x,y) = xy^\eta, 0 < x, y \le 1$, and if $U_V(V) = \ln(V)$, then

$$U(x,y) = \ln\left(xy^\eta\right) = \ln x + \eta \ln y = U_1(x) + U_2(y),$$

which is an additive function of two utility functions.

12.4 ARCHIMEDEAN COMBINATIONS OF UTILITY FUNCTIONS: GENERAL UTILITY FUNCTIONS OVER ADDITIVE PREFERENCE FUNCTIONS

In this section, we will see that a utility function over an Archimedean preference or value function results in an Archimedean multiattribute utility function that is an Archimedean combination of individual utility functions. For ease of expression, we start with the case of two attributes and then extend the results to multiple attributes.

12.4.1 Utility Function over a Two-Attribute Archimedean Preference or Value Function

Suppose you have an Archimedean value function, where more of any attribute is preferred to less, such that

$$V(x,y) = \phi^{-1}\left[\phi(V_1(x))\phi(V_2(y))\right].$$

The purpose of this section is to show that if

$$U(x,y) = U_V\left(V(x,y)\right) = U_V\left(\phi^{-1}\left[\phi(V_1(x))\phi(V_2(y))\right]\right),$$

then we can express $U(x,y)$ as

$$\boxed{U(x,y) = g^{-1}\left[g(U(x,y^*))g(U(x^*,y))\right].}$$

We outline this proof below. For simplicity, we shall assume (for now) that the value function is normalized and so the functions $V_1(x), V_2(y)$ have a maximum value of one. We shall relax this requirement in Section 12.4.2, which will show that it merely adds to normalizing terms. Assume further that $\phi(1) = 1$.

By direct substitution, we can determine the boundary values of this value function as

$$V(x,y^*) = \phi^{-1}\left[\phi(V_1(x))\phi(V_2(y^*))\right] = \phi^{-1}\left[\phi(V_1(x))\phi(1)\right] = \phi^{-1}\left[\phi(V_1(x)).1\right]$$
$$= \phi^{-1}\left[\phi(V_1(x))\right] = V_1(x).$$

Similarly,

$$V(x^*,y) = \phi^{-1}\left[\phi(V_1(x^*))\phi(V_2(y))\right] = \phi^{-1}\left[\phi(1)\phi(V_2(y))\right] = \phi^{-1}\left[1.\phi(V_2(y))\right]$$
$$= \phi^{-1}\left[\phi(V_2(y))\right] = V_2(y).$$

This observations implies that the functions $V_1(x), V_2(y)$ are the curves traced on the value function when the other attribute is maximum.

Now suppose you apply a normalized one-dimensional utility function, U_V, over this value function to construct a multiattribute utility function. The multiattribute utility function has the form

$$U(x,y) = U_V\left(V(x,y)\right) = U_V\left(\phi^{-1}\left[\phi(V_1(x))\phi(V_2(y))\right]\right) \tag{6}$$

Define the composite transformation

$$g \triangleq \phi \circ U_V^{-1}.$$

The inverse of this composite transformation is by definition equal to

$$g^{-1} = U_V \circ \phi^{-1}.$$

Substituting into (6) gives

$$U(x,y) = g^{-1}\left[\phi(V_1(x))\phi(V_2(y))\right]. \tag{7}$$

Further define the composite transformations

$$U_1 = U_V \circ V_1 \text{ and } U_2 = U_V \circ V_2.$$

Note that $U_1(x^*) = U_V\left(V_1(x^*)\right) = 1$ and $U_2(y^*) = U_V\left(V_2(y^*)\right) = 1$.
Furthermore, this implies that

$$V_1 = U_V^{-1} \circ U_1 \text{ and } V_2 = U_V^{-1} \circ U_2.$$

Therefore,

$$\phi \circ V_1 = \phi \circ U_V^{-1} \circ U_1 = g \circ U_1 \text{ and } \phi \circ V_2 = \phi \circ U_V^{-1} \circ U_2 = g \circ U_2$$

Direct substitution into (7) gives

$$U(x,y) = g^{-1}\left[g(U_1(x))g(U_2(y))\right]. \tag{8}$$

This implies that the value function that results from applying a monotone utility function to an Archimedean preference or value function also has an Archimedean form.

Each of the individual functions $U_1(x), U_2(y)$ in this new form is equal to the utility function over the value curves, i.e.,

$$U_1(x) = U_V(V_1(x)) \text{ and } U_2(y) = U_V(V_2(y)).$$

The generating function of the new form is $g(t) = \phi(U_V^{-1}(t))$, which satisfies

$$g(1) = \phi(U_V^{-1}(1)) = 1.$$

By direct substitution into (8),

$$U(x,y^*) = g^{-1}\left[g(U_1(x))g(U_2(y^*))\right] = g^{-1}\left[g(U_1(x))g(1)\right] = U_1(x).$$

Similarly,

$$U(x^*, y) = g^{-1}\left[g(U_1(x^*))g(U_2(y))\right] = g^{-1}\left[g(1)g(U_2(y))\right] = U_2(y).$$

Therefore, we can write:

$$\boxed{U(x, y) = g^{-1}\left[g(U(x, y^*))g(U(x^*, y))\right].}$$

The extension of these results to multiple attributes is straightforward. Two important observations need to be made:

1. A utility function over an Archimedean preference or value function results in an Archimedean combination of utility functions.
2. We can construct the utility function by assessing individual utility functions at the boundary value of the surface, $U(x, y^*)$ and $U(x^*, y)$, and then using an appropriate generating function. We shall discuss this method in detail in Chapter 30.

Note: We have already seen in Chapter 8 how additive preference functions can be expressed as Archimedean preference functions. In Chapter 9, we also saw that monotone transformations of Archimedean preference functions result in Archimedean value functions. Because a utility function over an Archimedean value function is just a monotone transformation applied to the Archimedean preference function, it should not have come as a surprise at this point that a utility function over an Archimedean value function results in an Archimedean combination of individual utility functions. This is an important result that enables us to interpret (and assess) the functions f_i of an additive preference function using utility lottery assessments.

Dialogue between Instructor (I) and Student (S)

I: Do you see what is happening here?

S: Yes, whenever you have an additive preference function, and you need to construct a multiattribute utility surface, you can do so using an Archimedean combination of individual utility assessments for each attribute at the maximum values of the remaining attributes, as well as a generating function.

I: Correct. Note that this also solves the problem of having to assess the functions $f_i(x_i)$ in the preference function. We now have a clear interpretation for how to assess them for the Archimedean combination using lottery assessments and individual utility functions for each attribute.

S: Can we use this method even if we did not have uncertainty in the problem, just to construct a preference function?

I: Yes, preference functions, value functions, and utility functions have the same isopreference contours. Therefore you can, in principle, construct a multiattribute utility function and use it as a preference function. This is in fact what most of the literature does. People construct multiattribute utility functions and use them to solve deterministic decision problems. But if you can construct a value function, your life will be a lot easier. Remember, this Archimedean conversion from the additive form requires functions $f_i(x_i)$ to be strictly increasing. As we have seen, not all additive preference functions need to satisfy this property. What ends up happening in many applications is that people formulate the attributes in a way that makes the utility function increasing with each attribute, and often end up using arbitrary scales in the process. Remember the peanut butter and jelly sandwich? Preferences do not need to be increasing with each attribute. Think of reconstructing the attributes for things like taste and convenience that are monotone. They become difficult to reason about.

S: But why we do we need to think about this Archimedean combination if we can assign a utility function over the additive preference or additive value function directly?

I: Well, remember: If the value function has a logical meaning or a structural model, like net present value, then you can just do that. But if you have an additive preference function, where you do not know the functions $f_i(x_i)$ or even their domain and range, then it might be easier to think in terms of the Archimedean form, where all such functions are normalized and have a clear method for their assessment using indifference assessments.

12.4.2 Utility Function over an Additive Preference or Value Function

For educational purposes, we now derive the multiattribute extension of the two-attribute case using the additive Archimedean combination. We consider the general case of utility functions over value functions that can be converted into an additive form using a monotone transformation.

$$V(x_1,...,x_n) = m\left(\sum_{i=1}^{n} f_i(x_i)\right),$$

where m is a monotone function. We shall outline this proof below and then formally assert it. Then we shall explain the implications this result has for multiattribute utility elicitation.

By assigning a utility function over an additive preference function, we have

$$U(x_1,...,x_n) = U_V\left(V(x_1,...,x_n)\right) = U_V\left(m\left(\sum_{i=1}^{n} f_i(x_i)\right)\right).$$

Define $\theta^{-1} \triangleq U_V \circ m$, and therefore $\theta = m^{-1} \circ U_V^{-1}$. Direct substitution gives

$$U(x_1,...,x_n) = \theta^{-1}\left(\sum_{i=1}^{n} f_i(x_i)\right).$$

Further define $f_i(x_i) = \theta(U_i(x_i))$ and substitute to get

$$U(x_1,...,x_n) = \theta^{-1}\left(\sum_{i=1}^{n} \theta(U_i(x_i))\right). \tag{9}$$

Equation (9) is the additive Archimedean combination of utility functions. It has an equivalent multiplicative form as we illustrate below.

Define $\theta = \ln\varphi$ and so $\theta^{-1}(t) = \varphi^{-1}(e^t)$. Direct substitution gives

$$U(x_1,...,x_n) = \varphi^{-1}\left(e^{\left(\sum_{i=1}^{n}\ln\varphi_i(U_i(x_i))\right)}\right) = \varphi^{-1}\left(e^{\ln\left(\prod_{i=1}^{n}\varphi_i(U_i(x_i))\right)}\right) = \varphi^{-1}\left(\left(\prod_{i=1}^{n}\varphi_i(U_i(x_i))\right)\right) \quad (10)$$

which is the multiplicative Archimedean functional form (Abbas 2009).

Furthermore, if $\varphi(1) = 1$ and $U_i(x_i^*) = 1$ (and we show that this is indeed the case in the appendix to this chapter), then we get by direct substitution,

$$U(x_1, x_2^*, x_3^*,..., x_n^*) = \varphi^{-1}\left(\varphi_1(U_1(x_1))\prod_{i\neq 1}\varphi_i(U_i(x_i^*))\right) = \varphi^{-1}\left(\varphi_1(U_1(x_1))\right) = U_1(x_1).$$

Similarly, by direct substitution, $U_i(x_i) = U(x_i, \bar{x}_i^*)$ is the curve traced on the surface of the utility function for x_i when all other attributes are at their maximum values. We use the notation $\bar{x}_i$ to represent an instantiation of the remaining attributes (the complement attributes to attribute X_i).

The functional form (10) can therefore be expressed as

$$U(x_1,...,x_n) = \varphi^{-1}\left[\prod_{i=1}^{n}\varphi(U(x_i, \bar{x}_i^*))\right]. \quad (11)$$

Theorem 12.1 (Abbas and Sun 2015)

If $U(x_1,...,x_n)$ is continuous and strictly increasing with each argument at the maximum value of the complement attributes, then the following two statements are equivalent:

1. $U(x_1,...,x_n) = U_V\left(V(x_1,...,x_n)\right)$ with $V(x_1,...,x_n) = m\left(\sum_{i=1}^{n}f_i(x_i)\right)$, where m is a contin
    uous monotonic function, and f_i are continuous and strictly increasing functions.

2. $U(x_1,...,x_n) = \varphi^{-1}\left[\prod_{i=1}^{n}\varphi(U(x_i, \bar{x}_i^*))\right]$,   where $\varphi^{-1}(V) = U_V\left(m\left(\ln V + \sum_{i=1}^{n}f_i(x_i^*)\right)\right)$

    and U_V is normalized to range from 0 to 1 as the value function traces its minimum and maximum values.

Why Is This Important?

The importance of this result is that it characterizes the family of multiattribute utility functions that correspond to additive preference functions. As we have discussed for the case of two attributes, for example, it asserts that you can assess only the utility curves $U(x, y^*)$ and $U(x^*, y)$ on the surface of the two-attribute utility function $U(x, y)$ and then combine them into the Archimedean combination of the form

$$\varphi^{-1}\left[\varphi(U(x, y^*))\varphi(U(x^*, y))\right]$$

using some function φ to derive the surface of the two attribute utility function $U(x, y)$.

Therefore, three univariate functions, $U(x, y^*)$, $U(x^*, y)$, and φ, enabled the construction of the whole two-attribute utility surface. The same analysis extends to multiple attributes using the form $U(x_1, ..., x_n) = \varphi^{-1} \left[\prod_{i=1}^{n} \varphi(U(x_i, \bar{x}_i^*)) \right]$.

Another fundamental implication of this result is that it provides another mechanism for assessing the functions $f_i(x_i)$ of an additive preference function, besides the value function approach that we discussed in Chapter 9 and the various methods of constructing a preference function we discussed in Chapter 7. Because the preference function and the multiattribute utility function will have the same isopreference contours, it suffices to assess univariate utility function assessments for each attribute of the form $U(x_i, \bar{x}_i^*)$ and a generating function to determine the contours of the preference function. We have seen in Chapter 10 how to assess univariate utility functions using lottery assessments. Furthermore, the isopreference contours of the preference function are the same as those of the multiattribute utility function because they are related by a monotone transformation. Therefore, we now have a clear interpretation for assessing the individual functions of an additive preference function. In Chapters 13 through 18, we shall provide more detail about properties of single attribute utility functions. What remains for the Archimedean combination then is to determine the generating function ϕ, and it would be great if it can also be determined using lottery assessments. Chapter 30 illustrates how to do this.

The following example illustrates the results of Theorem 12.1 and revisits the logarithmic utility function over net present value.

EXAMPLE 12.4: Expressing the Logarithmic Utility Function over Net Present Value Function as a Multiplicative Archimedean Combination

---

Recall the additive value function,

$$V(x, y) = x + \beta y = m(f_1(x) + f_2(y)), \quad x, y \in [1, 2].$$

By inspection, this implies that

$$f_1(x) = x, \ f_2(y) = \beta y, \ m(t) = 1.$$

Suppose we assign a logarithmic utility function over value in such a way that the resulting utility surface is normalized to range from 0 to 1 when x, y are at their minimum and maximum values, respectively. Here, we have $x^0 = y^0 = 1$ and $x^* = y^* = 2$.

Direct substitution shows that the maximum values of the functions $f_1(x), f_2(y)$ are, respectively,

$$f_1(x^*) = x^* = 2, \quad f_2(y^*) = \beta y^* = 2\beta,$$

and also that the maximum and minimum values of the value function are, respectively,

$$V^* = V(x^*, y^*) = 2 + \beta.2 = 2(1 + \beta) \text{ and } V^0 = V(x^0, y^0) = 1 + \beta.1 = 1 + \beta.$$

Suppose that $U_V(V)$ is normalized from zero to one, as will be the case in future chapters. We have

$$U_V(V) = \frac{\ln(V) - \ln(V^0)}{\ln(V^*) - \ln(V^0)} = \frac{\ln(V) - \ln(1+\beta)}{\ln(2(1+\beta)) - \ln(1+\beta)} = \frac{\ln(V) - \ln(1+\beta)}{\ln(2)}.$$

The corresponding multiattribute utility function is then

$$U(x,y) = U_V(V(x,y)) = \frac{\ln(V(x,y)) - \ln(V^0)}{\ln(V^*) - \ln(V^0)} = \frac{\ln(x + \beta y) - \ln(1+\beta)}{\ln(2(1+\beta)) - \ln(1+\beta)}.$$

We can express the corresponding multiattribute utility function in terms of an Archimedean multiattribute utility function

$$\varphi^{-1}\left[\varphi(u)\varphi(v)\right] = \varphi^{-1}\left[\prod_{i=1}^{n} \varphi(U(x,y^*))\varphi(U(x^*,y))\right]$$

using Theorem 12.1 as follows.

Step 1: Determine the curves $u = U(x,y^*)$ and $v = U(x^*,y)$

By direct substitution,

$$u \triangleq U(x,y^*) = U(x,2) = \frac{\ln(x + 2\beta) - \ln(1+\beta)}{\ln(2)} \quad \text{and}$$

$$v \triangleq U(x^*,y) = U(2,y) = \frac{\ln(2 + \beta y) - \ln(1+\beta)}{\ln(2)}.$$

The inverse of the generating function is

$$\varphi^{-1}(V) = U_V\left(m\left(\ln V + \sum_{i=1}^{n} f_i(x_i^*)\right)\right).$$

Step 2: Determine the inverse of the generating function

Direct substitution shows that

$$\varphi^{-1}(V) = U_V\left(m\left(\ln V + \sum_{i=1}^{n} f_i(x_i^*)\right)\right).$$
$$= U_V\left(\ln V + 2(1+\beta)\right)$$

As we can see, there is a direct relation between the inverse of the generating function, φ^{-1}, and the utility function over value, U_V. To derive the exact relation, define

$$\varphi^{-1}(V) \triangleq U_V(r(V))$$

where

$$r(V) = m\left(\ln V + \sum_{i=1}^{n} f_i(x_i^*)\right) = \ln V + 2(1+\beta).$$

Furthermore, this implies that the generating function is given by the inverse composite transformation

$$\varphi(t) = r^{-1}\left(U_{\bar{V}}^{-1}(t)\right).$$

Direct substitution shows that

$$r^{-1}(t) = e^{t-2(1+\beta)}.$$

Furthermore, because $U_V(V) = \dfrac{\ln(V) - \ln(1+\beta)}{\ln(2)}$, its inverse is

$$U_{\bar{V}}^{-1}(t) = e^{t\ln(2) + \ln(1+\beta)} = 2^t(1+\beta).$$

Step 3: Determine the generating function $\varphi = r^{-1} \circ U_{\bar{V}}^{-1}$.
Therefore, the generating function is

$$\varphi(t) = r^{-1}\left(U_{\bar{V}}^{-1}(t)\right) = r^{-1}\left(2^t(1+\beta)\right) = e^{2^t(1+\beta) - 2(1+\beta)} = e^{(2^t - 2)(1+\beta)}.$$

We have now identified the three univariate functions φ, u, v needed to construct the utility surface.

As a consistency check, let us evaluate the term $\varphi^{-1}\left(\varphi(u)\varphi(v)\right)$:
We have

$$u = \frac{\ln(x + 2\beta) - \ln(1+\beta)}{\ln(2)}, v = \frac{\ln(2 + \beta y) - \ln(1+\beta)}{\ln(2)}, \text{ and}$$
$$\varphi(t) = e^{(2^t - 2)(1+\beta)}.$$

Therefore,

$$\varphi(u) = e^{(2^u - 2)(1+\beta)} = e^{x-2} \text{ and } \varphi(v) = e^{\beta(y-2)}.$$

Then their product

$$\varphi(u)\varphi(v) = e^{x-2}e^{\beta(y-2)} = e^{x + \beta y - 2(1+\beta)}$$

because

$$\varphi^{-1}(t) = U_V\left(\ln t + 2(1+\beta)\right).$$

Direct substitution shows

$$\varphi^{-1}\left(\varphi(u)\varphi(v)\right) = U_V\left(\ln e^{x + \beta y - 2(1+\beta)} + 2(1+\beta)\right) = U_V\left(x + \beta y\right) = \frac{\ln(x + \beta y) - \ln(1+\beta)}{\ln(2)},$$

which is the multiattribute utility function.

This example illustrates that the functional form of the utility function can indeed be expressed as the Archimedean combination of individual utility functions $\varphi^{-1}\left(\varphi(u)\varphi(v)\right)$. We leave it as an exercise to the reader to show that it can also be written as $\theta^{-1}\left(\theta(u) + \theta(v)\right)$. We shall refer back to the general idea of a utility copula function in more detail in Chapters 29–30.

EXAMPLE 12.5: Expressing the Exponential Utility Function over the Cobb-Douglas Function as a Multiplicative Archimedean Combination

Consider again the ordinal Cobb-Douglas function

$$V(x,y) = xy^\eta, 0 < x, y \leq 1.$$

Suppose the decision-maker assigns an exponential utility function over value, such that

$$U_V(V) = \frac{1 - e^{-\gamma V}}{1 - e^{-\gamma}},$$

and the resulting two-attribute utility function is

$$U(x,y) = \frac{1 - e^{-\gamma x y^\eta}}{1 - e^{-\gamma}}, 0 < x, y \leq 1. \tag{12}$$

To express (12) using the multiplicative Archimedean form

$$\boxed{C(u,v) = \varphi^{-1}\left[\varphi(u)\varphi(v)\right],} \tag{13}$$

we conduct the following steps:

Step 1: Determine the curves $u = U(x, y^*)$ and $v = U(x^*, y)$

Direct substitution gives

$$u \triangleq U(x, y^*) = U(x,1) = \frac{1 - e^{-\gamma x}}{1 - e^{-\gamma}} \text{ and } v \triangleq U(x^*, y) = U(1, y) = \frac{1 - e^{-\gamma y^\eta}}{1 - e^{-\gamma}}.$$

Step 2: Determine the inverse of the generating function

First we express this function in an additive form using a logarithmic transformation to get

$$V(x,y) = xy^\eta = m(f_1(x) + f_2(y)), \quad 0 < x, y \leq 1.$$

Therefore,

$$m(t) = e^t, f_1(x) = \ln x, f_2(y) = \eta \ln y.$$

Direct substitution shows that $f_1(x^*) = 0, f_2(y) = 0$.

The inverse of the generating function is

$$\varphi^{-1}(V) = U_V\left(m\left(\ln V + \sum_{i=1}^{n} f_i(x_i^*)\right)\right) = U_V(V) = \frac{1 - e^{-\gamma V}}{1 - e^{-\gamma}}.$$

Step 3: Determine the generating function

The generating function can now be determined from its inverse function as

$$\varphi(t) = -\frac{1}{\gamma}\log(1 - (1 - e^{-\gamma})t).$$

Note that the generating function is a continuous and strictly monotone function with $\varphi(1) = 1$.

To verify that the Archimedean combination of φ, u, v yields the multiattribute utility function, we get by direct substitution

$$\varphi(u) = -\frac{1}{\gamma}\log(1-(1-e^{-\gamma})u) = -\frac{1}{\gamma}\log(1-(1-e^{-\gamma})\frac{1-e^{-\gamma x}}{1-e^{-\gamma}}) = x.$$

Similarly, $\varphi(v) = y^{\eta}$.

Therefore, the product

$$\varphi(u)\varphi(v) = xy^{\eta}.$$

Hence,

$$\varphi^{-1}[\varphi(u)\varphi(v)] = \frac{1-e^{-\gamma xy^{\eta}}}{1-e^{-\gamma}}.$$

This example also highlights the idea of constructing multiattribute utility functions by direct utility assessments over the attributes instead of direct utility assessments over the value function when the value function is additive. Chapters 21 through 27 explain the calculus of assessing multiattribute utility functions using conditional utility assessments that do not necessarily have additive preference functions.

12.5 ASSESSING MULTIATTRIBUTE UTILITY FUNCTIONS WITH ADDITIVE ORDINAL PREFERENCES USING ARCHIMEDEAN COMBINATIONS OF UTILITY FUNCTIONS

We have shown that if the preference function is additive, then the multiattribute utility function is an Archimedean combination of functions, which requires the two boundary curves, $U(x, y^*)$ and $U(x^*, y)$, as well as the generating function φ to determine the whole surface.

In the previous examples, we determined these functions when a utility function was assigned over a given preference function. What is of interest, however, is the reverse direct: how do we assess the functions $U(x, y^*), U(x^*, y)$, and φ to construct the multiattribute utility function to match the decision-maker's preferences if we know that the ordinal preferences are additive? Chapter 31 addresses this question in detail. For now, let us highlight some of the items needed for this construction and explain the connections of this approach with future chapters.

Figure 12.1 shows an example of a two-attribute utility surface with the curves $U(x, y^*)$ and $U(x^*, y)$ at the boundary values of the domain. Note that the assessment $U(x, y^*)$ returns the utility value of points (x, y^*) on the upper bound of the domain as x varies from its minimum to its maximum value. Therefore, the range of this curve is $U(x^0, y^*)$ to $U(x^*, y^*)$. Figure 12.2 shows an example of this curve $U(x, y^*)$. Note that the term $U(x^0, y^*)$ need not be zero and $U(x^*, y^*)$ is equal to one because it is the most preferred prospect.

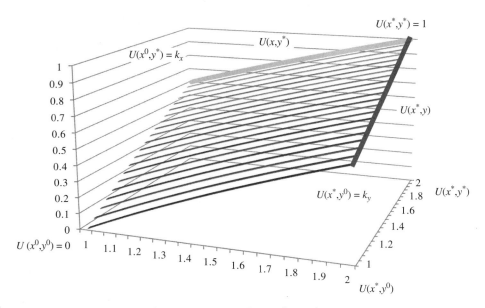

Figure 12.1. Two-attribute utility surface.

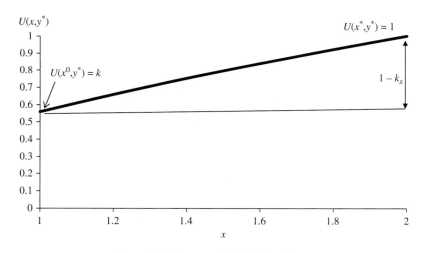

Figure 12.2. The curve $U(x, y^*)$ plotted vs x.

To assess the curve $U(x, y^*)$, we can use lottery assessments. If we pick n points at the boundary value, say $(x_1, y), (x_2, y), ..., (x_n, y)$, we can assess the utility values $U(x_1, y), U(x_2, y), ..., U(x_n, y)$ using lottery assessments as shown in Figure 12.3, where we assume that more of an attribute is strictly preferred to less. Note that these assessments are expressed in terms of the best and worst prospects on the entire domain.

We can also express the indifference assessments in terms of (x^0, y^*) and $U(x^*, y^*)$, which is normalized to range from zero to one as x varies from x^0 to x^*. Note that the value if y is equal to its maximum value, y^*, on both prospects of the right-hand side of the figure. Figure 12.4 shows this assessment. We refer to this normalized assessment as $U(x \mid y^*)$.

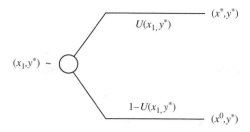

Figure 12.3. Assessment of $U(x, y^*)$ using lottery assessments.

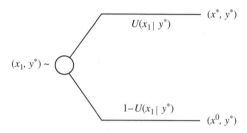

Figure 12.4. Normalized assessment of $U(x \mid y^*)$ using lottery assessments.

To convert from the normalized assessment $U(x \mid y^*)$ into the curve $U(x, y^*)$, we simply scale it using the equation

$$U(x, y^*) = k_x + [1 - k_x]U(x \mid y^*).$$

The same types of analyses apply to attribute Y, where

$$U(x^*, y) = k_y + [1 - k_y]U(y \mid x^*).$$

Chapters 13 through 18 present detailed properties of single-attribute utility functions that enable these assessments.

Having made the assessments $U(x, y^*)$ and $U(x^*, y)$, what remains for the construction of the utility surface is the generating function, φ, which is continuous, increasing, and passes by the point $(1,1)$. Chapter 30 presents an empirical method to determine the generating function directly from utility assessments. But we can also experiment with a library of generating functions to match the trade-offs through trial and error or sensitivity analysis. Different generating functions will lead to different contours for the given boundary assessments.

Note that if the generating function is the identity function, $\varphi(t) = t$, then the Archimedean combination is the product, where

$$\varphi^{-1}\big[\varphi(u)\varphi(v)\big] = uv.$$

It can also take on a more general linear form, for example, $\varphi(t) = at + b$, giving $\dfrac{[au + b][av + b] - b}{a}$. Many people use forms corresponding to a linear generating function for simplicity, but assuming a linear form means that you are also assuming the shape of the isopreference contours. We shall discuss the conditions that lead to a linear generating function in future chapters and the errors that might result when making this assumption if its conditions are not really satisfied.

Note: By constructing the utility surface of an additive preference function using a generating function and the two curves $U(x, y^*)$ and $U(x^*, y)$, we have in effect determined the functions $f_i(x_i)$ in the additive preference function $P(x_1,...,x_n) = \sum_{i=1}^{n} f_i(x_i)$. Therefore, we have converted the trade-off assessment into knowledge of an additive preference function and a set of utility assessments. In future chapters, we shall also discuss how to convert trade-off assessments into utility assessments even if the preference function is not additive.

Dialogue between Instructor (I) and Student (S)

S: This chapter has been really enlightening in the construction of multiattribute utility functions. Why do we need the other chapters in the book then?

I: Well, first I want to provide some technical derivations that will enable you to better understand the properties of single-attribute utility functions. We have seen how important they are, not only for decisions with a single attribute but also for decisions with multiple attributes. This is why we shall dedicate a whole section of this book for this discussion.

S: OK, what about the other sections?

I: People often rush into assigning multiattribute utility functions by combining individual utility assessments without using a value function. I would like to highlight that this combination need not be a simple addition and multiplication of individual utility functions, and sometimes not even an Archimedean combination. I will also show how using approximate methods may lead to errors in decision-making. Then, I will present the general calculus for building multiattribute functions without value functions. I will go beyond the construction of multiattribute utility functions of the Archimedean forms and show the general formulation for constructing multiattribute utility functions by direct utility assessment.

S: Will you present easy methods to assess those more general forms? We want to make sure that they can be assessed easily.

I: Yes. For example, the utility copula forms or other forms we shall discuss, such as one-switch independence, will require single-attribute utility assessments.

S: Are you sure we can't just multiply or add utility assessments over the attributes without using more complicated forms?

I: That is what most of the literature has done. In a future chapter we shall discuss the conditions under which this is valid, but you will see that it applies to only very special cases. We shall also discuss the errors introduced by rushing directly into doing that. It is difficult to interpret what we mean by adding utilities or by multiplying them. It is a lot easier to first think about your preferences for deterministic consequences using trade-offs and then assign a utility function over value or over an attribute. If you cannot do that for some reason, then you need to know the precise methods of combining individual utility assessments that we shall discuss in future chapters.

S: But what about the generating function φ ? You did not provide a clear interpretation for how to assess that.

I: True. This is coming. In Chapters 29 and 30, we shall revisit the Archimedean utility copula and we shall also provide methods to assess the function φ by direct utility assessments.

12.6 SUMMARY

1. For the special case of a utility function over an additive preference function (or over a value function that can be converted into an additive form using a monotone transformation), the resulting multiattribute utility function can be decomposed into individual utility functions over each of the attributes. The decomposition is not necessarily a simple product or addition, but a slightly more complex form known as an Archimedean combination of individual utility functions that also involves a generating function.

2. The idea of an Archimedean combination of utility functions provides some insights into the possibility of constructing multiattribute utility functions using individual utility assessments over the attributes and then combining them into some functional form. We shall revisit this idea, as well as properties and limitations of the corresponding functional forms, in Chapters 21 through 32.

ADDITIONAL READINGS

Abbas, A. E. 2003. *Entropy Methods in Decision Analysis*. PhD Thesis, Stanford University.

Abbas, A. E. 2009. Multiattribute utility copulas. *Operations Research* 57(6): 1367–1383.

Abbas, A. E. 2012. Utility copula functions matching all boundary assessments. *Operations Research* 61(2): 359–371.

Abbas, A. E. and R. A. Howard. 2005. Attribute dominance utility. *Decision Analysis* 2(4): 185–206.

Abbas, A. E. and Z. Sun. 2015. Multiattribute utility functions satisfying mutual preferential independence. *Operations Research* 63(3): 378–393.

Howard, R. A. 1984. On fates comparable to death. *Management Science*, 30(4), 407–422.

APPENDIX 12A CHANGE OF VARIABLES IN UTILITY ASSIGNMENT

The Archimedean form of multiattribute utility functions enables an analytical answer to the question posed in the previous chapter about determining the utility over force given the utility over acceleration and the utility over mass, which is important for change of variables in structural models as we illustrate below.

EXAMPLE 12.3: Utility over Force Given Utility over Mass and Utility over Acceleration

If the individual utility functions over mass and acceleration are assessed at the maximum values of the complement attributes, then the utility over force must be an

Archimedean combination of the utility functions over mass and acceleration. This is because the structural model,

Force = Mass x *Acceleration,*

can be converted into an additive form using a logarithmic transformation. Because we have the structural model fully specified, the individual utility functions are also related and the generating function is fully specified.

Consider the force $F_1 = m_1 a_1$ corresponding to a point (m_1, a_1) in the domain of the attributes. We have seen in the previous chapter that this implies that the utility over force can be related to the utility over mass at the maximum value of acceleration using the relation

$$U_F(F_1) = U_{m|a^*}\left(\frac{m_1 a_1}{a^*}\right) = U_{m|a^*}\left(\frac{F_1}{a^*}\right).$$

Similarly, the utility over force is related to the utility over acceleration at the maximum value of mass using

$$U_F(F_1) = U_{a|m^*}\left(\frac{m_1 a_1}{m^*}\right) = U_{a|m^*}\left(\frac{F_1}{m^*}\right).$$

Therefore, the individual utility functions over mass and acceleration are related by the structural model as

$$U_{m|a^*}\left(\frac{m_1 a_1}{a^*}\right) = U_{a|m^*}\left(\frac{m_1 a_1}{m^*}\right).$$

Because the structural model can be converted into an additive form, we also know that

$$U_F(F_1) = \varphi^{-1}\left[\varphi(u)\varphi(v)\right].$$

Substituting for the various utility functions gives

$$U_{m|a^*}\left(\frac{m_1 a_1}{a^*}\right) = \varphi^{-1}\left[\varphi\big(U_{m|a^*}(m_1)\big)\varphi\big(U_{a|m^*}(a_1)\big)\right].$$

Furthermore, substituting for $U_{m|a^*}\left(\dfrac{m_1 a_1}{a^*}\right) = U_{a|m^*}\left(\dfrac{m_1 a_1}{m^*}\right)$ and simplifying gives a relation between the utility function over mass and the generating function as

$$\varphi\left[U_{m|a^*}\left(\frac{m_1 a_1}{a^*}\right)\right] = \varphi\big(U_{m|a^*}(m_1)\big)\varphi\left(U_{m|a^*}\left(\frac{m^* a_1}{a^*}\right)\right). \tag{14}$$

To illustrate, consider a domain of mass and acceleration over a normalized domain of the unit square. Suppose that the utility over mass is exponential,

$$U_{m|a^*}(m) = \frac{1 - e^{-\gamma m}}{1 - e^{-\gamma}}.$$

Therefore,

$$U_{a|m^*}(a) = U_{m|a^*}\left(\frac{m^*a}{a^*}\right) = \frac{1-e^{-\gamma a}}{1-e^{-\gamma}}.$$

Furthermore, the generating function can be determined as per Theorem 12.1 or by solving (14).

From (14), the generating function needs to satisfy

$$\varphi\left(\frac{1-e^{-\gamma ma}}{1-e^{-\gamma}}\right) = \varphi\left(\frac{1-e^{-\gamma m}}{1-e^{-\gamma}}\right)\varphi\left(\frac{1-e^{-\gamma a}}{1-e^{-\gamma}}\right).$$

Direct substitution shows that

$$\varphi(t) = -\frac{1}{\gamma}\log(1-(1-e^{-\gamma})t)$$

does indeed satisfy this equation.

Three important learnings must be highlighted from assigning utility functions to structural or value models that are fully specified and can be converted into an additive form by a monotone transformation:

1. The utility over force is an Archimedean combination of the utility of mass and utility of velocity.
2. The utilities over mass and velocity are not independent of each other. They are related by the isopreference contours of the structural model. We shall revisit this in Chapter 19 and refer to it as utility transversality.
3. The generating function is not arbitrarily chosen. It is completely determined because the structural model is fully specified.

On the other hand, if all we know is that the value model can be converted into an additive, but we do not know the functions in the structural model, then we have the extra degree of freedom to assess the individual utility functions and the generating function.

APPENDIX 12B PROOF OF THEOREM 1

The Two-Attribute Case

To simplify the exposition, we shall first prove the results for two attributes. The results for multiple attributes follow the exact same analysis except that we add more arguments to the functions.

Necessity: Given the two-attribute utility function $U(x_1, x_2)$ constructed by assigning a utility function over an additive preference function, we have

$$U(x_1, x_2) = U_V\left(m\big(f_1(x_1) + f_2(x_2)\big)\right), (15)$$

where $f_1(x_1^*), f_2(x_2^*) < \infty$ because they are bounded.

Using properties of logarithms and exponential functions, we can rewrite (15) as

$$U(x_1, x_2) \triangleq U_V\left(m\left(f_1(x_1) + f_2(x_2)\right)\right) = U_V\left(m\left(\ln\left(e^{f_1(x_1)}e^{f_2(x_2)}\right)\right)\right).$$

Define the modified functions

$$\tilde{f}_1(x_1) = e^{(f_1(x_1) - f_1(x_1^*))} \text{ and } \tilde{f}_2(x_2) = e^{(f_2(x_2) - f_1(x_2^*))}.$$

Note that by direct substitution, $\tilde{f}_1(x_1^*) = \tilde{f}_2(x_2^*) = 1$. We can now write

$$U(x_1, x_2) = U_V\left(m\left(\ln\left(e^{\tilde{f}_1(x_1)}e^{\tilde{f}_2(x_2)}\right) + f_1(x_1^*) + f_2(x_2^*)\right)\right) \triangleq \tilde{U}_V\left(\tilde{f}_1(x_1)\tilde{f}_2(x_2)\right), (16)$$

where

$$\tilde{U}_V(v) = U_V\left(m\left(\ln v + f_1(x_1^*) + f_2(x_2^*)\right)\right),$$

which is a strictly increasing function.

When $x_2 = x_2^*$, equation (16) gives

$$U(x_1, x_2^*) = \tilde{U}_V\left(\tilde{f}_1(x_1)\tilde{f}_2(x_2^*)\right) = \tilde{U}_V\left(\tilde{f}_1(x_1) \cdot 1\right) = \tilde{U}_V\left(\tilde{f}_1(x_1)\right).$$

Define $\varphi \triangleq \tilde{U}_V^{-1}$, which is strictly increasing, and note that $\varphi(1) = 1$. Therefore,

$$\tilde{f}_1(x_1) = \tilde{U}_V^{-1}\left(U(x_1, x_2^*)\right) = \varphi\left(U(x_1, x_2^*)\right). (17)$$

Similarly,

$$\tilde{f}_2(x_2) = \varphi\left(U(x_1^*, x_2)\right). (18)$$

Substituting for (17), (18) into (16) with $\varphi \triangleq \tilde{U}_V^{-1}$ gives

$$U(x_1, x_2) = \varphi^{-1}\left(\varphi\left(U(x_1, x_2^*)\right) \cdot \varphi\left(U(x_1^*, x_2)\right)\right), (19)$$

which is the Archimedean utility form of (11).

Sufficiency: If $U(x_1, x_2)$ is expressed in terms of an Archimedean utility copula, then (19) holds. Applying a composite monotone transformation $\ln(\varphi(t))$ gives an additive form

$$\ln\left(\varphi\left(U(x_1, x_2^*)\right)\right) + \ln\left(\varphi\left(U(x_1^*, x_2)\right)\right). (20)$$

Because ordinal preferences are invariant to monotone transformations, the form in (20) is ordinally equivalent to the additive form $m(f_1(x_1) + f_2(x_2))$, where

$$m(t) = t, \ f_2(x_1) = \ln\left(\varphi\left(U(x_1^*, x_2)\right)\right) \text{ and } f_2(x_1) = \ln\left(\varphi\left(U(x_1^*, x_2)\right)\right).$$

We now generalize the results to multiple attributes.

Proof of Theorem 1 (Multiple Attributes)

Necessity: Following the same steps for the case of two attributes, with extensions to multiple attributes, we have

$$U(x_1,...,x_n) = U_V\left(m\left(\sum_{i=1}^{n} f_i(x_i) \right) \right) = \tilde{U}_V\left(\prod_{i=1}^{n} \tilde{f}_i(x_i) \right), \tag{21}$$

where $\tilde{f}_i(x_i) = e^{(f_i(x_i)-f_i(x_i^*))}, i = 1,...,n$ and $\tilde{U}_V(v) = U_V\left(m\left(\ln v + \sum_{i=1}^{n} f_i(x_i^*) \right) \right)$, which is a strictly increasing function.

Note that $\tilde{f}_i(x_i^*) = 1, i = 1,...,n$. When $\bar{x}_i = \bar{x}_i^*$, (21) gives

$$U(x_i, \bar{x}_i^*) = \tilde{U}_V\left(\tilde{f}_i(x_i) \prod_{j \neq i} \tilde{f}_j(x_j^*) \right) = \tilde{U}_V\left(\tilde{f}_i(x_i) \prod_{j \neq i} 1 \right) = \tilde{U}_V(\tilde{f}_i(x_i)).$$

Define $\varphi \triangleq \tilde{U}_V^{-1}$, then $\varphi(v)$ is strictly increasing with $\varphi(1) = 1$ and

$$\tilde{f}_i(x_i) = \tilde{U}_V^{-1}(U(x_i, \bar{x}_i^*)), \forall i = 1,...,n. \tag{22}$$

Substituting for $\varphi = \tilde{U}_V^{-1}$ and (22) into (21) gives

$$U(x_1,...,x_n) = \varphi^{-1}\left(\prod_{i=1}^{n} \varphi(U(x_i, \bar{x}_i^*)) \right),$$

which is the Archimedean utility form of (11).

Sufficiency: This is straightforward by applying a monotone transformation $\ln(\varphi(t))$ to (11).

PART IV   PROPERTIES OF SINGLE-ATTRIBUTE UTILITY FUNCTIONS

This part explores the properties of univariate utility functions defined for prospects characterized by a single direct value attribute. It is somewhat more mathematical than the previous parts, but the concepts themselves are simple and are presented in a methodical, step-by-step approach. The Appendix of the book also provides a tutorial on some of the mathematical foundations for readers who are not familiar with some of the mathematical content.

As we have seen in Chapter 11, a multiattribute utility function can be constructed using a univariate utility function over value or over an individual attribute. Therefore, it is important to understand the properties and implications of single-attribute utility functions. Furthermore, as demonstrated in Chapter 12, additive ordinal preferences correspond to a multiattribute utility function that is an Archimedean combination of univariate utility functions over the attributes. Therefore, single-attribute utility functions play an important role in constructing multiattribute utility functions corresponding to additive ordinal preferences.

As we shall see in future parts of this book, the multiattribute utility assessment can be decomposed into univariate conditional utility assessments at reference values of the remaining attributes. Therefore, univariate utility functions also enable the construction of appropriate multiattribute utility functions, without a preference or a value function, even when additive ordinal preferences are not present.

Many of the properties of single-attribute utility functions that we discuss in this part will transfer directly, using some change of variables, into concepts of multiattribute utility functions. For example, the "delta property" of Chapter 13 and the "invariant transformation" for single-attribute utility functions of Chapter 16 will translate into the concept of "utility independence" for multiattribute utility functions, as we shall see in Chapter 23. Likewise, the concept of "one-switch utility functions" of Chapter 18 will translate into the concept of "one-switch utility independence" for multiattribute utility functions in Chapter 27.

While there are many ways to characterize single-attribute utility functions, our focus in this part will be on two main themes:

1. The effects of the shape of the utility function on the valuation of lotteries.
2. The effects of the shape of the utility function on the change in valuation of lotteries when the outcomes are modified by monotone transformations.

These two themes provide several insights into the properties of utility functions and their characterizations. In this part we also discuss the concept of a risk aversion function and its extension to multivariate risk aversion using a value function. In Chapter 19, we also discuss the concept of utility transversality that relates the risk aversion functions over the individual attributes to the trade-off functions among them.

If you are already familiar with properties of single-attribute utility functions, or are not interested in the mathematical details of their properties, then you may skip this part and move directly to the next part of the book, in which we discuss multiattribute utility elicitation using expansion theorems and independence properties. However, I believe it will still be worth your time to gain a full understanding of the basic properties of single-attribute utility functions before moving on to multiple attributes.

CHAPTER 13 The Role of the Utility Function in Valuing
Uncertain Deals

Chapter Concepts

- Valuing a decision alternative with uncertain monetary outcomes
  ○ The certain equivalent
- Effects of the shape of the utility function on the certain equivalent
- Applying a linear transformation to the utility function
- Strategically equivalent utility functions
- Applying a concave or convex transformation to the utility function
  ○ Implications for the utility function when one certain equivalent is always lower (higher) than another
- Effects of increase in wealth on the valuation of lotteries:
  ○ The delta property
  ○ The sub-delta property
  ○ The super-delta property

13.1 INTRODUCTION

As we have seen in Chapter 11, a single-attribute utility function over value is sufficient to construct a multiattribute utility function. Therefore, it is important to understand the properties of single-attribute utility functions over value and their implications. This section examines these basic properties. The procedures we discuss in this chapter apply to utility functions over general value measures or other attributes. The value measure we shall focus on in our discussions is wealth. Our focus will be on valuation of decision alternatives with monetary outcomes and changes in valuation of alternatives when their outcomes are modified. Once the utility function is specified, the best decision alternative is the one with the highest expected utility. This chapter discusses examples of utility functions over value measures and discusses several of their properties.

13.2 BASIC DEFINITIONS AND NOTATION

In this and the following chapters, we shall use U to denote a utility function over a value measure or an attribute. We assume in this section that the value measure is the total wealth, and give it the particular symbol y,

$$\text{Total Wealth} = y = x + w,$$

where x is an outcome of a lottery and w is the initial wealth.

The value of a deterministic prospect is the personal indifference selling price of that prospect; i.e., it is the price that makes you just indifferent to owning the prospect or giving it up and receiving this particular indifference value. In this chapter, we shall see how this concept translates into an indifference selling price for a lottery.

Notation for Total Wealth Lotteries: We denote a total wealth lottery as a lottery that provides a p_i chance of total wealth y_i for possible prospects $i = 1, \dots, n$ using the notation

$$< p_1, y_1; p_2, y_2; \dots; p_n, y_n >.$$

Figure 13.1 shows a tree representation of a lottery with three wealth outcomes, y_1, y_2, y_3. This tree represents the wealth lottery $< p_1, y_1; p_2, y_2; p_3, y_3 >$.

Notation for Lottery Outcomes: To specify the outcomes of the lottery itself, x_i, $i = 1, \dots, n$, instead of the total wealth, we note that $x_i = y_i - w$ and write

$$< p_1, x_1; p_2, x_2; \dots; p_n, x_n >.$$

Notation for Expectation of a Lottery or a Function of a Lottery
We use E to represent the expectation of any function of the lottery. For example,

$$E[y] = \sum_{i=1}^{n} p_i y_i$$

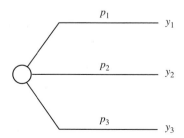

Figure 13.1. Tree representation of a three-outcome wealth lottery, $< p_1, y_1; p_2, y_2; p_3, y_3 >$.

is the expectation (mean) of the wealth lottery $< p_1, y_1; p_2, y_2; ...; p_n, y_n >$, and

$$E[U(y)] = \sum_{i=1}^{n} p_i U(y_i)$$

is the expectation of the utility function applied to the wealth outcomes of the lottery.

We will use the inverse of a utility function, U, denoted by U^{-1} frequently in the remaining chapters. Appendix A at the end of the book provides further details on the inverse function.

Definition: Inverse of a Function

To determine the inverse function of a utility function, U, set

$$U(y) = t$$

and solve for y in terms of t, i.e., $y = U^{-1}(t)$.
   For example, if the utility function is

$$U(y) = a + be^{-\gamma y},$$

then we set $U(y) = t$, and so

$$U(y) = t = a + be^{-\gamma y}.$$

Rearranging gives the relation between y and t as

$$y = \frac{-1}{\gamma} \ln \frac{t-a}{b} = U^{-1}(t).$$

13.3 ASSESSING UTILITY FUNCTIONS OVER VALUE MEASURES

By using a value measure for the prospects of a decision and then assigning a utility function over value, you get more than just the best decision alternative using the expected utility criterion: you can also calculate the ***certain equivalent*** of any alternative (or any uncertain deal) that you face.

Definition: Certain Equivalent

The certain equivalent of a lottery is the amount of money that makes you just indifferent to giving up a lottery you are facing or holding on to it; ***it is your personal indifferent selling price of the lottery.***

Calculating the certain equivalent enables us to assign monetary values to the decision alternatives we face. The certain equivalent, therefore, enables the translation of value statements from deterministic prospects into value statements for uncertain deals using the utility function and the characterization of the lottery.

Properties of Some Common Utility Functions

Of particular interest in our endeavor are utility functions on the domain of interest that are:

Continuous (you do not want any discontinuity in utility functions, otherwise you would value small increments in a deal by an abruptly large amount),

Bounded (this is not a major restriction in valuing lotteries: remember utility values are just indifference probabilities so they can be normalized to range from zero to one). In fact, if a utility function cannot be normalized over the domain of operation then there will be some violation of the Equivalence Rule over that domain.

Non-decreasing (because more money is preferred to less, the utility value increases as the value measure increases). If more is strictly preferred to less, then the utility function is strictly increasing.

Another condition that is widely used in many practical contexts is that the utility function be **twice continuously differentiable**. This condition enables a **smooth** curve for the utility function and also enables the calculation of what is known as the risk aversion function (Arrow 1965, Pratt 1964), which is discussed in Chapter 14. The risk aversion function provides many insights into the valuation of uncertain deals. Figure 13.2 shows an example of a continuous, bounded, strictly increasing, and twice continuously differentiable utility function.

In Chapter 10, we discussed how to assess utility values using lottery assessments. This method of direct assessment comprises one means of constructing the utility function. As we have seen, the method may become tedious, however, if numerous points are required on the curve.

One method for determining the utility function when the number of prospects is large is to assume a functional form for the utility function and then fit the parameters of this functional form to a few utility assessments. Chapter 10 presented an example of an exponential utility function and methods to assess its parameter.

It is clear that choosing a functional form and then assessing its parameters is easier than assessing several utility values and then fitting them. The main problem, of course, is how to choose that functional form. When choosing a particular functional form for the utility function, the decision-maker may wish to have it satisfy some desirable properties. For example,

1. The decision-maker may wish to value uncertain deals (or investments) by an amount that is lower than their mean (expected value); or alternatively,
2. The decision-maker may wish to value uncertain deals by a higher (or lower) amount as he gets richer; or alternatively,
3. The decision-maker may wish to provide statements that one investment is more desirable than another; or
4. The decision-maker may wish to provide qualitative (or quantitative) expressions about the change in value of a lottery or an investment when it gets modified.

To determine the functional form that best matches these preferences, it will be useful to understand the properties and implications of using different functional forms of utility functions. We shall discuss additional properties of utility functions in the next

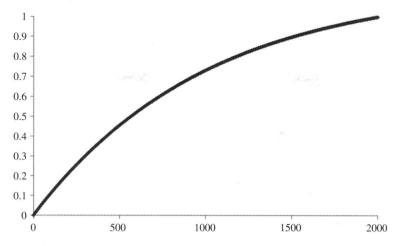

Figure 13.2. Example of a continuous, bounded, strictly increasing, and twice continuously differentiable utility function.

three chapters to help characterize the utility function based on these conditions. First, we provide some motivation for determining the utility function and describe how to use the utility function for valuing a decision alternative once it has been specified.

13.4 VALUING A DECISION ALTERNATIVE USING THE UTILITY FUNCTION (THE CERTAIN EQUIVALENT)

In Chapter 1, we defined the preference, value, and utility of a deterministic prospect. In Chapter 2, we demonstrated how to determine the preference order for uncertain deals or alternatives involving uncertainty using the expected utility criterion. Now we shall illustrate how to determine the value of an uncertain deal when we are uncertain about which prospect we shall get. We refer to the value of an uncertain deal as its *certain equivalent*.

The best decision alternative is the one with the highest expected utility. Using the Five Rules (Section 2.2), we should be indifferent between keeping a wealth lottery we own (or a decision alternative we own) and giving it up for a wealth equivalent that provides the same utility as the expected utility of the wealth lottery, i.e.,

$$E[U(y)] = U(\tilde{y}) = \sum_{i=1}^{n} p_i U(y_i),$$

where U is the utility function over total wealth, which is assumed to be continuous and monotone, and $\tilde{y}$ is the wealth equivalent of the lottery.

If the utility function over value is specified, we can do more than just determine the best decision alternative: we can also determine the monetary value that makes us indifferent to each alternative using the certain equivalent of the alternative. We explain this quantity in more detail below.

13.4.1 The Certain Equivalent of a Lottery and the Wealth Equivalent

Different people will value wealth lotteries differently based on their taste for risk. If we wish to value a lottery, X, that we own, and receive a deterministic amount instead, then it must be that the expected utility of the wealth lottery is equal to the utility of its certain equivalent plus the initial wealth w, i.e.,

$$U(\tilde{x}+w) = \sum_{i=1}^{n} p_i U(x_i + w), \tag{1}$$

where $\tilde{x}$ is the certain equivalent of the lottery – the amount of money that would make us just indifferent to selling the lottery or holding on to it. The certain equivalent is also the monetary value of a decision alternative described by this lottery. Recall that this amount is the personal indifferent selling price of the lottery.

Rearranging (1) gives an expression for the certain equivalent of the lottery, which we define below.

Definition
The Certain Equivalent of a Lottery, $\tilde{x}$, is equal to

$$\tilde{x} = U^{-1}\left(\sum_{i=1}^{n} p_i U(x_i + w)\right) - w \tag{2}$$

The wealth equivalent and certain equivalent are related by,

$$\boxed{\tilde{y} = \tilde{x} + w}.$$

EXAMPLE 13.1: Numerical Illustration: Calculating the Wealth Equivalent Using an Exponential Utility Function

---

Consider the wealth lottery

$$<0.5, 1000; 0.5, 5000>,$$

which provides a 50/50 chance of total wealth \$1,000 or \$5,000. When evaluated by a decision-maker with an exponential utility function,

$$U(y) = 1 - e^{-\gamma y},$$

with $\gamma = 0.001$.

The expected utility of total wealth is

$$E[U(y)] = 0.5U(1000) + 0.5U(5000) = 0.5(1 - e^{-\gamma 1000}) + 0.5(1 - e^{-\gamma 5000}) = 0.812.$$

The utility of the wealth equivalent of the lottery must be equal to the expected utility of total wealth, and so

$$U(\tilde{y}) = E[U(y)].$$

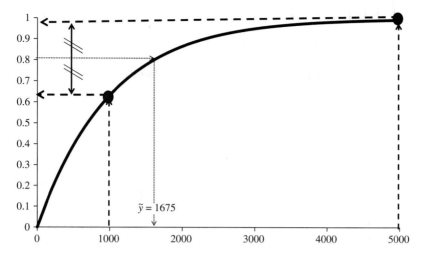

Figure 13.3. Wealth equivalent and expected utility.

For an exponential utility function, we get by direct substitution into its functional form, the utility of the wealth equivalent as

$$U(\tilde{y}) = 1 - e^{-\gamma\tilde{y}}.$$

Because the utility of the wealth equivalent is equal to the expected utility of total wealth, this implies that

$$1 - e^{-\gamma\tilde{y}} = E[U(y)].$$

Rearranging gives

$$\tilde{y} = \frac{-1}{\gamma}\ln\left(1 - E[U(y)]\right) = \frac{-1}{.001}\ln(1 - 0.812) = \$1675.$$

Figure 13.3 plots the utility function, the expected utility, and then illustrates the calculation of the wealth equivalent graphically.

EXAMPLE 13.2: Numerical Illustration: Wealth Equivalent Using a Logarithmic Utility Function

---

When the wealth lottery

$$<0.5, 1000; 0.5, 5000>$$

is evaluated by a decision-maker with a logarithmic utility function,

$$U(y) = \ln(y),$$

the expected utility of total wealth is

$$E[U(y)] = 0.5U(1000) + 0.5U(5000) = 0.5\ln(1000) + 0.5\ln(5000) = \ln\left(1000^{0.5} \times 5000^{0.5}\right).$$

The wealth equivalent must satisfy

$$U(\tilde{y}) = E[U(y)].$$

For a logarithmic utility function, we get by direct substitution

$$U(\tilde{y}) = \ln(\tilde{y}).$$

Rearranging gives

$$\tilde{y} = e^{E[U(y)]} = 1000^{0.5} \times 5000^{0.5} = \$2236.07.$$

The wealth equivalent includes the initial wealth and the certain equivalent of the lottery, $\tilde{y} = \tilde{x} + w$. Therefore, the certain equivalent of the actual lottery is the wealth equivalent less the initial wealth

$$\tilde{x} = \tilde{y} - w.$$

If the initial wealth is $500, then the lottery outcomes are <0.5, 500; 0.5, 4500> and the certain equivalent of this lottery is $2236.07 − $500 = $1736.07.

Dialogue between Instructor (I) and Student (S)

I: Do you have questions on how to calculate the wealth equivalent and the certain equivalent of a lottery? This is actually the response to the question you posed in Chapter 1 about how to place a dollar value on a decision alternative that involves uncertainty.

S: Yes, I have now seen how to make value statements for lotteries having monetary outcomes. But we have not discussed this for decision alternatives with monetary outcomes or outcomes that are expressed in terms of some value measure.

I: A decision alternative that you face is actually a lottery. When we say lottery, we do not mean that you are playing a lottery or that you are gambling; we simply mean you are facing some deal where you are uncertain about the prospect that you will get. In our earlier discussion in this chapter, the prospects of the decision alternative were represented in terms of a value measure, which was total wealth.

S: So if I want to place a value on a decision alternative, I calculate the certain equivalent of the alternative, and this involves the utility function. I can even determine the difference between the valuations of alternatives using the certain equivalent. And the wealth equivalent is just the certain equivalent plus the initial wealth.

I: Correct. You can calculate how much you need to be paid to give up one alternative for another.

S: But to determine the wealth equivalent or the certain equivalent, we need the utility function. We have not yet discussed how to assign a functional form for the utility function. It seems like this is important if we want to calculate the certain equivalent appropriately.

I: We do not necessarily need the full utility function; we just need the utility values for all prospects and the utility value of the wealth equivalent. So in effect, we can do this by direct utility assessments. But, of course, if we have numerous prospects, it is a lot more convenient to have a functional form for the utility function. We shall do this by first understanding the different properties of utility functions and their behavior as the lottery is modified.

S: We have seen that the logarithmic utility function and the exponential utility function gave different wealth equivalents for the same lottery. Are there utility functions that give the same wealth equivalents for all lotteries?

I: For a particular lottery, yes there are many. For example, we just need the utility functions to be equal at the prospects of the lottery and the utility and the expected utility. But if we would like the same wealth equivalents for all lotteries, then all such utility functions must be related, and we refer to those as "strategically equivalent utility functions."

S: What relates these strategically equivalent utility functions?

I: We will discuss this in detail. But first let us see what happens to the certain equivalent if we apply a linear transformation to the utility function.

13.4.2 Invariance of the Certain Equivalent to a Linear Transformation on the Utility Function

We shall now explore the effects of the shape of the utility function on the valuation of lotteries. First, we consider what happens to the certain equivalent of a lottery if we apply a linear transformation to the utility function. To illustrate, consider a utility function $U_1(y)$, which is a linear transformation of a utility function, $U(y)$, of the form,

$$U_1(y) = aU(y) + b, \ a \neq 0.$$

The inverse of this utility function can be expressed in terms of the inverse of U as

$$U_1^{-1}(t) = U^{-1}\left(\frac{t-b}{a}\right).$$

The certain equivalent calculated using this new utility function is

$$\tilde{x} = U_1^{-1}\left(\sum_{i=1}^{n} p_i U_1(x_i + w)\right) - w.$$

Substituting for $U_1 = aU(x) + b$ and its inverse U_1^{-1} from above gives

$$\tilde{x} = U^{-1}\left(\frac{\sum_{i=1}^{n} p_i[aU(x_i + w) + b] - b}{a}\right) - w = U^{-1}\left(\sum_{i=1}^{n} p_i U(x_i + w)\right) - w,$$

which is the same certain equivalent obtained in (2) using the utility function $U(y)$. Therefore, the certain equivalent is unaffected by applying a linear transformation to the utility function, even if the constant, a, of this linear transformation is negative.

The certain equivalent is invariant to a linear transformation on the utility function.

This result is in contrast to preference functions for deterministic consequences, which are invariant to a strictly monotone transformation. If a strictly increasing transformation is applied to the preference function, it does not change the deterministic preferences, and if we apply a strictly decreasing transformation, then the prospect with the lowest value of the preference function will be most preferred. The certain equivalent (or the wealth equivalent), on the other hand, is not invariant to an arbitrary monotone transformation (see Example 13.4). It is invariant only up to a linear transformation on the utility function. In general, if a property holds for preferences over deterministic prospects, the same property requires more stringent conditions when it applies to preferences over lotteries.

EXAMPLE 13.3: Numerical Illustration: Invariance of the Certain Equivalent (or Wealth Equivalent) to a Linear Transformation on the Utility Function

---

Consider again the wealth lottery

$$<0.5, 1000; 0.5, 5000>,$$

that provides a 50/50 chance of total wealth \$1,000 or \$5,000. As we have seen, the wealth equivalent, when evaluated by a decision-maker with an exponential utility function, $U(y) = 1 - e^{-\gamma y}$ with $\gamma = 0.001$, is $\tilde{y} = \$1675$.

If we apply a linear transformation to this utility function, we get the function

$$U_1(y) = ae^{-\gamma y} + b,$$

with $\gamma = 0.001$. Note that if a is positive, then the utility function is decreasing with y. Because the utility function is decreasing with wealth, when a is positive, we choose the alternative with the lowest expected utility in this case. On the other hand, if a is negative, we choose the alternative with the highest expected utility.

The expected utility of the lottery (in terms of a and b) is

$$E[U_1(y)] = 0.5U_1(1000) + 0.5U_1(5000)$$
$$= 0.5(ae^{-\gamma 1000} + b) + 0.5(ae^{-\gamma 5000} + b) = 0.5a[e^{-\gamma 1000} + e^{-\gamma 5000}] + b.$$

The wealth equivalent of the lottery after applying a linear transformation to the utility function is

$$\tilde{y} = \frac{-1}{\gamma} \ln\left(\frac{E[U(y)] - b}{a}\right) = \frac{-1}{.001} \ln\left(0.5[e^{-\gamma 1000} + e^{-\gamma 5000}]\right) = \$1675.$$

The wealth equivalent remains the same regardless of the value of a (positive or negative), and we still choose the alternative with the highest wealth equivalent.

EXAMPLE 13.4: Numerical Illustration: Applying a Monotone Transformation to the Utility Function

---

Consider again the wealth lottery $<0.5, 1000; 0.5, 5000>$, and the exponential utility function, $U(y) = e^{-\gamma y}$ with $\gamma = 0.001$. As we have seen, the wealth equivalent is $\tilde{y} = \$1675$.

What happens if we apply a monotone transformation to this utility function? For example, consider the monotone transformation

$$g(U) = U^3.$$

The new utility function becomes

$$U_1(y) = \left(e^{-\gamma y}\right)^3 = e^{-3\gamma y}.$$

In effect, we have tripled the value of the parameter, γ.

The wealth equivalent of the lottery using the modified utility function is

$$\tilde{y} = \frac{-1}{3\gamma} \ln\left(0.5[e^{-3\gamma 1000} + e^{-3\gamma 5000}]\right) = \$1231.05.$$

As opposed to the case of the linear transformation, the wealth equivalent of the lottery is no longer invariant, and it has decreased by applying a transformation $g(U) = U^3$ to the utility function.

Not every transformation applied to the utility function will preserve the certain equivalent. Our next example shows that if two decision-makers with the same initial wealth value all uncertain deals by the same amount (have the same certain equivalent for uncertain deals), they must have the same utility function up to a linear transformation.

EXAMPLE 13.5: When Two Decision-Makers Value Lotteries Equally

Consider two decision-makers with utility functions U_A, U_B. If they have the same initial wealth and their certain equivalents are the same for all lotteries, then

$$\tilde{y}_A = \tilde{y}_B \text{ for all lotteries.}$$

Or equivalently,

$$U_A^{-1}\left(\sum_{i=1}^{n} p_i U_A(y_i)\right) = U_B^{-1}\left(\sum_{i=1}^{n} p_i U_B(y_i)\right).$$

Rearranging gives

$$U_B\left(U_A^{-1}\left(\sum_{i=1}^{n} p_i U_A(y_i)\right)\right) = \sum_{i=1}^{n} p_i U_B(y_i). \tag{3}$$

We shall now use a trick that we shall incorporate quite frequently in this chapter and others.

Define t_i as the utility value of prospect y_i using the utility function U_A

$$U_A(y_i) = t_i \Rightarrow y_i = U_A^{-1}(t_i).$$

Further define $f(t)$, the composite function of U_B and U_A^{-1}:

$$f(t) = U_B\left(U_A^{-1}(t)\right).$$

Substituting reduces the complicated form of Equation (3) into the simpler equation

$$f\left(\sum_{i=1}^{n} p_i t_i\right) = \sum_{i=1}^{n} p_i f(t_i) \text{ for all lotteries.} \tag{4}$$

This is known as **Jensen's equality**, and if it applies for an interval of t, then it implies that f is a linear function, i.e.,

$$f(t) = kt + d.$$

Because $f(t) = U_B(U_A^{-1}(t))$, Jensen's equality implies that

$$U_B(U_A^{-1}(t)) = kt + d, \quad k \neq 0.$$

Substituting for $U_A(y) = t$ and $y = U_A^{-1}(t)$ gives

$$\boxed{U_B(y) = kU_A(y) + d},$$

proving that equality in the certain equivalents for all lotteries implies that U_A, U_B must be related by a linear transformation.

We have established that the certain equivalent invariant to a linear transformation on the utility function, i.e., if we apply a linear transformation to the utility function, it will result in the same certain equivalent. The reverse is also true: if two decision-makers, with the same initial wealth, value all uncertain deals by the same amount, then their utility functions must be related by a linear transformation. Utility functions that lead to the same certain equivalent (or wealth equivalent) are said to be ***strategically equivalent***.

13.5 EFFECTS OF THE SHAPE OF THE UTILITY FUNCTION ON THE CERTAIN EQUIVALENT

It is natural to ask about the relation between the certain equivalent of a lottery, $\tilde{x}$, and the mean value of the lottery, $E[X]$, or simply μ. Which is larger in magnitude? The shape of the utility function (concave, linear, or convex) determines the relative magnitude of the certain equivalent and the mean, i.e., whether the certain equivalent is above, below, or equal to the mean value of the lottery. Therefore, it is helpful to understand this relation in our assignment of the utility function to capture our preferences for lotteries appropriately.

13.5.1 Concave Utility Functions

If the utility function is concave, then the certain equivalent of the lottery, $\tilde{x}$, is less than its mean value, μ, i.e.,

$$\tilde{x} \leq \mu, \tag{5}$$

in which case, the decision-maker is said to be risk averse.

To see why this is true, observe that if a function is concave, then from **Jensen's inequality** (see Appendix B at the end of the book) the expectation of that function is less than or equal to the function of the expectation, i.e., if a function, U, is concave, then by Jensen's inequality,

$$E[U(x)] \leq U\left(E[x]\right).$$

Equivalently, we can write the expectations in terms of the probabilities, and so we get for a concave utility function

$$\sum_{i=1}^{n} p_i U(x_i + w) \leq U\left(\sum_{i=1}^{n} p_i(x_i + w)\right). \tag{6}$$

The certain equivalent is by definition equal to

$$\tilde{x} = U^{-1}\left(\sum_{i=1}^{n} p_i U(x_i + w)\right) - w.$$

Rearranging gives

$$U(\tilde{x} + w) = \sum_{i=1}^{n} p_i U(x_i + w). \tag{7}$$

Substituting from (7) into the left handside of (6) shows that if the utility function is concave, then we get from Jensen's inequality

$$U(\tilde{x} + w) \leq U\left(\sum_{i=1}^{n} p_i(x_i + w)\right)$$

Rearranging gives

$$\tilde{x} \leq U^{-1}\left(U\left(\sum_{i=1}^{n} p_i(x_i + w)\right)\right) - w$$

Because $U^{-1}\left(U(t)\right) = t$, we have $U^{-1}\left(U\left(\sum_{i=1}^{n} p_i(x_i + w)\right)\right) = \sum_{i=1}^{n} p_i(x_i + w)$ and so

$$\tilde{x} \leq \sum_{i=1}^{n} p_i(x_i + w) - w.$$

Simplifying gives

$$\tilde{x} \leq \sum_{i=1}^{n} p_i x_i = E[x] = \mu.$$

Therefore, the certain is less than or equal to the mean value. We summarize this result below.

> *The mean value of a lottery is an upper bound on the certain equivalent if the utility function is concave.*

13.5.2 Convex Utility Functions

On the other hand, if the utility function is convex, then Jensen's inequality implies that

$$E[U(x)] \geq U(E[x]).$$

We leave it as an exercise to the reader to show that this implies that the certain equivalent is greater than or equal to the mean value,

$$\tilde{x} \geq \sum_{i=1}^{n} p_i x_i.$$

We simply state this result.

> *The mean value of a lottery is the lower bound on the certain equivalent if the utility function is convex.*

13.5.3. Linear Utility Functions

If the utility function is linear (risk-neutral decision-maker), then from **_Jensen's Equality (eqn 4 in example 13.5)_**,

$$E[U(x)] = U(E[x]).$$

For example, if $U(x) = ax + b$, then the certain equivalent is equal to the mean value of the lottery. To illustrate, recall that the certain equivalent is

$$\tilde{x} = U^{-1}\left(\sum_{i=1}^{n} p_i U(x_i + w)\right) - w.$$

Substituting for a linear utility function $U(x) = ax + b$, and its inverse $U^{-1}(t) = \dfrac{t-b}{a}$ gives

$$\tilde{x} = \frac{\left(a\sum_{i=1}^{n} p_i x_i + aw + b\right) - b}{a} - w = \sum_{i=1}^{n} p_i x_i,$$

which is the mean value of the lottery.

> *The mean value of a lottery is equal to the certain equivalent if the utility function is linear.*

Dialogue between Instructor (I) and Student (I)

I: Let us recap what we have just discussed:

  If the utility function is linear, then the certain equivalent is equal to the mean.
  If the utility function is concave, then the certain equivalent is less than the mean.
  If the utility function is convex, then the certain equivalent is greater than the mean.

S: So we have a concave utility function for money?

I: Sometimes, linear utility functions are also effective if the monetary amounts are much smaller than your wealth.

S: Is the reverse direction also true? For example, if the certain equivalent is always less than the mean, does the utility function have to be concave?

I: If this is true for all lotteries, then the answer is yes. Let us show this reverse direction below by considering two utility functions U_A, U_B.

13.6 EFFECTS OF APPLYING A CONCAVE OR CONVEX TRANSFORMATION TO THE UTILITY FUNCTION

In this section, we continue to investigate the effects of the shape of the utility function on the valuation of lotteries. We have seen that applying a linear transformation to a utility function does not change the certain equivalent. What happens to the certain equivalent if we apply a concave or a convex transformation to the utility function?

Consider two decision-makers with utility functions U_A, U_B with the same wealth. Suppose that decision-maker A has a higher wealth equivalent than does decision-maker B for all lotteries. What can we say about their utility functions?

This condition can be expressed as

$$\tilde{y}_A \geq \tilde{y}_B \text{ for all lotteries.}$$

Substituting for the definitions of the wealth equivalent gives,

$$U_A^{-1}\left(\sum_{i=1}^{n} p_i U_A(y_i)\right) \geq U_B^{-1}\left(\sum_{i=1}^{n} p_i U_B(y_i)\right) \text{ for all lotteries.}$$

Rearranging gives

$$U_B\left(U_A^{-1}\left(\sum_{i=1}^{n} p_i U_A(y_i)\right)\right) \geq \sum_{i=1}^{n} p_i U_B(y_i).$$

Define $U_A(y_i) = t_i \Rightarrow y_i = U_A^{-1}(t_i)$. Further define $f(t) = U_B\left(U_A^{-1}(t)\right)$. Substituting gives

$$f\left(\sum_{i=1}^{n} p_i t_i\right) \geq \sum_{i=1}^{n} p_i f(t_i). \tag{8}$$

This is **Jensen's inequality**, and it implies that f is a concave function.

Recall that $f(t) = U_B\left(U_A^{-1}(t)\right)$. Hence $U_B\left(U_A^{-1}(t)\right) = kt + d$. Substituting for $t = U_A(y)$ gives

$$\boxed{U_B(y) = f[U_A(y)]}. \tag{9}$$

This implies that if U_B always provides a lower wealth equivalent than does U_A, then U_B must be a concave function of U_A.

Equation (9) is often referred to as "*concavifying the utility function.*"

The reverse direction – that if one utility function is a concave function of another, then it provides a lower wealth equivalent – is easy to verify, following the same steps as in the previous section, and so we leave it as an exercise to the reader to verify that if U_B is a concave function of U_A, then $\tilde{y}_A \geq \tilde{y}_B$. We have proved the following theorem:

Theorem 13.1

> *Given two decision-makers, A and B, with utility functions U_A and U_B (respectively) and the same initial wealth, decision-maker B values all uncertain deals lower than does decision-maker A if and only if*
>
> $$U_B(y) = f[U_A(y)], \qquad (10)$$
>
> *where f is a concave function.*

Special Case

If $U_A(t) = t$ in (10), i.e., it is a linear utility function, then U_B is a concave function. Because a decision-maker with a linear utility function values uncertain deals by their mean value, it must be that a decision-maker with a concave utility function must value deals by an amount that is less than their mean value. This decision-maker is said to be "*risk averse.*"

EXAMPLE 13.6:   Numerical Illustration: Concavifying the Utility Function Using a Square Root Transformation

---

Based on the previous results, we can tell right away that the utility function $U_1(y) = y$, $y \geq 0$ will have a higher certain equivalent for uncertain deals than the utility function $U_2(y) = \sqrt{y}$. This is because $U_2(y) = \sqrt{U_1(y)}$, and the square root function is a concave function on the interval $[0, \infty)$,

Similarly, the utility function $U_3(y) = y^2$ will have a higher certain equivalent for uncertain deals than the function $U_1(y) = y$ because $U_3(y) = U_1(y)^2$ and the square function is a convex function.

We leave it as an exercise to the reader to prove the following theorem:

Theorem 13.2

> *Given two decision-makers, A and B, with utility functions U_A and U_B (respectively) and the same initial wealth. Decision-maker B values all uncertain deals higher than A if and only if*

$$U_B(y) = f[U_A(y)],$$

where f is a convex function.

Dialogue between Instructor (I) and Student (S)

S: If we apply a concave transformation to the utility function, then the wealth equivalent (and certain equivalent) will be lower than those of the unmodified utility function, and if we apply a convex transformation, then the wealth equivalent (and certain equivalent) will be higher. What happens if we apply an S-shaped transformation to the utility function? How does the certain equivalent and wealth equivalent change?

I: There is no general answer. They could be higher or lower depending on the lottery and the transformation. If all wealth outcomes are such that only the convex portion of the transformation is invoked in the calculation, then they will be higher. If the wealth outcomes are such that only the concave portion is invoked then they will be lower. If the lottery spans a large range such that both concave and convex portions of the transformation are in invoked, then it will depend on the actual lottery itself.

S: You mentioned that utility functions also have implications on how the valuation of a lottery changes when the lottery is modified.

I: Yes, so far we have considered only the effects of modifying the utility function by a linear, concave, or convex utility function. Next let us discuss the effects of changing the lottery outcomes themselves and the effect that the utility function will have on the new valuation in this case.

13.7 EFFECTS OF INCREMENTING ALL LOTTERY OUTCOMES BY THE SAME SHIFT AMOUNT

We now consider the effects of the shape of the utility function on the change in valuation of lotteries. Sometimes a wealth lottery is modified by adding some fixed amount to all of its outcomes. This can happen for many reasons. For example, the decision-maker might gain some wealth, and so the total wealth lottery is now modified by this new wealth amount. Alternatively, the lottery itself might become better due to some cost saving, or some improved forecasts. It will be useful to understand how the wealth equivalent of the lottery changes when this shift amount is added to all lottery outcomes. This section considers the following problem:

Problem Formulation

Suppose we have a wealth lottery

$$< p_1, y_1; p_2, y_2; ...; p_n, y_n >$$

whose wealth equivalent, as calculated by some utility function, U, is $\tilde{y}$. Now suppose we apply a shift amount to all outcomes of the lottery to obtain the lottery

$$< p_1, y_1 + \delta; p_2, y_2 + \delta; ...; p_n, y_n + \delta >$$

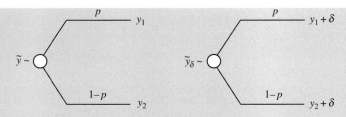

Figure 13.4. Example of a wealth lottery modified by a shift amount, δ.

and a wealth equivalent, calculated using the same utility function, equal to $\tilde{y}_\delta$.
Figure 13.4 shows a wealth lottery modified by a shift amount, δ.

What effect does this shift transformation have on the wealth equivalent of the original lottery?

Is the wealth equivalent of the modified lottery, $\tilde{y}_\delta$, equal to, higher than, or lower than the wealth equivalent of the unmodified lottery $\tilde{y}$ plus the shift amount δ? In other words, is

$$\tilde{y}_\delta = \tilde{y} + \delta, \ \tilde{y}_\delta = \tilde{y} > \delta, \ \tilde{y}_\delta = \tilde{y} < \delta?$$

13.7.1 The Delta Property

The delta property is the simplest form of observing change in certain equivalent when the lottery is modified by a shift amount. It assumes that the wealth equivalent of the modified lottery, $\tilde{y}_\delta$, is equal to the wealth equivalent of the unmodified lottery, $\tilde{y}$, plus the same shift amount, δ. In other words, the delta property assumes that for all lotteries,

$$\boxed{\tilde{y}_\delta = \tilde{y} + \delta}. \tag{11}$$

What does this equality imply about the utility function? By direct substitution into the definition of the wealth equivalent, (11) implies that

$$U^{-1}\left(\sum_{i=1}^{n} p_i U(y_i + \delta)\right) = U^{-1}\left(\sum_{i=1}^{n} p_i U(y_i)\right) + \delta.$$

Rearranging gives

$$U^{-1}\left(\sum_{i=1}^{n} p_i U(y_i)\right) = U^{-1}\left(\sum_{i=1}^{n} p_i U(y_i + \delta)\right) - \delta$$

Define

$$V_\delta(y) = U(y + \delta) \text{ and its inverse, } V_\delta^{-1}(t) = U^{-1}(t) - \delta.$$

Substitute to get

$$U^{-1}\left(\sum_{i=1}^{n} p_i U(y_i)\right) = V_\delta^{-1}\left(\sum_{i=1}^{n} p_i V_\delta(y_i)\right).$$

Applying the transformation V_δ to both sides gives

$$V_\delta \left(U^{-1} \left(\sum_{i=1}^{n} p_i U (y_i) \right) \right) = \sum_{i=1}^{n} p_i V_\delta(y_i).$$

Define $y_i = U^{-1}(t_i)$, and so $t_i = U(y_i)$. Further define $f = V_\delta(U^{-1})$, and substitute to get

$$f \left(\sum_{i=1}^{n} p_i t_i \right) = \sum_{i=1}^{n} p_i f(t_i),$$

which implies that f is a linear function.

Hence,

$$f(t) \triangleq V_\delta(U^{-1}(t)) = k(\delta)t + d(\delta) \quad \text{(a linear function of } t)$$

Now recall that $V_\delta(U^{-1}(t)) = U(U^{-1}(t) + \delta)$. Moreover, $y = U^{-1}(t)$, and so $y = U^{-1}(t)$. Therefore,

$$\boxed{U(y + \delta) = k(\delta)U(y) + d(\delta).}$$

To summarize:

> *If $\tilde{y}$ is the wealth equivalent of the lottery $< p_1, y_1; p_2, y_2; ...; p_n, y_n >$ and*
> *if $\tilde{y} + \delta$ is the wealth equivalent of the lottery $< p_1, y_1 + \delta; p_2, y_2 + \delta; ...; p_n, y_n + \delta >$,*
> *then the utility function satisfies*
>
> $$U(y + \delta) = k(\delta)U(y) + d(\delta).$$

Dialogue between Instructor (I) and Student (S)

I: We are interested in characterizing the effect of changes in the lottery on the certain equivalent. This will help us characterize the utility function. The simplest change is adding a shift transformation to the outcomes. We have provided a necessary condition that needs to be satisfied by all utility functions when the wealth equivalent increases by an amount δ when all outcomes increase by the same amount δ.

S: And we have shown that this condition implies that the utility function must satisfy $U(y + \delta) = k(\delta)U(y) + d(\delta)$.

I: Yes, this is a functional equation that characterizes the utility function. The math we have used to derive this equation will be used repeatedly. This is why we took the time to go through it. It is worth learning it. Also, do not get lost in the math and do remember the main result.

S: Got it. This is a clear necessary condition for satisfying the delta property. But does any utility function that satisfies this condition need to follow the delta property?

I: Yes. To illustrate, let us now prove sufficiency in the opposite direction, which is easier actually.

If the decision-maker's utility function satisfies the functional equation,

$$U(y + \delta) = k(\delta)U(y) + d(\delta),$$

then we can take the utility inverse of both sides to get

$$y + \delta = U^{-1}\left(k(\delta)U(y) + d(\delta)\right). \tag{12}$$

By definition, the wealth equivalent of the unmodified lottery is

$$\tilde{y} = U^{-1}\left(\sum_{i=1}^{n} p_i U\ (y_i)\right)$$

and the wealth equivalent of the modified lottery is

$$\tilde{y}_\delta = U^{-1}\left(\sum_{i=1}^{n} p_i U\ (y_i + \delta)\right). \tag{13}$$

Substituting for $U(y + \delta) = k(\delta)U(y) + d(\delta)$ into (13) gives

$$\tilde{y}_\delta = U^{-1}\left(\sum_{i=1}^{n} p_i [k(\delta)U(y_i) + d(\delta)]\right). \tag{14}$$

By direct substitution of $U(\tilde{y}) = \sum_{i=1}^{n} p_i U(y_i)$ into (14), we get

$$\tilde{y}_\delta = U^{-1}\left(k(\delta)U(\tilde{y}) + d(\delta)\right). \tag{15}$$

But from (12), if we substitute for the value of $y = \tilde{y}$, we get $\tilde{y} + \delta = U^{-1}\left(k(\delta)U(\tilde{y}) + d(\delta)\right)$. Substituting this into the right-hand side of (15) gives

$$\boxed{\tilde{y}_\delta = \tilde{y} + \delta}.$$

To summarize:

> If the utility function satisfies $U(y + \delta) = k(\delta)U(y) + d(\delta)$, and if
> the wealth equivalent of the lottery $< p_1, y_1; p_2, y_2; ...; p_n, y_n >$ is $\tilde{y}$, then
> the wealth equivalent of the lottery $< p_1, y_1 + \delta; p_2, y_2 + \delta; ...; p_n, y_n + \delta >$ is $\tilde{y} + \delta$.

We have proved the following theorem:

Theorem 13.3
A decision-maker follows the delta property if and only if the utility function satisfies

$$U(y + \delta) = k(\delta)U(y) + d(\delta).$$

We have converted the original problem of trying to identify the utility function that satisfies the delta property into a functional equation that provides a necessary and sufficient condition that the utility function must satisfy. What remains is to find the solutions of this functional equation. First note that both the linear and exponential utility functions satisfy (16).

EXAMPLE 13.7: Exponential Utility Functions Satisfy the Functional Equation

---

If $U(y) = a + be^{-\gamma y}$, then we get by direct substitution,

$$
\begin{aligned}
U(y+\delta) &= a + be^{-\gamma(y+\delta)} \\
&= a + be^{-\gamma y}e^{-\gamma\delta}[+ae^{-\gamma\delta} - ae^{-\gamma\delta}] \\
&= e^{-\gamma\delta}\left(a + be^{-\gamma y}\right) + a(1 - e^{-\gamma\delta}) \\
&= e^{-\gamma\delta}U(y) + a(1 - e^{-\gamma\delta}) \\
&= k(\delta)U(y) + d(\delta).
\end{aligned}
$$

Here, $k(\delta) = e^{-\gamma\delta}$ and $d(\delta) = a(1 - e^{-\gamma\delta})$.

EXAMPLE 13.8: Linear Utility Functions Satisfy the Functional Equation

---

If $U(y) = a + by$, then

$$
\begin{aligned}
U(y+\delta) = a + b(y+\delta) &= a + by + b\delta \\
&= U(y) + b\delta.
\end{aligned}
$$

Here, $k(\delta) = 1$ and $d(\delta) = b\delta$.

Dialogue between Instructor (I) and Student (S)

I: Is the idea of the delta property clear?

S: Yes, it is clear, both linear and exponential utility functions satisfy its invariance functional equation and therefore they must satisfy the delta property. But isn't it true that all utility functions must satisfy this functional equation? If we have a deal worth $10 and then we add $5 to all outcomes, shouldn't the new deal be worth $15?

I: Absolutely Not! This is the difference between deterministic prospects and uncertain deals. It is true that if you have a deterministic prospect worth $10 and that is someone sweetened the prospect by adding $5, then you would value the new prospect by $15 regardless of your utility function. But if you had an uncertain deal worth $10 and someone sweetened all possible outcomes by adding $5, it is not necessarily true that you will now value the sweetened deal by $15.

EXAMPLE 13.9: Logarithmic Utility Functions Do Not Satisfy This Functional Equation

---

If $U(y) = a + b\ln(y)$, then

$$
U(y+\delta) = a + b\ln(y+\delta),
$$

which cannot be expressed as $k(\delta)U(y) + d(\delta)$.

Therefore, the logarithmic utility function does not satisfy the delta property, and we now know that not all utility functions will satisfy this property.

13.7.2 Solutions to the Functional Equation

We now provide solutions to the functional equation

$$U(y+\delta) = k(\delta)U(y) + d(\delta). \tag{16}$$

There is an important subtlety that needs to be highlighted, however, and so we will consider two cases.

CASE I: THIS EQUATION IS SATISFIED FOR δ ON AN INTERVAL OF POSITIVE LENGTH This case is the simplest case and it has two possible solutions:

(a) **The function k is identically equal to one on its domain**

$$U(y+\delta) = U(y) + d(\delta)$$

This is a well-known **Pexider Equation** whose general continuous non-constant solution is the linear utility function

$$U(x) = cy + b, \qquad d(\delta) = c\delta, \quad c \neq 0.$$

(b) **The function k is not identically equal to one on its domain**

We solve this equation below assuming (for simplicity) that the utility function is defined at $y = 0$. For a more involved proof that does not make this assumption (and that arrives at the same solutions), see Abbas, Aczel, and Chudziak (2009).
    Put $y = 0$ into (16):

$$U(\delta) = k(\delta)U(0) + d(\delta).p \tag{17}$$

Subtracting (17) from (16) and adding the zero term $[U(0) - U(0)]$ gives

$$U(y+\delta) - U(\delta) = k(\delta)[U(y) - U(0)] + [U(0) - U(0)].$$

Rearranging gives

$$U(y+\delta) - U(0) = k(\delta)[U(y) - U(0)] + [U(\delta) - U(0)].$$

Equivalently, with

$$V(y) = U(y) - U(0), \tag{18}$$

we get

$$V(y+\delta) = k(\delta)V(y) + V(\delta). \tag{19}$$

We have converted the functional equation $U(y+\delta) = k(\delta)U(y) + d(\delta)$ into a simpler Equation (19) with one less function. Note further that the left-hand side of (19) is symmetric in y, δ. Therefore, the right-hand side must also be symmetric, and so we can write

$$k(\delta)V(y) + V(\delta) = k(y)V(\delta) + V(y). \tag{20}$$

Because $k(\delta)$ is not identically one on its interval, there exists δ_0 such that $k(\delta_0) \neq 1$. Substituting for δ_0 into (20) gives

$$k(\delta_0)V(y) + V(\delta_0) = k(y)V(\delta_0) + V(y).$$

This implies that

$$V(y)[k(\delta_0) - 1] = V(\delta_0)[k(y) - 1].$$

Rearranging gives

$$V(y) = D[k(y) - 1], \qquad (21)$$

where $D = \dfrac{V(\delta_0)}{k(\delta_0) - 1} \neq 0, \infty$.

Substituting from (21) into (19) gives

$$D[k(y + \delta) - 1] = k(\delta)D[k(y) - 1] + D[k(\delta) - 1].$$

Rearranging and simplifying gives

$$k(y + \delta) = k(\delta)k(y). \qquad (22)$$

We have now reduced the original functional equation into a functional equation with only one unknown function. The functional Equation (23) is the well-known **Cauchy Equation** whose solution is

$$k(y) = e^{cy}, c \neq 0. \qquad (23)$$

From (18), (21), and (23), we have

$$\boxed{U(y) = De^{cy} + b}, \qquad d, c \neq 0. \qquad (24)$$

We have proved the following theorem:

Theorem 13.4: *The general continuous non-constant solutions of (16) on an open interval I of positive length containing the origin for y and δ such that $y + \delta$ is also in I are given either by*

$$U(y) = ay + b, \quad k(y) = 1, d(y) = ay \qquad (25)$$

where $a \neq 0, b(= U(0))$ are arbitrary constants, or by

$$U(y) = \eta e^{\gamma y} + c, \quad k(y) = e^{\gamma y}, d(y) = c(1 - e^{\gamma y}), \qquad (26)$$

where $\eta \neq 0, \gamma \neq 0$ but otherwise arbitrary and c is an arbitrary constant ($U(0) = \eta + \gamma$).

We have now characterized the only two utility functions that satisfy the functional Equation (16) over an interval of positive length. This result was first established by Pfanzagl (1959) and is now referred to as the **Delta Property** (see, for example, Howard and Abbas 2015).

If a decision-maker wants to satisfy the delta property over an interval, then his utility function must be either linear, $U(y) = ay + b$, or exponential, $U(y) = \eta e^{\gamma y} + c$.

Since a utility function is invariant to a linear transformation, we do not need to assess the parameters a, b for the linear utility function. We can simply use $U(y) = y$. We also do not need to assess η or c. But we do need to assess γ. We shall discuss this assessment in detail, but let us first see if there are other solutions to the functional equation.

CASE 2: THE FUNCTIONAL EQUATION IS SATISFIED FOR A PARTICULAR SHIFT VALUE, δ_0 This case is less important than the previous case. We present it merely to show that testing with only a single value of a parameter (or even infinitely man discrete values) is not sufficient to assert the linear or exponential forms. Validation must be done on an interval for these cases.

In this case of a single parameter, additional periodic solutions occur. To illustrate a simple example, let us work on the derivative of the utility function. Taking the first derivative of both sides of (16) gives

$$U'(y + \delta_0) = k(\delta_0) U'(y). \tag{27}$$

To solve (27) for a particular value of δ_0, let $y = \delta_0 t$ and define

$$p(t) = \left(k(\delta_0)\right)^{-t} U'(\delta_0 t). \tag{28}$$

Dividing both sides of (27) by $k(\delta_0)^{t+1}$ gives

$$\left(k(\delta_0)\right)^{-(t+1)} U'(\delta_0(t+1)) = \left(k(\delta_0)\right)^{-t} U'(\delta_0 t). \tag{29}$$

Substituting for $p(t)$ from (28) gives

$$p(t+1) = p(t), \tag{30}$$

which implies that $p(t)$ is a periodic function with period equal to 1. From (28),

$$k^{-y/\delta_0} p(\frac{y}{\delta_0}) = U'(y). \tag{31}$$

Making the substitution $\gamma = -\dfrac{1}{\delta_0} \ln(k)$, gives

$$U'(y) = e^{-\gamma y} p(\frac{y}{\delta_0}). \tag{32}$$

As can be seen from (32), a periodic component has been added to the derivative of the utility function. This also implies that a decision-maker can be invariant to shift amounts $\delta_0, 2\delta_0, 3\delta_0$, etc. and still not have an exponential utility function. Appendix B provides more examples of periodic solutions for invariance with discrete values of transformation parameters.

EXAMPLE 13.10: Numerical Illustration with the Exponential Utility Function

---

Consider again the wealth lottery <0.5, 1000; 0.5, 5000> as evaluated by a decision-maker with an exponential utility function, $U(y) = 1 - e^{-\gamma y}$, where $\gamma = 0.001$. The wealth equivalent of this lottery is $\tilde{y} = \$1675$. If we modify the lottery by an amount $\delta = \$50$, we get the new lottery

$$<0.5, 1050; 0.5, 5050>$$

whose wealth equivalent is (by direct calculation) $\tilde{y}_\delta = \$1725$.

Note that

$$\tilde{y}_\delta = 1675, \ \tilde{y} + \delta = 1675 + 50 = 1725, \text{ and so } \tilde{y}_\delta = \tilde{y} + \delta \text{ as expected.}$$

Furthermore, if we modify the lottery by an amount $\delta = \$-50$, we get the new lottery

$$<0.5, 950; 0.5, 4950>$$

whose wealth equivalent is $\tilde{y}_\delta = \$1625$.

Once again,

$$\tilde{y}_\delta = 1675, \ \tilde{y} + \delta = 1675 - 50 = 1625, \text{ and so } \tilde{y}_\delta = \tilde{y} + \delta \text{ as expected.}$$

Dialogue between Instructor (I) and Student (S)

I: So now we have covered what happens to the wealth equivalent when all lotteries outcomes are modified by the same shift transformation.

S: Just to be sure. If the wealth equivalent increases by the same shift amount then the decision-maker must have a linear or an exponential utility function. Right?

I: Not precisely. Remember, we considered two cases. This must be true for shift amounts on an interval of positive length to assert the linear and exponential solutions.

S: Oh, yes, that is correct. It is not sufficient to test with just one shift amount δ_0. There can be many solutions to this formulation, particularly periodic solutions.

I: Indeed. And while testing with shift amounts on an interval determines the form of the utility function, the periodic solutions do not determine the form of the periodic function. Any periodic function with periodicity equal to one would work.

S: Now, is it sufficient to test with only one lottery, or do we need to test with multiple lotteries to assert the delta property?

I: You only need to test with one lottery (uncertain deal), but you need to test with shift amounts on an interval of positive length to assert linear and exponential utility functions on that domain. If the property holds for one lottery, then it is enough to assert linear or exponential utility functions. Now note that the delta property holds on some interval, but it need not hold on the entire domain.

S: Got it. But, why should we care about this delta property?

I: For a few reasons. You now have a test you can perform to validate the utility function if you are going to use either the linear or exponential. You simply apply the same shift amounts to all lottery outcomes (on an interval) and verify that your wealth equivalent does indeed change by the same shift amount.

S: Okay. That makes sense as a verification test for linear and exponential utility functions.

I: Another reason is that if you have a linear or exponential utility function, and you have already gone through the trouble of calculating the wealth equivalent of a lottery, then you can calculate the wealth equivalent of any lottery that is modified by a shift amount by simply adding a shift amount to the wealth equivalent that you have already calculated without having to redo the calculations from start.

S: That also makes sense, but what if the lottery is modified by other transformations besides shift?

I: We will cover that in Chapters 15 and 16.

S: Okay, what if we apply a shift amount, but the wealth equivalent of the modified lottery is not equal to the previous wealth equivalent plus the shift amount? What if it is higher or lower?

I: Let's discuss that right now.

13.7.3 The Sub-Delta Property

Of course, it is possible that the delta property need not hold either on an interval or for a particular shift amount. So it is natural to ask then, what happens if we add a shift amount to the lottery, and if the wealth equivalent of the modified lottery is always less than the wealth equivalent of the unmodified lottery plus the positive shift amount – that is, what if

$$\boxed{\tilde{y}_\delta < \tilde{y} + \delta}$$

for all lotteries, and for shift amounts defined on an interval of positive length. We refer to this property as the **sub-delta property** and discuss its implications on the utility function below.

By direct substitution into the definition of the wealth equivalent, the sub-delta property implies that

$$U^{-1}\left(\sum_{i=1}^{n} p_i U(y_i + \delta)\right) < U^{-1}\left(\sum_{i=1}^{n} p_i U(y_i)\right) + \delta.$$

Rearranging gives

$$U^{-1}\left(\sum_{i=1}^{n} p_i U(y_i + \delta)\right) - \delta < U^{-1}\left(\sum_{i=1}^{n} p_i U(y_i)\right)$$

Define

$$V_\delta(y) \triangleq U(y + \delta) \text{ and its inverse, } V_\delta^{-1}(t) = U^{-1}(t) - \delta.$$

Substitute to get

$$V_\delta^{-1}\left(\sum_{i=1}^{n} p_i V_\delta(y_i)\right) < U^{-1}\left(\sum_{i=1}^{n} p_i U(y_i)\right).$$

Applying the transformation V_δ to both sides gives

$$\sum_{i=1}^{n} p_i V_\delta(y_i) < V_\delta\left(U^{-1}\left(\sum_{i=1}^{n} p_i U\ (y_i)\right)\right)$$

Note: This step assumed that the utility function is strictly increasing. If it were strictly decreasing, then the sign of the inequality would be reversed.

Define $y_i = U^{-1}(t_i)$, and so $t_i = U(y_i)$. Further define $f = V_\delta(U^{-1})$, and substitute to get

$$\sum_{i=1}^{n} p_i f(t_i) < f\left(\sum_{i=1}^{n} p_i t_i\right).$$

This is again **Jensen's Inequality**, which implies that f is a concave function. Hence,

$$f(t) = V_\delta(U^{-1}(t)) = U\left(U^{-1}(t) + \delta\right) \quad \text{(is a concave function of } t\text{)}.$$

The reverse direction is straightforward to verify. Thus we have proved the following theorem:

Theorem 13.5: *A continuous and strictly increasing utility function, U, satisfies the sub-delta property if and only if*

$$U\left(U^{-1}(t) + \delta\right)$$

is a concave function of t, where t is defined over the range of the utility function.

If the utility function were strictly decreasing, then the results of this theorem would be reversed, and the utility function would satisfy the sub-delta property if and only if the function $U\left(U^{-1}(t) + \delta\right)$ were a convex function of t.

The delta property would require $f(t)$ to be a linear function of t, and so $U\left(U^{-1}(t) + \delta\right)$ is a linear function for utility functions satisfying the delta property.

An equivalent formulation for the sub-delta property that follows from the proof of Theorem 13.5 is that the utility function satisfies the sub-delta property for all lotteries if and only if

$$\boxed{U(y + \delta) = F(U(y), \delta)},$$

where F is a concave function of its first argument.

EXAMPLE 13.11: Logarithmic Utility Function Satisfying the Sub-Delta Property

---

Consider the case of a logarithmic utility function $U(y) = \ln(y)$ that is used to determine the certain equivalent of a lottery. Now suppose the lottery outcomes are incremented by a shift amount δ. To determine the relation between the certain equivalent of the original lottery, $\tilde{y}$, and the certain equivalent of the modified lottery, $\tilde{y}_\delta$, we first calculate the term $U\left(U^{-1}(t)+\delta\right)$. Since $U^{-1}(y)=e^y$, we have

$$U\left(U^{-1}(t)+\delta\right)=\ln\left(e^t+\delta\right),\ \ e^t+\delta>0.$$

The function $\ln\left(e^t+\delta\right)$ is concave when $\delta<0$. To illustrate, its first and second derivatives are $\dfrac{e^t}{e^t+\delta}$ and $\dfrac{\delta e^t}{(e^t+\delta)^2}$, respectively. The second derivative is negative when $z<0$, and so the term $\ln\left(e^t+\delta\right)$ is a concave function on this interval. Therefore, the logarithmic utility function satisfies the sub-delta property for negative shift amounts.

EXAMPLE 13.12: Numerical Illustration with the Logarithmic Utility Function

---

Consider again the wealth lottery <0.5, 1000; 0.5, 5000> as evaluated by a logarithmic decision-maker, with utility function $U(y)=\ln y$. The wealth equivalent of this lottery is $\tilde{y}=\$2236.07$.

If we modify the lottery by an amount $\delta=\$\text{-}50$, we get the new lottery

$$<0.5, 950; 0.5, 4950>$$

whose wealth equivalent is $\tilde{y}_\delta=\$2168.52$. Note that

$$\tilde{y}_\delta=2168.52,\ \tilde{y}+\delta=2236.07-50=2186.07,\text{ and so }\tilde{y}_\delta<\tilde{y}+\delta\text{ as expected.}$$

13.7.4 The Super-Delta Property

We can also define the **Super-Delta** property when the increase in the wealth equivalent is larger than the shift amount, i.e., the wealth equivalent of the modified lottery is larger than the wealth equivalent of the original lottery plus the shift amount,

$$\boxed{\tilde{y}_\delta > \tilde{y}+\delta}$$

and, using similar analysis, we state, without proof, the following.

Theorem 13.6: *A continuous and strictly increasing utility function satisfies the super delta property if and only if*

$$U\left(U^{-1}(t)+\delta\right)$$

is a convex function of t, where t is defined over the range of the utility function.

EXAMPLE 13.13: Logarithmic Utility Function Satisfying the Super-Delta Property

---

As we have seen in Example 13.11, $U\left(U^{-1}(t)+\delta\right)=\ln\left(e^t+\delta\right),\ e^t+\delta>0$ for a logarithmic utility function. This function is convex for $z>0$. Therefore, the logarithmic utility function satisfies the super-delta property for positive shift amounts.

EXAMPLE 13.14:   Numerical Illustration with the Logarithmic Utility Function

---

Consider again the wealth lottery <0.5, 1000; 0.5, 5000> as evaluated by a logarithmic decision-maker. As we have seen, the wealth equivalent of this lottery is $\tilde{y} = \$2{,}236.07$.

If we modify the lottery by an amount \$50, we get the new lottery

$$<0.5, 1050; 0.5, 5050>$$

whose wealth equivalent is $\tilde{y}_\delta = \$2{,}302.72$. Note that

$$\tilde{y}_\delta = 2302.72,\ \tilde{y} + \delta = 2236.07 + 50 = 2286.07,\ \text{and so } \tilde{y}_\delta > \tilde{y} + \delta \text{ as expected.}$$

Figure 13.5 plots the values of $\tilde{y}_\delta - \tilde{y}$ vs. δ for both the exponential and logarithmic utility functions for the lottery <0.5, 1000; 0.5, 5000>. Note how the $\tilde{y}_\delta - \tilde{y}$ curve corresponds to the straight line $y = \delta$ for the exponential utility function. For the logarithmic utility function, $\tilde{y}_\delta - \tilde{y} < \delta$ for $\delta < 0$ and $\tilde{y}_\delta - \tilde{y} > \delta$ for $\delta > 0$.

If the utility function is strictly increasing and if the change in wealth equivalent is always lower than the previous wealth equivalent plus the shift amount, $\tilde{y}_\delta < \tilde{y} + \delta$, then we have the sub-delta property, and this implies that $U\left(U^{-1}(t) + \delta\right)$ is a concave function. If it is higher, i.e., $\tilde{y}_\delta > \tilde{y} + \delta$, Then we have the super-delta property. For a strictly increasing utility function this means that $U\left(U^{-1}(t) + \delta\right)$ is a convex function. The reverse direction also holds. For a given utility function, you can determine the shape of the function $U\left(U^{-1}(t) + \delta\right)$. If it is concave, then $\tilde{y}_\delta < \tilde{y} + \delta$ for all lotteries and if it is convex, then $\tilde{y}_\delta > \tilde{y} + \delta$.

If $U\left(U^{-1}(t) + \delta\right)$ is S-shaped or of it is neither fully concave nor fully convex on the whole domain, the wealth equivalent will sometimes be higher or lower than the previous wealth equivalent plus the shift amount depending on the actual lottery.

In the next chapter we present additional properties of the utility function related to shift transformations on the lottery outcomes. The analysis in this chapter required only continuity and monotonicity of the utility function, but it did not require differentiability. In the next chapter, we shall revisit the concepts introduced here and relate

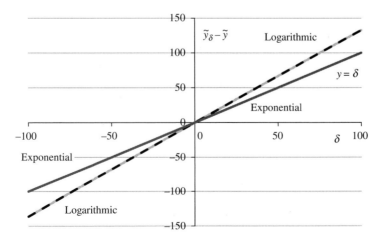

Figure 13.5. A plot of $\tilde{y}_\delta - \tilde{y}$ vs. δ for exponential and logarithmic utility functions.

them to the risk aversion function, where the utility function is assumed to be twice-continuously differentiable.

In Chapter 15 we discuss other transformations besides shift.

13.8 SUMMARY

The wealth equivalent of the wealth lottery $< p_1, y_1; ...; p_n, y_n >$ is

$$\tilde{y} = U^{-1}\left(\sum_{i=1}^{n} p_i U(y_i)\right).$$

The certain equivalent of the lottery $< p_1, x_1; ...; p_n, x_n >$ for a decision-maker with initial wealth w is

$$\tilde{x} = U^{-1}\left(\sum_{i=1}^{n} p_i U(x_i + w)\right) - w.$$

The certain equivalent and wealth equivalent are invariant to a linear transformation on the utility function.

Effects of the Shape of the Utility Function on the Certain Equivalent are

- If the utility function is concave, then the certain equivalent is below the mean value of the lottery (and the wealth equivalent is below the mean wealth level).
- If the utility function is linear an interval, then the certain equivalent is equal to the mean value of a lottery over this interval (and the wealth equivalent is equal to the mean wealth level).
- If the utility function is convex, then the certain equivalent is larger than the mean (and the wealth equivalent is larger the mean wealth level).

If one utility function, U_B, always provides a lower wealth equivalent than that obtained by a utility function, U_A, then U_B must be a concave function of U_A, $U_B(y) = f[U_A(y)]$, where f is a concave function.

The Delta Property

- Important Functional Equation: $U(x + \delta) = k(\delta)U(x) + d(\delta)$
- The solutions to this functional equation for values of δ on an interval are: linear and exponential
- The solutions to this functional equation for discrete values of δ
- An equivalent formulation for the delta property is that $U\left(U^{-1}(t) + \delta\right)$ is a linear function of t.

The Sub-Delta Property

- The term $U\left(U^{-1}(t) + \delta\right)$ is a concave function of t.
- An equivalent formulation for the Sub-Delta property:

$$U(y+\delta) = F(U(y),\delta),$$

where F is a concave function of its first argument.

The Super-Delta Property

- The term $U\left(U^{-1}(t)+\delta\right)$ is a convex function of t.
- An equivalent formulation for the Super-Delta property:

$$U(y+\delta) = F(U(y),\delta),$$

where F is a convex function of its first argument.

Food for Thought

Consider two extensions of what we have covered in this chapter:

1. Suppose we have a wealth lottery with wealth equivalent, $\tilde{y}$, and we modify the outcomes of this lottery by a general monotone transformation $g(y)$. Suppose further that the wealth equivalent of the new lottery is $g(\tilde{y})$. What does this imply about the utility function? We answer this question in Chapter 16.
2. What happens if we add more terms to the right hand side of the equation

$$U(x+\delta) = k(\delta)U(x)+d(\delta)?$$

For example, what if we have a new equation of the form

$$U(x+\delta) = k_1(\delta)U_1(x)+k_2(\delta)U_2(x)+d(\delta)?$$

What are the utility functions that satisfy this new functional equation? And what does it imply about the decision-maker's preferences? We answer this question in Chapter 18.

ADDITIONAL READINGS

Abbas, A. E. 2007. Invariant utility functions and certain equivalent transformations. *Decision Analysis* 4(3): 17–31.

Abbas, A. E. and J. Aczél. 2010. The role of some functional equations in decision analysis. *Decision Analysis* 7(2): 215–228.

Abbas, A. E., J. Aczél, and J. Chudziak, 2009. Invariance formulations for multiattribute utility functions under shift transformations. *Results in Mathematics* 54: 1–13.

Arrow, K. J. 1965. *The Theory of Risk Aversion. Lecture 2 in Aspects of the Theory of Risk-Bearing.* Yrjo Jahnssonin Saatio, Helsinki.

Howard, R. A. and A. E. Abbas. 2015. *Foundations of Decision Analysis.* Pearson, New York.

Pfanzagl, J. 1959. A general theory of measurement: Applications to utility. *Naval Research Logistics.* 6: 283–294.

Pratt, J. 1964. Risk aversion in the small and in the large. *Econometrica* 32: 122–136.

CHAPTER 14

The Risk Aversion Function

Chapter Concepts

- The absolute risk aversion function
- The risk tolerance
- The risk premium
- Implications of
  - The sign of the risk aversion function on the valuation of lotteries
  - The magnitude of the risk aversion function on the valuation of lotteries
  - The inclination of the aversion function (increasing or decreasing with wealth)
- Relating the shape of the risk aversion function to the delta, sub-delta, and super-delta properties
- Answering some of the questions posed in Chapter 13 when the utility function is twice continuously differentiable

14.1 INTRODUCTION

This chapter revisits several key concepts that were introduced in Chapter 13, but addresses them using a different notion: *the risk aversion function*. The idea of the risk aversion function was introduced by Kenneth Arrow (1965) and John Pratt (1964) (independently). The risk aversion function uses the second and first derivatives of the utility function for its calculation, and therefore we will focus on utility functions that are twice continuously differentiable. These additional "smooth" properties of the utility function enable simpler answers to some of the questions that were posed in the previous chapter. Recall that in the previous chapter, we merely required the utility function to be continuous and strictly monotonic. As we shall see, solutions that require only continuity are in general more complex than solutions that allow for differentiability.

The sign of the risk aversion function can answer the question we posed in Chapter 13: Which is higher, the mean or the certain equivalent? The magnitude of the risk aversion function provides an approximation for the difference between the certain equivalent and the mean value of a lottery. The inclination of the risk aversion function (increasing or decreasing with wealth) determines the behavior of the certain equivalent of a lottery when the outcomes of the lottery are incremented by a fixed shift amount. All of these questions were posed in the previous chapter. The risk aversion function can answer these questions when the utility function is twice continuously differentiable.

14.2 THE ABSOLUTE RISK AVERSION FUNCTION

Arrow (1965) and Pratt (1964) introduced the absolute risk aversion function, $\gamma(x)$, as the negative ratio of the second to the first derivative of the utility function.

Definition: The **risk aversion function** is defined as

$$\gamma(x) = -\frac{U''(x)}{U'(x)} = -\frac{d}{dx}\ln\left(U'(x)\right),$$

where $U'(x)$ is the first derivative of the utility function and $U''(x)$ is its second derivative.

This formulation provides two equivalent definitions of the risk aversion function:

(i)   the negative ratio of the second to the first derivative of the utility function, and
(ii) the negative derivative of the logarithm of the first derivative of the utility function.

Both of these definitions will be useful in various settings.

An important subtlety we need to observe here is that the risk aversion function can (theoretically) be infinite. For example, if the first derivative is zero and the second derivative is non-zero. If the utility function is reverse S-shaped, for example, then there will be an inflection point at which the first derivative can be zero and the second derivative may or may not be zero.

The units of the risk aversion function are those of $\frac{1}{x}$ because it is a ratio of the second derivative to the first derivative with respect to x.

Definition: The *risk tolerance* is the reciprocal of the risk aversion function

$$\rho(x) = \frac{1}{\gamma(x)} = -\frac{U'(x)}{U''(x)}.$$

The units of risk tolerance are those of x, the argument of the utility function. For example, if x is in dollars, then the risk tolerance has units of dollars.

The risk aversion function has many interesting properties.

14.2.1 Effects of Applying a Linear Transformation to the Utility Function

Like the certain equivalent, the risk aversion function is also invariant to a linear transformation on the utility function. To illustrate, note that a utility function $U_1(x)$, where

$$U_1(x) = aU(x) + b, \ a \neq 0$$

has the same risk aversion function as the utility function, $U(x)$, because $U_1'(x) = aU'(x)$ and $U_1''(x) = aU''(x)$ so

$$\Rightarrow -\frac{U_1''(x)}{U_1'(x)} = -\frac{aU''(x)}{aU'(x)} = -\frac{U''(x)}{U'(x)} = \gamma(x).$$

We highlight this result below.

> *Applying a linear transformation to a utility function does not change its risk aversion function.*

Because the risk aversion function will be used in approximating the certain equivalent of a lottery, and because the certain equivalent is invariant to a linear transformation applied to the utility function, it seems plausible that the risk aversion function be invariant to a linear transformation on the utility function.

14.2.2 Utility Functions Exhibiting Constant Absolute Risk Aversion

The following examples illustrate the calculation of the risk aversion function for exponential and linear utility functions.

EXAMPLE 14.1: Exponential Utility Function (Constant Absolute Risk Aversion)

Consider the exponential utility function

$$U(x) = a + be^{-\gamma x},$$

where a, γ are constants and $b \neq 0$ but otherwise constant.
     By direct derivatives, $U'(x) = -\gamma be^{-\gamma x}, U''(x) = \gamma^2 be^{-\gamma x}$. The risk aversion function is then

$$\gamma(x) = -\frac{U''(x)}{U'(x)} = -\frac{\gamma^2 be^{-\gamma x}}{-\gamma be^{-\gamma x}} = \gamma = \text{constant.}$$

This is why exponential utility functions are said to have constant absolute risk aversion (CARA). Constant absolute risk aversion means the risk aversion will not change with wealth.

Note: Because the risk aversion function is invariant to a linear transformation on the utility function, we could have calculated this function directly using the utility function $U(x) = e^{-\gamma x}$ to get $U'(x) = -\gamma e^{-\gamma x}, U''(x) = \gamma^2 e^{-\gamma x}$, and so $\gamma(x) = \gamma$ as predicted.

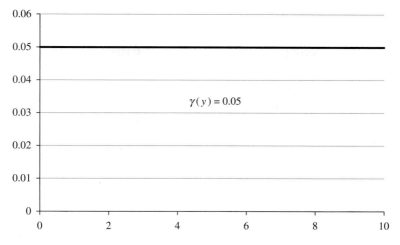

Figure 14.1. Constant absolute risk aversion $\gamma = 0.05$.

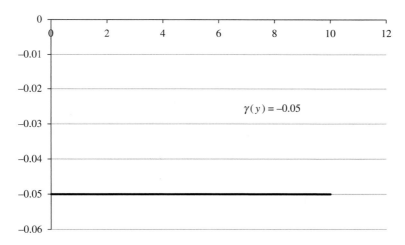

Figure 14.2. Constant absolute risk aversion $\gamma = -0.05$.

Figures 14.1 and 14.2 plot the absolute risk aversion functions for an exponential utility function with risk aversion coefficients $\gamma = 0.05$ and $\gamma = -0.05$, respectively.

EXAMPLE 14.2: Linear Utility Function (Constant Absolute Risk Aversion)

Consider the utility function

$$U(x) = a + bx,$$

where $a, b \neq 0$ are constants.

Taking the first and second derivatives, we get $U'(x) = b, U''(x) = 0$. The risk aversion function is then

$$\gamma(x) = -\frac{U''(x)}{U'(x)} = 0 = \text{constant}.$$

This is why linear utility functions are also said to have constant absolute risk aversion (CARA).

Note, again, that the constants a, b do not play a role in the risk aversion function calculation.

We will prove in Section 14.3 that there are no other functions that have constant absolute risk aversion. These two utility functions are also the only utility functions that satisfy the delta property on an interval of positive length. We shall see that there is a relation between the risk aversion function and the delta property. The risk aversion function plays a role in the valuation of lotteries and the calculation of the certain equivalent. Constant absolute risk aversion means that the risk aversion, and hence the certain equivalent, does not change with wealth. The delta property also meant that the wealth equivalent of a modified lottery is equal to the wealth equivalent of the unmodified lottery plus the wealth amount. Therefore, the certain equivalent is also constant with wealth.

Let us now calculate the risk aversion function for other utility functions.

14.2.3 Utility Functions Exhibiting Increasing or Decreasing Absolute Risk Aversion Functions

---

EXAMPLE 14.3: Logarithmic Utility Function (Decreasing Absolute Risk Aversion)

Consider the logarithmic utility function

$$U(x) = a + b\ln(x + w), \quad x > -w$$

where a, w are constants and $b \neq 0$ but otherwise constant.

By taking the first and second derivatives, we get $U'(x) = \dfrac{b}{x+w}, U''(x) = \dfrac{-b}{(x+w)^2}$.
The risk aversion function is then

$$\gamma(x) = -\frac{U''(x)}{U'(x)} = \frac{1}{x+w}.$$

Note that the risk aversion function is a decreasing function of the wealth, w. The logarithmic utility function is said to have decreasing risk aversion with wealth. As we shall see, this implies that the decision-maker will value uncertain deals more as he gets richer.

As in the earlier examples, neither of the constants a or b played a role in the risk aversion function, because a linear transformation of a utility function does not affect its risk aversion function.

Figure 14.3 plots the absolute risk aversion functions for a logarithmic utility function in terms of the total wealth.

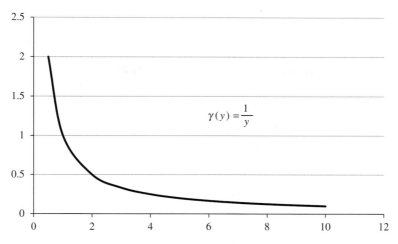

Figure 14.3. Decreasing absolute risk aversion for logarithmic utility function.

EXAMPLE 14.4: Power Utility Function

Consider the power utility function

$$U(x) = (x+w)^{\alpha}, \quad x > -w$$

where w is a constant and $\alpha \neq 0,1$.

By direct derivatives, $U'(x) = \alpha(x+w)^{\alpha-1}, U''(x) = \alpha.(\alpha-1)(x+w)^{\alpha-2}$.

The risk aversion function is then

$$\gamma(x) = -\frac{U''(x)}{U'(x)} = -\frac{\alpha.(\alpha-1)(x+w)^{\alpha-2}}{\alpha(x+w)^{\alpha-1}} = \frac{1-\alpha}{x+w}.$$

We have two cases:

Case 1: When $\alpha < 1$, the risk aversion function is a decreasing function of the wealth, w.

Case 2: When $\alpha > 1$, the risk aversion function is an increasing function of wealth.

EXAMPLE 14.5: Linear Risk Tolerance Utility Function

Consider the utility function $U(x) = (\rho + \eta x)^{1-\frac{1}{\eta}}$, where $\eta \neq 0,1$.

Taking the first derivative gives $U'(x) = (1 - \frac{1}{\eta})(\rho + \eta x)^{-\frac{1}{\eta}}$. The risk aversion function

is then

$$\gamma(x) = -\frac{d}{dx}\ln(U'(x)) = -\frac{d}{dx}\ln(1-\frac{1}{\eta}) + \frac{1}{\eta}\frac{d}{dx}\ln(\rho + \eta x) = \frac{1}{\rho + \eta x}.$$

This implies that the risk tolerance (the reciprocal of the risk aversion function) is a linear function

$$\rho(x) = \frac{1}{\gamma(x)} = \rho + \eta x.$$

This is why the utility function is known as linear risk tolerance utility function. For the special case when $\eta = 0$, we have $\gamma(x) = \frac{1}{\rho}$, which is the case of constant absolute risk aversion. When $\eta = 1$, we have $\gamma(x) = \frac{1}{\rho + x}$, which is the case of the logarithmic utility function.

EXAMPLE 14.6: Gaussian Utility Function

Gaussian utility functions have the same functional form as a Gaussian cumulative distribution function. Since the Gaussian cumulative distribution does not have a closed form expression, they are usually expressed in terms of their derivative,

$$U'(x) = \frac{1}{\sigma\sqrt{2\pi}} e^{-\frac{(x-\mu)^2}{2\sigma^2}}, \sigma \neq 0.$$

The risk aversion function can be calculated quite simply using the expression

$$\gamma(x) = -\frac{d}{dx}\ln\left(U'(x)\right) = -\frac{d}{dx}\left(\ln\frac{1}{\sigma\sqrt{2\pi}} - \frac{(x-\mu)^2}{2\sigma^2}\right) = \frac{x-\mu}{\sigma^2}.$$

The risk aversion function of a Gaussian utility function is therefore increasing with wealth. As we shall see, this implies that the decision-maker values uncertain deals by a smaller amount as he gets richer. Furthermore, it implies that the decision-maker changes from being risk seeking to risk averse at a wealth amount μ. These properties are somewhat undesirable and are why the Gaussian utility function has not found much popularity.

14.3 DETERMINING THE UTILITY FUNCTION FROM THE ABSOLUTE RISK AVERSION FUNCTION

The utility function can be determined uniquely (up to a linear transformation) from the risk aversion function. Recall that

$$\gamma(x) = -\frac{d}{dx}\ln\left(U'(x)\right).$$

Direct integration gives

$$\ln\left(U'(x)\right) = -\int \gamma(x)dx + c.$$

Taking the exponential of both sides relates the first derivative of the utility function to the risk aversion function as

$$U'(x) = ae^{-\int \gamma(x)dx}, \quad a = e^c \neq 0.$$

Integrating gives a relation between the utility function and the risk aversion function as

$$\boxed{U(x) = a\int e^{-\int \gamma(x)dx} dx + b}.$$

EXAMPLE 14.7: Utility Functions from Constant Absolute Risk Aversion

---

We consider two cases:

Case 1: $\gamma(x) = \gamma \neq 0$.

By direct substitution, $\int \gamma(x)dx = \int \gamma dx = \gamma x + c$, where c is an arbitrary constant.

$$U(x) = a\int e^{-(\gamma x+c)}dx + b = \alpha e^{-\gamma x} + b,$$

where $\alpha = \dfrac{ae^{-c}}{-\gamma} \neq 0$ but otherwise arbitrary, and b is an arbitrary constant.

Case 2: $\gamma(x) = 0$.

By direct substitution, $\int \gamma(x)dx = \int 0.\, dx = c$, where c is an arbitrary constant.

$$U(x) = a\int c\, dx + b = acx + b,$$

where $a \neq 0$ but otherwise arbitrary, and b is an arbitrary constant.

EXAMPLE 14.8: Logarithmic Utility Function from Its Risk Aversion Function

---

Consider the case where

$$\gamma(x) = \frac{1}{x+w}, \quad w > -x.$$

By direct substitution,

$$\int \gamma(x)dx = \int \frac{1}{x+w}dx = \ln(x+w) + c,$$

where c is a constant. This implies that

$$U(x) = a\int e^{-[\ln(x+w)+c]}dx + b = \alpha \int e^{\ln\left(\frac{1}{x+w}\right)}dx + b = \alpha\ln(x+w) + b,$$

where $\alpha = \dfrac{ae^{-c}}{-\gamma}$.

Dialogue between Instructor (I) and Student (S)

I: Is everyone clear on the definition of the risk aversion function and how it relates to the utility function?

S: Yes, we have discussed two equivalent definitions. The first is the negative ratio of the second to the first derivative of the utility function. The second is the negative derivative of the logarithm of the derivative of the utility function.

I: Yes. We have seen that one definition might be easier to use than the other in certain circumstances.

S: For example, when we related the risk aversion function to the form of the utility function, we used the second definition.

I: We have also seen that we can derive the form of the utility function from the risk aversion function up to a linear transformation on the utility function.

S: What does this risk aversion function mean?

I: As we shall see below, among other things, the risk aversion function determines the deviation of the mean value of a lottery from the certain equivalent. Remember the exponential utility function, where the risk aversion function was a constant? It had a meaning in terms of indifference lottery assessments. Think of the risk aversion function at any point as the answer to the same lottery question for small deviations around the initial wealth.

S: For exponential utility functions, it was useful to determine the risk aversion because it was a parameter that helped determine the form of the utility function. Why do we care about this interpretation for other utility functions?

I: Great question. Let's discuss this right now.

14.4 APPLICATIONS OF THE ABSOLUTE RISK AVERSION FUNCTION

Three characteristics of the risk aversion function are of interest, and each has its implications for the calculation of the certain equivalent:

1. its sign,
2. its magnitude, and
3. its inclination (constant, increasing or decreasing with wealth).

14.4.1 The Sign of the Risk Aversion Function

Definitions: Risk Attitude based on the Sign of the Risk Aversion Function
    The sign of the risk aversion function, $\gamma(x)$, determines the decision-maker's attitude toward risk: *he is said to be risk averse if the risk aversion function is positive; risk seeking if the risk aversion function is negative; and risk neutral if the risk aversion function is zero.*

Relating the Sign of the Risk Aversion Function to the Shape of the Utility Function
The sign of the risk aversion function also relates to the shape of the utility function. Consider a strictly increasing utility function. If, in addition:

1. The utility function is strictly concave, then the second derivative is negative and the first derivative is positive, therefore the risk aversion function is negative.
2. The utility function is strictly convex, then the second derivative is positive and so the risk aversion function is positive.
3. The utility function is linear, then the second derivative is zero and so the risk aversion function is zero.

Relating the Sign of the Risk Aversion Function to Bounds on the Certain Equivalent

Based on our discussions in the last chapter, where the shape of the utility function determines the relation between the certain equivalent of a lottery and its mean, we can deduce that:

1. If the utility function is concave, then the risk aversion function is positive and the certain equivalent is below the mean value (the mean is an upper bound on the certain equivalent if the risk aversion function is positive).
2. If the risk aversion function is zero at an interval, then the utility function is linear over that interval and the certain equivalent is equal to the mean value of a lottery over this interval.
3. If the risk aversion function is negative over an interval, then the utility function is convex over this interval and the certain equivalent is higher than the mean over this interval.

14.4.2 The Magnitude of the Risk Aversion Function

As we have seen, the sign of the risk aversion function is sufficient to characterize the relative magnitude between the certain equivalent and the mean value of the lottery. The actual magnitude of the risk aversion function can provide an approximation for the deviation between the certain equivalent of a lottery (of small variance) and its mean value.

Definition: Risk Premium

The difference between the mean value of a lottery and the certain equivalent,

$$\mu - \tilde{x}$$

is known as the **risk premium**.

Pratt (1964) showed that the risk premium of a lottery can be approximated in terms of the risk aversion function (for small risks) using the quantity,

$$\text{Risk premium} = \mu - \tilde{x} = \frac{\sigma^2}{2} \gamma(\mu),$$

where σ^2 is the variance of the lottery (which is non-negative) and $\gamma(\mu)$ is the magnitude of the risk aversion function evaluated at the mean value of the lottery, i.e., for lotteries of small variance, we can write as an approximation

$$\boxed{\tilde{x} \approx \mu - \frac{\sigma^2}{2} \gamma(\mu).}$$

The exact expression is

$$\boxed{\tilde{x} = \mu - \frac{\sigma^2}{2} \gamma(\mu) + o(\sigma^2) \text{ as } \sigma^2 \to 0,}$$

where $o(\sigma^2)$ is a function that approaches zero faster than σ^2 as $\sigma^2 \to 0$, i.e.,

$$\lim_{\sigma^2 \to 0} \frac{o(\sigma^2)}{\sigma^2} = 0.$$

Proof: see the appendix at the end of this chapter.

Recap: If $\gamma(x)$ is positive, then the decision-maker is said to be risk averse, and the approximate expression of the certain equivalent shows that the mean is larger than the expected value of the lottery (as expected); the mean is the upper bound on the certain equivalent of a lottery for concave utility functions.

If $\gamma(x)$ is negative, the decision-maker is said to be risk seeking, and the approximate expression shows that the mean is smaller than the expected value of the lottery; the mean is a lower bound on the certain equivalent for convex utility functions.

If $\gamma(x) = 0$, the decision-maker is risk neutral, and the approximate expression shows that the mean is equal to the expected value of the lottery. The mean is equal to the certain equivalent for linear utility functions.

Dialogue between Instructor (I) and Student (S)
I: So now you see the importance of the magnitude of the risk aversion function?
S: Yes, it can be used to approximate the risk premium and therefore to relate the certain equivalent to the mean value of the lottery. Just to verify, in the previous expression $\gamma(\mu)$ means the magnitude of the risk aversion function when calculated at the mean value of the lottery. Correct?
I: Correct. For lotteries of small enough variance, this is the case. As the variance gets closer to zero, the certain equivalent approaches the mean less half the risk aversion multiplied by the variance. This is why John Pratt called his paper "Risk Aversion in the Small and in the Large." In the small because lotteries that are local around the mean will have a small variance and would satisfy this relation.
S: What about in the large?
I: For this he was referring to the sign of the risk aversion on the whole domain, particularly the effects of concavity, convexity, and linearity of the utility function and how that provides bounds on the certain equivalent in relation to the mean.
S: Very interesting.
I: By the way, if you have a Gaussian utility lottery and an exponential utility function, the previous expression becomes exact. That is, $\tilde{x} = \mu - \frac{\sigma^2}{2}\gamma(\mu)$. We shall take a few examples now and also see that in more detail.

EXAMPLE 14.9: An Exponential Utility Function and an Exponential Lottery

---

Consider an exponential lottery with a density $f(y) = \lambda e^{-\lambda y}$, $y \geq 0$. The mean of this exponential lottery is $\mu = \frac{1}{\lambda}$, and the variance is $\sigma^2 = \left(\frac{1}{\lambda}\right)^2$. The exponential utility function with risk aversion coefficient, γ, is $U(y) = -e^{-\gamma y}$.

The expected utility of this exponential lottery is

$$E[U(y)] = \int_0^\infty f(y)U(y)dy = \int_0^\infty \lambda e^{-\lambda y}[-e^{-\gamma y}]dy = -\lambda \int_0^\infty e^{-(\lambda+\gamma)y}dy = \frac{-\lambda}{\lambda+\gamma}.$$

By definition, $E[U(y)] = U(\tilde{y})$, and so the exact expression for the certain equivalent is

$$\tilde{y} = U^{-1}(E[U(y)]) = \frac{-1}{\gamma}\ln\left(\frac{\lambda}{\lambda+\gamma}\right).$$

On the other hand, the approximate expression for the certain equivalent is

$$\tilde{y} \approx \mu - \frac{\sigma^2}{2}\gamma(\mu) = \frac{1}{\lambda} - \frac{1}{2}\left(\frac{1}{\lambda}\right)^2 \gamma = \frac{1}{\lambda}\left(1 - \frac{\gamma}{2\lambda}\right).$$

Tables 14.1 and 14.2 plot the exact and approximate certain equivalents as a function of the parameters λ and γ. Note that as λ increases, the variance decreases and the two expressions get closer. Table 14.3 displays the difference in wealth equivalents calculated by the different expressions.

Table 14.1. Exact wealth equivalent for different values of γ and λ

		λ									
		0.1	0.2	0.3	0.4	0.5	0.6	0.7	0.8	0.9	1
	0.01	9.531	4.879	3.279	2.469	1.980	1.653	1.418	1.242	1.105	0.995
	0.02	9.116	4.766	3.227	2.440	1.961	1.639	1.409	1.235	1.099	0.990
	0.03	8.745	4.659	3.177	2.411	1.942	1.626	1.399	1.227	1.093	0.985
	0.04	8.412	4.558	3.129	2.383	1.924	1.613	1.389	1.220	1.087	0.981
γ	0.05	8.109	4.463	3.083	2.356	1.906	1.601	1.380	1.212	1.081	0.976
	0.06	7.833	4.373	3.039	2.329	1.889	1.589	1.371	1.205	1.076	0.971
	0.07	7.580	4.287	2.996	2.304	1.872	1.576	1.362	1.198	1.070	0.967
	0.08	7.347	4.206	2.955	2.279	1.855	1.565	1.353	1.191	1.064	0.962
	0.09	7.132	4.128	2.915	2.255	1.839	1.553	1.344	1.185	1.059	0.958
	0.1	6.931	4.055	2.877	2.231	1.823	1.542	1.335	1.178	1.054	0.953

Table 14.2. Approximate wealth equivalent for different values of γ and λ

		λ									
		0.1	0.2	0.3	0.4	0.5	0.6	0.7	0.8	0.9	1
	0.01	9.500	4.875	3.278	2.469	1.980	1.653	1.418	1.242	1.105	0.995
	0.02	9.000	4.750	3.222	2.438	1.960	1.639	1.408	1.234	1.099	0.990
	0.03	8.500	4.625	3.167	2.406	1.940	1.625	1.398	1.227	1.093	0.985
	0.04	8.000	4.500	3.111	2.375	1.920	1.611	1.388	1.219	1.086	0.980
γ	0.05	7.500	4.375	3.056	2.344	1.900	1.597	1.378	1.211	1.080	0.975
	0.06	7.000	4.250	3.000	2.313	1.880	1.583	1.367	1.203	1.074	0.970
	0.07	6.500	4.125	2.944	2.281	1.860	1.569	1.357	1.195	1.068	0.965
	0.08	6.000	4.000	2.889	2.250	1.840	1.556	1.347	1.188	1.062	0.960
	0.09	5.500	3.875	2.833	2.219	1.820	1.542	1.337	1.180	1.056	0.955
	0.1	5.000	3.750	2.778	2.188	1.800	1.528	1.327	1.172	1.049	0.950

Table 14.3. Difference between approximate and exact wealth equivalent for different values of γ and λ

	$\boxed{\lambda}$									
	0.1	**0.2**	**0.3**	**0.4**	**0.5**	**0.6**	**0.7**	**0.8**	**0.9**	**1**
0.01	0.031	0.004	0.001	0.001	0.000	0.000	0.000	0.000	0.000	0.000
0.02	0.116	0.016	0.005	0.002	0.001	0.001	0.000	0.000	0.000	0.000
0.03	0.245	0.034	0.010	0.004	0.002	0.001	0.001	0.001	0.000	0.000
0.04	0.412	0.058	0.018	0.008	0.004	0.002	0.001	0.001	0.001	0.001
0.05	0.609	0.088	0.027	0.012	0.006	0.004	0.002	0.002	0.001	0.001
0.06	0.833	0.123	0.039	0.017	0.009	0.005	0.003	0.002	0.002	0.001
0.07	1.080	0.162	0.052	0.023	0.012	0.007	0.004	0.003	0.002	0.002
0.08	1.347	0.206	0.066	0.029	0.015	0.009	0.006	0.004	0.003	0.002
0.09	1.632	0.253	0.082	0.036	0.019	0.011	0.007	0.005	0.003	0.003
0.1	1.931	0.305	0.099	0.044	0.023	0.014	0.009	0.006	0.004	0.003

The row labels are indexed by $\boxed{\gamma}$.

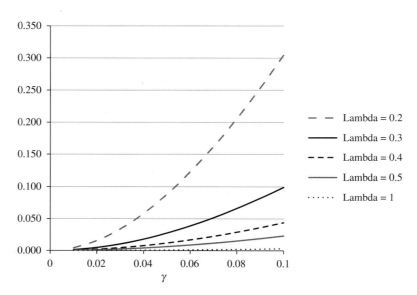

Figure 14.4. Difference between exact and approximate wealth equivalents.

Figure 14.4 plots the difference in wealth equivalents as a function of γ for different values of λ. Note that the error increases as γ increases and decreases as λ increases.

EXAMPLE 14.10: A Logarithmic Utility Function and an Exponential Lottery

Consider a decision-maker with a logarithmic utility function, $U(y) = \ln(y + \alpha)$, facing the exponential lottery with density $f(y) = \lambda e^{-\lambda y}$, $y \geq 0$. As we have seen, the mean of this lottery is $\mu = \dfrac{1}{\lambda}$, and its variance is $\sigma^2 = \left(\dfrac{1}{\lambda}\right)^2$. The risk aversion function is

$$\gamma(y) = -\frac{U''(y)}{U'(y)} = \frac{1}{y + \alpha}.$$ When evaluated at the mean value of the lottery,

$$\gamma(\mu) = \frac{1}{\mu + \alpha} = \frac{1}{1/\lambda + \alpha} = \frac{\lambda}{1 + \alpha\lambda}.$$

The approximate expression for the certain equivalent is

$$\tilde{y} \approx \mu - \frac{\sigma^2}{2}\gamma(\mu) = \frac{1}{\lambda} - \frac{1}{2}\left(\frac{1}{\lambda}\right)^2 \frac{\lambda}{1 + \alpha\lambda}.$$

EXAMPLE 14.11: Exponential Decision-Maker with a Gaussian Lottery

---

Consider a Gaussian lottery $f(y) = \frac{1}{\sigma\sqrt{2\pi}}e^{-\frac{(y-\mu)^2}{2\sigma^2}}, y \in R$ and a decision-maker with an exponential utility function, $U(y) = -e^{-\gamma y}$.

The expected utility of this lottery is

$$E[U(y)] = \int_{-\infty}^{\infty} f(y)U(y)dy = -\int_{-\infty}^{\infty} e^{-\gamma y}\frac{1}{\sigma\sqrt{2\pi}}e^{-\frac{(y-\mu)^2}{2\sigma^2}}dy = e^{-\gamma[\mu - \gamma\frac{\sigma^2}{2}]}.$$

Note that the previous integral is just the exponential transform of a Gaussian function. The wealth equivalent is then

$$\tilde{y} = \frac{-1}{\gamma}\ln(E[U(y)]) = \mu - \frac{\sigma^2}{2}\gamma.$$

Note that the approximate and exact expressions for the wealth equivalent match in this case.

Dialogue between Instructor (I) and Student (S)

S: So now we see some immediate applications of the risk aversion function. The magnitude really is important after all.

I: So is the sign, and we shall also discuss the inclination. But there are still more applications of the magnitude that we have not discussed.

S: Like what?

I: Remember in Chapter 13 when we asked the question about what it implies when one certain equivalent is always larger than the other for all lotteries?

S: Yes.

I: We can answer that using the risk aversion function now. The magnitude of one risk aversion function must be larger than the other. We will prove that now.

14.5 WHEN ONE CERTAIN EQUIVALENT IS ALWAYS LOWER THAN ANOTHER

If one decision-maker, call him A, values all uncertain deals higher than does another decision-maker, B, with the same initial wealth, what does this imply about his risk aversion function?

We have seen in Chapter 13 that this implies that the utility function that provides a lower valuation must be a concave function of the other, i.e.,

$$U_B(y) = f[U_A(y)], \tag{1}$$

where f is a concave function. We now investigate the implications this has on the risk aversion functions.

Taking the first and second derivatives of both sides of (1) gives

$$U'_B(y) = f'(U_A(y))U'_A(y)$$

and

$$U''_B(y) = f''(U_A(y))U'_A(y)^2 + f'(U_A(y))U''_A(y).$$

This implies that

$$\gamma_B(y) = -\frac{U''_B(y)}{U'_B(y)} = -\frac{f''(U_A(y))}{f'(U_A(y))}U'_A(y) + \gamma_A(y).$$

Rearranging gives

$$\gamma_B(y) - \gamma_A(y) = -\frac{f''(U_A(y))}{f'(U_A(y))}U'_A(y). \tag{2}$$

Because U_A, U_B are strictly increasing, $f'(U_A(y))$ and $U'_A(y)$ are almost everywhere positive. Furthermore, because f is a concave function, its second derivative is almost everywhere negative.

> Decision-maker A values all lotteries higher than does decision-maker B if and only if
>
> $$\gamma_B(y) \geq \gamma_A(y).$$

The higher the risk aversion function, the lower the certain equivalent. This result is consistent with that obtained from the approximate expression of the certain equivalent.

14.6 THE INCLINATION OF THE RISK AVERSION FUNCTION (INCREASING OR DECREASING WITH WEALTH)

In principle, the risk aversion function can be increasing, decreasing, or a constant, and can even assume various shapes when plotted as a function at the wealth level y. The inclination of the risk aversion function, and its behavior with wealth, determines how the certain equivalent of a lottery changes as the wealth increases (equivalently, it determines how the certain equivalent of a lottery changes as the outcomes of the lottery are modified by adding shift amount to each one).

To illustrate, we consider again the following problem:

If $\tilde{y}$ is the wealth equivalent of the lottery

$$< p_1, y_1; p_2, y_2; ...; p_n, y_n >,$$

and if $\tilde{y}_\delta$ is the wealth equivalent of the lottery

$$< p_1, y_1 + \delta; p_2, y_2 + \delta; ...; p_n, y_n + \delta >.$$

How does $\tilde{y}_\delta$ compare to $\tilde{y} + \delta$?

Is $\tilde{y}_\delta < \tilde{y} + \delta$? $\tilde{y}_\delta > \tilde{y} + \delta$? $\tilde{y}_\delta = \tilde{y} + \delta$?

What role does the shape of the risk aversion function play in this formulation?

This is the same formulation we discussed with the delta property, sub-delta property, and super-delta property in Chapter 13. The delta property implies that

$$U(y + \delta) = k(\delta)U(y) + d(\delta).$$

Equivalently, the delta property implies that

$$U(y + \delta) = F(U(y), \delta), \quad F \text{ is a linear function of its first argument.}$$

The Sub-Delta property implies that

$$U(y + \delta) = F(U(y), \delta), \quad F \text{ is a concave function of its first argument.}$$

The Super-delta property implies that

$$U(y + \delta) = F(U(y), \delta), \quad F \text{ is a convex function of its first argument.}$$

We now relate this formulation to the risk aversion function by taking the first and second derivatives of both sides of the equation

$$\boxed{U(y + \delta) = F(U(y), \delta)} \tag{3}$$

and calculating the risk aversion functions.

We have $U'(y + \delta) = U'(y)F'(U(y), \delta)$ and $U''(y + \delta) = U'(y)^2 F''(U(y), \delta) + U''(y)F'(U(y), \delta)$.

Assuming $U' \neq 0$, and therefore $F' \neq 0$, we can write

$$\gamma(y + \delta) \triangleq -\frac{U''(y + \delta)}{U'(y + \delta)} = -\frac{U'(y)^2 F''(U(y), \delta) + U''(y)F'(U(y), \delta)}{U'(y)F'(U(y), \delta)} = -U'(y)\frac{F''(U(y), \delta)}{F'(U(y), \delta)} + \gamma(y)$$

Rearranging gives the following equality for $F' \neq 0$, if either the delta property, sub-delta property, or super-delta property are satisfied,

$$\boxed{\gamma(y + \delta) - \gamma(y) = -U'(y)\frac{F''(U(y), \delta)}{F'(U(y), \delta)}.} \tag{4}$$

14.6.1 Interpreting the Delta Property with the Risk Aversion Function

For the delta property, F is a linear function and so $F'' = 0$. From (4) this implies that

$$\gamma(y + \delta) - \gamma(y) = 0 \Rightarrow \boxed{\gamma(y + \delta) = \gamma(y)}.$$

When this is satisfied for values of δ on an interval, this implies *constant absolute risk aversion*.

14.6.2 Interpreting the Sub-Delta Property with the Risk Aversion Function

For the sub-delta property, $\tilde{y}_\delta < \tilde{y} + \delta$, F is a concave function, and so $F'' \leq 0$. From (4) this implies that

$$\gamma(y + \delta) - \gamma(y) \geq 0 \Rightarrow \boxed{\gamma(y + \delta) \geq \gamma(y)}.$$

When this is satisfied for positive values of δ on an interval, this implies *increasing absolute risk aversion*.

14.6.3 Interpreting the Super-Delta Property with the Risk Aversion Function

For the super-delta property, $\tilde{y}_\delta > \tilde{y} + \delta$, F is a convex function, and so $F'' \geq 0$. From (4) this implies that

$$\gamma(y + \delta) - \gamma(y) \leq 0 \Rightarrow \boxed{\gamma(y + \delta) \geq \gamma(y)}.$$

When this is satisfied for positive values of δ on an interval, this implies *decreasing absolute risk aversion*.

To summarize:

> Consider a shift amount $\delta > 0$ applied to the lottery outcomes
>
> 1. If $\tilde{y}_\delta > \tilde{y} + \delta$, for all lotteries (super-delta property), then $\gamma(y)$ is a strictly decreasing function.
> 2. If $\tilde{y}_\delta < \tilde{y} + \delta$ for all lotteries (sub-delta property), then $\gamma(y)$ is a strictly increasing function.
> 3. If $\tilde{y}_\delta = \tilde{y} + \delta$ for all lotteries (delta property), then $\gamma(y)$ is a constant function.

We generalize these ideas in the next chapters and extend them to other transformations.

14.7 SUMMARY

In Section 14.1 we defined the risk aversion function, which requires the utility function to be twice differentiable.

Three important properties pertain to the risk aversion function: its sign, its magnitude, and its inclination (increasing or decreasing with wealth). The sign determines whether the certain equivalent is above or below the mean value of the lottery; the magnitude can be used to approximate the certain equivalent of a lottery; and the inclination determines the effects that applying a shift transformation to the lottery outcomes has on the wealth equivalent.

An approximate expression for the certain equivalent using the risk aversion function is

$$\tilde{x} \approx \mu - \frac{\sigma^2}{2}\gamma(\mu).$$

Decision-maker A values all lotteries higher than does decision-maker B if and only if $\gamma_B(y) \geq \gamma_A(y)$.

Relating the risk aversion function to the delta, sub-delta, and super-delta properties:

Consider a shift amount $\delta > 0$ applied to the lottery outcomes

4. If $\tilde{y}_\delta > \tilde{y} + \delta$, for all lotteries (super-delta property), then $\gamma(y)$ is a strictly decreasing function and the certain equivalent will increase with wealth.
5. If $\tilde{y}_\delta < \tilde{y} + \delta$ for all lotteries (sub-delta property), then $\gamma(y)$ is a strictly increasing function and the certain equivalent will decrease with wealth.
6. If $\tilde{y}_\delta = \tilde{y} + \delta$ for all lotteries (delta property), then $\gamma(y)$ is a constant function and the certain equivalent will not change with wealth. This implies constant absolute risk aversion and the linear and exponential utility functions.

ADDITIONAL READINGS

Abbas, A. E. 2007. Invariant utility functions and certain equivalent transformations. *Decision Analysis* 4(3): 17–31.

Abbas, A. E. 2012. Valuing changes in investment opportunities. *Operations Research* 60(6): 1451–1460.

Abbas, A. E. and J. Aczél. 2010. The role of some functional equations in decision analysis. *Decision Analysis* 7(2): 215–228.

Arrow, K. J. 1965. *The Theory of Risk Aversion. Lecture 2 in Aspects of the Theory of Risk-Bearing.* Yrjo Jahnssonin Saatio, Helsinki.

Pratt, J. 1964. Risk aversion in the small and in the large. *Econometrica* 32: 122–136.

APPENDIX 14A HOW PEOPLE GET IT WRONG

The terms "risk aversion" and "risk tolerance" are widely misused. Risk tolerance is often used in the context of the maximum amount of money that one can afford to lose, or some threshold of an attribute beyond which a person is not willing to accept any loss.

You may hear statements like, "Our risk tolerance for lives in this operation or mission is three." They imply that they are willing to lose three lives to get a certain mission accomplished. There are several issues with this. First, this is not the risk tolerance. As we have seen, the risk tolerance (like the risk aversion) has a specific meaning in terms of the derivatives of the utility function and also the risk premium. Using it in other contexts is misleading and can create confusion. Similarly, you might hear, "Our risk tolerance for this project is $2 million," and the person means that the maximum amount of money they are willing to lose before quitting is $2 million. Why would you specify that you can lose up to a certain amount? And why would you quit after some specified threshold value? What if investing more money would generate a much higher potential, such as the cure for cancer or Alzheimer's disease? Why don't you just make good decisions given your attitude toward risk?

You may also hear risk strategy statements in government security enterprises like "we are risk averse for loss of lives but are risk seeking for media criticism." They mean that they will continue to do certain actions even if criticized by the media, and they will not care about the criticism received. But they do care (and are concerned) about losing lives. The statement is again misleading because simply caring or not caring about an item does not mean you are risk averse or risk seeking for preferences over its lotteries. Use statements like that can incorrectly give the impression about some concave or convex utility function.

APPENDIX 14B PROOF OF THE APPROXIMATE EXPRESSION FOR THE CERTAIN EQUIVALENT

Step 1: First, we prove that

$$U(\tilde{y}) = U(\mu) + \frac{1}{2}\frac{\partial^2 U(\mu)}{\partial y^2}\sigma^2 + o(\sigma^2), \text{ as } \sigma^2 \to 0,$$

where $o(\sigma^2)$ is the "small o" notation, a function that goes to zero faster than σ^2 as $\sigma^2 \to 0$.

More formally, this "small o" notation $o(\sigma^2)$ implies that for every positive M, there exists σ_o^2 such that $o(\sigma^2) \le M\sigma^2$ for $\sigma^2 < \sigma_o^2$. The small o notation is stronger than the big O notation, which implies only the existence of a positive M. The former case applies to all positive M.

Proof. We prove this step by taking a Taylor expansion of $U(y)$ around $U(\mu)$. We have, as $y \to \mu$,

$$U(y) = U(\mu) + \frac{\partial U(y)}{\partial y}\Big|_{y=\mu}(y-\mu) + \frac{1}{2}\frac{\partial^2 U(y)}{\partial y^2}\Big|_{y=\mu}(y-\mu)^2 + o(|y-\mu|^2),$$

where $o(|y-\mu|^2)$ converges to zero faster than $(y-\mu)^2$ as $y \to \mu$.

The expected utility of the lottery obtained using the Taylor expansion is

$$U(\tilde{y}) \triangleq E[U(y)] = \int_{y_{\min}}^{y_{\max}} f(y)U(y)dy = U(\mu) + 0 + \frac{1}{2}\frac{\partial^2 U(\mu)}{\partial y^2}\sigma^2 + \int_{y_{\min}}^{y_{\max}} f(y)o(|y-\mu|^2)dy. \quad (5)$$

Note that $\lim\limits_{y\to\mu}\int_{y_{\min}}^{y_{\max}} f(y)o(|y-\mu|^2)dy = \lim\limits_{\sigma^2\to 0} o(\sigma^2)$. Therefore, (5) becomes

$$U(\tilde{y}) = U(\mu) + \frac{1}{2}\frac{\partial^2 U(\mu)}{\partial y^2}\sigma^2 + o(\sigma^2), \text{ as } \sigma^2 \to 0.$$

Step 2: We prove that

$$\boxed{\tilde{y} - \mu = O(\sigma^2), \text{ as } \tilde{y} \to \mu,}$$

where $O(\sigma^2)$ is the "big O" notation, which implies that there exists N such that $O(\sigma^2) \le N\sigma^2$ as $\sigma^2 \to 0$.

Proof: Using the results of Step 1,

$$U(\tilde{y}) - U(\mu) = \frac{1}{2}\frac{\partial^2 U(\mu)}{\partial y^2}\sigma^2 + o(\sigma^2) = O(\sigma^2), \text{ as } \sigma^2 \to 0,$$

where $N = \frac{1}{2}\frac{\partial^2 U(\mu)}{\partial y^2}$, and we observe that $O(\sigma^2) + o(\sigma^2) = O(\sigma^2)$ as $\sigma^2 \to 0$.

Note that $\tilde{y} - \mu = U^{-1}(U(\tilde{y})) - U^{-1}(U(\mu))$.

Expanding $U^{-1}(U(\tilde{y}))$ around the neighborhood of $U^{-1}(U(\mu))$ as $\tilde{y} \to \mu$ gives

$$\tilde{y} = U^{-1}(U(\mu)) + (U^{-1})'(U(\mu))[U(\tilde{y}) - U(\mu)] + o(U(\tilde{y}) - U(\mu)).$$

This implies that

$$\tilde{y} - \mu = (U^{-1})'(U(\mu))[U(\tilde{y}) - U(\mu)] + o(U(\tilde{y}) - U(\mu)), \text{ as } \tilde{y} \to \mu.$$

Since $U(\tilde{y}) - U(\mu) = O(\sigma^2)$, as $\sigma^2 \to 0$, we have

$$\tilde{y} - \mu = (U^{-1})'(U(\mu))O(\sigma^2) + o(O(\sigma^2)) = O(\sigma^2), \text{ as } \tilde{y} \to \mu.$$

Step 3: We prove that

$$\boxed{U(\tilde{y}) = U(\mu) + \frac{\partial U(\mu)}{\partial y}(\tilde{y} - \mu) + o(\sigma^2), \text{ as } \tilde{y} \to \mu.}$$

Proof: We first take a Taylor expansion of $U(\tilde{y})$ around $U(\mu)$ as $\tilde{y} \to \mu$ to get

$$U(\tilde{y}) = U(\mu) + \frac{\partial U(\mu)}{\partial y}(\tilde{y} - \mu) + o((\tilde{y} - \mu)), \quad \tilde{y} \to \mu. \tag{6}$$

Then we use the results of Step 2 and observe that $\tilde{y} - \mu = O(\sigma^2)$, as $\tilde{y} \to \mu$, to get

$$U(\tilde{y}) = U(\mu) + \frac{\partial U(\mu)}{\partial y}(\tilde{y} - \mu) + o(\sigma^2), \quad \tilde{y} \to \mu.$$

Step 4: We now use Steps 1, 2, and 3 to prove the Theorem. As $y_{\min}, y_{\max} \to \mu$, $\tilde{y} \to \mu$ at least as fast, and $\sigma^2 \to 0$. This allows us to equate the expressions of $U(\tilde{y})$ from Steps 1 and 3 and rearrange to get

$$\tilde{y} = \mu + \frac{\sigma^2}{2}\frac{\frac{1}{2}\frac{\partial^2}{\partial y^2}U(\mu)}{\frac{\partial}{\partial y}U(\mu)} + o(\sigma^2), \quad \sigma^2 \to 0.$$

Using the definition of the risk-aversion function gives

$$\tilde{y} = \mu - \frac{\sigma^2}{2}\gamma(\mu) + o(\sigma^2), \quad \sigma^2 \to 0. \quad Q.E.D.$$

Note that the derivation was conducted for the wealth equivalent and the mean of the wealth lottery. By subtracting the initial wealth from both sides, we get a related expression for the certain equivalent and the mean of the lottery itself.

CHAPTER 15 Scale Transformations Applied to a Lottery

Chapter Concepts

- The multiplicative delta property
- The multiplicative sub-delta and super-delta properties
- The relative risk aversion function

15.1 INTRODUCTION

In this chapter, we continue our discussion of the effects of modifying the lottery outcomes on the change in wealth equivalent. In particular, we consider modifying the wealth lottery by a scale transformation. By scaling a wealth lottery you receive either a multiple or a fraction of the wealth that you were going to receive originally. For example, you might increase your share in an investment (purchasing more equity), and so the amount you now receive would be a multiple of the previous amount. Alternatively, you might experience some delay in receiving the payoff from a lottery, and by receiving the prospects in the future you effectively scale all outcomes by a discount factor to yield their net present value.

The analysis in this chapter parallels that for shift transformations presented in earlier chapters. A new definition, the ***relative risk aversion function***, is introduced and parallels the applications of the absolute risk aversion function for shift amounts. Going through this chapter helps the reader solidify some of the concepts learned in the previous chapters. Furthermore, it prepares the reader for the next chapters where we generalize the results of both shift and scale transformations to more general monotone transformations. Therefore, the proofs of most theorems in this chapter have been omitted.

15.2 THE MULTIPLICATIVE DELTA PROPERTY

The delta property discussed in Chapters 13 and 14 is useful in many situations. When it applies on an interval of positive length, it helps us narrow down the utility function to either the linear or the exponential forms. It also helps us calculate the certain

293

equivalent of any lottery that is modified by a shift amount using the certain equivalent of the unmodified lottery. When the delta property holds, the shift amount can simply be added to the previous certain equivalent to determine the certain equivalent of the new lottery.

We now discuss another transformation that also appears a lot in practice: scale transformations applied to wealth outcomes of a lottery.

Problem Formulation

Let $\tilde{y}$ be the wealth equivalent of the wealth lottery

$$< p_1, y_1; p_2, y_2; ...; p_n, y_n >.$$

Now suppose that a scale amount, m, is applied to the wealth outcomes to yield the modified lottery

$$< p_1, my_1; p_2, my_2; ...; p_n, my_n >, \quad m > 0$$

whose wealth equivalent we denote as $\tilde{y}_m$.

What is the relation between $\tilde{y}_m$ and $m\tilde{y}$??

Figure 15.1 illustrates the original lottery and the modified lottery. This formulation parallels that of the delta property for shift transformations, and we therefore refer to it as the **_multiplicative delta property_**. This chapter discusses the implications of this formulation for both continuous utility functions and for twice-continuously differentiable utility functions. Following similar steps as those used in the delta property, we state without proof the following theorem, which we generalize (and prove) in the next chapter.

Theorem 15.1

A decision-maker with a continuous and strictly monotone utility function satisfies the multiplicative delta property, i.e.,

$$\tilde{y}_m = m\tilde{y}$$

for all lotteries, if and only if

$$U(my) = k(m)U(y) + d(m), \quad y > 0. \tag{1}$$

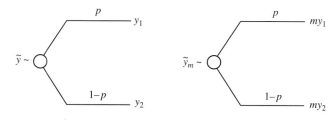

Figure 15.1. Scaling a wealth lottery to yield a new lottery with wealth equivalent $\tilde{y}_m$.

Dialogue between Instructor (I) and Student (S)

I: Once again, we are interested in utility functions that satisfy this functional equation. It is not surprising to note that, similar to the solutions of the functional equation for shift transformations, we can distinguish two cases:

1.   Solutions that arise when m is satisfied over an interval of positive length, and
2.   Solutions that arise when m is satisfied for only discrete values. In this case, additional periodic solutions arise. Because these periodic solutions do not determine the functional form of the utility function, we shall focus on solutions to the case where m is defined on an interval, and refer the reader to Abbas (2007) and Appendix C at the end of the book for the periodic solutions.

S: Are there utility functions that satisfy this functional equation?

I: Yes, by direct substitution, we can show that the logarithmic and power utility functions satisfy this functional equation over an interval for m.

EXAMPLE 15.1: Logarithmic Utility Function Satisfies the Scale Functional Equation

---

To illustrate that the logarithmic utility function satisfies this functional equation, note that if

$$U(y) = a + b\ln(y),$$

then

$$U(my) = a + b\ln(my) = a + b\ln(y) + b\ln(m) = U(y) + d(m),$$

where $k(m) = 1, d(m) = b\ln(m)$.

EXAMPLE 15.2: Power Utility Function Satisfies the Scale Functional Equation

---

For a power utility function,

$$U(y) = a + by^\alpha, \quad \alpha \neq 0,$$

we have

$$
\begin{aligned}
U(my) &= a + bm^\alpha y^\alpha = a + bm^\alpha y^\alpha + [am^\alpha - am^\alpha] \\
&= m^\alpha(a + by^\alpha) + a(1 - m^\alpha) \\
&= m^\alpha U(y) + a(1 - m^\alpha) \\
&= k(m)U(y) + d(m),
\end{aligned}
$$

where $k(m) = m^\alpha, \; d(m) = a(1 - m^\alpha)$.

We have now seen that the logarithmic and power utility functions satisfy the multiplicative delta property over an interval. As we prove below, these are the only utility functions that do so.

To solve (1), define

$$m = e^\delta, y = e^t, f(t) = U(e^t), k_1(t) = k(e^t), d_1(t) = d(e^t).$$

Substituting into (1) gives

$$f(\delta + t) = k_1(\delta)f(t) + d_1(\delta). \tag{2}$$

As we have seen, this implies one of the two possible solutions

1. $f(t) = a + bt$. This implies that $U(e^t) = a + bt$, which implies that $U(y) = a + b\ln(y)$.
2. $f(t) = a + be^{\alpha t}$. This implies that $U(e^t) = a + be^{\alpha t}$, which implies that $U(y) = a + by^\alpha$.

We summarize this result below.

Theorem 15.2

A decision-maker with a continuous and strictly monotone utility function satisfies the multiplicative delta property for values of m on an interval of positive length if and only if

$$U(y) = a + b\ln(y) \tag{3}$$

where $b \neq 0$ but otherwise arbitrary, and a is an arbitrary constant, or by

$$U(y) = a + by^\alpha, \tag{4}$$

where $\alpha, b \neq 0$ but otherwise arbitrary, and a is an arbitrary constant.

Theorem 15.2 asserts that the logarithmic and power utility functions are the only utility functions that satisfy the multiplicative delta property for scale intervals of positive length.

EXAMPLE 15.3: Numerical Illustration: Scale Transformations with a Logarithmic Utility Function

---

Consider again the wealth lottery <0.5, 1000; 0.5, 5000>, which provides a 50/50 chance of total wealth $1,000 or $5,000. When evaluated by a decision-maker with a logarithmic utility function, $U(y) = \ln(y)$, the wealth equivalent is

$$\tilde{y} = e^{E[U(y)]} = 1000^{0.5} \times 5000^{0.5} = \$2236.07.$$

Now consider a 20% share of this lottery that gives the new lottery

$$<0.5, 200; 0.5, 1000>.$$

The wealth equivalent of this modified lottery is

$$\tilde{y}_m = e^{E[U(y)]} = 200^{0.5} \times 1000^{0.5} = \$447.21$$

Note that the new wealth equivalent is equal to one-fifth (20%) of the previous wealth equivalent, i.e., (2236.07)/5 = 447.21, and so scaling the outcomes of the lottery has led to the same scaling of the wealth equivalent.

We also have an equivalent condition for the multiplicative delta property by setting $y = U^{-1}(t)$ into (1) to get

$$U(mU^{-1}(t)) = k(m)t + d(m).$$

Corollary: *A decision-maker with a continuous and strictly monotone utility function satisfies the multiplicative delta property for values of m on an interval of positive length if and only if*

$$U(mU^{-1}(t))$$

is a linear function of *t*.

Dialogue between Instructor (I) and Student (S)

I: We have now investigated the effects of modifying the lottery outcomes by scale transformations. We have discussed the multiplicative delta property and illustrated that it corresponds to logarithmic and power utility functions for scale changes on an interval.

S: Yes, and there is an interesting analogy between scale and shift transformations (delta property and multiplicative delta property), even though they correspond to different sets of utility functions.

I: Correct. In chapter 16, we shall also discuss even more general transformations.

S: What happens if the wealth equivalent of a lottery that is modified by a scale transformation that is less than (or greater than) than the prior wealth equivalent multiplied by the same scale amount? Do we have something corresponding to the sub-delta and super-delta properties, but for scale?

I: Yes, indeed. Let's discuss this below.

15.3 THE MULTIPLICATIVE SUB-DELTA PROPERTY

We can also consider the case where the increase in wealth equivalent due to scale transformations is less than the original wealth equivalent multiplied by the scale amount, i.e.,

$$\tilde{y}_m \leq m\tilde{y}.$$

When this occurs for all lotteries, we refer to it as the ***multiplicative sub-delta property***.

The proof follows the same steps as those for the additive sub-delta property. The multiplicative sub-delta property implies that

$$U^{-1}\left(\sum_{i=1}^{n} p_i U(my_i)\right) \leq mU^{-1}\left(\sum_{i=1}^{n} p_i U(y_i)\right).$$

Define $U(y_i) = t_i$ and substitute to get

$$U^{-1}\left(\sum_{i=1}^{n} p_i U(mU^{-1}(t_i))\right) \leq mU^{-1}\left(\sum_{i=1}^{n} p_i t_i\right).$$

Applying the transformation U to both sides gives

$$\sum_{i=1}^{n} p_i U(mU^{-1}(t_i)) \leq U\left(mU^{-1}\left(\sum_{i=1}^{n} p_i t_i\right)\right).$$

Note: This step assumes the utility function is strictly increasing. If it were strictly decreasing, then the sign of the inequality would be reversed in the previous equation.

Define $f(t_i) = U(mU^{-1}(t_i))$ and substitute to get

$$\sum_{i=1}^{n} p_i f(t_i) \leq f\left(\sum_{i=1}^{n} p_i t_i\right).$$

When this applies for all lotteries, it implies that f is a concave function. The reverse direction is easy to verify, and so we have proved the following.

Theorem 15.3

A continuous and strictly increasing utility function satisfies the sub-delta property, i.e.,

$$\tilde{y}_m \leq m\tilde{y} \text{ for all lotteries}$$

if and only if

$$U(my) = F(U(y), m),$$

where F is a concave function of its first argument.

If the utility function were strictly decreasing, then the results of this theorem would be reversed, and the utility function would satisfy the sub-delta property if and only if the function F was convex in its first argument.

Substituting for $y = U^{-1}(t)$ gives

$$U(mU^{-1}(t)) = F(t, m), \text{ which is concave in } t.$$

Corollary: *A continuous and strictly increasing utility function satisfies the sub-delta property, i.e.,*

$$\tilde{y}_m \leq m\tilde{y} \text{ for all lotteries}$$

if and only if $U(mU^{-1}(t))$ is a concave function.

EXAMPLE 15.4: Exponential Utility Function Satisfies the Multiplicative Sub-Delta Property for $m > 1$

Consider the case of an increasing exponential utility function and a lottery that is modified by applying a scale transformation. For an increasing exponential utility function,

$$U(y) = 1 - e^{-\gamma y}, \ \gamma > 0,$$

that is strictly increasing and whose range, is $[0,1]$, we have

$$U(mU^{-1}(t)) = 1 - e^{-\gamma\left[m.\frac{-1}{\gamma}\ln(1-t)\right]} = 1 - (1-t)^m, 0 \le t \le 1.$$

This function is concave when $m > 1$. Therefore, the exponential utility function satisfies the multiplicative sub-delta property on this interval of m.

Numerical Illustration: Exponential Utility Function Satisfying the Sub-Delta Property for $0 < m < 1$

Consider the wealth lottery <0.5, 1000; 0.5, 5000> that provides a 50/50 chance of total wealth $1,000 or $5,000. When evaluated by a decision-maker with an exponential utility function, $U(y) = 1 - e^{-\gamma y}$, with $\gamma = 0.001$, the wealth equivalent is $\tilde{y} = \$1675$ (as we have seen).

If we scale this wealth lottery by a multiple of 2, we get the modified lottery <0.5, 2000; 0.5, 10000>, whose new wealth equivalent is

$$\tilde{y}_m = \frac{-1}{\gamma}\ln\big(E[1 - U(y)]\big) = -1000 \times \ln\big(1 - 0.5(1 - e^{-2}) + 0.5(1 - e^{-10})\big) = \$2692.81$$

Note that $\tilde{y}_m = 2692.81, m\tilde{y} = 2 \times 1675 = 3350$ and so $\tilde{y}_m'' \ m\tilde{y}$ as expected.

15.4 THE MULTIPLICATIVE SUPER-DELTA PROPERTY

Finally, we consider the case where the new wealth equivalent is greater than the previous wealth equivalent multiplied by the scale amount for all lotteries, i.e.,

$$\tilde{y}_m \ge m\tilde{y} \ \text{for all lotteries.}$$

The proof of the following theorem should now be straightforward.

Theorem 15.4

A continuous and strictly increasing utility function satisfies the super-delta property, i.e.,

$$\tilde{y}_m \ge m\tilde{y} \ \text{for all lotteries}$$

if and only if

$$U(my) = F(U(y), m),$$

where F is a convex function of its first argument.

Equivalently, using the same substitution $U(y) = t$ as above, we get the following corollary.

Corollary: *A continuous and strictly increasing utility function satisfies the super-delta property if and only if*

$$U(mU^{-1}(t)) \text{ is a convex function.}$$

We now illustrate the implications of these results for twice-continuously differentiable utility functions.

EXAMPLE 15.5: Exponential Utility Function Satisfies the Multiplicative Super-Delta Property for $m > 1$

For an exponential utility function, $U(y) = 1 - e^{-\gamma y}$, we have seen that

$$U(mU^{-1}(t)) = 1 - (1-t)^m, 0 \le t \le 1.$$

This function is convex when $0 \le m \le 1$, and so the exponential utility function satisfies the multiplicative super-delta property on this interval of m.

EXAMPLE 15.6: Numerical Illustration

Consider again the wealth lottery <0.5, 1000; 0.5, 5000> that provides a 50/50 chance of total wealth $1,000 or $5,000. When evaluated by a decision-maker with an exponential utility function, $U(y) = 1 - e^{-\gamma y}$, with $\gamma = 0.001$, the wealth equivalent is $\tilde{y} = \$1675$ (as we have seen). If we scale this wealth lottery by a fraction of 0.2, we get the modified lottery <0.5, 200; 0.5, 1000>, whose new wealth equivalent is

$$\tilde{y}_m = \frac{-1}{\gamma} \ln\left(E[1 - U(y)]\right) = -1000 \times \ln\left(1 - 0.5(1 - e^{-0.2}) + 0.5(1 - e^{-1})\right) = \$522$$

Note that $\tilde{y}_m = 522$, $m\tilde{y} = 0.2 \times 1675 = 335$, and so $\tilde{y}_m > m\tilde{y}$ as expected.

Figure 15.2 plots the ratio $\frac{\tilde{y}_m}{\tilde{y}}$ vs m for both exponential utility function with $\gamma = 0.001$ and the logarithmic utility functions for the lottery <0.5, 1000; 0.5, 5000>. Note how $\frac{\tilde{y}_m}{\tilde{y}}$ corresponds to the curve $y = m$ for the logarithmic utility function. For the exponential utility function, $\frac{\tilde{y}_m}{\tilde{y}} > m$ for $0 < m < 1$ and $\frac{\tilde{y}_m}{\tilde{y}} < m$ for $m > 1$.

Dialogue between Instructor (I) and Student (S)
I: So now you see that we have a parallel phenomenon with the multiplicative sub-delta and multiplicative super-delta properties as we did with the sub-delta and super-delta properties.

We shall consider more general transformations in Chapter 16.

S: So we can see that these observations can be used to simplify and verify the analysis. If you have a logarithmic utility function, then you do not need to redo the whole wealth equivalent calculations if you scale the lottery. You simply multiply the previous calculated wealth equivalent by the scale amount.

I: Correct. But another use of these observations is verifying the use of the utility function itself. If a decision-maker says he has a logarithmic utility function, then you can verify whether his wealth equivalent will indeed be scaled if the wealth outcomes are scaled. This is a verification test. Likewise, if a decision-maker says he has an exponential utility function, then you can verify whether or not he satisfies the multiplicative sub-delta property for scale amounts m>1 and the multiplicative super-delta property for m<1.

S: Great. One more thought. In the last chapter, you illustrated that we could answer several questions about the delta property, and the sub- and super-delta properties for shift transformations using the risk aversion function.

I: Yes, when the utility function is twice continuously differentiable.

S: Do we have a similar function that can answer similar questions for scale transformations?

I: Indeed we do. Let's discuss that.

15.5 THE RELATIVE RISK AVERSION FUNCTION

As we illustrated in the previous chapter, the absolute risk aversion function provides many interpretations for the delta property when the utility function is twice continuously differentiable. In a similar manner, when the utility function is twice continuously differentiable, the relative risk aversion function provides many interpretations for the multiplicative delta property.

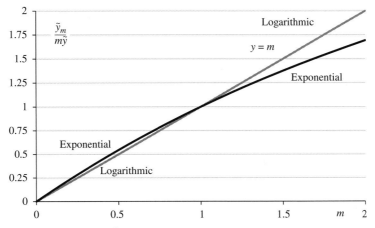

Figure 15.2. Plot of the ratio $\dfrac{\tilde{y}_m}{\tilde{y}}$ vs m for both exponential and logarithmic utility functions.

Definition

Arrow and Pratt introduced the **relative risk aversion function**, $r(y)$, as

$$r(y) = y\gamma(y), \quad y > 0.$$

The relative risk aversion function can be easily deduced from the absolute risk aversion function; it has the same sign as the risk aversion function, and it also defines the utility function uniquely up to a linear transformation.

EXAMPLE 15.7: Exponential Utility Function

Consider the exponential utility function

$$U(y) = a + be^{-\gamma y},$$

where a, γ are constants and $b \neq 0$ but otherwise constant. The relative risk aversion function is

$$r(y) = y\gamma(y) = \gamma y.$$

We have two cases:

  Case 1: If $\gamma < 0$, risk preferring exponential decision-makers, then the relative risk aversion function is decreasing with wealth.

  Case 2: If $\gamma > 0$, risk averse exponential decision-makers, then the relative risk aversion function is increasing with wealth.

Figures 15.3 and 15.4 illustrate the calculation of the relative risk aversion function from the absolute risk aversion graphically by multiplying it by the identity function $f(y) = y$ on the interval y>0.

EXAMPLE 15.8: Linear Utility Function

Consider the utility function

$$U(y) = a + by,$$

where $a, b \neq 0$ are constants.

The risk aversion function is

$$r(y) = y\gamma(y) = 0.$$

This is why linear utility functions are also said to have constant relative risk aversion (CRRA).

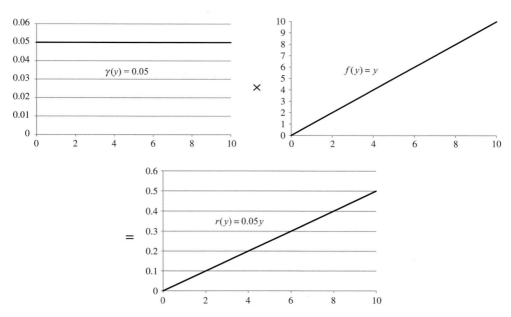

Figure 15.3. An exponential utility function with positive constant absolute risk aversion results in a relative risk aversion function that is increasing with wealth.

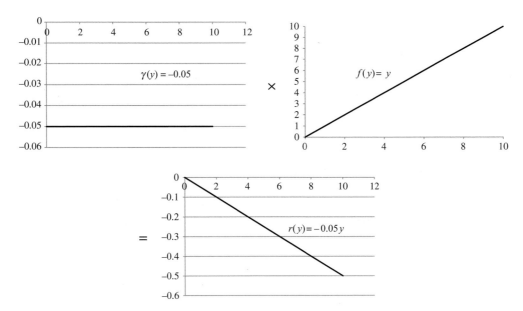

Figure 15.4. An exponential utility function with negative constant absolute risk aversion results in a relative risk aversion function that is decreasing with wealth.

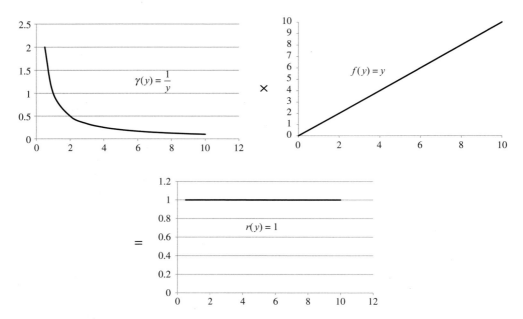

Figure 15.5. A logarithmic utility function with inverse absolute risk aversion results in a relative risk aversion function that is constant with wealth.

EXAMPLE 15.9: Logarithmic Utility Function

Consider the logarithmic utility function

$$U(x) = a + b\ln(y), \quad y > 0$$

where a is a constant and $b \neq 0$ but otherwise constant.

The absolute risk aversion function is $\gamma(y) = \dfrac{1}{y}$, and so the relative risk aversion function is

$$r(y) = y\gamma(y) = y.\frac{1}{y} = 1 = \text{Constant.}$$

The logarithmic utility function is said to have constant relative risk aversion with wealth.

Figure 15.5 illustrates the determination of the relative risk aversion function graphically.

EXAMPLE 15.10: Power Utility Function

Consider the power utility function

$$U(y) = y^{\alpha}, \quad y > 0$$

where $\alpha \neq 0, 1$.

The risk aversion function is then

$$r(y) = y\gamma(y) = y\frac{1-\alpha}{y} = 1 - \alpha = \text{constant.}$$

The power utility function (a special case of which is the linear utility function) also exhibits constant relative risk aversion with wealth.

15.5.1 The Behavior of the Relative Risk Aversion Function with a Scale Amount $m > 1$

The behavior of the relative risk aversion function with wealth determines how the wealth equivalent changes when the wealth outcomes of a lottery are modified by some scale transformation $m > 1$. To illustrate, consider the following scenario.

Problem Formulation for Scale Transformations

If $\tilde{y} = U^{-1}\left(\sum_{i=1}^{n} p_i y_i\right)$ is the wealth equivalent of the lottery

$$< p_1, y_1; p_2, y_2; ...; p_n, y_n >$$

how does the wealth equivalent of the lottery

$$< p_1, mx_1; p_2, mx_2; ...; p_n, mx_n >, m > 1,$$

which we refer to as $\tilde{y}_m$, compare to $m\tilde{y}$?

We can again characterize three situations:

$$\tilde{y}_m > m\tilde{y}, \ \tilde{y}_m = m\tilde{y}, \ \text{and} \ \tilde{y}_m < m\tilde{y}.$$

Because this formulation is the same as the super-/sub-/multiplicative delta properties, we start with the equation $U(my) = F(U(y), m)$, and determine its first and second derivatives as

$$mU'(my) = U'(y)F'(U(y), m), \ m^2 U''(my) = U''(y)F'(U(y), m) + U'(y)^2 F''(U(y), m).$$

With $U' \neq 0$, and so $F' \neq 0$, this implies that

$$\gamma(my) = -\frac{U''(my)}{U'(my)} = \frac{1}{m}[\gamma(y) - U'(y)\frac{F''(U(y), m)}{F'(U(y), m)}].$$

Multiplying both sides by my and rearranging gives

$$my\gamma(my) - y\gamma(y) = -yU'(y)\frac{F''(U(y), m)}{F'(U(y), m)}$$

or

$$r(my) - r(y) = -yU'(y)\frac{F''(U(y),m)}{F'(U(y),m)}.$$

Revisiting the Multiplicative Delta Property with the Relative Risk Aversion Function

If F is a linear function, then the right-hand side is zero, and so r(my) = r(y), which corresponds to the case of constant relative risk aversion. As we have seen, this also corresponds to the case of the multiplicative delta property.

If F is a concave function, then the right-hand side is positive, and so r(my) > r(y), for m>1, which corresponds to increasing relative risk aversion. This corresponds to the multiplicative sub-delta property.

If F is a convex function, then the right-hand side is negative, and so r(my) < r(y), for m>1, which corresponds to decreasing relative risk aversion. This implies that the wealth equivalent of a lottery whose wealth outcomes are modified by a scale transformation, $m > 1$, is greater than the wealth equivalent of the unmodified lottery multiplied by that scale amount. *This corresponds to the multiplicative super-delta property.*

15.6 SUMMARY

The multiplicative delta property implies the functional equation

$$U(my) = k(m)U(y) + d(m).$$

The multiplicative sub- (and super-)delta properties imply the equation

$$U(my) = F[U(y),m].$$

In particular,

1. $\tilde{y}_m \leq m\tilde{y}$ *for all lotteries if and only if*

    a. $U(my) = F(U(y),m)$, *where F is a concave function of its first argument.*
    b. $U(mU^{-1}(t))$ *is a concave function*

2. $\tilde{y}_m \geq m\tilde{y}$ *for all lotteries if and only if*

    a. $U(my) = F(U(y),m)$, *where F is a convex function of its first argument.*
    b. $U(mU^{-1}(t))$ *is a convex function*

3. $\tilde{y}_m = m\tilde{y}$ *for all lotteries if and only if*

    a. $U(my) = F(U(y),m)$, *where F is a linear function of its first argument.*
    b. $U(mU^{-1}(t))$ *is a linear function*

ADDITIONAL REFERENCES

Abbas, A. E. 2007. Invariant utility functions and certain equivalent transformations. *Decision Analysis* 4(3): 17–31.

Abbas, A. E. 2012. Valuing changes in investment opportunities. *Operations Research* 60(6): 1451–1460.

Abbas, A. E. and J. Aczél. 2010. The role of some functional equations in decision analysis. *Decision Analysis* 7(2): 215–228.

Arrow, K. J. 1965. *The Theory of Risk Aversion. Lecture 2 in Aspects of the Theory of Risk-Bearing.* Yrjo Jahnssonin Saatio, Helsinki.

Pratt, J. 1964. Risk aversion in the small and in the large. *Econometrica* 32: 122–136.

APPENDIX 15A LINEAR RISK TOLERANCE UTILITY FUNCTIONS

Related to the concept of modifying the lottery with scale transformations is the following transformation applied to the lottery outcomes.

Suppose that a decision-maker does not satisfy the multiplicative delta property for any scale amounts. However, he states that if his initial wealth is modified by a fixed amount, λ_0, (which could be positive or negative), then he would satisfy the multiplicative delta property on an interval of positive length of m. To achieve this new wealth level, the decision-maker can borrow a fixed amount, λ_0, and pay it back after the scaled prospect is revealed. This situation is represented by the following modified scale transformation:

$$g(y, m; \lambda_0) = m(y + \lambda_0) - \lambda_0, \, y > -\lambda_0. \tag{5}$$

Note that the amount λ_0 is fixed, while m is satisfied on an interval of positive length. This also implies that the wealth equivalent of the modified lottery, $\tilde{y}_m$, is related to the wealth equivalent of the unmodified lottery, $\tilde{y}$, by

$$\tilde{y}_m = m(\tilde{y} + \lambda_0) - \lambda_0.$$

We derive the utility function(s) whose wealth equivalent follows this transformation when it is applied to the lottery outcomes below.

Proposition: *A decision-maker with a continuous and strictly monotonic utility function satisfies invariance with (5) at a given value of λ_0, for $y > -\lambda_0$ if and only if he has one of the following utility functions:*

$$U(y) = a(\rho + \eta y)^{-\frac{1}{\eta}} + b \text{ or } U(y) = a \ln(y + \lambda_0) + c$$

where $a \neq 0$, but otherwise arbitrary, $\dfrac{\rho}{\eta} = \lambda_0$, and b, c are arbitrary constants.

Proof: As we show in the next chapter (and it should not come as a surprise to the reader at this point), this equality is satisfied if and only if

$$U(m(y + \lambda_0) - \lambda_0) = k(m)U(y) + d(m).$$

Define $t = y + \lambda_0$, and substitute to get

$$U(mt - \lambda_0) = k(m)U(t - \lambda_0) + d(m).$$

Further define $f(t) = U(t - \lambda_0)$, and substitute to get

$$f(mt) = k(m)f(t) + d(m).$$

This implies (as we have seen with the multiplicative delta property) that either

$$\text{(i) } f(t) = at^b + c \Rightarrow U(t - \lambda_0) = at^b + c \Rightarrow U(y) = a(y + \lambda_0)^b + c$$

or

$$\text{(ii) } f(t) = a\ln(t) + c \Rightarrow U(t - \lambda_0) = a\ln(t) + c \Rightarrow U(y) = a\ln(y + \lambda_0) + c$$

The utility function $U(y) = a(y + \lambda_0)^b + c$ is known as the linear risk tolerance utility function or the Hyperbolic Absolute Risk Aversion (HARA) utility function. This is because it provides a linear risk tolerance, as we illustrate below:

$$U(y) = a(y + \lambda_0)^b + c \Rightarrow U'(y) = ab(y + \lambda_0)^{b-1}, U''(y) = ab(b-1)(y + \lambda_0)^{b-2}.$$

From the definition of the risk aversion function,

$$\gamma(y) = -\frac{U''(y)}{U'(y)} = \frac{1-b}{y + \lambda_0},$$

which is a hyperbolic function. Alternatively, the risk tolerance is

$$\rho(y) = \frac{1}{\gamma(y)} = \frac{y + \lambda_0}{1-b},$$

which is a linear function.

Note also that the logarithmic utility function $U(y) = a\ln(y + \lambda_0) + c$ also provides linear risk tolerance because $U'(y) = \dfrac{a}{y + \lambda_0}, U''(y) = \dfrac{-a}{(y + \lambda_0)^2}$. Therefore,

$$\gamma(y) = -\frac{U''(y)}{U'(y)} = \frac{1}{y + \lambda_0},$$

which is a hyperbolic function, and the risk tolerance is

$$\rho(y) = \frac{1}{\gamma(y)} = y + \lambda_0,$$

which is a linear function.

CHAPTER 16

The Invariant Transformation of a
Utility Function

Chapter Concepts

- What happens if we modify all the outcomes of a lottery by a monotone transformation, g, and if the wealth equivalent follows the same transformation?
- The invariant transformation of a utility function U
- Characterizing a utility function by its invariant transformation
- The implications of the functional equation

$$U(g(y)) = kU(y) + d$$

- The g-risk aversion function

16.1 INTRODUCTION

This chapter generalizes some concepts from several previous chapters. In Chapter 13, we discussed valuation of lotteries, using the certain equivalent, and the changes in valuation of lotteries when a shift transformation is applied to the lottery outcomes. If the wealth equivalent of a lottery increases by the same shift amount, then the utility function is either linear or exponential.

In Chapter 14, we discussed the notion of a risk aversion function. Risk aversion is also concerned with the valuation of lotteries. The magnitude of the risk aversion function characterizes the relation between the certain equivalent of a lottery and its expected value. If the risk aversion function is constant, then the decision-maker's valuation of deals does not depend on the wealth (the wealth equivalent of any deal increases by an amount exactly equal to δ when all lottery outcomes, or the wealth, increase by δ), and the utility function is either linear or exponential.

In Chapter 15, we extended these valuation ideas to scale transformations, and we also introduced the relative risk aversion function. We illustrated how the logarithmic and power utility functions exhibit constant relative risk aversion.

How do these concepts of change in wealth equivalent extend to more general transformations? What happens if we modify the outcomes of a lottery by an arbitrary monotone transformation and if the wealth equivalent follows this same transformation?

This chapter illustrates that this property provides convenient characterizations for the utility function. In the next chapter, we shall discuss the implications to utility functions when the wealth equivalent of the modified lottery does not necessarily follow this transformation.

16.2 THE INVARIANT TRANSFORMATION OF A UTILITY FUNCTION (ABBAS 2007)

Before we proceed with the general formulation, let us first recap the results for shift and scale transformations.

Problem Formulation: Shift Transformations
Suppose you have a wealth lottery

$$< p_1, y_1; p_2, y_2; ...; p_n, y_n >$$

whose wealth equivalent is $\tilde{y}$. Now suppose we apply a shift amount to all outcomes of the lottery to obtain the lottery

$$< p_1, y_1 + \delta; p_2, y_2 + \delta; ...; p_n, y_n + \delta >$$

and a wealth equivalent, $\tilde{y}_\delta$.
  When does $\tilde{y}_\delta$ equal to $\tilde{y} + \delta$?

A utility function satisfies this property for all lotteries if and only if

$$\boxed{U(y+\delta) = k(\delta)U(y) + d(\delta)}, \tag{1}$$

for some functions k and d.
  Let us also recap the formulation for scale transformations:

Problem Formulation: Scale Transformations
Suppose we have a wealth lottery

$$< p_1, y_1; p_2, y_2; ...; p_n, y_n >$$

whose wealth equivalent is $\tilde{y}$. Now suppose we apply a scale transformation to all outcomes of the lottery to obtain the lottery

$$< p_1, my_1; p_2, my_2; ...; p_n, my_n >$$

and a wealth equivalent $\tilde{y}_m$.
  When does $\tilde{y}_m$ equal to $m\tilde{y}$ for all lotteries?

A utility function satisfies this property for all lotteries if and only if

$$\boxed{U(my) = k(m)U(y) + d(m)}. \tag{2}$$

The Invariant Transformation of a Utility Function

We now generalize the previous formulations to monotone transformations as follows.

Problem Formulation: General Monotone Transformations
Suppose you have a wealth lottery

$$< p_1, y_1; ...; p_n, y_n >$$

and you value this lottery by a wealth equivalent, $\tilde{y}$, where

$$\tilde{y} = U^{-1}\left(\sum_{i=1}^{n} p_i U(y_i) \right).$$

Now suppose that all the lottery outcomes are modified by a monotone transformation, g, to yield the modified lottery

$$< p_1, g(y_1); ...; p_n, g(y_n) >,$$

whose wealth equivalent we denote as $\tilde{g}$, where

$$\tilde{g} = U^{-1}\left(\sum_{i=1}^{n} p_i U(g(y_i)) \right).$$

When does $\tilde{g}$, equal to $g(\tilde{y})$ for all lotteries?
What is the most general form of g for a given U?

Figure 16.1 illustrates the idea of modifying the wealth outcomes of a lottery.

Definition
If we modify all wealth outcomes of a lottery by a transformation g, and if $g(\tilde{y}) = \tilde{g}$ for all lotteries, then we refer to the transformation g as the ***invariant transformation*** of the utility function, U, and we refer to the utility function, U, as the ***invariant utility function*** of the transformation g.

Note that by definition, $g(\tilde{y}) = \tilde{g}$ implies that

$$g\left(U^{-1}\left(\sum_{i=1}^{n} p_i U(y_i) \right) \right) = U^{-1}\left(\sum_{i=1}^{n} p_i U(g(y_i)) \right).$$

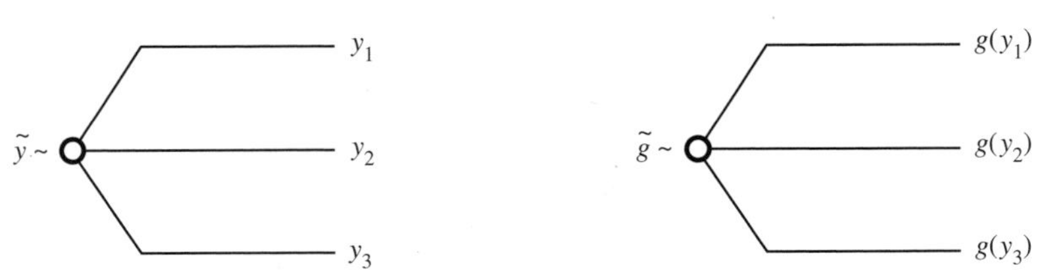

Figure 16.1. If $g(\tilde{y}) = \tilde{g}$ for all lotteries, what must be true about U and g?

Why Do We Care About This Formulation?

There are many reasons why we care about the change in value of a lottery when it is modified by a monotone transformation. As we shall see, this formulation helps us characterize the utility function uniquely up to a linear transformation, and so it is a validation test for the functional form of the utility function. Furthermore, as we shall see in Chapter 17, this formulation will enable us to determine the certain equivalent of a modified lottery using the certain equivalent of the prior lottery using simple substitutions. In addition, this formulation will help us understand many concepts in multiattribute utility functions such as utility independence and one-switch utility independence.

If the lottery is deterministic, then $\tilde{g} = g(\tilde{y})$ for any transformation, g.

To illustrate, if your wealth is deterministic at \$1,000, then the wealth equivalent is $\tilde{y} = 1000$. If suddenly your wealth becomes a deterministic amount of \$1,000,000 by a square transformation, the new value of the wealth equivalent is also \$1,000,000 by a square transformation, and so $g(\tilde{y}) = \tilde{g}$ for deterministic lotteries. Furthermore, for the trivial case, where the transformation is the identity, $g(y) = y$ (no change in the lottery outcomes), then $\tilde{g} = g(\tilde{y}) = \tilde{y}$.

Our main focus will therefore be situations where the deal is uncertain and where $g(y) \neq y$.

Theorem 16.1

For a continuous and strictly monotone utility function, U, and a strictly monotone transformation g applied to the wealth outcomes of a lottery, the following are equivalent:

1. *$\tilde{g} = g(\tilde{y})$ for all lotteries.*
2. *The utility function and the monotone transformation satisfy the equation*

$$U(g(y)) = kU(y) + d,$$

where $k \uparrow 0$ but otherwise arbitrary and d is an arbitrary constant.

We have provided a general equation that characterizes the utility functions that are invariant to a transformation, g. If we can assert equality of $\tilde{g} = g(\tilde{y})$ *for all lotteries, then we can determine the form of the utility function by solving the functional equation $U(g(y)) = kU(y) + d$.*

Dialogue between Instructor (I) and Student (S)

I: Does everyone see what is happening?

S: Yes, the linear and exponential utility functions were invariant to a shift transformation, and therefore they satisfied the delta property functional equation. And the logarithmic and power utility functions were invariant to a scale transformation, and therefore they satisfied the multiplicative delta property functional equation.

I: Correct.

S: But now, we are generalizing these concepts to other utility functions and more general transformations.

I: Precisely.

S: So given a wealth lottery with wealth equivalent, $\tilde{y}$, and an arbitrary monotone transformation, g, that is applied to the outcomes of the lottery, if the wealth equivalent of the new lottery is $g(\tilde{y})$, and if this identity holds for all lotteries, then it must be that the utility function satisfies the equation $U(g(y)) = kU(y) + d$.

I: Correct! This equation is known as the invariance functional equation for a utility function, U, and transformation, g. All we need to do to find the corresponding U for a given g is to solve this equation.

It is important to make two connections here. First, we can specify a particular parameter (or parameters) of the transformation as we did with shift and scale. In this case, we can write the transformation as $g(y, \delta)$. For a shift transformation, $g(y, \delta) = y + \delta$; for a scale transformation, $g(y, \delta) = \delta y$, and for a power transformation, $g(y, \delta) = y^\delta$. If the utility function is invariant to this transformation, then we can write

$$\boxed{U(g(y, \delta)) = k(\delta)U(y) + d(\delta)}.$$

This equation generalizes the shift formulation when $g(y, \delta) = y + \delta$. We simply solve the functional equation,

$$U(y + \delta) = k(\delta)U(y) + d(\delta),$$

which, as we have seen, results in a linear or exponential utility function for values of δ on an interval.

It also generalizes the scale formulation when $g(y, \delta) = \delta y$, where we solve the equation

$$U(\delta y) = k(\delta)U(y) + d(\delta).$$

We can also consider power transformations of the form $g(y, \delta) = y^\delta$ and find utility functions that satisfy this property by solving

$$U(y^\delta) = k(\delta)U(y) + d(\delta).$$

The second connection is that if $k(\delta)$ does not change sign, or, for simplicity, if $k(\delta) > 0$, then preferences for lotteries over attribute Y will not change as we apply the transformation $g(y, \delta)$ to the lottery outcomes, and as we change the parameter δ. In a sense we can say preferences for lotteries over attribute Y are independent of δ. We refer back to this idea in Chapter 23 and illustrate why this is the case below.

Take two lotteries over y, call them A and B, and suppose that the expected utility of lottery A is higher than that of lottery B and so $E_A[U(y)] > E_B[U(y)]$. Now suppose we modify the lotteries by a transformation $g(y, \delta)$. The expected utilities of the modified lotteries are $E_A[U(g(y, \delta))]$ and $E_B[U(g(y, \delta))]$. If the utility function satisfies $U(g(y, \delta)) = k(\delta)U(y) + d(\delta)$, then the expected utility of the modified lotteries is, respectively,

$$E_A[U(g(y, \delta))] = E_A[k(\delta)U(y) + d(\delta)] = k(\delta)E_A[U(y)] + d(\delta)$$

and

$$E_B[U(g(y,\delta))] = E_B[k(\delta)U(y) + d(\delta)] = k(\delta)E_B[U(y)] + d(\delta).$$

If $k(\delta) > 0$, and if $E_A[U(y)] > E_B[U(y)]$, then it is straightforward to see that $E_A[U(g(y,\delta))] > E_B[U(g(y,\delta))]$ because

$$E_A[U(g(y,\delta))] - E_B[U(g(y,\delta))] = k(\delta)\big[E_A[U(y)] - E_B[U(y)]\big] > 0.$$

This implies that preferences for lotteries do not changes as we change the parameter δ.

16.3 DETERMINING THE INVARIANT TRANSFORMATION OF A UTILITY FUNCTION

We now ask the opposite question. For a given utility function, U, what is the most general form of its invariant transformation g for which $g(\tilde{y}) = \tilde{g}$? We know that the invariance functional equation,

$$U(g(y)) = kU(y) + d,$$

must hold.

Rearranging gives *the invariant transformation*, g, of the form

$$g(y,k,d) = U^{-1}\big(kU(y) + d\big), \tag{3}$$

where k, d are variables that can be varied on their respective domains and still satisfy this invariance condition. The domain of feasibility of k, d is determined by the requirement that $kU(y) + d$ lies on the range of U. Equation (3) provides a general form for the invariant transformation of a given utility function.

A special case of (3) is when each of k, d is varied on an interval while keeping the other fixed. For example, the transformations $g(y,k) = U^{-1}\big(kU(y)\big)$ and $g(y,d) = U^{-1}\big(U(y) + d\big)$ would both need to be satisfied.

16.3.1 The Invariant Transformation of a Linear Utility Function

For a linear utility function, say $U(y) = y$, we substitute into (3) to get

$$g(y,k,d) = U^{-1}\big(kU(y) + d\big) = ky + d, \tag{4}$$

which is a shift-scale transformation. This implies that if we have a wealth lottery and a wealth equivalent calculated for a linear utility function, we can modify all the outcomes of the lottery by any linear transformation and the wealth equivalent will follow the same transformation. As a special case, if we modify the lottery first by a shift transformation $g(y,d) = y + d$ and then by a scale transformation $g(y,k) = ky$ (or the reverse order), both would satisfy transformation invariance.

EXAMPLE 16.1: Numerical Illustration: Invariant Transformation of an Exponential Utility Function

---

Consider again the wealth lottery

$$<0.5, 1000; 0.5, 5000>$$

and a linear utility function.

The wealth equivalent of this lottery is just the mean value, which is

$$\tilde{y} = 0.5 \times 1000 + 0.5 \times 5000 = \$3000$$

Now suppose we apply a linear transformation

$$g(y, k, d) = ky + d.$$

For any arbitrary values of k and d, say $k = 0.1$ and $d = 100$, we get the new lottery

$$<0.5, 200; 0.5, 600>$$

The wealth equivalent of this modified lottery is

$$\tilde{g} = 0.5 \times 200 + 0.5 \times 600 = \$400$$

Note that if we apply the same linear transformation to the wealth equivalent of the original lottery, we get

$$g(\tilde{y}) = k\tilde{y} + d = 0.1(3000) + 100 = \$400 = \tilde{g}$$

as expected.

Note: This equality holds for all values of k and d. For example, if we pick other values, $k = 0.2$ and $d = 50$, we get the lottery $<0.5, 250; 0.5, 1050>$, whose wealth equivalent is

$$\tilde{g} = 0.5 \times 250 + 0.5 \times 1050 = \$650.$$

If we apply the same transformation to the prior wealth equivalent, we get

$$g(\tilde{y}) = k\tilde{y} + d = 0.2(3000) + 50 = \$650 = \tilde{g}.$$

16.3.2 The Invariant Transformation of an Exponential Utility Function

We know that an exponential utility function, $U(y) = -e^{-\gamma_0 y}$, is invariant to a shift transformation because it satisfies the delta property. But is this the most general transformation for which a given exponential utility function with risk aversion coefficient γ_0 is invariant?

Substituting into (3) gives the invariant transformation for an exponential utility function

$$g(y, k, d) = -\frac{1}{\gamma_0} \ln(ke^{-\gamma_0 y} - d), \ \forall ke^{-\gamma_0 y} > d. \tag{5}$$

Equation (5) defines an invariant transformation for an exponential utility function with risk aversion coefficient γ_0. Both k and d can be varied across their feasible domains, and the resulting transformation will still result in transformation invariance. The transformation in (5) is specific to the value of the risk aversion coefficient γ_0. Therefore, knowledge of transformation invariance with the transformation (5) asserts both the exponential form of the utility function and the value of its risk aversion coefficient.

Special Cases

1.   If $d = 0$ and we vary only k in (5): **The delta property**

When $d = 0$, (5) becomes $g(y,k,0) = -\dfrac{1}{\gamma_0}\ln(ke^{-\gamma_0 y}) = U^{-1}(kU(y))$, which reduces to the shift transformation,

$$g_1(y,k_1) = y + k_1,$$

where $k_1 = -\dfrac{1}{\gamma_0}\ln(\delta)$.

Because δ is arbitrary, the variable k_1 no longer depends on the particular value of the risk aversion coefficient γ_0. As such, the shift transformation $g_1(y,k_1) = y + k_1$ is satisfied by all exponential utility functions and not just those with risk aversion coefficient γ_0. Therefore, the shift transformation alone would not characterize the utility function uniquely.

2.   **If $k = 1$ and we vary only d in (5)**

When $k = 1$, (5) becomes $g(y,1,d) = -\dfrac{1}{\gamma_0}\ln(e^{-\gamma_0 y} + d) = U^{-1}(U(y) + d)$. Equation (5) asserts that this transformation must satisfy invariance with this exponential utility function.

To characterize the exponential utility function, $U(y) = -e^{-\gamma_0 y}$, up to a linear transformation, we will see that we must satisfy invariance with values of k, d each on an interval and not just a fixed value.

Dialogue between Instructor (I) and Student (S)

S: We learned in previous chapters that exponential utility functions are invariant to shift transformations. What is this new transformation $g(y,k,d) = -\dfrac{1}{\gamma_0}\ln(ke^{-\gamma_0 y} - d), \forall ke^{-\gamma_0 y} > d$?

I: Indeed, all exponential utility functions are invariant to shift transformations, and therefore they satisfy the delta property. But the question we are asking is whether a given exponential utility function, say $U(y) = -e^{-\gamma_0 y}$, can be invariant to more general transformations besides the shift transformation.

S: And what we have seen is that the answer is yes; that a given exponential utility function with risk aversion coefficient γ_0 is invariant to the more general transformation,

$$g(y,k,d) = -\dfrac{1}{\gamma_0}\ln(ke^{-\gamma_0 y} - d), \forall ke^{-\gamma_0 y} > d.$$

> **I:** Correct. And a special case of this transformation is the shift transformation. Not all exponential utility functions with arbitrary risk aversion coefficients satisfy invariance with this transformation. Only an exponential utility function with risk aversion coefficient γ_0 would satisfy invariance with this transformation as we vary k and d over intervals of positive length. To illustrate, recall that if $d = 0$ (and so it is not defined on an interval of positive length), then it becomes the shift transformation, and all exponential utility functions would satisfy invariance with it.
>
> **S:** So does the general invariant transformation of a utility function $g(y,k,d) = U^{-1}(kU(y)+d) = ky+d$, where both k and d are defined on intervals of positive length, characterize the utility function up to a linear transformation?
>
> **I:** Great question. Let's discuss this below.

16.3.3 Characterizing the Utility Function

For values of k,d on an interval, the transformation (6) defines a utility function up to a linear transformation. We assert this below.

Proposition 16.1: *Two continuous and strictly monotonic utility functions, $U_1(y)$ and $U_2(y)$ have the same invariant transformations satisfying*

$$g_1(y,k,d) = g_2(y,k,\frac{d+b(k-1)}{a})$$

if and only if they are related by a linear transformation, $U_2(y) = aU_1(y)+b, \quad a \neq 0$.

To characterize the utility function up to a linear transformation, the invariant transformation must be satisfied for values of y,k,d on an interval.

To illustrate, we now show that if we apply a linear transformation to a linear utility function, we again get a linear (shift-scale) transformation. If $U(y) = y$, we substitute into (3) to get

$$g(y,k,d) = U^{-1}(kU(y)+d) = ky+d.$$

On the other hand, if $U(y) = ay+b$, and substitute into (3), we get

$$g(y,k,d) = U^{-1}(kU(y)+d) = \frac{1}{a}[k(ay+b)+d]-b = ky+d', \tag{6}$$

where $d' = \dfrac{d+b(k-1)}{a}$.

Equation (6) shows that the general invariant transformation for the modified linear utility function is again a shift-scale transformation. The following sections illustrate these results using other utility functions.

16.3.4 Applying a Linear Transformation to the Utility Function

Applying a linear transformation to the exponential utility function of the previous example yields $U_2(y) = b - ae^{-\gamma_0 y}$ whose invariant transformation is

$$g_2(y,k,d) = -\frac{1}{\gamma_0}\ln(ke^{-\gamma_0 y} - \frac{b(k-1)+d}{a}). \tag{7}$$

Defining $d^{eff} = \dfrac{b(k-1)+d}{a}$ expresses the invariant transformation as

$$g_2(y,k,d) = g_1(y,k,d^{eff}) = -\frac{1}{\gamma_0}\ln(ke^{-\gamma_0 y} - d^{eff}), ke^{-\gamma_0 y} > d^{eff}. \tag{8}$$

Comparing (7) and (8) shows the equivalence of the invariant transformations for two linearly related utility functions.

EXAMPLE 16.2: Numerical Illustration: Invariant Transformation of an Exponential Utility Function

---

Consider again the wealth lottery <0.5, 1000; 0.5, 5000> and an exponential utility function $U(y) = -e^{-\gamma_0 y}$, with risk aversion coefficient $\gamma_0 = 0.001$. The wealth equivalent of this lottery using this exponential utility function is (as we have seen in Example 13.1) $\tilde{y} = \$1675$.

We can now apply the invariant transformation (5) to this lottery for different values of k and d that lie on the domain of feasibility $ke^{-\gamma_0 x} > d$ and see what happens to the wealth equivalent.

Different values of k and d provide different invariant transformations that satisfy the invariance relation. As we apply each of these transformations to the outcomes of the lottery, the certain equivalent follows the same transformation.

To illustrate, the values $k = 2, d = -0.1$ yield the invariant transformation

$$g(y,k,d) = -1000\ln(2e^{-0.001y} + 0.1)$$

and the modified lottery

$$<0.5, 179.4; 0.5, 2176.16>.$$

The wealth equivalent of the modified lottery is $\tilde{g} = 745.25$. Moreover, the invariant transformation applied to the original wealth equivalent gives

$$g(\tilde{y},k,d) = -1000\ln(2e^{-0.001(1675)} + 0.1) = 745.25$$

as expected.

To further demonstrate the invariance relationship, we repeat the analysis with $k = 3, d = 0.01$ yield the invariant transformation

$$g(y,k,d) = -1000\ln(3e^{-0.001y} - 0.01)$$

and the modified lottery

$$<0.5, -89.51; 0.5, 4584.01>.$$

The wealth equivalent of the modified lottery is $\tilde{g} = 594.34$. Moreover, the invariant transformation applied to the original wealth equivalent gives

$$g(\tilde{y},k,d) = -1000\ln(3e^{-0.001(1675)} - 0.01) = 594.34$$

as expected.

16.3.5 Invariant Transformation of a Logarithmic Utility Function

For a logarithmic utility function, $U(y) = \ln(y)$, we substitute into (3) to get

$$\boxed{g(y,k,d) = U^{-1}\left(kU(y) + d\right) = e^{k\ln(y)+d} = e^d y^k = my^k}, \qquad (9)$$

where $m = e^d > 0$.

This implies that the invariant transformation for a logarithmic utility function is a composite power-scale transformation for values of m and k on an interval of positive length. When applying a power-scale transformation to a wealth lottery, the wealth equivalent follows the same transformation. Once again, this transformation characterizes a logarithmic utility function up to a linear transformation.

EXAMPLE 16.3: Numerical Illustration: Invariant Transformation of a Logarithmic Utility Function

---

Consider again the wealth lottery <0.5, 1000; 0.5, 5000> and the logarithmic utility function

$$U(y) = \ln(y).$$

The wealth equivalent of this lottery using this logarithmic utility function is (as we have seen) $\tilde{y} = \$2236.07$. We now apply the power scale transformation (9) to the lottery outcomes and observe the change in the wealth equivalent.

Consider the values $m = 3$, $k = 1.5$ that yield the invariant transformation

$$g(y,m,k) = 3y^2$$

and the modified lottery

$$<0.5, 94{,}868.33; 0.5, 1{,}060660>.$$

The wealth equivalent of the modified lottery is $\tilde{g} = \$317{,}211.38$. Moreover, the invariant transformation applied to the original wealth equivalent gives

$$g(\tilde{y},k,d) = 3(2236.07)^{1.5} = 317{,}211.38$$

as expected.

Once again, this transformation applies to power-scale values on interval. To further demonstrate the invariance relationship, we repeat the analysis with $m = 2, k = 0.01$ yield the invariant transformation

$$g(y,m,k) = 2y^{0.01}$$

and the modified lottery <0.5, 12.86; 0.5, 13.07>. The wealth equivalent of the modified lottery is $\tilde{g} = \$12.96$. Moreover, the invariant transformation applied to the original wealth equivalent gives

$$g(\tilde{y}, k, d) = 2(2236.07)^{0.01} = \$12.96 = \tilde{g}$$

as expected.

16.3.6 The Invariant Transformation of a Power Utility Function

It is now straightforward to determine the invariant transformation for any continuous utility function that has a closed-form expression for its inverse.

For a power utility function,

$$U(y) = y^{\alpha},$$

we substitute into (3) to get

$$g(y, k, d) = \left(k y^{\alpha} + d \right)^{\frac{1}{\alpha}}. \tag{10}$$

We leave it as an exercise to the reader to verify that this transformation does indeed satisfy invariance with a power utility function for values of k and d on an interval.

16.3.7 The Invariant Transformation Requires Invariance with k and d on an Interval

As we have seen, shift transformations alone do not characterize the utility function up to a linear transformation: they allow for all exponential utility functions (with all risk aversion coefficients) as well as linear utility functions. Likewise, scale transformations alone are not sufficient to specify the utility function uniquely up to a linear transformation: they specify all power utility functions as well as the logarithmic utility function.

To characterize the utility function up to a linear transformation, it is not sufficient to merely pick a particular value of each of k or d and verify invariance with the transformation (3). Rather, the general invariant transformation requires invariance with both k and d each on an interval of positive length

---

EXAMPLE 16.4: The Invariant Transformation Requires k and d Each on an Interval

---

To further illustrate that invariance is required for both k and d on an interval, suppose we set $k = 1$ and vary only d in (3). By satisfying invariance with only d, we characterize the utility function up to a linear and exponential transformation of the utility function as we demonstrate below.

The invariant transformation with $k = 1$ as we vary d on its feasible domain is

$$g(y, k, d) = U^{-1}\left(U(y) + d \right). \tag{11}$$

We now show that both functions $aU(y)+b$ and $ae^{cU(y)}+b$ satisfy invariance with this transformation:

1. Invariance with $aU(y)+b$.

We get by direct substitution from (11),

$$aU(g(y,k,d))+b = aU\left(U^{-1}[U(y)+d]\right)+b = a[U(y)+d]+b = aU(y)+\beta,$$

where $\beta = ad+b$. This implies that the transformation applied to the argument y, i.e., $U(g(y,k,d))$, results in a linear transformation of $U(y)$. Therefore, $U(y)$ is invariant to (11) (as expected).

2. Invariance with $f(y) = ae^{cU(y)}+b$

Now we show that an exponential form of $U(y)$ also satisfies invariance. We get by direct substitution from (11),

$$f(g(y,k,d)) = ae^{cU(g(y,k,d))}+b = ae^{c[U(y)+d]}+b = ae^{cd}e^{cU(y)}+b = e^{cd}f(y)+b(1-e^{cd}),$$

which implies that $f(g(y,k,d))$ is a linear transform of $f(y)$, and so it is also invariant to (11).

However, because it is required that both k and d be variables on an interval, and not just a particular value of k, it is not sufficient to test with only a single value of $k = 1$. The term $f(y) = ae^{cU(y)}+b$ does not satisfy invariance for values of k on an interval and is thus excluded. If we include k, we get

$$f(g(y,k,d)) = ae^{cU(g(y,k,d))}+b = ae^{c[kU(y)+d]}+b = ae^{cd}e^{ckU(y)}+b = e^{cd}\left(f(y)\right)^{k}+b(1-e^{cd}),$$

which is not a linear transformation of $f(y)$.

16.4 THE g-RISK AVERSION FUNCTION

By now, you might be considering the relation between the risk aversion function and the invariant transformation $g(y,k,d) = U^{-1}\left(kU(y)+d\right)$, because they both define the utility function up to a linear transformation. Moreover,

1. For shift transformations (the delta property), the invariant transformation characterized utility functions satisfying constant absolute risk aversion.
2. For scale transformations (the multiplicative delta property), the invariant transformation characterized utility functions satisfying constant relative risk aversion.

It seems natural that there be a general form of risk aversion function that is characterized by utility functions satisfying invariance with a given transformation. Indeed, this is the case. It will now be useful to define a modified risk aversion function that generalizes both the risk aversion function and the relative risk aversion function, and that sheds light into how the wealth equivalent of a lottery changes when its outcomes are modified by a monotone transformation, g. We refer to this function as the g-risk aversion function.

Define the composite transformation

$$U_g(y) \triangleq U(g(y)).$$

Definition
The g-risk aversion function, $\gamma_g(y)$, for the composite function $U_g(y)$ is:

$$\gamma_g(y) = -\frac{U_g''(y)}{U_g'(y)}.$$

The g-risk aversion will play a significant role in the next chapter when valuing changes in the wealth equivalent of lotteries, and it will also play a role in charactering the change in risk aversion function for multiattribute utility functions as we deviate from the boundary values.

Proposition 16.2: The g-risk aversion function, $\gamma_g(y)$, is related to the absolute risk aversion function, $\gamma(y)$, as follows:

$$\gamma_g(y) = \gamma(g(y))g'(y) - \frac{g''(y)}{g'(y)}.$$

The following examples illustrate the calculation of the g-risk aversion function.

1. **Shift Transformations:**

$$g(y) = y + \delta \Rightarrow g'(y) = 1, \ g''(y) = 0.$$

By definition,

$$\gamma_g(y) = \gamma(y + \delta).$$

2. **Scale Transformations:**

$$g(y) = my \Rightarrow g'(y) = m, \ g''(y) = 0.$$

By definition,

$$\gamma_g(y) = m\gamma(my). \tag{12}$$

3. **Linear Transformations:**

Linear transformations can be interpreted in terms of receiving a fraction of the lottery (or a delayed lottery) and a receiving a deterministic monetary amount in return. For a linear transformation,

$$g(y) = z_1 y + z_2, \ z_1 > 1, z_2 > 0 \Rightarrow g'(y) = z_1 \quad \text{and} \quad g''(y) = 0.$$

By definition,

$$\gamma_g(y) = z_1 \gamma(z_1 y + z_2).$$ (13)

16.5 SUMMARY

This chapter focused on the following problem:

Suppose you have a wealth lottery $< p_1, y_1; ...; p_n, y_n >$ and you value this lottery by a wealth equivalent, $\tilde{y}$, where $\tilde{y} = U^{-1}\left(\sum_{i=1}^{n} p_i U(y_i)\right)$. Now suppose that all the lottery outcomes are modified by a monotone transformation, g, to yield the modified lottery $< p_1, g(y_1); ...; p_n, g(y_n) >$, whose wealth equivalent we denote as $\tilde{g}$, where $\tilde{g} = U^{-1}\left(\sum_{i=1}^{n} p_i U(g(y_i))\right)$. When does $\tilde{g}$, equal to $g(\tilde{y})$ for all lotteries? And what is the most general form of g for a given U?

We showed that $\tilde{g}$ is equal to $g(\tilde{y})$ for all lotteries if and only if $U(g(y)) = kU(y) + d$.

We referred to the equation $U(g(y)) = kU(y) + d$ as the invariance functional equation of a utility function U and a monotone transformation g.

We then considered the opposite question: What is the most general invariant transformation, g, for a given utility function U? We showed that the invariant transformation of a utility function is $g(y, k, d) = U^{-1}(kU(y) + d)$.

These functional equations and the idea of the invariant transformation will help us better understand the following concepts in future chapters:

- One-switch utility functions
- Utility independence
- Valuing changes in investment opportunities
- One-switch utility independence

ADDITIONAL READINGS

Abbas, A. E. 2007. Invariant utility functions and certain equivalent transformations. *Decision Analysis* 4(3): 17–31.

Abbas, A. E. and J. Aczél. 2010. The role of some functional equations in decision analysis. *Decision Analysis* 7(2): 215–228.

Pfanzagl, J. 1959. A general theory of measurement: Applications to utility. *Naval Research & Logistics Quarterly* 6: 283–294.

Pratt, J. 1964. Risk aversion in the small and in the large. *Econometrica* 32: 122–136.

APPENDIX

THEOREM 16.1: THE INVARIANT TRANSFORMATION OF A
UTILITY FUNCTION

Sufficiency: First we show that if $U(g(y)) = kU(y) + d$, then $\tilde{g} = g(\tilde{y})$.. The wealth equivalent of the modified lottery is equal to the transformation g applied to the wealth equivalent of the unmodified lottery. The proof is as follows:

If $U(g(y)) = kU(y) + d$, then the wealth equivalent of the modified lottery is

$$\tilde{g} = U^{-1}\left(\sum_{i=1}^{n} p_i U(g(y_i))\right) = U^{-1}\left(\sum_{i=1}^{n} p_i [kU(y_i) + d]\right) = U^{-1}\left(k\sum_{i=1}^{n} p_i U(y_i) + d\right)$$

But $\sum_{i=1}^{n} p_i U(y) = U(\tilde{y})$ because it is the expected utility of the lottery, which is also the utility of the wealth equivalent. Therefore,

$$\tilde{g} = U^{-1}\left(kU(\tilde{y}) + d\right). \tag{14}$$

But if $U(g(y)) = kU(y) + d$, then

$$g(y) = U^{-1}\left(kU(y) + d\right). \tag{15}$$

Substituting for $y = \tilde{y}$, this implies that

$$g(\tilde{y}) = U^{-1}\left(kU(\tilde{y}) + d\right). \tag{16}$$

From (14) and (16),

$$\tilde{g} = g(\tilde{y}).$$

Necessity: Now we show the reverse direction: if $g(\tilde{y}) = \tilde{g}$ for all lotteries, then we must have $U(g(y)) = kU(y) + d$. The proof goes as follows:

If $g(\tilde{y}) = \tilde{g}$, then by definition,

$$g\left(U^{-1}\left(\sum_{i=1}^{n} p_i U(y_i)\right)\right) = U^{-1}\left(\sum_{i=1}^{n} p_i U(g(y_i))\right).$$

Rearranging gives

$$U^{-1}\left(\sum_{i=1}^{n} p_i U(y_i)\right) = g^{-1}\left(U^{-1}\left(\sum_{i=1}^{n} p_i U(g(y_i))\right)\right) \tag{17}$$

Define $V(y) = U(g(y))$ and its inverse, $V^{-1}(t) = g^{-1}(U^{-1}(t))$.

Substitute into (17) to get

$$U^{-1}\left(\sum_{i=1}^{n} p_i U\ (y_i)\right) = V^{-1}\left(\sum_{i=1}^{n} p_i V(y_i)\right).$$

Equivalently,

$$V\left(U^{-1}\left(\sum_{i=1}^{n} p_i U\ (y_i)\right)\right) = \sum_{i=1}^{n} p_i V(y_i).$$

Define $y_i = U^{-1}(t_i)$, and so $t_i = U(y_i)$. Further define $f = V(U^{-1})$, and substitute to get

$$f\left(\sum_{i=1}^{n} p_i t_i\right) = \sum_{i=1}^{n} p_i f(t_i),$$

Because this equation must be valid for all lotteries, it must be satisfied for values of t_i on an interval. This implies that f is a linear function. Hence,

$$f(t) \triangleq V(U^{-1}(t)) = kt + d \quad \text{(a linear function of } t\text{)}.$$

Note that $V(U^{-1}(t)) = U(g(U^{-1}(t)))$. Moreover, $y = U^{-1}(t)$, and so $t = U(y)$. Therefore,

$$U(g(y)) = kU(y) + d.$$

PROPOSITION 16.1: THE INVARIANT TRANSFORMATION CHARACTERIZES A UTILITY FUNCTION UP TO A LINEAR TRANSFORMATION

Sufficiency: If U_1, U_2 have the same invariant transformation, then

$$U_1^{-1}\left(\kappa U_1(y) + \tau\right) = U_2^{-1}\left(\alpha U_2(y) + \beta\right),$$

where $\kappa, \tau, \alpha, \beta$ are variables that on the domain of feasibility of their respective transformations.

Rearranging gives

$$U_2\left(U_1^{-1}\left(\kappa U_1(y) + \tau\right)\right) = \alpha U_2(y) + \beta, \tag{18}$$

Define $U_1(y) = t \Rightarrow y = U_1^{-1}(t)$, and further define $h(t) = U_2\left(U_1^{-1}(t)\right)$ with $h^{-1}(t) = U_1\left(U_2^{-1}(t)\right)$. Substituting into (18) gives

$$h(\kappa t + \tau) = \alpha h(t) + \beta \tag{19}$$

for some variable t and values of $\kappa, \tau, \alpha, \beta$ each satisfied on an interval. The solution of this functional equation is

$$h(t) = at + b, \quad a \neq 0. \tag{20}$$

Recall that $h(t) = U_2\left(U_1^{-1}(t)\right)$ and $U_1(y) = t$. Substituting into (20) gives

$$U_2(y) = aU_1(y) + b.$$

Furthermore, we can relate the variables $\kappa, \tau, \alpha, \beta$ by substituting from (20) into (19) to get

$$a(\kappa t + \tau) + b = \alpha[at + b] + \beta,$$

which implies that $\kappa = \alpha$ and that $a\tau + b = \alpha b + \beta \Rightarrow \tau = \dfrac{\alpha b(\alpha - 1) + \beta}{a}.$

Necessity: If U_1, U_2 are continuous monotone utility functions related by a linear transformation, then

$$U_2(y) = aU_1(y) + b, \quad a \neq 0. \tag{21}$$

The characteristic transformations are determined by the two equations,

$$U_1(g_1(y, \kappa, \tau)) = \kappa U_1(y) + \tau \Rightarrow g_1(y, \kappa, \tau) = U_1^{-1}\left(\kappa U_1(y) + \tau\right) \tag{22}$$

and

$$U_2(g_2(y, \alpha, \beta)) = \alpha U_2(y) + \beta. \tag{23}$$

Substituting from (21) into (23) gives

$$aU_1(g_2(y, \alpha, \beta)) + b = \alpha[aU_1(y) + b] + \beta. \tag{24}$$

From (24),

$$g_2(y, \alpha, \beta) = U_1^{-1}\left(\frac{\alpha[aU_1(y) + b] + (\beta - b)}{a}\right) = U_1^{-1}\left(\alpha U_1(y) + \frac{b(\alpha - 1) + \beta}{a}\right) \tag{25}$$

Comparing (22) and (25) shows that $g_1(y, \kappa, \tau) = g_2(y, \alpha, \beta)$ with $\alpha = \kappa$ and $\tau = \dfrac{b(\alpha - 1) + \beta}{a}.$

Proof of Proposition 16.2: We can prove this by direct substitution. Taking the first partial derivative of $U_g(y)$ gives

$$U_g'(y) \triangleq \frac{\partial}{\partial y} U_g(y) = \frac{\partial}{\partial y} U(g(y)) = \frac{\partial}{\partial g} U(g(y)) \frac{\partial}{\partial y} g(y),$$

Taking the second partial derivative gives

$$U_g''(y) \triangleq \frac{\partial^2}{\partial y^2} U_g(y) = \frac{\partial^2}{\partial y^2} U(g(y)) = \frac{\partial^2}{\partial g^2} U(g(y)) \left(\frac{\partial}{\partial y} g(y)\right)^2 + \frac{\partial}{\partial g} U(g(y)) \frac{\partial^2}{\partial y^2} g(y).$$

Therefore,

$$\boxed{\gamma_g(y) = -\frac{U_g''(y)}{U_g'(y)} = \gamma(g(y))g'(y) - \frac{g''(y)}{g'(y)}.}$$

CHAPTER 17

Valuing Changes in Uncertain Deals
(In the Small and in the Large)

Chapter Concepts

- The characteristic transformation of a utility function U and a transformation g
- The effects of the shape of the characteristic transformation on the change in value of a lottery
- A measure of change in the certain equivalent of a lottery when it is modified by a transformation, g. We denote this measure of change as η_g.
- A formula that uses the measure of change, η_g, to determine the change in value of a lottery when it is modified by a transformation, g

17.1 INTRODUCTION

In the previous chapters, we discussed properties of utility functions over single attributes and the notions of the risk aversion function and invariant transformation. The invariant transformation provided insights into the change in the wealth equivalent of a lottery when its outcomes are modified by a monotone transformation. If the outcomes are modified by the invariant transformation of a utility function, we simply apply that same invariant transformation to the wealth equivalent to get the wealth equivalent of the modified lottery.

In this chapter, we extend these results and quantify the changes in the wealth equivalent when lottery outcomes are modified by a transformation that need not be the invariant transformation of the utility function. Can we characterize some relations for the new wealth equivalent when the utility function is not invariant to this monotone transformation? This characterization could be useful for many purposes: first, if the new wealth equivalent is less than (or greater than) the same transformation applied to the wealth equivalent of the original lottery, then we can infer certain properties about the utility function. Second, this analysis enables us to determine relatively easily upper or lower bounds on the wealth equivalent of modified lotteries when the utility function is not invariant to the modifying transformation. For example, we might be interested in the value gained due to cost savings in an investment project (which corresponds to a shift transformation on the lottery outcomes), but the utility function is neither linear nor exponential. Alternatively, cost savings might result in

327

some project delay corresponding to a discount factor, which results in a linear (or even non-linear) transformation on the lottery outcomes depending on the cost structure of the firm. It would be convenient to have a method that determines the bounds on the certain equivalent of the modified lottery for general utility functions.

This chapter introduces two measures for characterizing the change in the wealth equivalent of a lottery when its outcomes are modified by a monotone transformation. The first is a ***characteristic transformation*** of a utility function U and a monotone transformation g. The shape of the characteristic transformation determines an upper bound, lower bound, or equality on the certain equivalent of the modified lottery. The second is *a **measure of change in certain equivalent**,η_g*, whose sign determines whether the certain equivalent of the modified lottery is above or below some reference value. For lotteries with small variance, the magnitude of η_g relates the new certain equivalent to the mean, variance, and certain equivalent of the unmodified lottery.

17.2 PROBLEM FORMULATION: BOUNDS ON THE WEALTH EQUIVALENT

Our focus will again be continuous and strictly increasing utility functions U and continuous and strictly increasing transformations g. For convenience, we define utility functions and lotteries over total wealth, y, where $y = x + w$. Our development will consider the following problem.

Problem Formulation: *Suppose you face an uncertain investment whose prospects are characterized by the wealth lottery,* $< p_1, y_1; ...; p_n, y_n >$, *and you value this lottery by a wealth equivalent,$\tilde{y}$, where*

$$\tilde{y} = U^{-1}\left(\sum_{i=1}^{n} p_i U(y_i) \right).$$

Now suppose that, due to some changes in this investment project, all the lottery prospects are modified by a transformation, g, to yield the modified lottery $< p_1, g(y_1); ...; p_n, g(y_n) >$. How does the wealth equivalent of the modified lottery, which we denote as $\tilde{g}$, where

$$\tilde{g} = U^{-1}\left(\sum_{i=1}^{n} p_i U(g(y_i)) \right),$$

compare to $g(\tilde{y})$?

Is $\tilde{g} > g(\tilde{y})$ or $\tilde{g} < g(\tilde{y})$ or $\tilde{g} = g(\tilde{y})$?

Figure 17.1 illustrates the problem formulation.

We are interested in the change in value of a lottery in relation to the reference value $g(\tilde{y})$. In other words,

Is $\tilde{g} > g(\tilde{y})$ or $\tilde{g} < g(\tilde{y})$ or $\tilde{g} = g(\tilde{y})$?

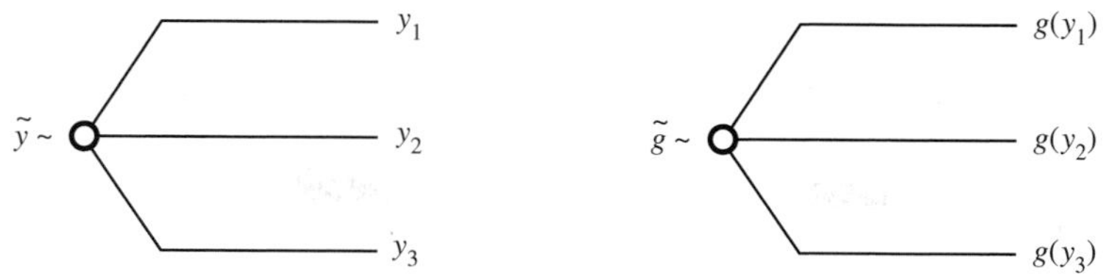

Figure 17.1. What is the relation between $\tilde{g}$ and $g(\tilde{y})$? Is $\tilde{g} > g(\tilde{y})$ or $\tilde{g} < g(\tilde{y})$ or $\tilde{g} = g(\tilde{y})$?.

Substituting for the values of $\tilde{g}$ and $\tilde{y}$ helps rephrase this question as:

$$\text{Is } U^{-1}\left(\sum_{i=1}^{n} p_i U(g(y_i))\right) >, <, = g\left(U^{-1}\left(\sum_{i=1}^{n} p_i U(y_i)\right)\right)?$$

It is straightforward to see that if $g(y) = y$, then there is no change in the lottery, and so

$$\tilde{g} = g(\tilde{y}) = \tilde{y}.$$

Our main focus will therefore be transformations where $g(y) \neq y$. We consider three scenarios:

1. $\tilde{g} = g(\tilde{y})$ for all lotteries:
   We saw in Chapter 16 that this condition is satisfied for a given utility function U and for all lotteries if and only if the transformation is the invariant transformation of the utility function, which implies that $U(g(y)) = kU(y) + d$ or

   $$g(y,k,d) = U^{-1}(kU(y) + d),$$

   where k and d are variables whose domain is defined by the condition that $kU(y) + d$ lies on the range of U.
2. If $\tilde{g} > g(\tilde{y})$ for all lotteries, i.e., if the wealth equivalent of the modified lottery is greater than or equal to the same transformation applied to the wealth equivalent of the unmodified lottery, and if this applies to all lotteries, then $g(\tilde{y})$, which is straightforward to calculate, provides a ***lower bound*** on the new certain equivalent. To calculate this lower bound, we simply apply a monotonic transformation to the previously calculated certain equivalent. But how can we tell whether or not $\tilde{g} > g(\tilde{y})$ from the utility function and the monotone transformation? We answer this question in this chapter.
3. If $\tilde{g} < g(\tilde{y})$ for all lotteries, then $g(\tilde{y})$ provides an ***upper bound*** on the new certain equivalent. Once again, this upper bound is straightforward to calculate. But how can we assert whether $g(\tilde{y})$ is an upper bound? We answer this question below.

17.3 THE CHARACTERISTIC TRANSFORMATION OF A UTILITY FUNCTION, U, AND A TRANSFORMATION, g

Definition: The ***characteristic transformation*** of a utility function U and a monotone transformation g is

$$m(t) = U(g(U^{-1}(t))). \tag{1}$$

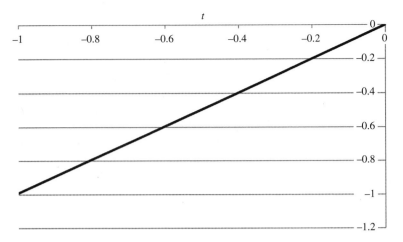

Figure 17.2. Characteristic transformation for an exponential utility function and shift transformation.

Note that the domain of t is defined on the range of U. The shape of the characteristic transformation determines the relative magnitude of $\tilde{g}$ and $g(\tilde{y})$ as we shall demonstrate in this section. First, we illustrate some examples of characteristic transformations for common transformations and common utility functions.

EXAMPLE 17.1: Shift Transformation with Exponential Utility Function

---

Consider the exponential utility function, $U(y) = -e^{-\gamma y}, \gamma = 0.01$, and a shift transformation, $g(y) = y + \delta, \delta = 4$. The characteristic transformation is

$$m(t) = -e^{-\gamma\left(\delta + \left(-\frac{1}{\gamma}\ln(-t)\right)\right)} = -te^{-\gamma\delta},$$

which is a linear function of t. Figure 17.2 plots this curve for $t = -1$ to 0.

EXAMPLE 17.2: Scale Transformation with Exponential Utility Function

---

Consider the exponential utility function, $U(y) = -e^{-\gamma y}, \gamma = 0.01$, and a scale transformation, $g(y) = \delta y, \delta = 3$. The characteristic transformation is

$$m(t) = -e^{-\gamma\left(\delta\left(-\frac{1}{\gamma}\ln(-t)\right)\right)} = -(-t)^{\delta} = t^3,$$

which is a non-linear function of t. Figure 17.3 plots this curve for $t = -1$ to 0.

EXAMPLE 17.3: Power Transformation with Exponential Utility Function

---

Consider the exponential utility function, $U(y) = -e^{-\gamma y}, \gamma = 0.01$, and a scale transformation, $g(y) = y^{\delta}, \delta = 1.3$. The characteristic transformation is

$$m(t) = -e^{-\gamma\left(-\frac{1}{\gamma}\ln(-t)\right)^{\delta}},$$

which is a linear function of t. Figure 17.4 plots this curve for $t = -1$ to 0.

EXAMPLE 17.4: Logarithmic Utility Function with Shift, Scale, and Power Transformations

---

In a similar manner Figures 17.5, 17.6, and 17.7 show the characteristic transformation for a logarithmic utility function $U(y) = \ln y$ with shift, scale, and power transformations.

Examples 17.1 through 17.4 demonstrate that the characteristic transformation can have many shapes depending on the utility function and the monotone transformation.

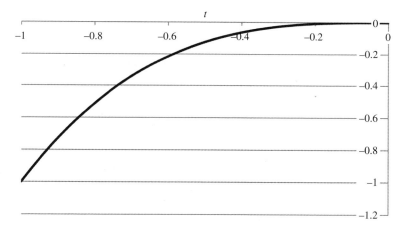

Figure 17.3. Characteristic transformation for an exponential utility function and scale transformation.

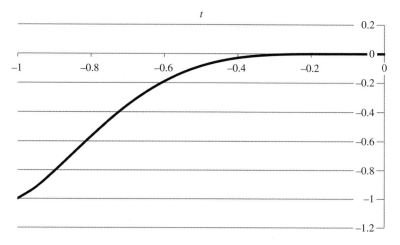

Figure 17.4. Characteristic transformation for an exponential utility function and a power transformation.

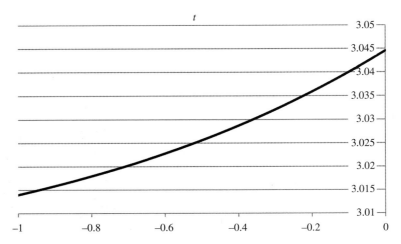

Figure 17.5. Characteristic transformation for a logarithmic utility function and shift transformation $g(y) = y + 20$.

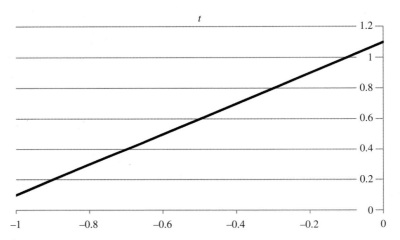

Figure 17.6. Characteristic transformation for a logarithmic utility function and scale transformation $g(y) = 3y$.

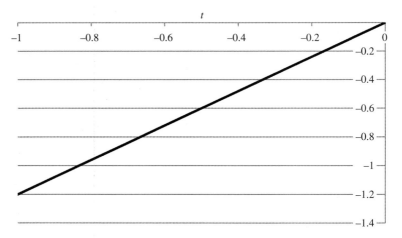

Figure 17.7. Characteristic transformation for a logarithmic utility function and a power transformation $g(y) = y^{1.2}$.

The following Theorem 17.1 explains the importance of the shape of the characteristic transformation.

Theorem 17.1: The Shape of the Characteristic Transformation

Given a continuous and strictly increasing utility function U, and a monotone transformation g applied to the wealth outcomes of a lottery, the following statements hold:

a) $\tilde{g} = g(\tilde{y})$ *for all lotteries if and only if $m(t)$ is a linear function of t.*
b) $\tilde{g} > g(\tilde{y})$ *for all lotteries if and only if $m(t)$ is a strictly convex function of t.*
c) $\tilde{g} < g(\tilde{y})$ *for all lotteries if and only if $m(t)$ is a strictly concave function of t.*

Theorem 17.1 explains the importance of the characteristic transformation: its shape determines whether we have an upper bound, lower bound, or equality on the value of the new certain equivalent in relation to the reference value $g(\tilde{y})$.

Theorem 17.1 also shows that the deviation of the characteristic transformation from linearity drives the inequality between $\tilde{g}$ and $g(\tilde{y})$. When the characteristic transformation is linear, we get $\tilde{g} = g(\tilde{y})$. As the characteristic transformation deviates from linearity, either strictly concave or convex, $\tilde{g}$ deviates from $g(\tilde{y})$. Theorem 17.1 also enables us to determine the bounds on the new certain equivalent in this case. We illustrate some applications of Theorem 17.1 through the following examples.

17.4 MODIFYING LOTTERIES BY SHIFT TRANSFORMATIONS

Suppose that the outcomes of a lottery are modified by a shift transformation

$$g(t) = t + z.$$

By definition, the characteristic transformation associated with this shift transformation is

$$m(t) = U\left(U^{-1}(t) + z\right),$$

for values of t that lie on the range of the utility function.

Note: This characteristic transformation, $m(t)$, is the general solution to the functional equation

$$m(m(t, z_1), z_2) = m(t, z_1 + z_2),$$

which is commonly known as the translation equation (see, for example, Aczel 1966). This equation implies that applying the characteristic transformation n times to a variable t is equivalent to applying it only once, using the parameter $z_1 + z_2 + \ldots + z_n$.

EXAMPLE 17.5: The Shape of the Characteristic Transformation with Shift Transformations

---

We now calculate $m(t)$ for various utility functions to determine the relation between $\tilde{g}$ and $g(\tilde{y})$ for shift transformations directly.

1. For linear and exponential utility functions,

$$Linear: U(y) = cy + w \Rightarrow m(t) = t + cz,$$

and

$$Exponential: U(y) = a + be^{-y} \Rightarrow m(t) = te^{-z} + a(1 - e^{-z}),$$

both of which are *linear* in t. From Theorem 17.1, we can assert that

$$\tilde{g} = g(\tilde{y}) = \tilde{y} + z,$$

with these utility functions.

2. For a logarithmic utility function,

$$U(y) = \ln(y) \Rightarrow m(t) = \ln(e^t + z), \ z > 0,$$

which is a strictly *convex* function of t. From Theorem 17.1, we can assert that

$$\tilde{g} > g(\tilde{y}) = \tilde{y} + z, \ z > 0,$$

and therefore $\tilde{y} + z$ presents a lower bound on the new certain equivalent.

3. For a power utility function,

$$U(x) = (x + w)^{\alpha} \Rightarrow m(t) = \left(t^{\frac{1}{\alpha}} + z \right)^{\alpha}, \ z > 0,$$

We consider three cases:

(i) $\alpha > 1$: $m(t)$ is a strictly *concave* function of t and so $\tilde{g} < \tilde{y} + z$;
(ii) $\alpha < 1$: $m(t)$ is a strictly *convex* function of t and so $\tilde{g} > \tilde{y} + z$,
(iii) $\alpha = 1$: $m(t)$ is a *linear* function of t and so $\tilde{g} = \tilde{y} + z$.

Thus we can get either an upper bound, lower bound, or equality, depending on the value of α.

Dialogue between Instructor (I) and Student (S)

I: Does everyone see what is happening here?

S: Yes, with the characteristic transformation, we have generalized the ideas of the delta property and sub-delta and super-delta properties to more general monotone transformations besides just shift or scale. We have also interpreted this phenomenon in terms of the shape of the characteristic transformation.

I: Yes, we construct the characteristic transformation for a utility function and a monotone transformation and its shape tells us the relation between the wealth equivalent of the modified lottery and the same transformation applied to the previous wealth equivalent. Special cases of this formulation include the additive transformation and the scale transformation that we discussed earlier.

S: One piece of the puzzle is still missing.

I: What is that?

S: With the risk aversion function, we were able to characterize the magnitude of the difference between the mean value of a lottery and the certain equivalent. Can we define some function that characterizes the magnitude of the change in wealth equivalent when the prospects are modified by monotone transformations? For example, is there a function that can quantify the difference between $\tilde{g}$ and $g(\tilde{y})$?

I: Yes, we can! Let's do that now.

17.5 A MEASURE OF CHANGE IN WEALTH EQUIVALENT

The shape of the characteristic transformation $m(t)$ determines the relative magnitude of $\tilde{g}$ and $g(\tilde{y})$. This section derives a new measure whose sign defines the relative magnitude of $\tilde{g}$ and $g(\tilde{y})$, and whose magnitude helps quantify the difference between them.

Our focus will now be utility functions and transformations that are twice differentiable. We assume that the first derivative of the utility function is either positive or negative, while the first derivative of the monotone transformation is positive.

Define the composite transformation of the utility function and the monotone transformation as

$$U_g(y) = U(g(y)),$$

and recall the definition of the *g-risk aversion function* from Chapter 16,

$$\gamma_g(y) = -\frac{U_g''(y)}{U_g'(y)} = \gamma(g(y))g'(y) - \frac{g''(y)}{g'(y)}. \qquad (2)$$

Definition

The ***measure of change in wealth equivalent*** $\eta_g(y)$ is the difference between the g-risk aversion function and the absolute risk aversion function:

$$\eta_g(y) = \gamma_g(y) - \gamma(y).$$

EXAMPLE 17.6: Calculating the Measure of Change $\eta_g(y)$ for an Exponential Utility Function

We calculate $\eta_g(y)$ for an exponential utility $\gamma(y) = \gamma_0$. From (0), the g-risk aversion function is

$$\gamma_g(y) = \gamma(g(y))g'(y) - \frac{g''(y)}{g'(y)} = \gamma_0 g'(y) - \frac{g''(y)}{g'(y)}.$$

Hence, the measure of change in wealth equivalent is

$$
\boxed{
\begin{aligned}
\eta_g(y) &= \gamma_g(y) - \gamma(y) \\
&= \gamma_0[g'(y) - 1] - \frac{g''(y)}{g'(y)}.
\end{aligned}
}
$$

For a shift transformation,

$$g(y) = y + z, g'(y) = 1, g''(y) = 0.$$

Hence, for an exponential utility function and a shift transformation, $\eta_g(y) = 0$.
For a scale transformation,

$$g(y) = my, g'(y) = m, g''(y) = 0.$$

Hence, for an exponential utility function and a scale transformation,

$$\eta_g(y) = \gamma_0(m - 1).$$

The following Theorem 17.2 presents an application of the measure $\eta_g(y)$.

Theorem 17.2

The measure of change $\eta_g(y)$ is invariant under a linear transformation on the utility function, and it satisfies the following conditions:

$$
\mathrm{sgn}(\eta_g(y)) =
\begin{cases}
+1 & \forall y \Leftrightarrow \tilde{g} < g(\tilde{y}) \text{ for all lotteries} \\
0 & \forall y \Leftrightarrow \tilde{g} = g(\tilde{y}) \text{ for all lotteries} \\
-1 & \forall y \Leftrightarrow \tilde{g} > g(\tilde{y}) \text{ for all lotteries}
\end{cases}
$$

Theorem 17.2 shows that the sign of $\eta_g(y)$ determines the relative magnitude of $\tilde{g}$ in relation to the reference value $g(\tilde{y})$. This result is invariant to a linear transformation on the utility function. From Theorems 17.1 and 17.2, we can also see how the sign of $\eta_g(y)$ determines the shape of the characteristic transformation.

We have seen that an exponential utility function and a shift transformation have a measure of change $\eta_g(y) = 0$. This asserts that $\tilde{g} = g(\tilde{y})$ for an exponential utility function and a shift transformation for all lotteries. This assertion is in fact further validation that the exponential utility function satisfies the delta property.

The following examples illustrate an application of Theorem 17.2 for some common transformations.

EXAMPLE 17.7: Shift Transformations and General Utility Functions

---

$$g(y) = y + z, \ z > 0 \Rightarrow g'(y) = 1, \ g''(y) = 0.$$

By definition,

$$\eta_g(y) = \gamma(y + z) - \gamma(y). \tag{3}$$

Equation (3) shows that for a shift transformation, $\eta_g(y)$ is zero when $\gamma(y+z) = \gamma(y)$, which can only be true for constant absolute risk aversion (linear or exponential utility functions) because the function γ does not depend on z. This implies that for constant absolute risk aversion, $\tilde{g} = g(\tilde{y})$ for shift transformations (a verification of what we have proved earlier).

Equation (3) shows that $\eta_g(y)$ is negative when $\gamma(y+z) < \gamma(y)$, which can only be the case for decreasing absolute aversion because $z > 0$. This would apply to logarithmic utility functions or power utility functions when $\alpha < 1$.

Moreover, $\eta_g(y)$ is positive when $\gamma(y+z) > \gamma(y)$, which can only be true for increasing absolute aversion such as power utility functions when $\alpha > 1$.

EXAMPLE 17.8: Scale Transformations and General Utility Functions

---

$$g(y) = zy,\ z > 1 \ \Rightarrow \ g'(y) = z,\ g''(y) = 0.$$

By definition,

$$\eta_g(y) = z\gamma(zy) - \gamma(y) = \frac{r(zy) - r(y)}{y}, \tag{4}$$

where $r(y) = y\gamma(y)$ is the relative risk aversion function.

For scale transformations, $\eta_g(y)$ is zero when $r(zy) = r(y)$, which implies constant relative risk aversion, such as power or logarithmic utility functions.

$\eta_g(y)$ is negative when $r(zy) < r(y)$, $z > 1$, which implies decreasing relative risk aversion with wealth, and is positive when $r(zy) > r(y)$, $z > 1$, which implies increasing relative risk aversion with wealth.

EXAMPLE 17.9: Linear Transformations and General Utility Functions

---

Linear transformations can be interpreted in terms of receiving a fraction of the lottery (or a delayed lottery) and a receiving a deterministic monetary amount in return. For a linear transformation,

$$g(y) = z_1 y + z_2,\ z_1 > 1, z_2 > 0 \Rightarrow g'(y) = z_1 \text{ and } g''(y) = 0.$$

By definition,

$$\eta_g(y) = z_1 \gamma(z_1 y + z_2) - \gamma(y). \tag{5}$$

For linear transformations, $\eta_g(y)$ is zero when $z_1\gamma(z_1 y + z_2) = \gamma(y)$, it is negative when $z_1\gamma(z_1 y + z_2) < \gamma(y)$, which implies that the certain equivalent of a lottery modified by a linear transformation will be greater than $z_1\tilde{y} + z_2$, and it is positive when $z_1\gamma(z_1 y + z_2) > \gamma(y)$, which implies that the certain equivalent of a lottery modified by a linear transformation will be less than $z_1\tilde{y} + z_2$.

The next section presents further applications of the magnitude of the measure η_g and an interpretation of its magnitude for small risks.

17.6 VALUING CHANGES IN THE SMALL

How does the magnitude of $\tilde{g}$ compare to $g(\tilde{y})$ for lotteries whose support forms a small neighborhood around the mean? These lotteries also have a small variance. We now quantify the difference between $\tilde{g}$ and $g(\tilde{y})$ using the magnitude of η_g.

Theorem 17.3: Valuing Changes in the Small

$$\tilde{g} = g\left(\tilde{y} - \frac{\sigma^2}{2}\eta_g(\mu) + o(\sigma^2)\right) \text{ as } \sigma^2 \to 0. \tag{6}$$

Theorem 17.3 interprets the magnitude of η_g in terms of the change in certain equivalent. For lotteries whose support is small, the term $\dfrac{\sigma^2}{2}\eta_g(\mu)$ fills in the inequality gap up to an order of $o(\sigma^2)$ in the argument of g. In the limit, when $\sigma^2 = 0$, we get $\tilde{g} = g(\tilde{y})$, which must be satisfied for deterministic lotteries.

EXAMPLE 17.10: Numerical Illustration: Valuing Changes for a Logarithmic Utility Function

Consider again the wealth lottery of Example 13.2, <0.5, 1000; 0.5, 5000> and the logarithmic utility function $U(y) = \ln(y)$. The wealth equivalent of this lottery using this logarithmic utility function is (as we have seen) $\tilde{y} = \$2236.07$.

We now apply the linear transformation

$$g(y) = 0.1y + 200$$

to the lottery outcomes and observe the change in wealth equivalent. The new lottery is

$$<0.5, 300; 0.5, 700>$$

We now calculate the g-risk aversion and the measure of change. We know that for a logarithmic utility function, $\gamma(y) = \dfrac{1}{y}$. As we have seen, the g-risk aversion function is

$$\gamma_g(y) = z_1\gamma(z_1 y + z_2) = 0.1\gamma(0.1y + 200) = \frac{0.1}{0.1y + 200}.$$

Therefore, the measure of change is

$$\eta_g(y) = z_1\gamma(z_1 y + z_2) - \gamma(y) = \frac{0.1}{0.1y + 200} - \frac{1}{y} = \frac{0.1y - (0.1y + 200)}{y(0.1y + 200)} = -\frac{200}{y(0.1y + 200)}.$$

Because the measure of change is negative, we know that $\tilde{g} > g(\tilde{y})$ for all lotteries, and therefore we also have a lower bound on the wealth equivalent of the modified lottery, which is $g(\tilde{y})$.

Let us verify this result. The wealth equivalent of the modified lottery is

$$\tilde{g} = \$458.26$$

On the other hand, the same transformation applied to the wealth equivalent of the original lottery is

$$g(\tilde{y}) = z_1\tilde{y} + z = 0.1 \times 2236.07 + 200 = \$423.61$$

This verifies that $\tilde{g} > g(\tilde{y})$ as expected.

Furthermore, let us examine the accuracy of the approximation

$$\tilde{g} \approx g\left(\tilde{y} - \frac{\sigma^2}{2}\eta_g(\mu)\right),$$

even though this lottery has a large variance. Direct calculations give $\mu = 3000$, $\sigma^2 = 4000000$, and therefore $\eta_g(\mu) = -0.00013$. Substituting into the approximate expression gives

$$\tilde{g} \approx g\left(\tilde{y} - \frac{\sigma^2}{2}\eta_g(\mu)\right) = 0.1\left(2236.07 - \frac{4000000}{2} \times -0.00013\right) = \$450.28$$

Compare this to the exact expression of $\$458.26$, which is close given the large variance of this lottery.

17.7 JENSEN'S INEQUALITY GAP FOR SMALL VARIANCE

We have previously discussed the use of Jensen's inequality as it determines the relative magnitude of the expectation of a function to the function of the expectation, i.e., it determines the relative magnitude of $E[g(y)]$ and $g(\infty)$. We can now use the measure of change to quantify the difference between them. We illustrate this below.

As a simple application of Theorem 17.3, consider the special case of the linear utility function. As we have seen, the certain equivalent in this case is equal to the expected value of the lottery. For a linear utility function, $\gamma(y) = 0$. Direct substitution gives the measure of change as

$$\eta_g(y) = -\frac{g''(y)}{g'(y)}$$

for a linear utility function.

From Theorem 17.3, with $\tilde{y} = \mu$, writing $\tilde{g} = E[g(y)]$, we get

$$\boxed{E[g(y)] = g\left(\mu + \frac{\sigma^2}{2}\frac{g''(y)}{g'(y)} + o(\sigma^2)\right), \text{ as } \sigma^2 \to 0.} \tag{7}$$

Equation (2) fills in Jensen's inequality gap for small risks.

Observe the following:

If g is strictly concave, then $g'' < 0$, and so $E[g(y)] < g(\mu)$, an indication of Jensen's inequality.

If g is strictly convex, then $g'' > 0$, and so $E[g(y)] > g(\mu)$. The inclusion of the term $\dfrac{\sigma^2}{2} \dfrac{g''(y)}{g'(y)}$ on the right hand side fills in the Jensen's inequality gap for non-linear transformations and small risks.

As an approximation, we can also write for small risks,

$$E[g(y)] \approx g\left(\mu + \frac{\sigma^2}{2} \frac{g''(\mu)}{g'(\mu)} \right).$$

EXAMPLE 17.11: Numerical Illustration: Jensen's Inequality Gap: Linear Utility Function

---

Consider again the wealth lottery <0.5, 1000; 0.5, 5000> and the linear utility function $U(y) = y$. The wealth equivalent of this lottery using this linear utility function is its mean, $\tilde{y} = \$3000$. Its standard deviation is $\sigma = 2000$.

We now apply the power-scale transformation $g(y,m,k) = 3y^2$ to the lottery outcomes and observe the change in wealth equivalent. Note that this transformation is convex.

The modified lottery is

<0.5, 3,000,000; 0.5, 75,000,000>.

The wealth equivalent of the modified lottery is

$$0.5 \times 3,000,000 + 0.5 \times 75,000,000 = \$39,000,000$$

If we apply the same transformation to the previous wealth equivalent, we get $3\tilde{y}^2 = 3(3000)^2 = 27,000,000$. We can do better using the approximate expression for the change in certain equivalent.

The measure of change is $\eta_g(y) = -\dfrac{g''(y)}{g'(y)} = -\dfrac{6}{6y} = -\dfrac{1}{y}$, which is negative. The approximate expression using the measure of change is

$$E[g(y)] \approx g\left(\mu + \frac{\sigma^2}{2} \frac{g''(\mu)}{g'(\mu)} \right) = 3\left(3000 + \frac{(2000)^2}{2} \times \frac{1}{3000} \right)^2 = 40,333,333.$$

17.8 SUMMARY

The characteristic transformation of a utility function and monotone transformation is

$$m(t) \triangleq U(g(U^{-1}(t))).$$

The shape of the characteristic transformation detertmines the change in value of a lottery when its outcomes are modified by a monotone transformation.

- The g-risk aversion function is

$$\gamma_g(y) = -\frac{U_g''(y)}{U_g'(y)} = \gamma(g(y))g_y(y) - \frac{g_{yy}(y)}{g_y(y)},$$

where $U_g(y) = U(g(y))$ and $\gamma_g(y) = -\dfrac{U_g''(y)}{U_g'(y)}$.

- A *measure of change in wealth equivalent*, $\eta_g(y)$, is

$$\eta_g(y) = \gamma_g(y) - \gamma(y).$$

- Implications of the sign of the measure of change are

$$\text{sgn}(\eta_g(y)) = \begin{cases} +1 & \forall y \Leftrightarrow \tilde{g} < g(\tilde{y}) \text{ for all lotteries} \\ 0 & \forall y \Leftrightarrow \tilde{g} = g(\tilde{y}) \text{ for all lotteries} \\ -1 & \forall y \Leftrightarrow \tilde{g} > g(\tilde{y}) \text{ for all lotteries} \end{cases}$$

- Implications of the magnitude of the measure of change are

$$\tilde{g} = g\left(\tilde{y} - \frac{\sigma^2}{2}\eta_g(\mu) + o(\sigma^2)\right) \quad \text{as } \sigma^2 \to 0.$$

ADDITIONAL READINGS

Abbas, A. E. 2007. Invariant utility functions and certain equivalent transformations. *Decision Analysis* 4(3): 17–31.

Abbas, A. E. 2012. Valuing changes in investment opportunities. *Operations Research* 60(6): 1451–1460.

Abbas, A. E. and J. Aczél. 2010. The role of some functional equations in decision analysis. *Decision Analysis* 7(2): 215–228.

Aczel, J. 1966. *Lectures on Functional Equations and Their Applications*. Academic Press, New York.

Arrow, K. J. 1965. *The Theory of Risk Aversion. Lecture 2 in Aspects of the Theory of Risk-Bearing*. Yrjo Jahnssonin Saatio, Helsinki.

Pratt, J. 1964. Risk aversion in the small and in the large. *Econometrica* 32: 122–136.

APPENDIX 17A

Proof of Theorem 17.1

Sufficiency: If $m(t) = U(g(U^{-1}(t)))$ is a strictly convex function of t, then by Jensen's inequality, $m(\sum_{i=1}^{n} p_i t_i) < \sum_{i=1}^{n} p_i m(t_i)$, for $\sum_{i=1}^{n} p_i = 1, p_i > 0$. Put $t_i = U(y_i)$.

This implies that $U(g[U^{-1}\left(\sum\limits_{i=1}^{n} p_i U(y_i)\right)]) < \sum\limits_{i=1}^{n} p_i U(g(y_i))$. Rearranging gives

$g[U^{-1}\left(\sum\limits_{i=1}^{n} p_i U(y_i)\right)] < U^{-1}\left(\sum\limits_{i=1}^{n} p_i U(g(y_i))\right)$, or $g(\tilde{y}) < \tilde{g}$. The proof of strict concavity

follows the exact same analysis by reversing the inequality sign, and the proof of linearity follows from Jensen's equality instead of the inequality.

Necessity: If $g(\tilde{y}) < \tilde{g}$, then $g[U^{-1}\left(\sum\limits_{i=1}^{n} p_i U(y_i)\right)] < U^{-1}\left(\sum\limits_{i=1}^{n} p_i U(g(y_i))\right)$.

Define $V_g(y) = U(g(y))$, and substitute to get $V_g^{-1}\left(\sum\limits_{i=1}^{n} p_i V_g(y_i)\right) > U^{-1}\left(\sum\limits_{i=1}^{n} p_i U\ (y_i)\right)$.

Define $y_i = U^{-1}(t_i)$, $f = V_g(U^{-1})$, and substitute to get $\left(\sum\limits_{i=1}^{n} p_i f(t_i)\right) > f\left(\sum\limits_{i=1}^{n} p_i t_i\right)$, which

implies that f is a strictly convex function. Hence, $U(g(U^{-1}(y))) \triangleq m(y)$ is a strictly convex function of y. The proof of strict concavity follows by reversing the inequality sign, and the proof of linearity follows from Jensen's equality.

Proof of Theorem 17.2

Define $d(y) = -\dfrac{m''(y)}{m'(y)}$. The following Lemma will be useful in our development.

Lemma: Relating the measure of change to the negative ratio of the second to the first derivative of the characteristic transformation.

$$\eta_g(y) = -d(U(y))U'(y)$$

Proof of Lemma: By definition, $m(t) = U(g(U^{-1}(t)))$. Hence, $U(g(y)) = m(U(y))$. Taking the first and second derivatives with respect to y gives

$$U'(g(y))g'(y) = m'(U(y))U'(y).$$
$$U''(g(y))\left(g'(y)\right)^2 + U'(g(y))g''(y) = m''(U(y))\left(U'(y)\right)^2 + m'(U(y))U''(y).$$

Therefore,

$$\frac{U''(g(y))}{U'(g(y))} g'(y) + \frac{g''(y)}{g'(y)} = \frac{m''(U(y))}{m'(U(y))} U'(y) + \frac{U''(y)}{U'(y)}.$$

Rearranging gives

$$-\frac{U''(g(y))}{U'(g(y))} g'(y) + \frac{U''(y)}{U'(y)} - \frac{g''(y)}{g'(y)} = \frac{m''(U(y))}{m'(U(y))} U'(y).$$

By definition of the risk aversion function, we can write

$$\gamma(g(y))g'(y) - \gamma(y) - \frac{g''(y)}{g'(y)} = \frac{m''(U(y))}{m'(U(y))} U'(y).$$

From the definition of the g-risk aversion function, $\gamma_g(y) = \gamma(g(y))g'(y) - \dfrac{g''(y)}{g'(y)}$

This implies that

$$\gamma_g(y) - \gamma(y) = -\frac{m_{UU}(U(y))}{m_U(U(y))}U_y(y) = d(U(y))U_y(y) \triangleq \eta_g(y).$$

Now, we prove the theorem. First, we show that $\eta_g(y)$ is invariant under a linear transformation on the utility function. The proof goes as follows:

If $m(t)$ is the characteristic transformation of U, g, then the characteristic transformation of $a + bU$ is by definition equal to $a + bm(\dfrac{t-a}{b})$. By direct differentiation, the negative ratio of the second to first derivative of $a + bm(\dfrac{t-a}{b})$ is $\dfrac{1}{b}d(\dfrac{t-a}{b})$ and the first derivative of $a + bU$ is bU_x. Hence, the product of the two terms is $\dfrac{1}{b}d(\dfrac{t-a}{b})bU_x(x) = d(\dfrac{t-a}{b})U_x(x)$. Substituting for $t = a + bU$ gives $d(U(x))U_x(x)$, which is independent of both a and b. Therefore, we can assume without loss of generality that U is increasing.

Now if $d(t)$ is negative ($m(t)$ is a strictly convex function of t), then from Theorem 17.1, we know that $g(\tilde{y}) < \tilde{g}$, and if $d(t)$ is positive ($m(t)$ is a strictly concave function of t), then $g(\tilde{y}) > \tilde{g}$. Finally, if $d(t) = 0$ is zero ($m(t)$ is a linear function of t), then $g(\tilde{y}) = \tilde{g}$.

Proof of Theorem 17.3

We prove this theorem in four steps.

Step 1: First, we prove that

$$\boxed{U(\tilde{g}) = m(U(\mu)) + \frac{1}{2}\frac{\partial^2 m(U(\mu))}{\partial y^2}\sigma^2 + o(\sigma^2), \text{ as } \sigma^2 \to 0,}$$

where $o(\sigma^2)$ is the "small o" notation, a function that goes to zero faster than σ^2 as $\sigma^2 \to 0$.

More formally, this "small o" notation $o(\sigma^2)$ implies that for every positive M, there exists σ_o^2 such that $o(\sigma^2) \leq M\sigma^2$ for $\sigma^2 < \sigma_o^2$. The small o notation is stronger than the big O notation, which implies only the existence of a positive M. The former case applies to all positive M.

Proof. We prove this step by taking a Taylor expansion of $U(g(y))$ around $U(g(\infty))$. The region of convergence of the Taylor expansion is feasible if the domain of the lottery $U(g(\infty))$ forms a sufficiently small neighborhood around ∞. We have, as $y \to \mu$,

$U(g(y)) = m(U(y))$
$\quad = m(U(\mu)) + \dfrac{\partial m(U(y))}{\partial y}\Big|_{y=\mu}(y-\mu) + \dfrac{1}{2}\dfrac{\partial^2 m(U(y))}{\partial y^2}\Big|_{y=\mu}(y-\mu)^2 + o(|y-\mu|^2),$

where $o(|y-\mu|^2)$ converges to zero faster than $(y-\mu)^2$ as $y \to \mu$. More precisely, $o(|y-\mu|^2)$ implies that for every positive M, there exists y_0 in the neighborhood of ∞ such that $o(|y-\mu|^2) \leq M|y-\mu|^2$ as $y \to \mu$ (within this neighborhood defined by $\|y-\mu\| < \|y_o - \mu\|$).

The expected utility of the modified lottery obtained using the Taylor expansion for lotteries whose domain forms a small neighborhood around $\propto$, i.e., $y_{\min}, y_{\max} \to \mu$, and so $y \to \mu \ \forall y \in [y_{\min}, y_{\max}]$ gives

$$U(\tilde{g}) = \int_{y_{\min}}^{y_{\max}} f(y)U(g(y))dy = \int_{y_{\min}}^{y_{\max}} f(y)m(U(y))dy$$

$$= m(U(\mu)) + 0 + \frac{1}{2}\frac{\partial^2 m(U(\mu))}{\partial y^2}\sigma^2 + \int_{y_{\min}}^{y_{\max}} f(y)o(|y-\mu|^2)dy.$$

(8)

Note that as $\lim\limits_{y \to \mu} \int_{y_{\min}}^{y_{\max}} f(y)o(|y-\mu|^2)dy = \lim\limits_{\sigma^2 \to 0} o(\sigma^2)$. Therefore, (2) becomes

$$U(\tilde{g}) = m(U(\mu)) + \frac{1}{2}\frac{\partial^2 m(U(\mu))}{\partial y^2}\sigma^2 + o(\sigma^2), \text{ as } \sigma^2 \to 0.$$

Now define $\tilde{x}$ as

$$U(\tilde{g}) = U(g(\tilde{x})) \Rightarrow \tilde{x} = g^{-1}(\tilde{g}).$$

The transformation g is invertible as per our assumptions.

Step 2: We prove that

$$\boxed{\tilde{x} - \mu = O(\sigma^2), \text{ as } \tilde{x} \to \mu,}$$

where $O(\sigma^2)$ is the "big O" notation, which implies that there exists N such that $O(\sigma^2) \leq N\sigma^2$ as $\sigma^2 \to 0$.

Proof: Using the definition of $U^{Mod}(y) = U(g(y))$, and the results of Step 1, with the observation that $m(U(\mu)) = U(g(\mu)) = U^{Mod}(\mu)$, gives

$$U^{Mod}(\tilde{x}) = U^{Mod}(\mu) + \frac{1}{2}\frac{\partial^2 m(U(\mu))}{\partial y^2}\sigma^2 + o(\sigma^2), \text{ as } \sigma^2 \to 0.$$

This implies that

$$U^{Mod}(\tilde{x}) - U^{Mod}(\mu) = \frac{1}{2}\frac{\partial^2 m(U(\mu))}{\partial y^2}\sigma^2 + o(\sigma^2) = O(\sigma^2), \text{ as } \sigma^2 \to 0,$$

where $N = \frac{1}{2}\frac{\partial^2 m(U(\mu))}{\partial y^2}$, and we observe that $O(\sigma^2) + o(\sigma^2) = O(\sigma^2)$ as $\sigma^2 \to 0$.

Define the function $F(.)$ as the inverse function of $U^{Mod}(.)$. Therefore, $\tilde{x} - \mu = F(U^{Mod}(\tilde{x})) - F(U^{Mod}(\mu))$. Expanding $F(U^{Mod}(\tilde{x}))$ around the neighborhood of $F(U^{Mod}(\propto))$ as $\tilde{x} \to \mu$ gives

$\tilde{x} = F(U^{Mod}(\mu)) + F'(U^{Mod}(\mu))[U^{Mod}(\tilde{x}) - U^{Mod}(\mu)] + o(U^{Mod}(\tilde{x}) - U^{Mod}(\mu)).$

This implies that

$\tilde{x} - \mu = F'(U^{Mod}(\mu))[U^{Mod}(\tilde{x}) - U^{Mod}(\mu)] + o(U^{Mod}(\tilde{x}) - U^{Mod}(\mu)),$ as $\tilde{x} \to \mu.$

Since $U^{Mod}(\tilde{x}) - U^{Mod}(\mu) = O(\sigma^2)$, as $\sigma^2 \to 0$, we have

$$\tilde{x} - \mu = F'(U^{Mod}(\mu))O(\sigma^2) + o(O(\sigma^2)) = O(\sigma^2), \text{ as } \tilde{x} \to \mu.$$

Step 3: We prove that

$$\boxed{U(\tilde{g}) = m(U(\mu)) + \frac{\partial m(U(\mu))}{\partial y}(\tilde{x} - \mu) + o(\sigma^2), \text{ as } \tilde{x} \to \mu.}$$

Proof: We first take a Taylor expansion of $U(g(\tilde{x}))$ around $U(g(\infty))$ as $\tilde{x} \to \mu$ to get

$$U(\tilde{g}) = U(g(\tilde{x})) = U(g(\mu)) + \frac{\partial U(g(\mu))}{\partial y}(\tilde{x} - \mu) + o((\tilde{x} - \mu)), \ \tilde{x} \to \mu$$

$$= m(U(\mu)) + \frac{\partial m(U(\mu))}{\partial y}(\tilde{x} - \mu) + o((\tilde{x} - \mu)), \ \tilde{x} \to \mu. \quad (9)$$

Then we use the results of Step 2 and observe that $\tilde{x} - \mu = O(\sigma^2)$, as $\tilde{x} \to \mu$, to get

$$U(\tilde{g}) = m(U(\mu)) + \frac{\partial m(U(\mu))}{\partial y}(\tilde{x} - \mu) + o(\sigma^2), \ \tilde{x} \to \mu.$$

Step 4: We now use Steps 1, 2, and 3 to prove the Theorem. As $y_{\min}, y_{\max} \to \mu, \tilde{x} \to \mu$ at least as fast, and $\sigma^2 \to 0$. This allows us to equate the expressions of $U(\tilde{g})$ from Steps 1 and 3 and rearrange to get

$$\tilde{x} = \mu + \frac{\sigma^2}{2} \frac{\frac{1}{2} \frac{\partial^2}{\partial y^2} m(U(\mu))}{\frac{\partial}{\partial y} m(U(\mu))} + o(\sigma^2), \quad \sigma^2 \to 0.$$

We have

$$\frac{\partial m(U(\mu))}{\partial y} = \frac{\partial m(U(y))}{\partial U(y)} \frac{\partial U(y)}{\partial y}\Big|_{y=\mu}.$$

$$\frac{\partial^2 m(U(\mu))}{\partial y^2} = \frac{\partial^2 m(U(y))}{\partial^2 U(y)} \left(\frac{\partial U(y)}{\partial y}\right)^2\Big|_{y=\mu} + \frac{\partial m(U(y))}{\partial U(y)} \frac{\partial^2 U(y)}{\partial y^2}\Big|_{y=\mu}$$

Dividing gives

$$\frac{\frac{\partial^2 m(U(\mu))}{\partial y^2}}{\frac{\partial m(U(\mu))}{\partial y}} = \frac{\frac{\partial^2 m(U(y))}{\partial^2 U(y)}}{\frac{\partial m(U(y))}{\partial U(y)}} \left(\frac{\partial U(y)}{\partial y}\right)\Big|_{y=\mu} + \frac{\frac{\partial^2 U(y)}{\partial y^2}}{\frac{\partial U(y)}{\partial y}}\Big|_{y=\mu} = -\eta_g(\mu) - \gamma(\mu).$$

Hence,

$$\tilde{x} = \mu - \frac{\sigma^2}{2}[\eta_g(\mu) + \gamma(\mu)] + o(\sigma^2), \quad \sigma^2 \to 0.$$

By definition, $\tilde{x} = g^{-1}(\tilde{g})$. Therefore,

$$\tilde{g} = g\left(\mu - \frac{\sigma^2}{2}[\eta_g(\mu) + \gamma(\mu)] + o(\sigma^2) \right), \quad \sigma^2 \to 0.$$

But, from Pratt (1964),

$$\tilde{y} = \mu - \frac{\sigma^2}{2}\gamma(\mu) + o(\sigma^2), \quad \sigma^2 \to 0.$$

Therefore,

$$\tilde{g} = g\left(\tilde{y} - \frac{\sigma^2}{2}\eta_g(\mu) + o(\sigma^2) \right), \quad \sigma^2 \to 0.$$

CHAPTER 18

One-Switch Utility Functions

Chapter Concepts

- A property about changes in preferences between lotteries called the "one-switch property" (Bell 1988)
- The functional forms of utility functions that satisfy the one-switch property
- A functional equation that characterizes the one-switch property
- Insights into how the one-switch property extends to multiple attributes

18.1 INTRODUCTION

This chapter builds on several concepts that we have discussed earlier, and extends them to a new property called the one-switch property. Suppose that you face any two wealth gambles, A and B, and that you prefer A to B. Suppose that your preference for the gambles does not change as your wealth increases. As we have seen, this implies that your utility function is either linear or exponential.

Now, suppose that your preferences for the gambles can change with wealth, but can only change once, and this is true for all lotteries. What does this imply about your utility function? We refer to this property as the ***one-switch property*** because it characterizes a maximum of one change in preferences between lotteries as wealth increases. We refer to utility functions satisfying this one-switch property as one-switch utility functions. The idea of one-switch change in preferences is an interesting concept that may narrow the search of the utility function to only a few forms, and was first introduced in Bell (1988).

It is clear that all zero-switch utility functions (both the linear and exponential functions) satisfy the one-switch property, because if no switch in preferences occurs, then the condition that a maximum of one switch change in preferences is automatically satisfied. In addition to the linear and exponential forms, there are other utility functions that allow for a maximum of one-switch change in preferences.

In this chapter, we discuss the formulation of both ***zero-switch*** and ***one-switch*** change in preferences. We also show how these concepts extend to any number of switches between lotteries. We also provide a new approach for the deriving the one-switch utility functions using functional equations that will help extend the concept of one-switch preferences to multiple attributes and to more general monotone transformations besides the shift transformations. We start by exploring the relationship to the delta property and extensions of the delta property and to invariant transformations.

18.2 RELATING THE DELTA PROPERTY TO ZERO-SWITCH UTILITY FUNCTIONS

Recall from Chapter 13 that the delta property implies that if $\tilde{y}$ is the wealth equivalent of the lottery $< p_1, y_1; p_2, y_2; ....; p_n, y_n >$ then $\tilde{y} + \delta$ is the wealth equivalent of the lottery $< p_1, y_1 + \delta; p_2, y_2 + \delta; ....; p_n, y_n + \delta >$, that is, that preferences between lotteries do not change as our wealth increases.

We demonstrated that only the linear and exponential utility functions satisfy the delta property for all lotteries and for values of δ on an interval of positive length, i.e., a decision-maker who follows the delta property has either

(i)   a linear utility function $U(y) = ay + b$, or
(ii) an exponential utility function, $U(y) = ae^{\gamma y} + b$.

Furthermore, we demonstrated that his utility function must satisfy the functional equation,

$$U(y + \delta) = k(\delta)U(y) + d(\delta), \tag{1}$$

where $k(\delta) = 1$ for a linear utility function and $k(\delta) = e^{\gamma \delta}$ for an exponential utility function. Note that in both of these cases, $k(\delta)$ does not change sign. In Chapter 14, we illustrated that these utility functions are the only utility functions that exhibit constant absolute risk aversion, and therefore related the delta property to absolute risk aversion for values of δ on an interval.

Relating the Delta Property to Zero-Switch Preferences
Suppose that a decision-maker faces any two lotteries A and B, and if he prefers A to B and follows the Five Rules, then

$$E_A[U(y)] > E_B[U(y)],$$

where $E_A[.]$ and $E_B[.]$ represent the expectation operators with respect to lotteries A and B, respectively.

What happens if we increment the outcome of each lottery by a shift amount δ? Will the decision-maker's preferences for the lotteries change? The difference in expected utility of the modified lotteries is by definition equal to the difference

$$E_A[U(y + \delta)] - E_B[U(y + \delta)]. \tag{2}$$

If the decision-maker's utility function satisfies the delta property, then we can substitute from (1) into (2) to get

$$E_A[U(y + \delta)] - E_B[U(y + \delta)] = k(\delta)\big(E_A[U(y)] - E_B[U(y)]\big). \tag{3}$$

Equation (3) shows that when the delta property holds, we can interpret the difference in expected utility of the modified lotteries $E_A[U(y+\delta)] - E_B[U(y+\delta)]$ in terms of the difference in expected utility of the unmodified lotteries $(E_A[U(y)] - E_B[U(y)])$ and some function $k(\delta)$ that does not change sign.

Because $k(\delta)$ does not change sign when the delta property holds, then the difference in expected utilities of the modified lotteries will not change sign with δ. Therefore, the decision-maker's preference for the modified lotteries will not change as we increment the outcome of each lottery by an amount δ. If she prefers lottery A to lottery B, then she will always prefer the modified lottery A to the modified lottery B for any value of δ.

We summarize this **sufficient** result below.

Sufficient Condition for Zero-Switch Utility: A utility function satisfies zero switch change in preferences with shift transformations if it satisfies the functional Equation (1) where $k(\delta)$ does not change sign.

The linear and exponential utility functions are the only utility functions that satisfy (1) over an interval of positive length for δ. But can you satisfy the zero-switch condition without satisfying (1)? The answer is negative. As we show below, this is also a necessary condition, and so the linear and exponential utility functions are the only zero-switch utility functions for all lotteries.

The proof of **necessity** is as follows: If a utility function satisfies the zero-switch condition, then there cannot exist δ_1, δ_2 such that

$$E_A[U(y+\delta_1)] - E_B[U(y+\delta_1)] > 0$$
$$E_A[U(y+\delta_2)] - E_B[U(y+\delta_2)] < 0.$$

We now observe that the expected utilities of any two lotteries, A and B, can be expressed as the expected utility of binary lotteries having the same probabilities but with different outcomes (we adjust the outcomes such that the expected utility of the equivalent lottery is equal to the expected utility of the original lottery).

The difference in expected utility between lotteries A and B is

$$E_A[U(y+\delta)] - E_B[U(y+\delta)] = \sum_{i=1}^{2} p_i U(y_{Ai} + \delta) - \sum_{i=1}^{2} p_i U(y_{Bi} + \delta) = \sum_{i=1}^{2} p_i V_i(\delta) = \theta(\delta),$$

where $V_i(\delta) = U(y_{Ai} + \delta) - U(y_{Bi} + \delta)$.

For zero-switch, there cannot exist δ_1, δ_2 such that $\theta(\delta_1) = +\text{ve}$, $\theta(\delta_2) = -\text{ve}$. This implies that the system

$$\begin{pmatrix} V_1(\delta_1) & V_2(\delta_1) \\ V_1(\delta_2) & V_2(\delta_2) \end{pmatrix} \begin{pmatrix} p_1 \\ p_2 \end{pmatrix} = \begin{pmatrix} \theta(\delta_1) \\ \theta(\delta_2) \end{pmatrix}$$

cannot have a solution, and so the matrix

$$\begin{pmatrix} V_1(\delta_1) & V_2(\delta_1) \\ V_1(\delta_2) & V_2(\delta_2) \end{pmatrix}$$

must be singular. This implies that

$$V_2(\delta) = \alpha V_1(\delta),$$

where α may depend on the outcomes $y_{A1}, y_{B1}, y_{A2}, y_{B2}$ but does not depend on δ. This in turn implies that

$$U(y_{A2} + \delta) - U(y_{B2} + \delta) = \alpha(y_{A1}, y_{B1}, y_{A2}, y_{B2})[U(y_{A1} + \delta) - U(y_{B1} + \delta)]. \qquad (4)$$

Equation (4) applies to any arbitrary fixed values $y_{A1}, y_{B1}, y_{A2}, y_{B2}$ and relates four assessments at the corresponding value of δ. Let $y_{A2} = y, y_{B2} = y_0$. This implies that

$$U(y + \delta) - U(y_0 + \delta) = \alpha(y_{A1}, y_{B1}, y, y_{B2})[U(y_{A1} + \delta) - U(y_{B1} + \delta)], \qquad (5)$$

which in turn implies that

$$U(y + \delta) = k(\delta)g(y) + d(\delta), \qquad (6)$$

where $k(\delta) = U(y_{A1} + \delta) - U(y_{B1} + \delta)$, $g(y) = \alpha(y_{A1}, y_{B1}, y, y_{B2})$, $d(\delta) = U(y_0 + \delta)$. We summarize this result below.

> **Necessary Condition for Zero-Switch Utility:** A necessary condition for a utility function to satisfy zero-switch change in preferences for all shift amounts and for all lotteries is that the utility function satisfy the functional Equation (1) where $k(\delta)$ does not change sign.

The general continuous solutions of (6) are the linear and exponential utility functions. Therefore, a utility function satisfies zero-switch change in preferences with wealth for all lotteries if and only if it is linear or exponential.

18.2.1 Relating the Invariant Transformation to Zero-Switch Change in Preferences

We can generalize the idea of zero-switch utility functions with shift transformations using the concept of an invariant transformation of the utility function. If a utility function is invariant to an increasing monotone transformation, then the preference between any two lotteries will not change when their outcomes are modified by the same transformation. To illustrate, suppose that g is an invariant transformation of a utility function, U. Then

$$U(g(y, \delta)) = k(\delta)U(y) + d(\delta)$$

for some strictly increasing transformation g.

Note that if the invariant transformation $g(y, \delta)$ is strictly increasing $\forall \delta$, then $k(\delta) > 0$. To see why this is true, consider two lotteries, X_a, X_b such that that wealth equivalent of X_a, namely $\tilde{y}_A$, is higher than the certain equivalent of X_b, $\tilde{y}_B$. What happens if we modify all the wealth outcomes of lotteries X_a and X_b with the same invariant strictly increasing transformation $g(y, \delta)$?

If the wealth equivalents of the unmodified lotteries are $\tilde{y}_A$ and $\tilde{y}_B$, respectively, and if $\tilde{y}_A > \tilde{y}_B$, then the wealth equivalents of the modified lotteries will be $g(\tilde{y}_A, \delta)$ and $g(\tilde{y}_B, \delta)$. But since g is strictly increasing $\forall \delta$, it must be the case that $g(\tilde{y}_A, \delta) > g(\tilde{y}_B, \delta)$ $\forall \delta$. Therefore, the modified version of X_a will always be more preferred than the modified version of X_b. Therefore, the difference in expected utilities between the modified versions of X_a and X_b will remain positive for all values of δ, and so

$$E_A[U(g(y, \delta))] - E_B[U(g(y, \delta))] = k(\delta)\big[E_A[U(y)] - E_B[U(y)]\big] > 0 \quad \forall \delta.$$

If $E_A[U(y)] - E_B[U(y)] > 0$ (the utility function is strictly increasing), it must be that $k(\delta) > 0$.

EXAMPLE 18.1: Zero-Switch Utility Functions with Scale Transformations

---

The concept of zero-switch change in preferences extends to other transformation. Consider, for example, the case where the lotteries are modified by a scale transformation. Zero-switch change in preferences implies that the difference

$$E_A[U(\delta y)] - E_B[U(\delta y)], \delta > 0$$

does not cross zero. It should not come as a surprise now that this equation is satisfied for all lotteries if and only if

$$U(\delta y) = k(\delta)U(y) + d(\delta),$$

where $k(\delta)$ does not cross zero. As we have seen, this equation is satisfied for power and logarithmic utility functions.

Dialogue between Instructor (I) and Student (S)

I: Is everyone now clear on the relation between the invariant transformation and the idea of zero-switch change in preferences for lotteries?

S: Yes. It seems very interesting. For the special case of shift transformations, it means that as we do a sensitivity analysis to the wealth equivalent of any two lotteries versus wealth, the wealth equivalent curves will not cross.

I: Correct, and for the invariant transformation, it means that as you modify the lottery using the invariant transformation parameters and conduct a sensitivity analysis versus those parameters, then the wealth equivalents of the modified lotteries will not cross.

S: What happens if preferences between lotteries can change?

I: Then, they will not satisfy the invariance functional equation. In the next section we analyze one-switch changes in preferences and see how this changes things.

18.3 ONE-SWITCH UTILITY FUNCTIONS

Suppose that your preference for two investments might change as you get richer. At a poorer wealth level, you prefer one investment, and as you get richer, you prefer another (riskier) investment, but you do not revert back to preferring the first investment as you get even richer. What is the functional equation that needs to be satisfied if such one-switch preference structures hold? What is the form of the utility functions that satisfy this one-switch property? We answer these questions in this section.

Problem Formulation

Suppose you face two wealth gambles, A and B, and you prefer A to B. Suppose that your preference for the gambles can change, but only once, as your wealth increases. If this is true for all lotteries, what does this imply about your utility function?

Consider two lotteries, A and B, such that that the expected utility of lottery A is higher than that of lottery B. This implies that

$$E_A[U(y)] - E_B[U(y)] > 0.$$

What happens if we increment the outcome of each lottery by a shift amount, δ, or equivalently that our wealth increases by an amount δ? The difference in expected utility is now

$$E_A[U(y+\delta)] - E_B[U(y+\delta)].$$

If preferences between lotteries A and B can change only once, then the difference in expected utility of the modified lotteries can cross zero only once.

This implies that the shape of the sensitivity analysis curves of the difference in expected utilities $E_A[U(y+\delta)] - E_B[U(y+\delta)]$ with respect to δ can cross zero only once. What if this property were to hold for all lotteries A and B?

Bell (1988) characterized four utility functions that uniquely satisfy this property. We can also extend this property to other transformations besides the shift transformations. For example, we may scale the outcomes by an amount δ (such as if you get a fraction of the lottery). In the case of one-switch preferences, the difference

$$E_A[U(\delta y)] - E_B[U(\delta y)]$$

would cross zero only once as we increase δ.

Finally, we can envision modifying all the wealth outcomes of lotteries X_a and X_b with the some general monotone transformation $g(y, \delta)$ and require that preferences between these lotteries can switch at most once as we increase δ. What must the functional form of the utility function be?

18.3.1 Characterizing the One-Switch Property by Extending the Invariance Functional Equation

We shall now provide a new functional equation that characterizes utility functions satisfying the one-switch property. We start with the following theorem.

Theorem 18.1: Functional Equation for One-Switch Utility Functions

A utility function, U, satisfies one-switch change in preferences for all lotteries when a shift transformation $g(y) = y + \delta$ is applied to the wealth outcomes as we increase δ if and only if

$$U(y + \delta) = k_0(\delta) + k_1(\delta)[f_1(y) + \phi(\delta)f_2(y)], \tag{7}$$

where $k_1(\delta)$ does not change sign and the $\phi(\delta)$ is a monotone function.

The functional Equation (7) can be expanded as

$$U(y + \delta) = k_0(\delta) + k_1(\delta)f_1(y) + k_2(\delta)f_2(y),$$

where $k_2(\delta) = \phi(\delta)k_1(\delta)$ simply adds a product term $k_2(\delta)f_2(y)$ to the invariance functional Equation (6).

If $k_2(\delta)f_2(y)$ were equal to zero, then this would be the invariance functional equation $U(y + \delta) = k_0(\delta) + k_1(\delta)f_1(y)$, which, as we have seen, characterizes the delta property and the linear and exponential utility functions. Therefore, these utility functions must satisfy this functional equation. We shall consider the general solutions of this equation in this chapter.

To gain more insight into this formulation, let us first demonstrate why a utility function that satisfies this functional equation will satisfy one-switch preferences for lotteries.

Illustration of Sufficiency: Consider two lotteries, X_a, X_b. What happens if we modify all the wealth outcomes of lotteries X_a and X_b with the same amount δ such that

$$U(y + \delta) = k_0(\delta) + k_1(\delta)[f_1(y) + \phi(\delta)f_2(y)]?$$

The difference in expected utility of the modified lotteries is then

$$E_A[U(y + \delta)] - E_B[U(y + \delta)]$$

$$= k_1(\delta)\{(E_A[f_1(y)] - E_B[f_1(y)]) + \phi(\delta)(E_A[f_2(y)] - E_B[f_2(y)])\}.$$

We have expressed the difference in expected utility in terms of a function $k_1(\delta)$ that does not change sign, and a term $(E_A[f_1(y)] - E_B[f_1(y)]) + \phi(\delta)(E_A[f_2(y)] - E_B[f_2(y)])$ that involves a monotone function $\phi(\delta)$. As we now show, this difference can equal zero at only one value of δ. To see why this is true, we set the difference in expected utility to zero, and divide by $k_1(\delta)$, to get

$$E_A[f_1(y)] - E_B[f_1(y)] + \phi(\delta)(E_A[f_2(y)] - E_B[f_2(y)]) = 0.$$

If $E_A[f_2(y)] - E_B[f_2(y)]$ is not zero, then this implies that the difference in expected utility of the modified lotteries crosses zero when

$$\phi(\delta) = -\frac{E_A[f_1(y)] - E_B[f_1(y)]}{E_A[f_2(y)] - E_B[f_2(y)]},$$

which can occur only once as we increase δ, because $\phi(\delta)$ is a monotone function.

The proof of **necessity** is in the appendix to this chapter and follows the same type of necessity argument (although a bit more extensive) that we conducted for zero-switch preferences.

It suffices to solve the functional Equation (7) to determine the one-switch utility functions with shift transformations. Before we do that, however, let us first generalize (7) to a monotone transformation $g(y,\delta)$ applied to the wealth outcomes and show that it does indeed satisfy the one-switch condition.

EXAMPLE 18.2: Extensions of the One-Switch Property to More General Transformations

---

Of course it is now simple to see intuitively the extension of this idea to more general transformations of the form $g(y,\delta)$ if the utility function satisfies

$$\boxed{U(g(y,\delta)) = k_0(\delta) + k_1(\delta)[f_1(y) + \phi(\delta)f_2(y)]}.$$

To illustrate, consider again two lotteries, X_a, X_b. What happens if we modify all the wealth outcomes of lotteries X_a and X_b with the same strictly increasing transformation $g(y,\delta)$ and we increase δ such that

$$U(g(y,\delta)) = k_0(\delta) + k_1(\delta)[f_1(y) + \phi(\delta)f_2(y)]?$$

The difference in expected utility of the modified lotteries is

$$E_A[U(g(y,\delta))] - E_B[U(g(y,\delta))]$$

$$= k_1(\delta)\big(E_A[f_1(y)] - E_B[f_1(y)]\big) + k_1(\delta)\phi(\delta)\big(E_A[f_2(y)] - E_B[f_2(y)]\big).$$

This difference can equal zero at only one value of δ,

$$E_A[f_1(y)] - E_B[f_1(y)] + \phi(\delta)\big(E_A[f_2(y)] - E_B[f_2(y)]\big) = 0$$

$$\Rightarrow \phi(\delta) = -\frac{E_A[f_1(y)] - E_B[f_1(y)]}{E_A[f_2(y)] - E_B[f_2(y)]},$$

which can occur only once as we increase δ because $\phi(\delta)$ is a monotone function.

18.3.2 One-Switch Utility Functions with Shift Transformations

Bell (1988) characterized the four one-switch utility functions for lotteries with shift transformations (or with increase In wealth). In his work, he used a Taylor expansion to derive a differential equation whose solution leads to the four one-switch forms below. These utility functions are also the general continuous solutions of (7).

The four one-switch utility functions are the

1. Linear plus exponential $U(y) = ay + be^{-\gamma y} + c$
2. Sumex $U(y) = ae^{-\lambda y} + be^{-\gamma y} + c$
3. Linear times exponential $U(y) = (ay + b)e^{-\gamma y} + c$
4. Quadratic $U(y) = ay^2 + by + c$

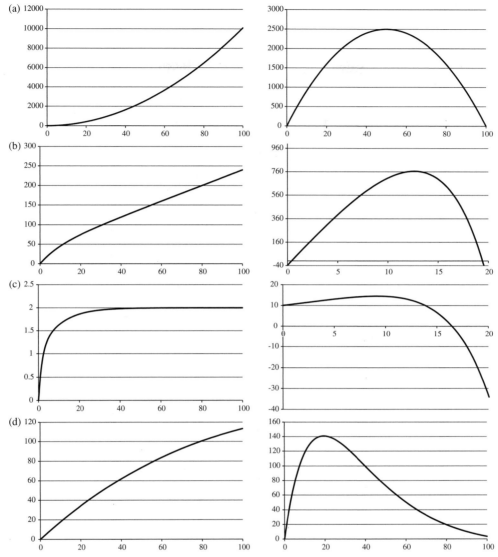

Figure 18.1. Examples of one-switch utility functions.
Quadratic $ay^2 + by + c$. Left: $a = b = c = 1$. Right $a = -1$, $b = 100$, $c = 0$.
Linear plus exponential $ay + be^{-\gamma y} + c$. (Left) $a = 2$, $b = -40$, $\gamma = .1$, $c = 40$
(Right) $a = 2$, $b = -40$, $\gamma = .1$, $c = 40$
Sumex $ae^{-by} + ce^{-dy} + f$.
(Left) $a = -1$, $b = 0.1$, $c = -1$, $d = 0.6$, $f = 2$ (Right) $a = 10$, $b = -0.1$, $c = -2$, $d = -0.2$, $f = 2$
Linear times exponential $(ay + b)e^{-\gamma y} + c$.
(Left) $a = 2$, $b = 20$, $\gamma = .005$, $c = -20$ (Right) $a = 20$, $b = 10$, $\gamma = .05$, $c = -10$.

Figure 18.1 shows examples of the four one-switch utility functions with different parameters.

There is an important feature that these four one-switch utility functional forms have in common:

The four one-switch utility functions all have at most one peak when plotted versus y no matter how you set the parameters.

Refer back to our discussion of ordinal one-switch preferences for deterministic prospects characterized by a single direct value attribute in Chapter 6. We illustrated that an ordinal one-switch preference function over one attribute can have at most one peak. It is not surprising that the four one-switch utility functions over lotteries have at most one peak.

Not every function with at most one peak will satisfy one-switch changes in preferences for lotteries, however. As we discussed, similar properties over lotteries provide more specificity than properties for deterministic consequences. The condition of a one-switch change in preferences with wealth over monetary lotteries allows for only the four functional forms described above.

Of course each of these four forms also provides one-switch preferences over deterministic consequences because they have at most one peak. Their applications to utility functions over money will set the parameters such that the utility functions are strictly increasing on their relevant domain.

Zero-Switch Utility Functions Are One-Switch Utility Functions
A zero-switch utility function guarantees zero-switch in preferences. A one-switch utility function requires at most one switch change in preferences. Therefore, a zero-switch utility function is by definition a one-switch utility function.

EXAMPLE 18.3: Linear Plus Exponential Utility Function Satisfies the Functional Equation

---

To show that the linear plus exponential utility function $U(y) = ay + be^{-\gamma x} + c$, satisfies (7), we get by direct substitution

$$U(y + \delta) = ay + a\delta + be^{-\gamma y}e^{-\gamma\delta} + c + [be^{-\gamma y} - be^{-\gamma y}]$$
$$= \{ay + be^{-\gamma y} + c\} + be^{-\gamma y}(e^{-\gamma\delta} - 1) + a\delta = f_1(y) + \phi(\delta)f_2(y) + k_0(\delta),$$

where $k_1(\delta) = 1$, $\phi(\delta) = b(e^{-\gamma\delta} - 1)$, $k_0(\delta) = a\delta$. Note that $k_1(\delta) = 1$ does not change sign and that $\phi(\delta) = b(e^{-\gamma\delta} - 1)$ is indeed a monotone function.

We leave it as an exercise to the reader to show that all other one-switch utility functions satisfy (7).

EXAMPLE 18.4: $U(y) = \sin(y)$ **Utility Function Does Not Satisfy the One-Switch Functional Equation**

Note that the function $U(y) = \sin(y)$ also satisfies

$$U(y + \delta) = \sin(y)\cos(\delta) + \cos(y)\sin(\delta)$$
$$= f_1(y)k_1(\delta) + f_2(y)k_2(\delta),$$

which may appear at first to satisfy (7), with $k_0(\delta) = 0$. However, closer examination shows that it does not satisfy the monotonicity and sign change conditions. Therefore, it cannot be expressed as (7) because

$$U(y + \delta) = \cos(\delta)[\sin(y) + \cos(y)\tan(\delta)].$$

First, $\cos(\delta)$ changes sign and, furthermore, $\tan(\delta)$ is a *not* a continuous monotone function of δ. Therefore, $U(y) = \sin(y)$ and even $U(y) = \cos(y)$ are not one-switch utility functions.

The following numerical example demonstrates a sensitivity analysis to wealth and shows that the difference in expected utility for two lotteries crosses zero at most once.

EXAMPLE 18.5: Adding a Shift Amount to the Lottery Outcomes

Consider the linear plus exponential utility function $U(y) = ay + be^{-\gamma y} + c$, with $a = 2$, $b = -40$, $\gamma = .1$, and $c = 40$. Suppose a decision-maker with this utility function faces two wealth lotteries, A and B, such that

Lottery A: 50/50 chance of $0 and $90
Lottery B: 50/50 chance of $30 and $60.

Based on this calculation, the decision-maker prefers lottery B to A.

Now suppose that the outcomes of each lottery are modified by a shift amount δ. The linear plus exponential one-switch utility function asserts that preferences between these lotteries can shift only once as we add an amount δ to each outcome. Indeed Figure 18.2 plots the sensitivity analysis of the expected utility of each lottery to δ. The figure shows that switching occurs at a value of δ approximately equal to 4.865.

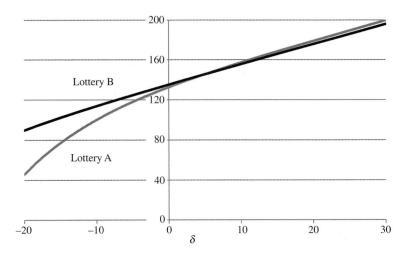

Figure 18.2. Sensitivity analysis to the value of the shift amount δ.

18.3.3 Risk Aversion Function Implied by the Four One-Switch Utility Functions

It is natural to ask about the absolute risk aversion function that results from a one-switch utility function.

1. **Linear Plus Exponential**

   The linear plus exponential utility function has the form

   $$U(y) = ay + be^{-\gamma y} + c.$$

Taking the first and second derivatives gives $U'(y) = a - \gamma be^{-\gamma y}$, $U''(y) = \gamma^2 be^{-\gamma y}$. Therefore the absolute risk aversion function is

$$\gamma(y) = -\frac{U''(y)}{U'(y)} = -\frac{\gamma^2 be^{-\gamma y}}{a - \gamma be^{-\gamma y}}.$$

Note that if $a = 0$, then $\gamma(y) = \gamma$, which has constant absolute risk aversion.

2. **Sumex**

   The sumex utility function has the form

   $$U(y) = ae^{-\lambda y} + be^{-\gamma y} + c.$$

Taking the first and second derivatives gives $U'(y) = -\lambda ae^{-\lambda y} - \gamma be^{-\gamma y}$, $U''(y) = \lambda^2 ae^{-\lambda y} + \gamma^2 be^{-\gamma y}$

   Therefore, the absolute risk aversion function is

   $$\gamma(y) = -\frac{U''(y)}{U'(y)} = \frac{\lambda^2 ae^{-\lambda y} + \gamma^2 be^{-\gamma y}}{\lambda ae^{-\lambda y} + \gamma be^{-\gamma y}}.$$

3. **Linear times exponential**

   The linear times exponential utility function has the form

   $$U(y) = (ay + b)e^{-\gamma y} + c.$$

Taking the first and second derivatives gives

$$U'(y) = -a\gamma ye^{-\gamma y} + ae^{-\gamma y} - b\gamma e^{-\gamma y} = e^{-\gamma y}(a - b\gamma - a\gamma y)$$

$$U''(y) = e^{-\gamma y}(-a\gamma) - \gamma e^{-\gamma y}(a - b\gamma - a\gamma y) = \gamma e^{-\gamma y}(b\gamma + ay).$$

Therefore, the absolute risk aversion function is

$$\gamma(y) = -\frac{U''(y)}{U'(y)} = -\frac{\gamma e^{-\gamma y}(b + ay)}{e^{-\gamma y}(a - b\gamma - a\gamma y)} = -\frac{\gamma(b\gamma + ay)}{(a - b\gamma - a\gamma y)}.$$

4. **Quadratic**

   The quadratic utility function has the form

   $$U(y) = ay^2 + by + c.$$

Taking the first and second derivatives gives

$$U'(y) = 2ay + b, \ U''(y) = 2a.$$

Therefore, the absolute risk aversion function is

$$\gamma(y) = -\frac{U''(y)}{U'(y)} = -\frac{2a}{2ay + b}.$$

18.4 ADDITIONAL CHARACTERIZATION OF ONE-SWITCH UTILITY FUNCTIONS

We now provide further characterizations for the functional Equation (7) to provide additional insights about its implications and its extension to more general transformations and to multiple attributes.

Proposition 18.1: *A utility function is a one-switch utility function with shift transformations if and only if it satisfies*

$$U(y + \delta) = k_1(\delta)U(y) + k_2(\delta)U_0(y) + k_0(\delta),$$

where $k_1(\delta)$ does not change sign, $U_0(y) = \begin{cases} ae^{\alpha y} + b \\ ay + b \end{cases}$ is a zero-switch utility function with shift transformations, and the ratio $\dfrac{k_1(\delta)}{k_2(\delta)}$ is monotone.

Proposition 18.1 provides more specificity for the functions $f_1(y), f_2(y)$ in (7). It shows that the effects of adding an amount δ to lottery outcomes for a one-switch utility function replaces the utility of the modified outcome $U(y + \delta)$ with a weighted combination of the utility of the unmodified outcome, $U(y)$, and a zero-switch utility of that outcome, $U_0(y)$ and some constants.

18.4.1 Expressing One-Switch Utility Functions in terms of Zero-Switch Utility Functions

A further characterization of (7) that will help us extend the concept of one-switch utility functions with shift to other transformations and to multiple attributes is the following proposition.

Proposition 18.2: *A utility function is a one-switch utility function with shift transformations if and only if it can be expressed as one of the following forms:*

$$U(y) = (aU_A(y) + b)(eU_A(y) + f), \tag{8}$$

$$U(y) = (aU_A(y) + b)U_M(y), \tag{9}$$

$$U(y) = (aU_A(y)+b)+(eU_M(y)+f), \qquad (10)$$

$$U(y) = (aU_{M_1}(y)+b)+(eU_{M_2}(y)+f), \qquad (11)$$

where

$$U_A(y+\delta) = U_A(y)+U_A(\delta), \quad \forall \delta \in \Delta,$$

and $U_M(y), U_{M_1}(y), U_{M_2}(y)$ are solutions to the equation

$$U_M(y+\delta) = U_M(y)U_M(\delta), \quad \forall \delta \in \Delta,$$

and Δ is an interval of positive length.

The following corollary, which we state without proof, extends the notion of one-switch utility functions with shift transformations to transformations of the form $g(y,\delta)$.

Corollary 18.1: *A utility function is a one-switch utility function for a transformation $g(y,\delta)$ as we increase δ if and only if it can be expressed as one of the following forms:*

$$U(y) = (aU_A(y)+b)(eU_A(y)+f), \qquad (12)$$

$$U(y) = (aU_A(y)+b)U_M(y), \qquad (13)$$

$$U(y) = (aU_A(y)+b)+(eU_M(y)+f), \qquad (14)$$

$$U(y) = (aU_M(y)+b)+(eU_M(y)+f), \qquad (15)$$

where $U_A(g(y,\delta)) = U_A(y)+U_A(\delta)$, and $U_M(g(y,\delta)) = U_M(y)U_M(\delta), \quad \forall \delta \in \Delta$, and Δ is an interval of positive length.

Corollary 18.1 enables the determination of one-switch utility functions for general transformations by solving the corresponding functional equations for zero-switch transformations. We demonstrate this below.

18.4.2 One-Switch Utility Functions with Scale Transformations

We can now characterize one-switch utility functions with scale transformations by first solving the simpler zero-switch equations.

EXAMPLE 18.6: One-Switch Utility Function for a Transformation, $g(y,\delta) = \delta y$

The additive zero-switch equation for scale takes the form

$$U_A(\delta y) = U_A(y)+U_A(\delta), \quad \forall \delta \in \Delta,$$

whose solution is the logarithmic form

$$U_A(y) = a\ln(y)$$

and the multiplicative zero-switch equation for scale

$$U_M(\delta y) = U_M(y)U_M(\delta), \quad \forall \delta \in \Delta,$$

whose solution is the power form

$$U_{M_i}(y) = y^{\alpha_i}.$$

Substituting for the logarithmic and power forms into (12) through (15) gives the four one-switch utility functions for scale transformations as

a) $U(y) = (a\ln(y) + b)(e\ln(y) + f)$
b) $U(y) = (a\ln(y) + b)y^{\alpha_j}, \forall j$
c) $U(y) = (a\ln(y) + b) + (ey^{\alpha_j} + f), \forall j$
d) $U(y) = (ay^{\alpha_i} + b) + (ey^{\alpha_j} + f).$

18.4.3 One-Switch Utility Functions with Power Transformations

Example 18.7: For a power transformation, $g(y, \delta) = y^{\delta}$, we first solve the additive and multiplicative zero-switch equations,

$$U_A(y^{\delta}) = U_A(y) + U_A(\delta) \text{ and } U_M(y^{\delta}) = U_M(\delta)U_M(y),$$

to get

$$U_A(y) = a\ln(\ln(y)) \text{ and } U_{M_j}(y) = (\ln(y))^{\alpha_j} \tag{16}$$

and then substitute into (12) through (15) to get

a) $U(y) = (a\ln(\ln(y)) + b)(e\ln(\ln(y)) + f)$
b) $U(y) = (a\ln(\ln(y)) + b)(\ln(y))^{\alpha_j}, \forall j$
c) $U(y) = (a\ln(\ln(y)) + b) + (e(\ln(y))^{\alpha_j} + f), \forall j$
d) $U(y) = (a(\ln(y))^{\alpha_i} + b) + (e(\ln(y))^{\alpha_j} + f).$

Dialogue between Instructor (I) and Student (S)

I: Is the idea of one-switch preferences now clear?

S: Yes. It means that preferences for any two lotteries can switch at most once as their outcomes are modified by some transformation, and a parameter varies from low to high.

I: Correct.

S: For shift transformations, it means that if we plotted a sensitivity analysis of the certain equivalents of any two lotteries with respect to the shift parameter (which can be interpreted as the wealth), the certain equivalent curves can cross at most once.

I: Correct. And what is really happening with one-switch transformations is that the utility of a modified prospect is replaced by a combination of the utility before modification and the zero-switch utility of the prospect. This observation enables us to derive the one-switch utility functions for other transformations.

S: We have one more question: What if preferences can switch only twice or thrice or a maximum of any fixed number of times?

I: What an intelligent question! Let's discuss this.

18.5 N-SWITCH UTILITY FUNCTIONS

It is natural to ask about the functional forms that lead to a maximum of n-switch changes in preferences. Using the same matrix argument in the proof of zero-switch and one-switch (or by simple intuition of adding more terms to the invariance functional equation), it is clear that a necessary condition for a utility function to satisfy n-switch change in preferences as we add an amount δ to lottery outcomes is that the utility function satisfy

$$U(y+\delta) = k_0(\delta) + \sum_{i=1}^{n+1} f_i(y)k_i(\delta).$$

Aczél (1966) solved this functional equation by converting it into a differential equation, and Aczél and Chung (1982) showed that the only continuous solutions to this functional equation are also infinitely differentiable. An n-switch utility function has the form

$$U(y) = f_0(y) + \sum_{i=1}^{k} f_i(y)e^{c_i y},$$

where $f_i(y)$ is a polynomial of order n_i such that $\sum_{i=0}^{k} n_i \leq n+1-k$.

18.6 EXTENSIONS TO TWO ATTRIBUTES

In Chapters 13 through 16, we discussed the preference for two univariate lotteries as the initial wealth, or some other parameter changes. In principle, the parameter δ can be another attribute that need not be additive. This basic idea will enable us to extend the notion of zero-switch and one-switch preferences to construct multiattribute utility functions. It should not come as a surprise that these utility functions will take the following forms:

Zero-Switch preferences with another attribute

$$U(y,z) = g_0(z) + g_1(z)f(y)$$

where $g_1(z)$ does not change sign, which implies that preferences for lotteries over y do not change as we change z.

One-switch preferences with another attribute

$$U(y,z) = g_0(z) + g_1(z)[f_1(y) + \phi(z)f_2(y)],$$

where $g_1(z)$ does not change sign and $\phi(z)$ is monotone implies that preferences for lotteries over y can change, but only once, as we change z.

18.7 SUMMARY

- The concept of one-switch preferences with monotone transformations means that a decision-maker with a specific utility function can change preferences at most once (this includes zero switches) when the wealth outcomes are modified by monotone transformations.
- A functional equation that characterizes one-switch change in preferences is $U(y + \delta) = k_0(\delta) + k_1(\delta)[f_1(y) + \phi(\delta)f_2(y)]$.
- The four one-switch utility functions satisfying one-switch change in preferences with wealth are: linear plus exponential, sumex, linear times exponential, and quadratic.
- We can express one-switch utility functions in terms of zero-switch utility functions.
- The notion of one-switch preferences over lotteries can be extended to two attributes.

ADDITIONAL READINGS

Abbas, A. E. 2007. Invariant utility functions and certain equivalent transformations. *Decision Analysis* 4(3): 17–31.

Abbas, A. E. and D. E. Bell. 2011. One-switch independence for multiattribute utility functions. *Operations Research* 59(3): 764–771.

Abbas, A. E. and D. E. Bell. 2012. One-switch conditions for multiattribute utility functions. *Operations Research* 60(5): 1199–1212.

Abbas, A. E. and D. E. Bell. 2012. Ordinal one-switch utility functions. *Operations Research* 63(6): 1411–1419.

Aczél, J. 1966. *Lectures on Functional Equations and Their Applications*. Academic Press, New York.

Aczél, J. and J. K. Chung. 1982. Integrable solutions of functional equations of a general type. *Studia Sci. Math. Hungar* 17: 51–67.

Bell, D. E. 1988. One-switch utility functions and a measure of risk. *Management Science* 34(12): 1416–1424.

APPENDIX

Proof of Theorem 18.1

Necessity: We start the proof by considering two discrete uncertain lotteries, $\tilde{x}_A$ and $\tilde{x}_B$, having the same probabilities but with different values of the outcomes. We assume that the two lotteries are

$$< p_1, x_{A1}; p_2, x_{A2}; ....; p_m, x_{Am} > \text{ and } < p_1, x_{B1}; p_2, x_{B2}; ....; p_m, x_{Bm} >.$$

The difference in expected utility of two lotteries over X when another attribute (or parameter), y, is fixed, can be written as

$$Eu_A(y) - Eu_B(y) = \sum_{i=1}^{m} p_i[u(x_{Ai}, y) - u(x_{Bi}, y)] = \theta(y) = \sum_{i=1}^{m} p_i V_i(y),$$

where $V_i(y) = u(x_{Ai}, y) - u(x_{Bi}, y)$.

For fixed values $y_0, y_1, ..., y_m$, we can write in matrix form

$$\begin{pmatrix} V_1(y_0) & V_2(y_0) \, ... & V_m(y_0) \\ V_1(y_1) & V_2(y_1) \, ... & V_m(y_1) \\ ... & & \\ V_1(y_m) & V_2(y_m) \, ... & V_m(y_m) \end{pmatrix} \begin{pmatrix} p_1 \\ p_2 \\ ... \\ p_m \end{pmatrix} = \begin{pmatrix} \theta(y_0) \\ \theta(y_1) \\ ... \\ \theta(y_m) \end{pmatrix}$$

$X \, 1S \, Y \Rightarrow$ there does not exist $y_0 < y_1 < y_2$ such that $\theta(y_0) = +ve, \theta(y_1) = -ve, \theta(y_2) = +ve$.

For this to occur, any 3x3 portion of the V matrix above must be singular, which implies that

$$V_i(y) = \alpha V_j(y) + \beta V_k(y), \forall i, j, k, \tag{17}$$

where α, β do not depend on y (or else the columns would not be linearly dependent).

Referring back to the definition of $V_i(y) = u(x_{Ai}, y) - u(x_{Bi}, y)$, gives

$$u(x_{Ai}, y) - u(x_{Bi}, y) = \alpha[u(x_{Aj}, y) - u(x_{Bj}, y)] + \beta[u(x_{Ak}, y) - u(x_{Bk}, y)]. \tag{18}$$

Equation (18) applies to any arbitrary fixed values $x_{A_i}, x_{B_i}, x_{A_j}, x_{B_j}, x_{A_k}, x_{B_k}$ and relates the six functional assessments of y at these instantiations. The parameters α and β depend on the particular instances of $x_{A_i}, x_{B_i}, x_{A_j}, x_{B_j}, x_{A_k}, x_{B_k}$.

Denote

$$x_{Bi} = x_0, x_{Ai} = x_0 + t, u(x_0, y) = g_0(y),$$
$$[u(x_{Aj}, y) - u(x_{Bj}, y)] = g_1(y), [u(x_{Ak}, y) - u(x_{Bk}, y)] = g_2(y).$$

Rearranging gives

$$u(x_0 + t, y) = \alpha(x_0, t, x_{A_j}, x_{B_j}, x_{A_k}, x_{B_k}) g_1(y) + \beta(x_0, t, x_{A_j}, x_{B_j}, x_{A_k}, x_{B_k}) g_2(y) + g_0(y). \tag{19}$$

Further define $x = x_0 + t$, $\delta^1_{x_0, x_{A_j}, x_{B_j}, x_{A_k}, x_{B_k}}(t) = \alpha(x_0, t, x_{A_j}, x_{B_j}, x_{A_k}, x_{B_k})$, $f_1(x) = \delta^1_{x_0, x_{A_j}, x_{B_j}, x_{A_k}, x_{B_k}}(x - x_0)$, $\delta^2_{x_0, x_{A_j}, x_{B_j}, x_{A_k}, x_{B_k}}(t) = \beta(x_0, t, x_{A_j}, x_{B_j}, x_{A_k}, x_{B_k})$, $f_2(x) = \delta^2_{x_0, x_{A_j}, x_{B_j}, x_{A_k}, x_{B_k}}(x - x_0)$. Substituting into (19) gives

$$u(x, y) = f_1(x) g_1(y) + f_2(x) g_2(y) + g_0(y). \tag{20}$$

Define $\phi(y) = \dfrac{g_2(y)}{g_1(y)}, g_1(y) \neq 0$. Substituting into (15) gives

$$Eu_A(y) - Eu_B(y) = g_1(y)\left([f_1(x_A) - f_1(x_B)] + [f_2(x_A) - f_2(x_B)]\phi(y)\right). \tag{21}$$

We distinguish two cases: Case (1) $g_1(y)$ does not change sign: for (21) to change sign only once, with two given arbitrary lotteries $\tilde{x}_A$ and $\tilde{x}_B$, and constant terms $f_1(\tilde{x}_A), f_1(\tilde{x}_B), f_2(\tilde{x}_A), f_2(\tilde{x}_B)$, then $\phi(y)$ must be monotonic or else, we may have different points of indifference for arbitrary lotteries. Case (2) $g_1(y)$ does change sign: This would imply that the term $\left([f_1(\tilde{x}_A) - f_1(\tilde{x}_B)] + [f_2(\tilde{x}_A) - f_2(\tilde{x}_B)]\phi(y)\right)$ either does not change sign or it changes sign at the same value of y for which $g_1(y)$ changes sign. This is impossible for arbitrary functions non-constant f_1, f_2 and arbitrary lotteries.

Sufficiency: For two arbitrary lotteries, $\tilde{x}_A$, $\tilde{x}_B$ and a monotonic $\phi(y)$ we have only one possible indifference point which occurs at y_1

$$Eu(\tilde{x}_A, y) - Eu(\tilde{x}_B, y) = g_1(y_1)\big([f_1(\tilde{x}_A) - f_1(\tilde{x}_B)] + [f_2(\tilde{x}_A) - f_2(\tilde{x}_B)]\phi(y_1)\big) = 0,$$

$$\phi(y_1) = \frac{f_1(\tilde{x}_B) - f_1(\tilde{x}_A)}{f_2(\tilde{x}_B) - f_2(\tilde{x}_A)},$$

below and after which the rank order of the lotteries must reverse.

Since X 1S Y must apply for lotteries having either the same or different probabilities, the general solution for X 1S Y must be contained by this solution (since it needs to satisfy both constraints; same and different probabilities). Using the same sufficiency argument discussed above, it is straightforward to show that the form (20) does indeed satisfy one-switch preferences for any two lotteries even if they have different probabilities. Consequently the functional form (20) must be the general solution for X 1S Y.

Proof of Proposition 18.1: Consider the functional equation

$$U(y+\delta) = k_0(\delta) + k_1(\delta)[U(y) + \phi(\delta)U_0(y)], \tag{22}$$

where $U_0(y)$ is a zero-switch utility function.

Aczél and Chung (1982) have shown that the only continuous solutions to the equation $U(y+\delta) = \sum_{i=1}^{n} f_i(y)k_i(\delta)$ are also infinitely differentiable. Therefore, we can solve this problem by taking its derivative without loss of generality to continuous solutions. If we take the first partial derivative with respect to y, we get

$$U'(y+\delta) = k_1(\delta)U'(y) + k_2(\delta)U_0'(y). \tag{23}$$

We now consider the two possible cases for $U_0(y)$:

Case 1: The *additive* solution, $U_0(y) = ay + b$, and so $U_0'(y) = a$. Equation (22) becomes

$$U'(y+\delta) = k_1(\delta)U'(y) + k_2(\delta)a. \tag{24}$$

Recall that this is the invariance functional equation for shift transformations whose solutions are

$$U'(y) = \begin{cases} ay + b \\ ce^{\gamma y} + d \end{cases}. \tag{25}$$

Integrating (25) with respect to y gives the two solutions

$$U(y) = \begin{cases} \alpha y^2 + \beta y + \mu & \text{(quadratic)} \\ \kappa e^{\gamma y} + \varepsilon y + \lambda & \text{(linear plus exponential)} \end{cases} \tag{26}$$

Case 2: The *multiplicative* solution, $U_0(y) = ae^{\alpha y} + b$ and so $U_0'(y) = a\alpha e^{\alpha y}$. To solve the functional equation in this case, we first take the first derivative of (23) with respect to y and equate it to the derivative with respect to δ (due to symmetry) to get

$$U''(y+\delta) = k_1(\delta)U''(y) + k_2(\delta)U_0''(y) = k_1'(\delta)U'(y) + k_2'(\delta)U_0'(y). \tag{27}$$

Substituting for $U_0'(y), U_0''(y)$ and rearranging gives the linear differential equation

$$U''(y) + lU'(y) = me^{\alpha y}. \tag{28}$$

The solution to this equation is

$$U'(y) = e^{-\int l dy}\left(\int e^{\int l dy}me^{\alpha y}dy + s\right) \tag{29}$$

We have three possible cases:

1. $-l \neq \alpha, l \neq 0$. Equation (29) yields the solution $U'(y) = ae^{\beta x} + be^{\lambda x}$, which integrates to the sumex form $U(y) = \varepsilon e^{\beta y} + \kappa e^{\lambda y} + e$ (*Sumex*).
2. $-l = \alpha$. Equation (29) yields $U'(y) = (my + s)e^{\alpha y}$, which integrates for some constants a, b, c, d to $U(y) = aye^{\alpha y} + be^{\alpha y} + d = e^{cy}(ay + b) + d$ (*linear times* exp*onential*).
3. $l = 0$. Equation (29) yields $U'(y) = \dfrac{m}{\alpha}e^{\alpha y} + s$, which integrates to $U(y) = ay + be^{\alpha y} + d$ (linear plus exponential).

We observe that there is another case of (29) that arises when the exponents, α, are imaginary and leads to a solution of the form $U(y) = e^{-\gamma y}\cos\beta y$; however, this solution violates the monotonicity constraint of $\phi(\delta)$ and is thus excluded.

Thus the general continuous (and also infinitely differentiable) solutions to (29) are the four functional forms of the one-switch solution.

Sufficiency: It is simple to show that the solutions satisfy the functional form (22) by direct substitution.

Proof of Proposition 18.2. First, we observe that the solution of

$$U_A(y+\delta) = U_A(y) + U_A(\delta), \quad \forall \delta \in \Delta \text{ is } U_A(y) = ay$$

and the solution of

$$U_{M_i}(y+\delta) = U_{M_i}(y)U_{M_i}(\delta), \quad \forall \delta \in \Delta, \text{ is } U_{M_i}(y) = e^{m_i y}.$$

This result should not come as a surprise, as they are both invariant utility functions to shift transformations. See Aczél (1966) for a further derivation. Substituting for $U_A(y) = ay$ and $U_{M_i}(y) = e^{m_i y}$ into (8), (9), (10), and (11) shows that these functional forms are indeed the four one-switch utility functions.

CHAPTER 19

Utility Transversality

Chapter Concepts

- Meaning and implications of the transversality relation
- Relating the risk aversion function over value to the risk aversion over any of the attributes
- Relating the risk aversion functions of the individual attributes
- Incorporating time and risk preference – discounting value not utility

19.1 INTRODUCTION

We have discussed in Chapter 11 that when the value function is specified, all that is needed to construct the multiattribute utility function is a one-dimensional utility function over value. The purpose of this chapter is to explore the relation between the risk aversion function over value and the risk aversion function over any of the individual attributes. We also explore the relation between the risk aversion functions across the different attributes and the trade-off functions between them. We call this relation *utility transversality*.

19.2 DERIVING THE CONDITIONAL UTILITY FUNCTION AT DIFFERENT LEVELS

Let $V(x,y)$ be a value function that returns the value of a prospect (x,y) in terms of some value measure, and let U_V be the utility function over value.

As we discussed in Chapter 12, the utility $U(x,y)$ of any prospect (x,y) is equal to the composite transformation of utility over value,

$$U(x,y) = U_V(V(x,y)). \tag{1}$$

The value function and the utility function over value completely characterize the multiattribute utility function and the individual utility functions over the individual attributes. Consequently, we do not have another degree of freedom to assign individual utility functions over the attributes once the value function and the utility function over value have been specified.

For example, if we have assigned a value function, $V(x, y)$, and a utility function over value, $U_V(V)$, then the utility function of attribute X at the maximum value of attribute Y, must be

$$U(x, y^*) = U_V(V(x, y^*)).$$

Similarly, the utility function of X at any other instantiation of Y, say y_1, can be obtained by direct substitution,

$$U(x, y_1) = U_V(V(x, y_1)).$$

The curve $U(x, y_1)$ is a single-attribute utility assessment for X at the instantiation y_1. This curve is the path traced by the intersection of the surface $U(x, y)$ with the plane $Y = y_1$. There is no room for arbitrarily assigning $U(x, y_1)$ once the value function and the utility over value have been assigned.

The topic we investigate now is the relation between the risk aversion function we assign over value, i.e., the risk aversion function for U_V, and the risk aversion function over any attribute.

It is clear that the risk aversion function over an attribute X need not be constant as the level of another attribute Y varies, and so we need to specify this risk aversion for X at a reference value, say y_1, i.e., the risk aversion function for the curve $U(x, y_1)$. Section 19.3 relates the risk aversion functions of these two utility functions.

19.3 THE CHAIN RULE FOR RISK AVERSION

Let us start with the general formulation where we have a single-attribute utility function over value and a value function, and so

$$U(x, y) = U_V(V(x, y)). \tag{2}$$

Definition: Attribute Risk Aversion in a Multiattribute Decision
The risk aversion function for attribute, X, in multiattribute decisions is equal to the negative ratio of the second to the first partial derivative of the multiattribute utility function with respect to x, i.e.,

$$\gamma_x^U(x, y) \triangleq -\frac{\dfrac{\partial^2 U(x, y)}{\partial x^2}}{\dfrac{\partial U(x, y)}{\partial x}}. \tag{3}$$

Compare this definition to the definition of risk aversion in the single-attribute case, from Chapter 14 that involved the total derivative for the single attribute.

Using the chain rule for partial derivatives, we now take the first derivative of (2) with respect to x, to get

$$\frac{\partial U(x,y)}{\partial x} = \frac{\partial U_V(V(x,y))}{\partial x} = \frac{\partial U_V(V(x,y))}{\partial V}\frac{\partial V(x,y)}{\partial x} = U_V'\frac{\partial V(x,y)}{\partial x}, \qquad (4)$$

where $U_V' = \frac{\partial}{\partial V}U_V$.

Equation (4) relates the partial derivative, $\frac{\partial U(x,y)}{\partial x}$, to the derivative, $U_V' = \frac{\partial}{\partial V}U_V$, and a unit conversion term, $\frac{\partial V(x,y)}{\partial x}$.

Now we use the chain rule again and evaluate the partial derivative of (4) with respect to x

$$\frac{\partial^2 U(x,y)}{\partial x^2} = \frac{\partial^2 U_V(V(x,y))}{\partial V^2}\left(\frac{\partial V(x,y)}{\partial x}\right)^2 + \frac{\partial U_V(V(x,y))}{\partial V}\frac{\partial^2 V(x,y)}{\partial x^2}$$

$$= U_V''\left(\frac{\partial V(x,y)}{\partial x}\right)^2 + U_V'\frac{\partial^2 V(x,y)}{\partial x^2}. \qquad (5)$$

Substituting (4) and (5) into (3) gives the risk aversion function for attribute X as

$$\gamma_x^U(x,y) = -\left(\frac{U_V''}{U_V'}\right)\frac{\partial V(x,y)}{\partial x} - \left(\frac{\frac{\partial^2 V(x,y)}{\partial x^2}}{\frac{\partial V(x,y)}{\partial x}}\right). \qquad (6)$$

Note that the term $-\left(\frac{U_V''}{U_V'}\right)$ in (6) is the risk aversion function for the utility function over value. For notational convenience, we also define γ_x^V as

$$\gamma_x^V(x,y) \triangleq -\frac{\frac{\partial^2 V(x,y)}{\partial x^2}}{\frac{\partial V(x,y)}{\partial x}}. \qquad (7)$$

Note that (7) is not a risk aversion function as it applies to the value function not the utility function. It is determined by the properties of the value function. However, it is the same operator as the risk aversion function but applied to the value function so we give it the notation $\gamma_x^V(x,y)$.

Substituting Equation (7) into Equation (3) gives *the **chain rule for risk aversion.***

Relating the Risk Aversion of an Attribute to the Risk Aversion over Value
The chain rule for risk aversion implies that

$$\gamma_x^U(x,y) = \gamma_V^U\frac{\partial V(x,y)}{\partial x} + \gamma_x^V(x,y). \qquad (8)$$

For simplicity of expression, we will write the chain rule as

$$\gamma_x^U = \gamma_V^U \frac{\partial V}{\partial x} + \gamma_x^V, \tag{9}$$

while keeping in mind that these terms are in fact functions of both X and Y.

Equation (8) expresses the risk aversion function of an attribute X as a sum of two terms:

1. The first term, $\gamma_V^U \frac{\partial V}{\partial x}$, is a product of the risk aversion function of the utility function over value and the partial derivative of the value function with respect to the attribute of interest. The partial derivative, $\frac{\partial V}{\partial x}$, is in effect a unit conversion factor.

2. The second term, γ_x^V, is the value function contribution to the risk aversion function for the attribute of interest. It is defined in (7).

Historical Perspective

Dyer and Sarin (1982) used a similar equation as (8) for value functions over single attributes and defined the difference

$$\gamma_x^U - \gamma_x^V = \gamma_V^U \frac{\partial V}{\partial x}$$

as the ***relative risk version*** of that attribute. I have also seen handwritten notes by Jim Matheson representing the multiattribute version of this equation that dated back to 1968, as well as a typed presentation at Stanford Research Institute expressing this relation for two attributes and dating back to 1974. It is included in the appendix to this chapter.

The Chain rule for risk aversion has numerous implications as we illustrate below.

EXAMPLE 19.1: Discounting Value, Not Utility (Revisited)

Let us refer back to the two-period cash flow and a net present value function with a discount factor, β,

$$V(x, y) = x + \beta y,$$

where x is the amount received in the current year (no discounting) and y is the amount received in one year (with a discount factor). The decision-maker assigns an exponential utility function over value, $U_V(V) = -e^{-\gamma V}$.

Direct substitution gives $U_V' = \gamma e^{-\gamma V}, U_V'' = -\gamma^2 e^{-\gamma V}, \dfrac{\partial V(x,y)}{\partial x} = 1$ and $\dfrac{\partial^2 V(x,y)}{\partial x^2} = 0$. Therefore,

$$\gamma_V^U = -\left(\frac{U_V''}{U_V'}\right) = \gamma \text{ and } \gamma_x^V = 0$$

Therefore, <u>the risk aversion function for attribute, X, money received today</u>, is

$$\gamma_x^U = \gamma.1 + 0 = \gamma.$$

On the other hand, if we repeat the same analysis for attribute Y, we get

$$\frac{\partial V(x,y)}{\partial y} = \beta \text{ and } \frac{\partial^2 V(x,y)}{\partial x^2} = 0.$$

The risk aversion function for attribute, Y, money received in the future, is

$$\gamma_y^U = \gamma.\beta + 0 = \gamma\beta$$

The risk aversion for future cash flows is equal to the risk aversion over present cash flows multiplied by the time preference discounting factor, β. Or equivalently,

$$\boxed{\rho_y = \rho(1+R)},$$

where $\rho_y = 1/\gamma_y^U$ is the risk tolerance for future cash flows, $\rho = 1/\gamma$ is the risk tolerance for present cash flows, and $(1+R) = 1/\beta$ is the compounding rate.

This important result implies that the risk tolerance compounds at the same rate as the time preference rate.

If you have a net present value function with a compound rate $(1+R)=1/\beta$, then your risk tolerance for future cash flows is equal to your risk tolerance over present cash flows multiplied by $(1+R)$.

Equivalently, we may write

Risk aversion function for money received today =
Risk aversion function for money received in the future multiplied by the discount factor.

19.4 UTILITY TRANSVERSALITY (MATHESON AND ABBAS 2009)

We have assessed a utility function over the value measure and related it to the utility function over an individual attribute. We now relate the risk aversion functions across the individual attributes using the value function. We refer to this relation as "utility transversality". This relation will further demonstrate that the risk aversion functions over the attributes are related by properties of the value function and that, once the value function is specified, no additional degree of freedom is needed to assign individual utility functions over the attributes.

The risk aversion function for attribute, X, is

$$\gamma_x^U = \gamma_V^U \frac{\partial V}{\partial x} + \gamma_x^V. \tag{10}$$

Similarly for any other attribute Y, we get

$$\gamma_y^U = \gamma_V^U \frac{\partial V}{\partial y} + \gamma_y^V. \tag{11}$$

From (10) and (11),

$$\gamma_x^U - \gamma_x^V = \left[\gamma_y^U - \gamma_y^V\right]\left(\frac{\partial V}{\partial x} \Big/ \frac{\partial V}{\partial y}\right). \tag{12}$$

Equation (12) may seem difficult to interpret at first, but we can simplify this equation by observing that the value function is constant along an "isopreference contour", i.e.

$$V(x,y) = \text{Constant along an isopreference contour.} \tag{13}$$

As a consequence, the total derivative of the value function across an isopreference contour is zero. Hence

$$dV(x,y) \triangleq \frac{\partial V(x,y)}{\partial x}dx + \frac{\partial V(x,y)}{\partial y}dy = 0. \tag{14}$$

Re-arranging (14) gives the tradeoff between the two attributes x and y as

$$\frac{\dfrac{\partial V(x,y)}{\partial x}}{\dfrac{\partial V(x,y)}{\partial y}} = -\frac{dy}{dx}\Big|_{\text{isopreference contour}} \triangleq t(x,y). \tag{15}$$

where $t(x,y)$ is the deterministic trade-off function between attributes y and x along an isopreference contour.

Substituting (15) into (12) yields

$$\gamma_x^U(x,y) - \gamma_x^V(x,y) = \left[\gamma_y^U(x,y) - \gamma_y^V(x,y)\right]t(x,y). \tag{16}$$

Re-arranging gives

$$t(x,y) = \frac{\gamma_x^U(x,y) - \gamma_x^V(x,y)}{\gamma_y^U(x,y) - \gamma_y^V(x,y)}. \tag{17}$$

Equation (17) is a fundamental expression that relates the risk aversion functions across the different attributes to the trade-off function between them. Matheson and Abbas introduce this relation and refer to it as *"Utility Transversality"*.

Definition
*The **utility transversality relation** asserts that the ratio of*
    the difference $\gamma_x^U(x,y) - \gamma_x^V(x,y)$ to the difference $\gamma_y^U(x,y) - \gamma_y^V(x,y)$
    is equal to the trade-off function between X and Y (the slope of the isopreference contour).
    This is further demonstration that the risk aversion functions are related by properties of the value function.

EXAMPLE 19.2: On Fates Comparable to Death (Revisited)

Suppose a decision-maker has a value function of the form

$$V(x, y) = xy^\eta.$$

Suppose also that he assigns an exponential utility function over value, $U_V(V) = 1 - e^{-\gamma V}$.

Using the chain rule for risk aversion of Equation (8), we can now deduce the risk aversion function toward each of the attributes of the decision problem. For example, in the case of attribute X, we have

$$\frac{\partial V}{\partial x} = y^\eta \text{ and } \gamma_x^V = 0.$$

Equation (8) gives

$$\gamma_x^U = \gamma_V^U \frac{\partial V}{\partial x} + \gamma_x^V = \gamma y^\eta. \tag{18}$$

For attribute Y we have

$$\frac{\partial V}{\partial y} = \eta x y^{\eta-1} \text{ and } \gamma_y^V = -\frac{\eta - 1}{y}.$$

Equation (8) gives

$$\gamma_y^U = \gamma_V^U \frac{\partial V}{\partial y} + \gamma_y^V = \gamma \eta x y^{\eta-1} - \frac{\eta - 1}{y}. \tag{19}$$

Note that while the utility function over value exhibited constant absolute risk aversion, neither of the risk aversion expressions for any of the attributes in (18) or (19) is a constant.

The transversality relation also relates the risk aversion functions over the attributes to the slope of the isopreference contours.

In particular, for attribute X,

$$\gamma_x^U - \gamma_x^V = \gamma_V^U \frac{\partial V}{\partial x} = \gamma y^\eta.$$

For attribute Y,

$$\gamma_y^U - \gamma_y^V = \gamma_V^U \frac{\partial V}{\partial y} = \gamma \eta x y^{\eta-1}.$$

The ratio of these quantities is

$$\boxed{\frac{\gamma_x^U - \gamma_x^V}{\gamma_y^U - \gamma_y^V} = \frac{\gamma y^\eta}{\gamma \eta x y^{\eta-1}} = \frac{y}{\eta x}.}$$

On the other hand, the slope of the isopreference contour can be set by setting the change in value to zero to get

$$dV(x,y) = 0 = \frac{\partial V}{\partial x}dx + \frac{\partial V}{\partial y}dy \Rightarrow y^{\eta}dx + \eta xy^{\eta-1}dy = 0.$$

Rearranging gives

$$\boxed{t(x,y) = -\frac{dy}{dx} = \frac{y}{\eta x}}.$$

Therefore, the ratio of $\dfrac{\gamma_x^U - \gamma_x^V}{\gamma_y^U - \gamma_y^V}$ is equal to the negative slope of the isopreference contour.

Dialogue between Instructor (I) and Student (S)

I: Do we all understand what we have just covered?

S: Yes, so the first thing is that if you have a value function, and if you assign a utility over value, then you can effect determine the utility function over each attribute at any level of the remaining attributes. This is consistent with the ideas covered in Chapter 11 about constructing a multiattribute utility function using a utility function over value and a value function.

I: Correct. We can also determine the relation between the risk aversion over the value measure and the risk aversion over any individual attribute in the problem. We refer to this relation as the chain rule for risk aversion.

S: So we really have no further degree of freedom to assess individual utility functions over the attributes if we are given a value function or the set of isopreference contours.

I: Correct, and as we have seen, the ratio of the risk aversion functions for X and Y, actually determines the slope of the isopreference contours and the trade-offs between the two attributes, i.e., the ratio $\gamma_x^U - \gamma_x^V$ to $\gamma_y^U - \gamma_y^V$ is equal to $-\dfrac{dy}{dx}\Big|_{\text{isopreference contour}}$.

S: And conversely, it must be that if we chose the utility funcitons over the attributes arbitrarily and combine them arbitrarily, then we are in effect making assumptions about the slopes of the isopreference contours and the trade-offs between the attributes.

I: Precisely! This is one of the common mistakes that can occur in multiattribute decisions, particularly when there are logical relations or structural models that relate the attributes. We shall discuss further implications of utility transversality below.

19.5 IMPLICATIONS OF UTILITY TRANSVERSALITY

Perhaps the biggest implication of utility transversality is that once you have established a preference or a value function, there is no additional degree of freedom for providing individual utility functions for each of the attributes independently: they are all related. Likewise, by assuming the functional forms of all individual utility functions and then combining them arbitrarily, you are in effect making assumptions about the trade-offs among the attributes and the shapes of the isopreference contours of the value function. This can be a problem if there is a natural relationship between them such as a logical relation or a structural model based on engineering, physical, or accounting principles.

The utility transversality relation relates the risk aversion functions of the attributes and the trade-offs among them. An immediate application of this result appears in cash flows when trade-offs occur between present and future payoffs. The utility transversality relation relates the risk aversion function for present cash flows and those for different years in the future, using the deterministic time preference (or time trade-offs) for receiving money in different years. Time preference is a deterministic relation that is determined by deterministic trade-offs among the attributes based on contours of the net present value function. Risk aversion is one-dimensional assessment over any of the attributes. The utility transversality relation also enables the calculation of the risk aversion over all the attributes from the risk aversion over any of the attributes using the trade-off function. Therefore, there is a relation between time preference and the risk aversion over the individual years.

The beauty of utility transversality is that it enables us to think about risk aversion over value and the deterministic trade-offs among the attributes separately.

19.6 SUMMARY

- The chain rule for risk aversion relates the risk aversion function of an attribute to the risk aversion function of some function of the attribute for a univariate function $V(x)$ is

$$\gamma_x^U = \gamma_V^U \frac{dV(x)}{dx} + \gamma_x^V$$

- For a bivariate function $V(x, y)$, the chain rule for risk aversion is

$$\gamma_x^U(x, y) = \gamma_V^U(V(x, y)) \frac{\partial V(x, y)}{\partial x} + \gamma_x^V(x, y)$$

- The utility transversality relation relates the risk aversion functions of two attributes to the slope of the isopreference contours. In particular, the ratio $\dfrac{\gamma_x^U - \gamma_x^V}{\gamma_y^U - \gamma_y^V}$ is equal to the negative slope of the contour $t(x, y) = -\dfrac{dy}{dx}$.

- Many people rush to incorporate them both using some arbitrary forms of utility functions. One of the most common methods of building multiattribute utility functions is when people immediately assess individual utility functions over each of the attributes and then combine them into some aggregate form. By doing so, they are making explicit assumptions about the trade-offs among the attributes.

- If there are structural models that capture the logical or physical relations among the attributes, they should first be incorporated and the corresponding individual risk aversion functions over the attributes can be determined using the transversality relation.

ADDITIONAL READINGS

Dyer, J. S. and R. K. Sarin. 1982. Relative risk aversion. *Management Science* 28: 875–886.
Howard, R. A. 1984. On fates comparable to death. *Management Science* 30(4): 407–422.
Howard, R. A. and A. E. Abbas. 2015. *Foundations of Decision Analysis*. Pearson. New York.

Matheson J. E. and A. E. Abbas. 2005. Utility transversality: A value-based approach. *Journal of Multi-criteria Decision Analysis* 13: 229–238.

Matheson J. E. and R. A. Howard. 1968. An introduction to decision analysis. In R. A. Howard and Matheson J. E., eds. *The Principles and Applications of Decision Analysis*, vol. I. Strategic Decisions Group: Menlo Park, CA.

APPENDIX

Historical Notes: An early version of the transversality relation for an additive value function and an exponential utility function over value, as well as notes by Ronald Howard on the linear risk tolerance utility functions and its assessment. These notes were transcribed from a research seminar at Stanford Research Institute (SRI). Used with Permission.

```
To:       Distribution
From:     Lee Merkhofer
Subject:  Minutes of the Decision Analysis Research Seminar, February 28, 1974
```

```
Jim Matheson led a discussion on normative risk preference.  Ron Howard followed
with a short talk on an interesting family of utility functions.  The following
is a brief summary in outline form of some of the ideas covered.
```

Normative Risk Preference
Can we tell a person what his utility function should look like?

A.   Possible additional axioms of rationality
   1.   Bernoulli: $du = dx/x$
   2. $u(x)$ ought to be risk averse
   3.   delta property
   4.   no zero illusion
   5.   non-increasing risk aversion
   6.   long-run stability

B.   Interorganizational-interpersonal comparison
   1.   An individual's risk tolerance is invariably measured to be approximately \$5,000
   2.   For small bets one should behave as if $u(x)$ were linear; for medium-sized bets one should behave as if $u(x)$ were exponential
   3.   Rules of thumb:
      a.   risk tolerance should be proportional to your income, e.g., $1/10\ I < \rho < I$
      b.   to determine an organization's desired behavior scale according to budget :: your salary

C.   Induced risk preference
   1.   *Can we deduce the risk attitude of an organization from higher considerations?*
      a.   Let utility = $u(v)$, value = $v(x,y)$, risk aversion coefficients

$$r_x^u(x,y) = \frac{-d^2u/dx^2}{du/dx} \qquad r_x^v(x,y) = \frac{-d^2v/dx^2}{dv/dx} \qquad r_v(v) = \frac{-d^2u/dv^2}{du/dv}$$

then

$$r_x^u(x,y) = -r_v(v)\frac{dv}{dx} + r_x^v(x,y)$$

In other words, the risk aversion coefficient induced by u on x is made up of two terms: a term induced by u on v and a term induced by v on x . The coefficient dv/dx represents a trade-off factor. Suppose the value function is linear:

$$v(x,y) = ax + by + k$$

Then the relationship between the risk aversion coefficient
induced on x and that induced on y is

$$\frac{r_x^u}{r_y^u} = \frac{a}{b} \quad ,$$

and is independent of the utility function. Also note that

$$r_x^u(x,y) = -ar_v(v)$$

These results allow us to determine the risk attitude of an
individual profit center within a corporation from the risk
attitude of the corporation. Notice that if, for example,
a represents the fraction of our ownership in company x ,
then the smaller a is the more we should be inclined
toward risk-taking in that enterprise.

Distribution (* denotes attendance)

Steve Barrager*	Dale Nesbitt*	Dean Boyd	George Murray
Bruce Judd*	Ken Oppenheimer	Ed Cazalet	Warner North
Al Grum*	Tom Rice*	Ron Howard*	Jacques Pezier*
Tom Keelin*	Steve Tani*	Jim Matheson*	Dick Smallwood*
Lee Merkhofer*	Hans Wynholds*	Al Miller	Ed Sondik
			Carl Spetzler

HAND-WRITTEN NOTES BY RONALD HOWARD (1974)

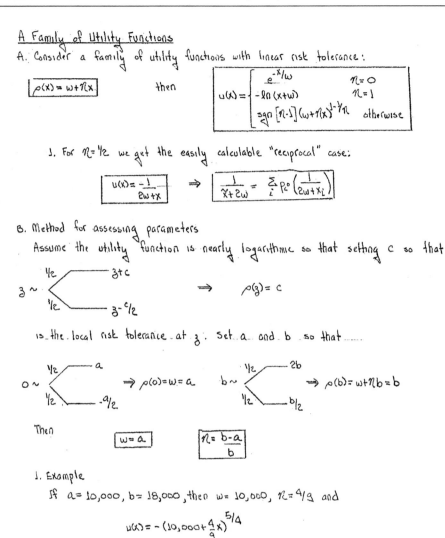

A Family of Utility Functions

A. Consider a family of utility functions with linear risk tolerance:

$$\rho(x) = \omega + \eta x \qquad \text{then}$$

$$u(x) = \begin{cases} e^{-x/\omega} & \eta = 0 \\ -\ln(x+\omega) & \eta = 1 \\ \text{sgn}[\eta-1](\omega+\eta x)^{1-1/\eta} & \text{otherwise} \end{cases}$$

1. For $\eta = 1/2$ we get the easily calculable "reciprocal" case:

$$u(x) = \frac{-1}{2\omega + x} \qquad \Rightarrow \qquad \frac{1}{\bar{x}+2\omega} = \sum_i p_i \circ \left(\frac{1}{2\omega + x_i}\right)$$

B. Method for assessing parameters

Assume the utility function is nearly logarithmic so that setting c so that

$$3 \sim \begin{cases} 1/2 & 3+c \\ 1/2 & 3 - c/2 \end{cases} \qquad \Rightarrow \qquad \rho(3) = c$$

is the local risk tolerance at 3. Set a and b so that

$$0 \sim \begin{cases} 1/2 & a \\ 1/2 & -a/2 \end{cases} \Rightarrow \rho(0) = \omega = a \qquad b \sim \begin{cases} 1/2 & 2b \\ 1/2 & b/2 \end{cases} \Rightarrow \rho(b) = \omega + \eta b = b$$

Then

$$\boxed{\omega = a} \qquad \boxed{\eta = \frac{b-a}{b}}$$

1. Example

If $a = 10,000$, $b = 18,000$, then $\omega = 10,000$, $\eta = 4/9$ and

$$u(x) = -\left(10,000 + \frac{4}{9}x\right)^{5/4}$$

$$5.28 \sim \begin{cases} 1/2 & 10,000 \\ 1/2 & -5,000 \end{cases} \qquad 17,930.50 \sim \begin{cases} 1/2 & 36,000 \\ 1/2 & 9,000 \end{cases}$$

$\begin{pmatrix} \text{instead} \\ \text{of } 0 \end{pmatrix}$ $\begin{pmatrix} \text{instead of} \\ 18,000 \end{pmatrix}$

Note accuracy of approximation

C. A Comparison: $n = 0, \frac{1}{2}, 1$

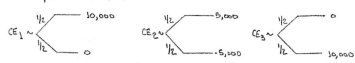

CASES			
ω	10,000	10,000	10,000
n	0	$\frac{1}{2}$	1
Name	exponential	reciprocal	logarithmic
$u(x)$	$e^{-x/10,000}$	$-\dfrac{1}{20,000 + x}$	$\ln(10,000 + x)$
CE_1	3798.85	4000	4142.14
CE_2	-1201.15	-1250	-1333.75
CE_3	-6801.15	-6666.67	$-\infty$

CHAPTER 20

Multiattribute Risk Aversion

Chapter Concepts

- Implications of the sign of the cross-derivative of
  - A preference function and
  - A multiattribute utility function
- The concept of multivariate risk aversion and its implications
- Relating the multivariate risk aversion concept to the risk aversion function over value
- Relating multivariate risk aversion to the concept of correlation aversion
- Implications of having a multivariate risk neutral (additive) utility function

20.1 INTRODUCTION

The main purpose of this chapter is to discuss the implications of the sign of the ***cross-derivative*** of a two-attribute preference function and a two-attribute utility function on the decision-maker's preferences for bivariate lotteries. In so doing, this chapter introduces the concept of ***multivariate risk aversion*** and its relation to preference for lotteries including correlated vs. uncorrelated variables. A second objective is to relate the multivariate risk aversion concept to the absolute risk aversion function over value and to properties of the value function.

20.2 THE SIGN OF THE CROSS-DERIVATIVE OF A PREFERENCE FUNCTION

Before we begin our discussion of the cross-derivative of the multiattribute utility function for lotteries, we should understand the implications of this cross-derivative of a preference function for deterministic prospects, i.e., what does the cross-derivative $\dfrac{\partial^2 P(x, y)}{\partial x \partial y}$ imply about preferences for the order of prospects as the level of an attribute varies.

Implications of the Sign of the Cross-Derivative of a Preference Function
If the cross-derivative of the preference function is either strictly positive or negative, i.e., if

$$\frac{\partial^2 P(x,y)}{\partial x \partial y} > 0 \text{ or } \frac{\partial^2 P(x,y)}{\partial x \partial y} < 0,$$

then preferences for any two prospects can change at most one time as the level of another attribute increases.

A constant sign of the cross-derivative of a preference function (either strictly positive or strictly negative on the entire domain) therefore implies that

$P(x_1, y) - P(x_2, y)$ can change sign at most once as y increases.

Because the cross-derivative is symmetric, this also implies that

$P(x, y_1) - P(x, y_2)$ can change sign at most once as x increases.

To see why this is true, note that if the cross-derivative has a constant sign (either positive or negative), then for any $x_1 > x_2, y_1 > y_2$, the difference

$$[P(x_1, y_1) - P(x_1, y_2)] - [P(x_2, y_1) - P(x_2, y_2)] = \int_{x_1}^{x_2} \int_{y_1}^{y_2} \frac{\partial^2 P(x,y)}{\partial x \partial y} dx dy$$

does not change sign. If the cross-derivative is either always positive or always negative, this implies that the difference

$$\Delta(y) = P(x_1, y) - P(x_2, y)$$

is strictly monotone and so it can change sign at most once.

Because utility functions have the same ordinal ranking as preference functions, the same implication applies to multiattribute utility functions for the case of preferences for deterministic prospects. That is, if we have a two-attribute utility function, $U(x, y)$, whose cross-derivative,

$$\frac{\partial^2 U(x,y)}{\partial x \partial y} > 0$$

does not change sign, then preferences for any two prospects, having any fixed attribute levels x_1, x_2 (or y_1, y_2), can change at most once as the level of the other attribute changes from low to high.

As we now show, the sign of the cross-derivative of a utility function has further implications for preferences for multivariate lotteries.

20.3 THE SIGN OF THE CROSS-DERIVATIVE OF A MULTIATTRIBUTE UTILITY FUNCTION: IMPLICATIONS FOR LOTTERIES

We now discuss the implications of the sign of the cross-derivative of a multiattribute utility function $\frac{\partial^2 U(x,y)}{\partial x \partial y}$ for choices between lotteries. Consider the following example, where you have a choice between two deals.

EXAMPLE 20.1: A Choice between Two Bivariate Lotteries

Consider a choice between two bivariate lotteries. The prospects of each lottery are characterized by two attributes, X and Y. You prefer more of each attribute to less. For example, the attributes could represent consumption and health state. Assume that $x_2 > x_1$ represent two levels of attribute X and that $y_2 > y_1$ represent two levels of attribute Y. We can think of four prospects

$$(x_1, y_1), (x_1, y_2), (x_2, y_1), (x_2, y_2).$$

It is clear that preferences for prospects satisfy the relation

$$(x_1, y_1) \prec (x_1, y_2) \prec (x_2, y_2)$$

because more is preferred to less, and also that

$$(x_1, y_1) \prec (x_2, y_1) \prec (x_2, y_2).$$

But the preference order for (x_1, y_2) and (x_2, y_1) is not clear.
Now suppose you face two uncertain deals as follows:

Deal 1: Lottery $< 0.5, (x_1, y_1); 0.5, (x_2, y_2) >$.

This deal provides a 50/50 chance of receiving either (x_1, y_1) or (x_2, y_2), i.e., you get either the best or the worst prospect.

Deal 2: Lottery $< 0.5, (x_1, y_2); 0.5, (x_2, y_1) >$.

This deal provides a 50/50 chance of receiving either (x_1, y_2) or (x_2, y_1), i.e., you will get either the second best or the third best.
Figure 20.1 represents both deals graphically. Which deal do you prefer?

We consider three cases:

Case 1: Indifference between the Two Deals
If you are indifferent between the two deals, and if this equality holds $\forall x_1 > x_2, y_1 > y_2$ on the domain, then the expected utilities of both sides of Figure 20.1 must be equal and so this implies that

$$0.5 \times U(x_1, y_1) + 0.5 \times U(x_2, y_2) = 0.5 \times U(x_1, y_2) + 0.5 \times U(x_2, y_1) \quad \forall x_1 > x_2, y_1 > y_2.$$

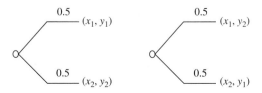

Figure 20.1. Choice between two bivariate lotteries.

Dividing both sides by 0.5 and rearranging gives

$$U(x_1, y_1) - U(x_1, y_2) - U(x_2, y_1) + U(x_2, y_2) = 0 \quad \forall x_1 > x_2, y_1 > y_2.$$

This implies that the cross-derivative

$$\frac{\partial^2 U(x, y)}{\partial x \partial y} = 0, \quad \forall x, y$$

on the domain of interest, which further implies that the decision-maker is indifferent between a deal where you get either the best of both attributes or the worst with a 50/50 chance, and a deal that gives you the best of one and the worst of the other with a 50/50 chance.

If we think of the two attributes as random variables, then this implies that the decision-maker is indifferent between deals where the attributes are positively correlated (Deal 1), where he gets either the best of both or the worst of both, and deals where the attributes are negatively correlated (Deal 2), where he always gets the best of one and the worst of the other.

Case 2: Prefer Deal 1 to Deal 2

If you prefer Deal 1 to Deal 2 $\forall x_1 > x_2, y_1 > y_2$ on the domain of interest, then the expected utility of Deal 1 is greater than that of Deal 2, and so

$$0.5 \times U(x_1, y_1) + 0.5 \times U(x_2, y_2) > 0.5 \times U(x_1, y_2) + 0.5 \times U(x_2, y_1) \quad \forall x_1 > x_2, y_1 > y_2$$

or equivalently,

$$U(x_1, y_1) - U(x_1, y_2) - U(x_2, y_1) + U(x_2, y_2) > 0 \quad \forall x_1 > x_2, y_1 > y_2.$$

This implies that the cross-derivative

$$\frac{\partial^2 U(x, y)}{\partial x \partial y} > 0 \quad \forall x_1 > x_2, y_1 > y_2.$$

This implies that the decision-maker prefers a deal that provides either the best of both attributes or the worst of both attributes to a deal that provides the best of one attribute and the worst of the other. If we think of the two attributes as random variables, then the decision-maker has a preference for deals where the attributes are positively correlated (Deal 1) to those where the attributes are negatively correlated (Deal 2).

Recall that even if more of an attribute was not necessarily preferred to less, then his preference for any two prospects (x_1, y) and (x_2, y) can change only once as y increases.

Case 3: Prefer Deal 2 to Deal 1

If you prefer Deal 2 to Deal 1 $\forall x_1 > x_2, y_1 > y_2$, then the expected utility of Deal 2 is greater than that of Deal 1, and so

$$U(x_1, y_1) - U(x_1, y_2) - U(x_2, y_1) + U(x_2, y_2) < 0 \quad \forall x_1 > x_2, y_1 > y_2.$$

This implies that the cross-derivative

$$\frac{\partial^2 U(x, y)}{\partial x \partial y} < 0 \quad \forall x_1 > x_2, y_1 > y_2.$$

Note that this implies that the decision-maker prefers a deal where you get the best of one attribute and the worst of another in either case to a deal where you get the best of both attributes or the worst of both attributes. If we think of the two attributes as random variables, then the decision-maker has a preference for games that are negatively correlated than those that are correlated positively.

The following definition of risk aversion was proposed by Richard (1975).

Definition: Two-Attribute Risk Aversion

A decision-maker is said to exhibit two-attribute risk aversion if

$$\frac{\partial^2 U(x, y)}{\partial x \partial y} < 0.$$

He is two-attribute risk seeking if

$$\frac{\partial^2 U(x, y)}{\partial x \partial y} > 0,$$

and is two-attribute risk neutral if

$$\frac{\partial^2 U(x, y)}{\partial x \partial y} = 0.$$

20.4 RELATING TWO-ATTRIBUTE RISK AVERSION AND CORRELATION AVERSION

As we have seen in the previous example, the sign of the cross-derivative provides an indication of whether or not a decision-maker prefers deals whose attributes are positively (or negatively) correlated. To further investigate this property, consider the following example.

EXAMPLE 20.2: Certain Equivalent of a Bivariate Gaussian Distribution

A decision-maker faces an uncertain cash flow for two years, whose payoffs are determined by a bivariate Gaussian distribution,

$$f(x, y) = \frac{1}{2\pi\sigma_x\sigma_y\sqrt{1-\rho^2}} \exp\left(-\frac{1}{2(1-\rho^2)}\left[\frac{(x-\mu_x)^2}{\sigma_x^2} + \frac{(y-\mu_y)^2}{\sigma_y^2} - \frac{2\rho(x-\mu_x)(y-\mu_y)}{\sigma_x\sigma_y}\right]\right).$$

The decision-maker has a net present value function for the cash flow, with a discount factor, β,

$$V(x,y) = x + \beta y.$$

Suppose the decision-maker has an exponential utility function over value and is risk averse, i.e.,

$$U_V(V) = 1 - e^{-\gamma V}.$$

Her two-attribute utility function is then

$$U(x,y) = 1 - e^{-\gamma(x+\beta y)}.$$

The cross-derivative of this two-attribute function is negative (she is two-attribute risk averse), since

$$\frac{\partial U(x,y)}{\partial x} = \gamma e^{-\gamma(x+\beta y)}, \quad \frac{\partial^2 U(x,y)}{\partial x \partial y} = -\gamma^2 \beta e^{-\gamma(x+\beta y)} < 0.$$

The certain equivalent of net present value for the bivariate Gaussian lottery is

$$\tilde{v} = \bar{v} - \frac{\gamma}{2}\sigma_v^2 = (\mu_x + \beta\mu_y) - \frac{\gamma}{2}\left(\sigma_x^2 + \beta^2\sigma_y^2 + 2\rho_{xy}\beta\sigma_x\sigma_y\right). \tag{1}$$

The mean of the net present value is

$$\bar{v} = \mu_x + \beta\mu_y$$

and the variance of net present value is

$$\sigma_v^2 = \sigma_x^2 + \beta^2\sigma_y^2 - 2\rho_{xy}\beta\sigma_x\sigma_y.$$

It is clear that the variance decreases as the correlation coefficient decreases. Because the decision-maker is risk averse, he prefers less variance and so he prefers a lower correlation coefficient.

This decision-maker prefers less correlation to more because the certain equivalent increases as ρ decreases. This implies that the decision-maker prefers lotteries whose attributes are negatively correlated to those that are positively correlated (he is correlation averse).

On the other had, if the decision-maker were risk seeking over net present value in this example, then $\gamma = -ve$, and we can set $U_V(V) = e^{-\gamma V}$ to be increasing with V,

$$\frac{\partial U(x,y)}{\partial x} = -\gamma e^{-\gamma(x+\beta y)}, \quad \frac{\partial^2 U(x,y)}{\partial x \partial y} = \gamma^2 \beta e^{-\gamma(x+\beta y)} > 0.$$

The decision-maker is multivariate risk seeking. From (1) the certain equivalent increases as the variance of net present value increases, and so the decision-maker prefers positive values of ρ to negative values. This implies that the decision-maker prefers lotteries whose attributes are positively correlated to those that are negatively correlated (he is correlation seeking).

Finally, if the decision-maker is risk neutral over net present value in this example, then $\gamma \to 0$, and so the certain equivalent is determined by the mean of the net present value, $\bar{v} = \mu_x + \beta\mu_y$. This implies that the decision-maker is indifferent to whether or not the attributes are correlated. She is said to be correlation neutral. Note that a decision-maker who is risk-neutral over net present value has an additive utility function $U(x, y) = x + \beta y$. We shall discuss the implications of having an additive utility function in Section 20.6.

The notion of correlation aversion (which stems from variance aversion in this example) makes perfect sense in a stock portfolio problem with Gaussian structure, where the dependence is captured entirely by the correlation coefficients, and where we prefer less variance to more.

It is also straightforward to extend this example to a multivariate Gaussian distribution where the certain equivalent will be equal to the mean less half the risk aversion coefficient multiplied by the variance of the portfolio. Because less variance is preferred to more in the case of risk aversion toward money, we can derive the effects that the pairwise correlation coefficients will have on the overall variance and the certain equivalent of the portfolio.

Dialogue between Instructor (I) and Student (S)

I: Is everyone clear on the implications of the sign of the cross-derivative?

S: For preference or value functions, it means that preferences for any two deterministic prospects can change at most once as the level of the second attribute varies.

I: Correct. This is a sufficient condition for a maximum of one-switch change in preferences for deterministic prospects. And what about the implications of the sign of the cross-derivative of a utility function?

S: As we have seen, if the sign is negative, then the decision-maker is multivariate risk averse, and would prefer a 50/50 lottery that provides the best of one attribute and the worst of another in either case to one that provides the best of both attributes or the worst with equal probabilities.

I: Correct, in other words, the decision-maker prefers lotteries where the variables are negatively correlated to those that are positively correlated. And what about a positive sign?

S: This implies the decision-maker prefers the lottery that provides the best or the worst levels of both attributes with equal probability to one that provides the best of one and the worst of the other. The decision-maker is said to be correlation seeking. And likewise, if the cross-derivative is zero, then the decision-maker is indifferent to correlation between the variables in these lotteries.

I: Would you want to have this correlation aversion property for any attributes and any decision you face?

S: Hmmm, not really. As we just saw in Example 20.2, if you were risk neutral over money, i.e., $\gamma = 0$, then you would not be correlation averse because you would not care about the variance in the first place. You just care about the mean value of the attributes in this case.

I: Correct. There is a relation between the risk aversion over net present value, the actual net present value function, and the implied multivariate risk aversion. In general, this property of correlation aversion will depend on the value function and the utility function over value as well as the types of attributes in the problem. As the value function becomes more complex (losing monotonicity), it will be more difficult to reason about this bivariate risk aversion property or its desirability in a given problem. It is a lot simpler to think about correlation aversion by thinking about the deterministic value function over the attributes, and then thinking about the risk aversion over value.

S: What if the prospects are characterized by non-monetary factors, would you not then prefer the best of one and the worst of the other to the best of both or the worst of both?

I: Not necessarily. When the attributes become more general, such as one of the attributes representing the "left shoe" and the other representing the "right shoe," most people would prefer a gamble providing a "pair of shoes" or nothing to a lottery that provides a single shoe (left or right). This is due to the value obtained by having both shoes together than having just one. This is why it is often preferable to think in terms of the value of the prospect combined using a value function.

The next section relates multivariate risk aversion to the absolute risk aversion function over value and to properties of the value function.

20.5 RELATING RISK AVERSION AND TWO-ATTRIBUTE RISK AVERSION

It is now natural to ask the following question:

> *Does a person who is risk averse over value have a multiattribute utility function that exhibits multivariate risk aversion?*

We answer this question below and provide insights into the signs of the cross-derivatives in terms of the value function.

Consider a value function constructed using a utility function over the value function

$$U(x, y) = U_V(V(x, y)). \tag{2}$$

We assume that U_V has a positive first derivative and a negative second derivative, i.e., $U_V' > 0$ and $U_V'' < 0$.

Taking the first partial derivative of (2) with respect to x and using the chain rule for derivatives gives

$$U_x(x, y) = U_V'(V)V_x(x, y), \tag{3}$$

where $U_x(x, y) = \dfrac{\partial}{\partial y}U(x, y)$, $U_V' = \dfrac{d}{dV}U_V$, $V_x(x, y) = \dfrac{\partial}{\partial x}V(x, y)$.

Taking the partial derivative of (3) with respect to y gives

$$U_{xy}(x, y) = U_V'(V)V_{xy}(x, y) + U_V''(V)V_x(x, y)V_y(x, y), \tag{4}$$

where $U_V'' = \dfrac{d^2}{dV^2}U_V$, $V_y(x, y) = \dfrac{\partial}{\partial y}V(x, y)$, $V_{xy}(x, y) = \dfrac{\partial^2}{\partial x \partial y}V(x, y)$.

The effects of the sign of the first derivatives on the cross-derivative are now clear. If the decision-maker prefers more value (and more of any attribute) to less, then $U_V'(V) \geq 0, V_x(x, y) \geq 0, V_y(x, y) \geq 0$. If any of these derivatives is zero (such as when there are inflection points on the utility curve), then they set their corresponding product term to zero. In particular, if there is an inflection point in the utility function, then $U_V'(V) = 0, U_V''(V) = 0$ at that point and, from (4) the cross-derivative will be zero regardless of the value function.

We now consider the case where $U_V'(V) > 0, V_x > 0, V_y > 0$ (the utility function over value is strictly increasing, and more of every attribute is strictly preferred to less). By rearranging (4), we obtain the following.

Proposition 20.1: Interpreting the multivariate risk aversion in terms of the value function

$$U_{xy}(x,y) = -U_V'(V)V_x(x,y)V_y(x,y)\left[\gamma(V) - \frac{V_{xy}(x,y)}{V_x(x,y)V_y(x,y)}\right], \quad U_V'(V) > 0, V_x > 0, V_y > 0,$$

where $\gamma(V) = -\dfrac{U_V''(V)}{U_V'(V)}$ is the absolute risk aversion function over value.

Proposition 20.1 relates the sign of the cross-derivative to the Arrow-Pratt absolute risk aversion function over value and the derivatives of the value function. Note that this proposition does not make assumptions about a particular functional form for either the value function or the utility function over value. We have merely assumed differentiability of the second order and also that the first derivatives are positive.

Two terms are now important in determining the sign of the mixed partial derivative of the utility function:

a) **Risk Aversion Function of the Single-Attribute Utility Function over Value**

   The *risk aversion function over value*, $\gamma(V)$: Based on our assumptions, $\gamma(V) \geq 0$.

b) **Cross-derivatives of the Value Function**

   The *term* $\dfrac{V_{xy}(x,y)}{V_x(x,y)V_y(x,y)}$: Note that this term has the same units as the risk

aversion function over value, namely $\dfrac{1}{V}$. Since $V_x(x,y) > 0, V_y(x,y) > 0$, then the

sign of this term depends on the sign of the mixed partial derivative of the value function, $V_{xy}(x,y)$.

A positive value for $V_{xy}(x,y)$ across the domain of the attributes implies that for any $x_0 < x_1$ and $y_0 < y_1$,

$$V(x_1,y_1) - V(x_0,y_1) - V(x_1,y_0) + V(x_0,y_0) > 0.$$

Rearranging gives

$$[V(x_1,y_1) - V(x_0,y_1)] > [V(x_1,y_0) - V(x_0,y_0)],$$

which implies that for the chosen dimension of value provided by the value function, the decision-maker values increments in one attribute, x, higher, when the level of attribute y is higher, and by symmetry, he values increments in attribute y higher for larger values of x.

Examples of attributes that may satisfy this property are Health and Wealth, when the decision-maker values changes in wealth more if he has a better health state because he can enjoy the benefits of wealth (such as engaging in extreme sports or travelling) more than if he had a lower health state that prevented him from travel or engaging in such activities.

A negative value for $V_{xy}(x,y)$ across the domain implies that for any $x_0 < x_1$ and $y_0 < y_1$,

$$V(x_1, y_1) - V(x_0, y_1) - V(x_1, y_0) + V(x_0, y_0) < 0,$$

or

$$[V(x_1, y_1) - V(x_0, y_1)] > [V(x_1, y_0) - V(x_0, y_0)],$$

which implies that the decision-maker values increments in one attribute, x, lower when the level of attribute y is higher (and the same applies to increments in y at levels of x).

The sign of the cross-derivative of the value function depends on the particular problem and the units that we would like to use for assessing the utility function over value. As we have seen, it may be positive, negative, or even change from positive to negative across the domain. Furthermore, the sign of the value function asserts one-switch change in preferences for deterministic prospects at a fixed level of one attribute as the level of the other attribute increases.

Observations about the Sign of the Cross-Derivative of the Utility Function
Assuming

$$U_V' > 0, \gamma(V) \geq 0, V_x(x,y) > 0, V_y(x,y) > 0.$$

1. If $V_{xy}(x,y) < 0$, then any person who is risk averse or risk neutral over value will also be multivariate risk averse, $U_{xy}(x,y) < 0$, and will choose Deal 1 over Deal 2.
2. If $V_{xy}(x,y) > 0$, then the sign of the cross-derivative can be positive or negative depending on the relative magnitude of $\dfrac{V_{xy}(x,y)}{V_x(x,y)V_y(x,y)}$ and $\gamma(V)$.
3. If $V_{xy}(x,y) = 0$, then $U_{xy}(x,y) < 0$ if $\gamma(V) > 0$, and $U_{xy}(x,y) = 0$ if $\gamma(V) = 0$.

Table 20.1 tabulates the sign of the cross-derivative of the utility function for different signs of $V_{xy}(x,y), \gamma(V)$ under the assumptions that $U_V' > 0, \gamma(V) \geq 0, V_x(x,y) > 0, V_y(x,y) > 0$. As shown in Table 20.1, a negative cross-derivative can be achieved by a positive or negative value of $\gamma(V)$.

The following examples present an application of Proposition 20.1 and several insights for reasoning about the sign of the mixed partial derivative for some given value functions and single-attribute utility functions over value.

Table 20.1. Sign of cross-derivative of the utility function versus sign of $V_{xy}(x, y)$, $\gamma(V)$.

	$V_{xy}(x, y)$		
	+	**0**	**−**
+	+/o/−	−	−
$\gamma(V)$ **0**	+	0	−
−	+	+	+/o/−

EXAMPLE 20.3: Additive Value Functions

Suppose that a decision-maker has an additive value function over two attributes (such as NPV for cash flows in two years). Here, more of any attribute is preferred to less and the value function is given by

$$V(x, y) = ax + by, \quad a, b > 0.$$

Here $V_x, V_y > 0$. Suppose also that the decision-maker is risk averse over value, i.e., $\gamma(V) > 0$, and therefore $U_V' > 0$. For an additive value function, we have

$$V_{xy}(x, y) = 0$$

and from Proposition 20.1,

$$U_{xy}(x, y) = -\gamma(V)U_V'(V)V_x(x, y)V_y(x, y). \tag{5}$$

From (5), a decision-maker who strictly prefers more of any attribute to less and has an additive value function will have a negative cross-derivative for the utility function if and only if the decision-maker is risk averse.

The concepts of the Arrow-Pratt risk aversion and the multivariate risk aversion coincide for linear value functions with $V_x(x, y) > 0, V_y(x, y) > 0$.

EXAMPLE 20.4: Multiplicative Value Functions with Constant Elasticity of Substitution

---

Thinking in terms of the value function and a single-attribute utility function over value can help us determine the sign of the cross-derivative for a variety of value functions. Suppose, for example, that the value function is a product of two functions, x and y^η, on a domain of the unit square,

$$V(x, y) = xy^\eta, \ 0 < x, y \leq 1. \tag{6}$$

This value function represents constant elasticity of substitution where the fractional increase in x needed to compensate for a fractional decrease in y is a constant,

$$\eta = -\frac{dx/x}{dy/y}.$$

When $\eta = 1$, we get the product value function $V(x,y) = xy$, and when $\eta \neq 1$, we have the more general Cobb-Douglas value function that we discussed in Chapters 7, 8, and 12. As we have seen, the value function (6) implies that more of any attribute is preferred to less. It has also appeared in medical decision-making contexts where the attributes are consumption and health state (see, for example, Howard 1984).

From (4), we have $V_x(x,y) = y^\eta > 0, V_y(x,y) = \eta xy^{\eta-1} > 0, V_{xy}(x,y) = \eta y^{\eta-1} > 0$, and so

$$\frac{V_{xy}(x,y)}{V_x(x,y)V_y(x,y)} = \frac{\eta y^{\eta-1}}{y^\eta(\eta xy^{\eta-1})} = \frac{1}{xy^\eta} = \frac{1}{V} > 0.$$

The sign of the cross-derivative for this value function is positive. Assume that the decision-maker is risk averse over value, i.e., $\gamma(V) > 0$.

Substituting into (4) gives

$$U_{xy}(x,y) = -U_V'(V)V_x(x,y)V_y(x,y)\left[\gamma(V) - \frac{1}{V}\right].$$

We now consider the sign of the cross-derivative for several utility functions belonging to S_1 that may be assigned over this value function:

1. If the decision-maker has a logarithmic utility function over value, then

$$U_V = \ln(V) \Rightarrow \gamma(V) = \frac{1}{V}, \text{ and } U_{xy}(x,y) = 0.$$

The mixed partial derivative of the utility function is zero on the entire domain of the utility function.

2. If the decision-maker has an exponential utility function and is risk averse, then $\gamma(V) = \gamma = \text{constant} > 0$, and the mixed partial derivative is positive when $\gamma V < 1$, and negative when $\gamma V > 1$. Generally speaking, it is more desirable to have a situation where $\gamma V < 1$ (and hence a positive mixed partial derivative) than $\gamma V > 1$. The rationale is that $\gamma V > 1$ includes the region of high-risk aversion and the saturation effects of the exponential utility function.

3. Finally, if the decision-maker is risk neutral over value, then from (2),

$$U_{xy}(x,y) = U_V'(V)V_{xy}(x,y) = U_V'(V)\frac{1}{V} > 0.$$

We will refer back to the role of linear utility functions over value in developing a pattern for the sign of the cross-derivative.

EXAMPLE 20.4: Multivariate Risk Averse but Risk Seeking over Value

---

Being multivariate risk averse does not guarantee that the decision-maker is risk averse over value. We have seen that you can be risk averse and have a cross- derivative for the utility function. We will now show that you can be risk seeking over value and yet be multivariate risk averse (have a negative cross-derivative). To illustrate, suppose that the value function is multiplicative and has the form

$$V(x,y) = -e^{-(\alpha x + \beta y)}, \ a,b > 0, \tag{7}$$

which also implies that more of any attribute is strictly preferred to less.

Suppose, further, that the decision-maker is risk seeking over value and has an exponential utility function. This implies that $\gamma(V) = \gamma = $ constant < 0. This implies (from Proposition 20.1) that

$$U_{xy}(x,y) = -U_V'(V)V_x(x,y)V_y(x,y)[\gamma + e^{\alpha x + \beta y}]. \tag{8}$$

From (6), it is straightforward to see that if the decision-maker is risk seeking over value, and if the absolute value of the risk aversion function is less than $e^{\alpha x + \beta y}$, then the cross-derivative is negative indicating multivariate risk aversion.

NOTE

1. Using a value function and thinking in terms of a single-attribute utility function over value can help us think about the sign of a cross-derivative for a given problem.
2. It is quite plausible that a utility function that is risk averse over value results in a multiattribute utility function with a cross- derivative that is positive or negative (and vice versa).

20.6 THE IMPLICATIONS OF ADDITIVE UTILITY FUNCTIONS

Additive utility functions have a cross-derivative equal to zero. Because they are widely used models of multiattribute utility functions, we dedicate this section to the implications of using such additive utility functions in practice.

As we have seen, additive preference functions (or monotone transformations thereof) allow for a wide variety of trade-offs including those representing constant elasticity of substitution.

Additive value functions (on an absolute monetary scale) apply in many business and accounting situations but need not apply in many other applications such as all attributes associated with high-speed milling.

Additive utility functions (correlation neutral) have more serious implications than additive preference and value functions, and can be problematic if used arbitrarily. We illustrate this in the following example.

EXAMPLE 20.5: Implications of Additive Utility Functions for Cash Flows

You are given a choice between two deals the two deals of Figures 20.2 and 20.3.

Deal 1: We toss a coin. If it lands Heads, you get $10,000 every year for the next 10 years. If it lands Tails, you get $0 every year for the next 10 years.

Deal 2: We toss a coin every year for the next 10 years (i.e., the coin is tossed 10 times, once every year). If the coin lands Heads in a given year, you get $10,000 in that year, and if it lands Tails in that year, you get $0 that year. The coin tosses are independent for each year.

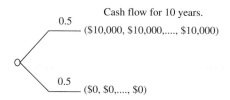

Deal 1: Coin is tossed only once. Pay-off for 10 years is determined by a single toss.

Figure 20.2. Deal 1. Ten-year payoff determined by a single toss.

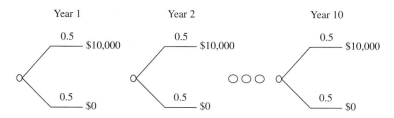

Deal 2: Coin is tossed each year.
Pay-off for each year is determined by its corresponding toss.

Figure 20.3. Deal 2. Payoff for each year is determined by its coin toss.

Which deal do you prefer?

Note that the payoffs in Deal 1 are 100% positively correlated (if you know the payoff in one year, you will know the payoffs in all years), but in Deal 2, they are uncorrelated, and you have a 50/50 chance of receiving a payoff in any given year.

Most people would prefer Deal 2 to Deal 1 because they feel there is a higher chance that they will get something out of this deal, whereas there is a 50/50 chance that they will get nothing with Deal 1. The probability distributions for the payoffs of these deals are in fact very different: although they have the same mean of $50,000, the variance is significantly higher for Deal 1.

If you have a net present value function for the cash flows, $X_i, i = 1,...,n$, with corresponding discount factors β_i, then

$$NPV = \sum_{i=1}^{10} \beta_i X_i,$$

which is an additive value function. This net present value function has a cross-derivative equal to zero for any two attributes.

If you are risk averse over this net present value function, then, as we have seen in Table 20.1, you are correlation averse, and you should prefer Deal 2 to Deal 1. In fact, if you assigned an exponential utility function, $-e^{-\gamma V}$, over this additive value function, then you get a multiplicative utility function of the form

$$U(X_1,...,X_{10}) = -e^{-\gamma \sum_{i=1}^{10} \beta_i X_i} = -\prod_{i=1}^{10} e^{-\gamma \beta_i X_i}.$$

The cross-derivative of this multiplicative utility function is negative for any two attributes, thereby exhibiting multivariate risk aversion.

But let us investigate what happens if instead you used an additive utility function of the form

$$U(X_1,...,X_{10}) = \sum_{i=1}^{10} U_i(X_i).$$ (6)

The expected utility of both deals would be equal in this case:

The additive utility function implies that you are indifferent between the two deals.

This is because an additive utility function implies correlation-neutral behavior, and thus it incorporates the marginal distribution and not the dependence relations between the variables, and so a decision-maker with such an additive utility function will be indifferent between the two deals.

Additive utility functions are simple, and there is a wealth of literature that uses them for their simplicity, but, as this example demonstrates, they are much more restrictive than using an additive preference function or an additive value function. You can have an additive value function and still be correlation averse.

A better approach to modeling the utility function over future cash flows would be to first construct a value function and then assign a utility function over value. If this value function is the net present value function, then the utility function would take the form

$$U(X_1,...,X_n) = U_V\left(V(X_1,...,X_n)\right) = U_V\left(\sum_{i=1}^{n} \beta_i X_i\right).$$

This requires only a single risk aversion assessment over value and discount factors instead of having to assess individual utility functions independently for each year.

This example highlights again the power of a structural model and why it is important to think about your value of a prospect first and then assign your utility function over value, instead of assigning an individual utility function over each attribute and then combining the individual utility assessments using some arbitrary form (albeit simple).

Dialogue between Instructor (I) and Student (S)

I: Is everyone clear about the implications of multivariate risk aversion?

S: Yes. It relates to the preference for correlation among the attributes. It is really surprising to see that additive utility functions that are widely used in practice have such severe limitations when it comes to indifference between whether or not attributes are probabilistically correlated.

I: I agree. The additive utility function is one of the most widely used (and widely misused) forms of multiattribute utility functions. People often construct them because of their simplicity without thinking about their implications. And, of course, simplicity is not an excuse for using a wrong approach.

20.7 SUMMARY

If the sign of the cross-derivative of a preference or a value function is either always positive or always negative, then this is a sufficient condition to guarantee that preferences for prospect characterized by the levels of any of the attributes can only switch once as the other attribute increases.

Implications of the cross-derivatives for multiattribute utility functions relate to preferences for correlation between the attributes.

This chapter also discussed relations between

(i)   multiattribute risk aversion and correlation aversion, and
(ii) multiattribute risk aversion and the risk aversion function over value and properties of the value function.

The additive utility function is much more restrictive than the additive preference function.

ADDITIONAL READINGS

Abbas, A. E. 2011. Decomposing the cross-derivatives of a multiattribute utility function into risk attitude and value. *Decision Analysis* 8(2): 103–116.

Howard R. A. 1984. On fates comparable to death. *Management Science* 30(4): 407–422.

Matheson, J. E. and A. E. Abbas. 2005. Utility transversality: A value-based approach. *Journal of Multi-Criteria Decision Analysis* 13(5–6): 229–238.

Richard, S. 1975. Multivariate risk aversion, utility independence and separable utility functions. *Management Science* 22(1): 12–21.

PART V

MULTIATTRIBUTE UTILITY FUNCTIONS WITHOUT PREFERENCE OR VALUE FUNCTIONS

In the previous chapters, we illustrated how to construct a multiattribute utility function using a univariate utility assessment over a value measure or over an attribute of the preference function. This approach reduces the multiattribute indifference assessment into a univariate utility assessment, but it also requires the identification of a value function (or the determination of the isopreference contours of a preference function that define the deterministic trade-offs among the attributes). Either the value function or the preference function is sufficient to determine the preference order for the prospects in correspondence with the Order Rule.

It is natural to consider whether we can also construct a multiattribute utility function directly using the Equivalence Rule indifference assessments, for prospects characterized by multiple attributes, without having to first consider the deterministic trade-offs among the attributes and the preference order of the prospects. In this part we see how to construct multiattribute utility functions by direct utility elicitation over the attributes without the use of a preference or a value function.

Every point on the surface of a multiattribute utility function has a clear interpretation in terms of multiattribute indifference probability assessments. If we have prospects characterized by multiple attributes, and we have a strict preference order for any three prospects, then we can state a multiattribute indifference probability according to the Equivalence Rule.

To illustrate, let X be an attribute denoting wealth, with x^* being the maximum wealth and x^0 the minimum wealth. Further, let Y denote health as determined by equivalent remaining life years, with y^* being the best health and y^0 being the worst. It is reasonable to assume that more of each attribute is preferred to less in this case. Figure Introduction 5.1 shows a pictorial representation of the attributes and their corresponding domain.

We can think of three prospects in the domain using this characterization:

- Prospect A with maximum wealth and best health (x^*, y^*)
- Prospect C with minimum wealth and worst health (x^0, y^0)
- Prospect B characterized by (x, y), where $x^0 < x < x^*$ and $y^0 < y < y^*$

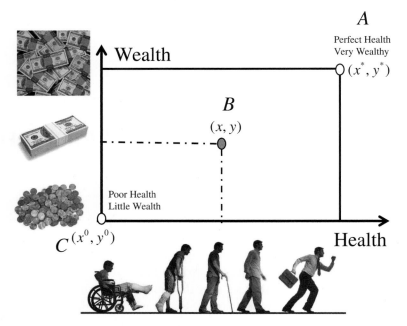

Figure Introduction 5.1. Domain of the two attributes Health and Wealth. © Netalieh/iStock/Getty Images Plus; © Burazin/Photographer's Choice/Getty Images; © Don Farrall/DigitalVision/Getty Images; © Echo/Cultura/Getty Images; © Izf/iStock/Getty Images Plus

Because more of these attributes is preferred to less, the deterministic order for these three prospects is clear,

$$C \prec B \prec A.$$

In other cases, of course, where we have more of one attribute and less of the other, the preference order needs explicit trade-off assessments. To state a multiattribute indifference assessment directly without the isopreference contours, we need the indifference probability that makes us indifferent to getting B for certain, or a binary deal that gives A (with a probability equal to this indifference probability) or C (with a probability equal to one minus this indifference probability). The indifference probability is in fact the multiattribute utility, $U(x, y)$, for prospect B. Figure Introduction 5.2 shows an example of this two-attribute indifference probability assessment.

It is clear from Figure Introduction 5.1 that we now have to think about the variation of both attributes when making this indifference assessment. Prospect B has levels (x, y), of the attributes, while prospects A and C have levels (x^*, y^*) and (x^0, y^0) (respectively).

If more of an attribute is preferred to less, as in most of the remaining chapters, then the utility of any prospect (any point in the domain of the attributes) has a clear meaning as the indifference probability that makes us indifferent to either receiving that prospect or receiving a binary deal that provides the **top right prospect** of the domain (with a probability equal to that indifference probability) or the **bottom left prospect** with a probability equal to one minus the indifference probability.

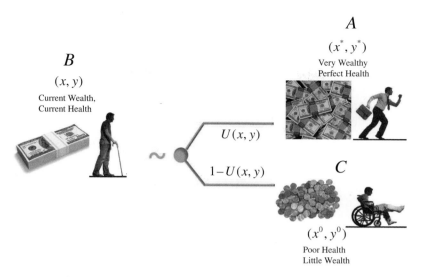

Figure Introduction 5.2. Two-attribute Indifference assessment. © Don Farrall/DigitalVision/Getty Images; © Izf/iStock/Getty Images Plus

While the concept of an indifference probability assessment is the same in both univariate and multiattribute cases, the extension to multiple attributes brings the following considerations:

1. There is an added cognitive complexity in the indifference probability assessment resulting from having to think about the variations in levels of multiple attributes instead of a single attribute.
2. The domain of the indifference probability assessment increases in dimensionality, thereby requiring assessments for a multidimensional domain, which is a large task.
3. The assessment of a utility value over or any individual attribute (such as an assessment over a vertical or horizontal line in the domain) requires the specification of the levels of the remaining attributes during the utility assessment.

To remedy some of these problems, this section illustrates how to

1. Decompose the multiattribute indifference assessment into multiple lower-order assessments where the level of only one attribute changes each time. We refer to this decomposition as the basic expansion theorem for multiattribute utility assessments.
2. Identify certain conditions under which an assessment is not needed on the entire domain, but rather on some lower-order subsets of the domain, such as its boundary values. We discuss this idea using notions of utility independence, interpolation independence, and boundary independence.
3. Identify independence conditions where the utility assessment over an attribute does not depend on the levels of the remaining attributes.

While the tools presented in this section can extend to situations where more of an attribute is not preferred to less, we find it difficult to apply in practice for these more general cases, where a preference or a value function or even a structural model could be very handy. We therefore recommend this multiattribute indifference assessment approach only when more of an attribute is preferred to less, or when you really do not want to construct a preference or a value function.

This section starts with a special class of multiattribute utility functions called Attribute Dominance Utility Functions. We show how to decompose this assessment into a product of conditional utility functions and derive several concepts that simplify the utility assessment. Next, we extend the results to more general utility functions that relax the attribute dominance condition.

CHAPTER 21

Attribute Dominance Utility Functions

Chapter Concepts

- A class of utility functions – called attribute dominance utility functions – that applies in many decisions including medical decision-making
- The attribute dominance condition and its implications on the shape of the utility function
- Methods for constructing attribute dominance utility functions
  - Using preference functions and a single-attribute utility assessment
  - Using marginal-conditional utility assessments
  - Using utility copulas and utility curves at the boundary values of the domain
- The product expansion formula for attribute dominance utility functions
- Utility inference: updating the conditional utility function of an attribute as the level of another attribute varies
- Trade-offs implied by assessments for attribute dominance utility functions
- Independence conditions for attribute dominance utility functions
- Graphical representations of attribute dominance utility functions
- Conducting sensitivity analysis to the product form of utility functions

---

21.1 INTRODUCTION

In Chapter 11, we discussed how to construct multiattribute utility functions using a single-attribute utility function over value (or over an attribute of the preference function). This chapter discusses the assessment of multiattribute utility functions using direct utility assessments over the attributes, without identifying a preference or a value function.

Our focus in this chapter is a particular class of multiattribute utility functions that satisfies the general property that more of an attribute is preferred to less, and that also has a specific feature: *the multiattribute utility function is zero if any one of the attributes falls below a minimum value.* We refer to any attribute that satisfies this condition as a ***utility-dominant attribute***. When all attributes satisfy this property, we refer to this condition as the ***attribute dominance utility condition*** or the ***grounding condition***. The term "attribute dominance utility function" is an indication that any single attribute dominates the remaining attributes when it falls below a certain level, and it sets the utility function to zero, regardless of the levels of the other attributes.

Examples of situations where the attribute dominance property holds are abundant. For example, consider a decision involving an individual's well-being characterized by health state and consumption. If either the health state or consumption falls below a minimum value, then that individual would die and so the utility function would be zero (if health state goes below a certain level they would die and so the utility would be zero, and if consumption goes below a certain level then they would die and so the utility would also be zero). Many other examples are available in our daily lives. In a car purchase decision, leg space and mileage might also be considered "utility dominant attributes" for some specified minimum levels because the value of the car would be zero if either leg space or mileage falls below a minimum value.

Figure 21.1 shows an example of an attribute dominance utility function and its corresponding isopreference contours. The attribute dominance condition is manifested by the fact that the function is zero on the axes of the figure (here the axes are the threshold values). We discussed the attribute dominance (grounding) property for preference functions in Chapters 8 and 11. The attribute dominance condition is also manifested in Figure 21.1 by the fact that all the isopreference contours intersect the axes $(x, y_{\max})$ and $(x_{\max}, y)$, i.e., for any x_1 and y_1, there exist points $(x_1, y_{\max})$ and $(x_{\max}, y_1)$ such that $U(x_1, y_{\max}) = U(x_{\max}, y_1)$. As we shall see, the attribute dominance condition provides significant simplifications for multiattribute utility assessment.

We start with a general discussion of attribute dominance utility functions for two attributes in Section 21.2, and then extend the analysis to multiple attributes.

21.2 PROPERTIES OF ATTRIBUTE DOMINANCE UTILITY FUNCTIONS (TWO ATTRIBUTES)

Our focus in this part of the book will be utility functions that satisfy the following conditions:

1. The function is **continuous** (to avoid sudden changes in preferences under uncertainty).
2. The function is **bounded.**
3. **Range:** The function is normalized to range from 0 to 1. This is because a utility assessment is in fact an indifference probability assessment.
4. **Domain**: The function is defined on a domain $[x_{\min}, x_{\max}] \times [y_{\min}, y_{\max}]$
5. The **attribute dominance** condition:

$$U(x_{\min}, y) = U(x, y_{\min}) = 0, \ \forall x, y.$$

This implies that if either x or y are at their minimum values, the utility function is zero.

6. Strictly increasing with each attribute except at the minimum boundary values.

$$U(x_1, y) > U(x_2, y) \qquad \forall x_1 > x_2 > x_{\min}, \ y \neq y_{\min}$$

$$U(x, y_1) > U(x, y_2) \qquad \forall y_1 > y_2 > y_{\min}, \ x \neq x_{\min}.$$

7. From the previous conditions, we can set $U(x_{\min}, y_{\min}) = 0$ and $U(x_{\max}, y_{\max}) = 1$.

From here on, we use a superscript d to indicate a utility function corresponding to utility dominant attributes, i.e., $U_{xy}^d(x, y)$.

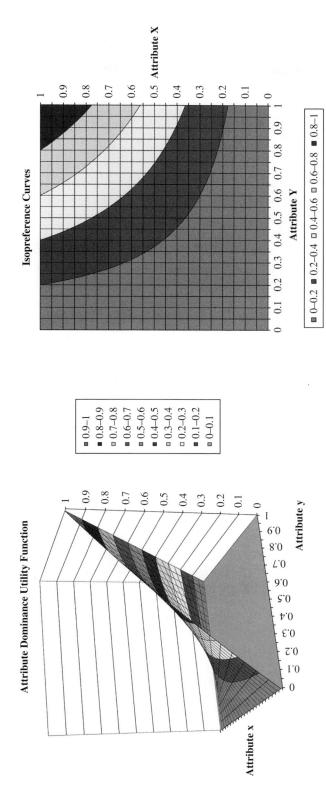

Figure 21.1. Example of an attribute dominance utility function and its corresponding isopreference contours.

Note: A Product of Single-Attribute Utility Functions is an Attribute Dominance Utility, But…

Before we proceed with the discussion of attribute dominance utility functions, we observe that the product of normalized single-attribute utility functions satisfies the attribute dominance conditions. For example, if $U_1(x)$ is the normalized utility over consumption and $U_2(y)$ is the normalized utility over health state measured in terms of equivalent healthy remaining lifetime, then the function

$$U^d(x, y) = U_1(x)U_2(y)$$

is an attribute dominance utility function over health and consumption. Note that the utility function $U^d(x, y)$ is zero if any of the attributes in the product are minimum. This is the simplest extension of univariate utility assessments into attribute dominance utility functions. As we shall see, however, this product form is limited in the types of trade-offs that it can provide. Furthermore, it makes an assertion that the multiattribute indifference assessment is a product of univariate assessments, but this need not be the case. Recall that the utility of a prospect, such as $U(x_1, y_1)$ is the indifference probability of

(i)   receiving the prospect (x_1, y_1) for certain or
(ii) receiving a binary deal that provides prospect (x^*, y^*) with probability $U(x_1, y_1)$, and prospect (x^0, y^0) with probability $1 - U(x_1, y_1)$.

The product form implies that this indifference probability is equal to the product of two univariate assessments:

1.   The indifference probability $U_1(x_1)$ of receiving the prospect x_1 for certain or a binary deal that provides either x^* or x^0 (and this indifference probability is the same regardless of the value of Y)
2.   The indifference probability $U_2(y_1)$ of receiving the prospect y_1 for certain or a binary deal that provides either y^* or y^0 (and this indifference probability is the same regardless of the value of X)

In principle, there is no reason why this indifference probability should be the product of two utility values or two indifference probabilities. The Five Rules do not require it. We shall therefore discuss a general procedure for constructing attribute dominance utility functions that is more general than the product form.

21.3 MARGINAL-CONDITIONAL EXPANSION OF ATTRIBUTE DOMINANCE UTILITY FUNCTIONS

21.3.1 Marginal (Boundary) Utility Function of an Attribute Dominance Utility Function

Definition
Marginal (Boundary) Utility Function: The marginal (boundary) utility function of an attribute, X, is the curve traced on the utility surface as we vary X, when all other

attributes are set at their maximum values. For two attributes, the marginal utility function of X is

$$U_x^d(x) = U_{xy}^d(x, y_{max}).\tag{1}$$

This curve occurs at the boundary (margin) of the domain, and hence the name – marginal utility. The marginal utility function $U_{xy}^d(x, y_{max})$ is normalized to range from zero to one for a utility-dominant attribute because by definition $U_{xy}^d(x_{min}, y_{max}) = 0$, and $U_{xy}^d(x_{max}, y_{max}) = 1$. As we shall see in future chapters, the curve $U_{xy}^d(x, y_{max})$ need not be normalized for general utility functions that do not satisfy the attribute dominance condition.

We can also define the marginal utility for attribute Y as

$$U_y^d(y) = U_{xy}^d(x_{max}, y).$$

21.3.2 Conditional Utility Function of an Attribute Dominance Utility Function

The curve traced by the intersection of the surface $U_{xy}^d(x, y)$ with the plane $X = x$ gives the variation of the utility surface with Y at a fixed value of X. If this curve is normalized to range from zero to one, then we refer to it as the normalized conditional utility function for attribute Y at that fixed value of X and denote it as $U_{y|x}^d(y \mid x)$.

CONDITIONAL UTILITY FUNCTION FOR ATTRIBUTE Y AT A GIVEN VALUE OF $X=x$
The conditional utility function for attribute Y is the curve traced on the surface $U_{xy}^d(x, y)$, as we vary y when X has a fixed value, x.

If Y is a utility-dominant attribute, then the curve traced on the surface $U_{xy}^d(x, y)$ by fixing $X = x$ and varying from Y from y_{min} to y_{max} is not normalized. It ranges from $U_{xy}^d(x, y_{min}) = 0$ to $U_{xy}^d(x, y_{max}) = U_x^d(x)$. We can normalize this curve to range from zero to one, however, by dividing by $U_x^d(x)$.

Definition
Normalized Conditional Utility Function for Utility-Dominant Attribute Y at a Fixed Value of $X = x$ The normalized conditional utility function, $U_{y|x}^d(y \mid x)$, for a utility-dominant attribute Y is

$$U_{y|x}^d(y \mid x) = \frac{U_{xy}^d(x, y)}{U_{xy}^d(x, y_{max})} = \frac{U_{xy}^d(x, y)}{U_x^d(x)}, \quad x \neq x_{min}.\tag{2}$$

Rearranging (2) gives the product expansion for attribute dominance utility functions.

Product Expansion Theorem for Attribute Dominance Utility Functions
$$U_{xy}^d(x, y) = U_x^d(x) U_{y|x}^d(y \mid x).\tag{3}$$

Equation (3) allows the construction of an attribute dominance utility (indifference probability) of a prospect (x, y) using a product expansion into marginal-conditional utility assessments analogous to the marginal-conditional approach for joint probability distributions.

Decomposition of an Indifference Probability Assessment for Utility-Dominant Attributes

Equation (3) asserts the equality of two different methods for assessing the multiattribute utility (indifference probability) of a prospect (x_1, y_1), for utility-dominant attributes:

1. **Direct Assessment of the Two-Attribute Utility**

This method uses indifference probability assessments of receiving (x_1, y_1) for certain or receiving a binary deal that provides (x^*, y^*) with probability $U(x_1, y_1)$ and (x^0, y^0) with probability $1 - U(x_1, y_1)$. This requires thinking about the variation in the levels of two attributes simultaneously.

2. **Decomposing the Utility Assessment**

With this method, we conduct the following:

1. First assess the utility value $U(x_1, y^*)$ using indifference assessments. This is the indifference probability of getting (x_1, y^*) for certain or a binary deal that gives (x^*, y^*) with probability $U(x_1, y^*)$ or (x^0, y^*) with probability $1 - U(x_1, y^*)$. This requires thinking about the variation in attribute X alone at a fixed value of $Y = y^*$.
2. Then we assess the utility $U(y_1 \mid x_1)$ also using lottery assessments, which is the indifference probability, $U(y_1 \mid x_1)$, of receiving (x_1, y_1) for certain or receiving a binary deal that provides either (x_1, y^*) with probability $U(y_1 \mid x_1)$ or (x_1, y^0) with probability $1 - U(y_1 \mid x_1)$. This requires thinking about the variations in attribute Y alone at a fixed value of $X = x_1$.
3. We then multiply the two assessments.

This decomposition approach requires thinking about only one variation in each attribute at a time but it requires two indifference assessments.

To construct the whole utility surface using (3), we need to repeat this expansion for all prospects on the domain of the attributes. To do this, we first assess the marginal utility function over attribute X conditioned at the maximum value of Y, i.e., $U_x^d(x)$. Then, we assess the conditional utility function for attribute Y at all fixed values of attribute X, $U_{y|x}^d(y \mid x)$. Observe that the $U_{y|x}^d(y \mid x)$ is quite complicated to assess because it needs to be assessed for all fixed values of X. Then we multiply the two functions.

21.3.3 Bayes' Rule for Attribute Dominance Utility

Using similar analysis as that conducted in the section 21.2.2, we can also define the conditional utility function, $U_{x|y}^d(x \mid y)$ as

$$U_{x|y}^d(x \mid y) = \frac{U_{xy}^d(x, y)}{U_y^d(y)}, \qquad y \neq y_{\min}. \tag{4}$$

Rearranging gives

$$U_{xy}^d(x, y) = U_y^d(y)U_{x|y}^d(x \mid y). \tag{5}$$

Equating (3) and (5) gives

$$U_{xy}^d(x, y) = U_x^d(x)U_{y|x}^d(y \mid x) = U_y^d(y)U_{x|y}^d(x \mid y).$$

Rearranging gives

$$\boxed{U_{x|y}^d(x \mid y) = \frac{U_{y|x}^d(y \mid x)U_x^d(x)}{U_y^d(y)}, \qquad x \neq x_{\min}, y \neq y_{\min}}. \tag{6}$$

Bayes' Rule for Utility Inference

It is important to pause here and think about the interpretation and implications of Equation (6). In the analogous world of probability inference, we update our state of information about the outcome of an uncertain event by conditioning on the outcome of another event. This updating of information can be made even if there is no decision to be made, and often helps us think more clearly about our degree of belief on the outcomes of uncertain events. In a similar manner, Equation (6) provides a method to update our utility values over an attribute when we are guaranteed a fixed amount of another attribute. It is natural to extend our analogy and refer to Equation (6) as *Bayes' rule for utility inference.*

Dialogue between Instructor (I) and Student (S)

I: Is the idea of an attribute dominance utility function clear?

S: Yes, it sounds like an important class of functions that may be especially useful in medical decisions, among others. But how do we assess it?

I: We shall discuss several methods. We have already presented the first method in Chapter 11 using a preference function and a utility over value or over an attribute. In this chapter, we also discussed a second method using direct utility elicitation of marginal-conditional assessments $U_x^d(x)$ and $U_{y|x}^d(y \mid x)$.

S: This seems more general than a simple product of functions $U_x^d(x)U_y^d(y)$.

I: Of course. It requires us to think about our utility for X, and our conditional utility for Y at this value of X. For example, if we would like the utility of (x_1, y_1), then we first need to think about our utility of x_1 at the maximum value of Y, namely $U(x_1 \mid y^*)$, and then multiply that by our utility of y_1 at x_1, namely $U(y_1 \mid x_1)$.

S: And will the utility for Y change with the value of X?

I: Yes. Wouldn't your utility function for investments, for example, change with your health?

S: Almost surely. But it sounds like the term $U_{y|x}^d(y \mid x)$ is somewhat difficult to assess. It requires the conditional utility assessments of Y at all values of X.

I: Yes. Every level x of attribute X might imply a different conditional utility function $U^d_{y|x}(y\,|\,x)$. Are you also clear on the interpretation and implications of Equation (6)?

S: In probability inference, we update our state of information about the outcome of an uncertain event by conditioning on the outcome of another event.

I: Yes, and this updating of information can be made even if there is no decision to be made. It often helps us think more clearly about our degree of belief on the outcomes of uncertain events. In a similar manner, Equation (6) provides a method to update our utility values over an attribute when we are guaranteed a fixed amount of another attribute.

Sometimes a functional form can be used for $U^d_{y|x}(y\,|\,x)$, and so its assessment would involve only parameters, as we illustrate in Example 21.1

EXAMPLE 21.1: Constructing an Attribute Dominance Utility Function for a Medical Decision

---

A patient undergoing a cancer treatment is deciding whether to have chemotherapy or radiotherapy. The two attributes involved are health state Y (measured by the remaining life years divided by the expected remaining life years, a normalized scale from zero to one) and consumption X (measured in millions of dollars on a normalized scale from zero to one). The patient's preferences are such that when any of these attributes has a minimum value, the prospect is considered a least preferred prospect and so the attribute dominance condition holds.

To assess the multiattribute utility function, we start by assessing the marginal utility function for consumption. This is the utility for consumption at the maximum health state.

Suppose that the decision-maker is risk neutral over the range of consumption levels considered when health state is maximum, i.e.,

$$U^d_x(x) = x, \ 0 \le x \le 1, \tag{7}$$

where x is in millions of dollars.

Now we assess the conditional utility function for health given consumption. The decision-maker states he is risk averse over the health state, with a risk aversion function that depends on the value of the consumption attribute. The decision-maker assigns an exponential conditional utility function for health given consumption as

$$U^d_{y|x}(y\,|\,x) = \frac{1 - e^{-(\gamma(x))y}}{1 - e^{-(\gamma(x))}} \ 0 \le y \le 1, 0 \le x \le 1, \tag{8}$$

where y is the health state and the risk aversion function for health depends on consumption level as

$$\gamma(x) = \frac{1}{\rho_0 + x},$$

where ρ_0 is a constant assessed to be \$0.3 million in this example.

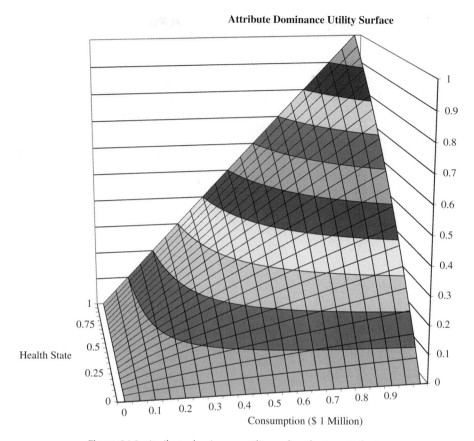

Attribute Dominance Utility Surface

Figure 21.2. Attribute dominance utility surface for two attributes.

This risk aversion function implies that the decision-maker becomes less risk averse over health as consumption increases. Of course, we can model any other type of behavior using an appropriate risk aversion function from our previous discussions in Chapter 14. For example, we can model increasing risk aversion over wealth as consumption increases using a function such as $\gamma(x) = x - \rho_0,\ x > \rho_0$.

The conditional utility function of Equation (8) is normalized to range from zero to one and can be assessed directly without the need for any further scaling of the utility function or assessing its boundary values. The only unknown in this equation is the single parameter ρ_0, which can be determined by fitting the curve to indifference probability assessments.

The attribute dominance utility function can now obtained using Equation (3) as

$$U_{xy}^d(x,y) = U_x^d(x)U_{y|x}^d(y\,|\,x) = \frac{x(1 - e^{-(\gamma(x))y})}{1 - e^{-(\gamma(x))}}, \quad 0 \le y \le 1, 0 \le x \le 1. \tag{9}$$

Figure 21.2 shows the multiattribute utility function for the decision-maker's prospects. Note that it is more involved than a simple product of two univariate utility functions.

The following Example 21.2 illustrates the idea of utility inference and shows how we can infer the utility assessments in reverse conditioning order. In this example, we infer $U(x\,|\,y)$ even though it was not previously assessed.

EXAMPLE 21.2: Utility Inference for an Attribute Dominance Utility Function in a Medical Decision

---

Having constructed the attribute dominance utility function in (9), the decision-maker would like some consistency check to see if the results are meaningful. To do this, we think of the implied utility assessments in the reverse order.

We can now determine the marginal utility function for health state as

$$U_y^d(y) = U_{xy}^d(x_{max}, y) = \frac{1 - e^{-\frac{y}{1.3}}}{1 - e^{-\frac{1}{1.3}}}, \quad 0 \le y \le 1, \tag{10}$$

where $\gamma(x_{max}) = \dfrac{1}{0.3 + 1} = \dfrac{1}{1.3}$.

Using Bayes' rule for utility inference (6), we can also determine the conditional utility function for consumption given health state as

$$U_{x|y}^d(x \mid y) = \frac{U_{xy}^d(x, y)}{U_y^d(y)} = \frac{x(1 - e^{-\frac{1}{1.3}})(1 - e^{-(\gamma(x))y})}{(1 - e^{-\frac{y}{1.3}})(1 - e^{-(\gamma(x))})}, \quad 0 \le y \le 1, 0 \le x \le 1. \tag{11}$$

Note that this assessment order was not elicited directly.

Figure 21.3 (a) shows the marginal utility function for remaining life obtained by utility inference, and Figure 21.3 (b) shows the conditional utility functions for consumption given different values of health states y using Bayes' rule for utility inference. The figure shows that the decision-maker is risk neutral for consumption only when the health state Y is maximum. For any other value of Y the decision-maker is risk averse over consumption.

Dialogue between Instructor (I) and Student (S)

S: This is really interesting what is going on here. A lot of concepts similar to those we have studied in probability.

I: I know. The basic idea is the same. The math does not know whether this function represents probability or utility.

S: So what does utility inference really mean?

I: It is a way to help us think about conditional utility assessments for an attribute as the level of another attribute varies.

S: In the previous example, the decision-maker provided $U_x^d(x)$ and $U_{y|x}^d(y \mid x)$. But doesn't the function $U_{y|x}^d(y \mid x)$ already enable us to think about our utility for y as the level of x varies?

I: Correct, but by using Bayes' rule for utility we are able to think about our conditional utility function $U_{x|y}^d(x \mid y)$ from these assessments. Even though we assessed only $U_x^d(x)$ at the maximum value of attribute Y, we were able to infer $U_{x|y}^d(x \mid y)$ at any value of Y.

S: Can we use this as a consistency check for utility assessment?

I: Precisely. It can also be used when the decision-maker is more comfortable conditioning on one attribute versus another. For example, the decision-maker might be more comfortable conditioning on health when making a utility assessment over consumption than conditioning on consumption when making a utility assessment over

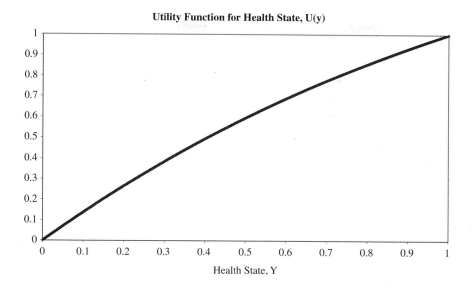

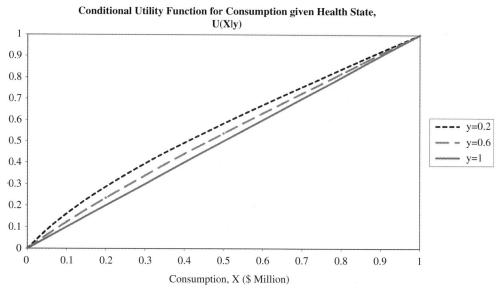

Figure 21.3. (a) Marginal utility function, $U_y^d(y)$; (b) conditional utility function, $U_{x|y}^d(x \mid y)$.

health. Whichever order the decision-maker provides can be used to infer the other and to construct the utility function.

S: The idea is very clear and quite novel. We have seen in Example 21.2 how the decision-maker may provide a conditional utility assessment such as $U_{y|x}^d(y \mid x)$ using a functional form. We now understand the difference between a single curve like $U_{y|x}^d(y \mid x_1)$ and the conditional assessment $U_{y|x}^d(y \mid x)$ that requires a conditional utility function for y at all possible values of x.

I: We shall also discuss other methods to construct attribute dominance utility functions without having to provide a complex assessment like $U_{y|x}^d(y \mid x)$. But even when we provide the simple method of construction, the idea of utility inference can serve as a very powerful consistency check.

21.3.4 Special Case: Product of Normalized Single-Attribute Utility Functions

The product of two (or more) normalized utility functions, $U_x^d(x)$ and $U_y^d(y)$, is an attribute dominance utility function. If we had used the product form of the marginal utility functions in our analysis in Example 21.2, then the utility function for consumption would not have updated with different values of health states. This is known as ***utility independence***. Likewise, because of the symmetry of the product operator in the decomposition, the utility for health state would not update with consumption.

To illustrate the implications of the product form, if

$$U_{y|x}^d(y\,|\,x) = U_y^d(y),$$

then from (6),

$$U_{x|y}^d(x\,|\,y) = \frac{U_{y|x}^d(y\,|\,x)U_x^d(x)}{U_y^d(y)} = \frac{U_y^d(y)U_x^d(x)}{U_y^d(y)} = U_x^d(x), \quad y \neq y_{\min}.$$

Therefore, the conditional utility function of any attribute would *not* depend on the level of the other attribute. We shall refer to the concept of utility independence in more detail in Chapters 23 and 25.

21.3.5 Different Attribute Dominance Utility Surfaces Having the Same Boundary Assessments

Attribute dominance utility functions enable more general utility functions than those prescribed by the product form. They have the same marginal utility functions as the product form, but they allow the normalized conditional utility functions to change as the level of another attribute(s) varies, thereby allowing for richer trade-offs in preferences.

EXAMPLE 21.3: The Same Utility Functions at the Margins but Different Isopreference Contours

---

Figure 21.4 shows attribute dominance utility functions having the same marginal utility functions at the boundaries. Each function, however, has a different set of trade-offs between the attributes as demonstrated by the shape of the isopreference contours.

Figure 21.4 (b) shows a surface constructed by the product of the two marginal utility functions; Figure 21.4 (a) shows an attribute dominance utility function with the same marginal utility functions but the surface is slightly lower. Figure 21.4 (c) also shows an attribute dominance utility function with the same marginal utility functions but the surface is raised higher than that of the product form.

As we can see, simply multiplying the marginal utility functions by each other results in a surface that implies a specific set of trade-offs between the attributes. It is therefore a very strong assumption in this case. Each of the graphsin Figure 21.4 has the same marginal utility function with different trade-offs. The surface of the utility

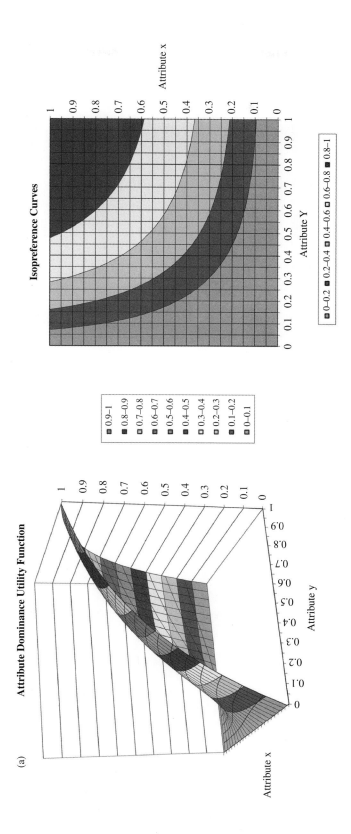

414

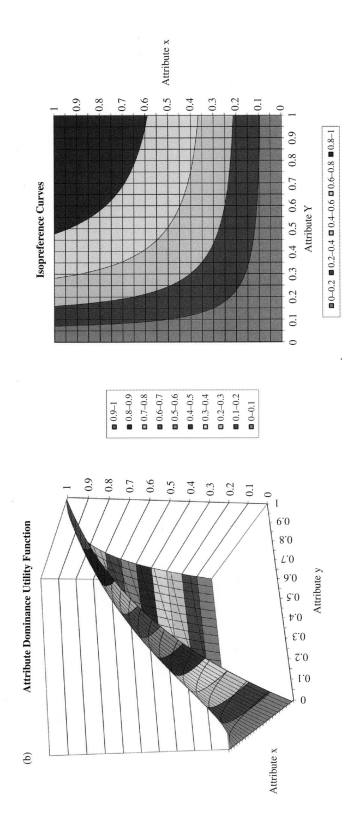

Isopreference Curves

Attribute x

Attribute Y

0–0.2 ■ 0.2–0.4 □ 0.4–0.6 □ 0.6–0.8 ■ 0.8–1

0.9–1
0.8–0.9
0.7–0.8
0.6–0.7
0.5–0.6
0.4–0.5
0.3–0.4
0.2–0.3
0.1–0.2
0–0.1

(b)

Attribute Dominance Utility Function

Attribute y

Attribute x

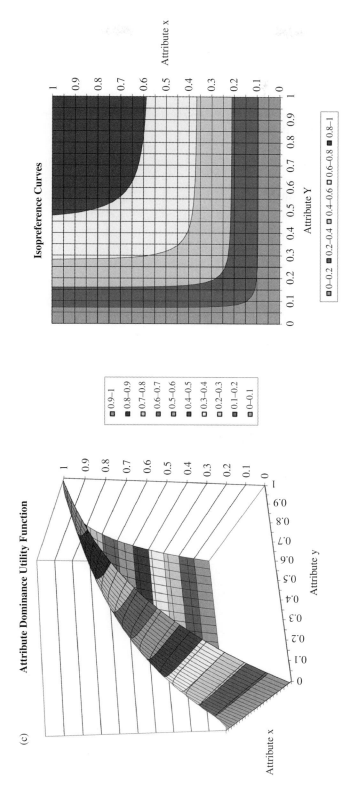

Figure 21.4. Attribute dominance utility functions having the same boundary utility functions but different shapes of isopreference contours.

functions in Figure 21.4 is raised as we move from (a) to (c), while maintaining the same marginal utility functions. This implies a higher utility value for any point in the domain. Attribute dominance utility functions provide the flexibility to model utility surfaces that match the boundary values but have an additional degree of freedom to vary the trade-offs or to provide higher (or lower) utility values across the domain.

21.4 TRADE-OFFS IMPLIED BY UTILITY-DOMINANT ATTRIBUTES

We have previously discussed a general method for constructing multiattribute utility functions using preference functions – that determine the deterministic trade-offs between health and consumption – and a one-dimensional utility function over consumption. The multiattribute utility function can be determined using these assessments, as we discussed in Chapter 11. Some people prefer not to make explicit trade-off assessments of these kinds, and therefore prefer to replace these trade-off assessments with direct utility assessments as we have seen in this chapter. Even though we have not made explicit trade-off assessments when constructing attribute dominance utility functions, it is important to observe that the resulting functions themselves do imply trade-off assessments and shapes of isopreference contours. There is no escaping form this fact. Any multiattribute utility function, no matter how it was constructed, will imply deterministic trade-offs. We illustrate this below to highlight that it is important to verify the implied trade-offs by the direct utility assessment approach.

21.4.1 Trade-Offs Implied by General Attribute Dominance Utility Functions

The trade-off function between two utility-dominant attributes x and y can be determined as we move along a constant isopreference contour, with utility value say U_i, where

$$U_{xy}^d(x, y) = U_i = Const. \tag{12}$$

The change in utility value along this path must therefore be zero

$$dU_{xy}^d(x, y) = \frac{\partial U_{xy}^d(x, y)}{\partial x} dx + \frac{\partial U_{xy}^d(x, y)}{\partial y} dy = 0. \tag{13}$$

Rearranging gives

$$\frac{dy}{dx} = -\frac{\dfrac{\partial U_{xy}^d(x, y)}{\partial x}}{\dfrac{\partial U_{xy}^d(x, y)}{\partial y}} = -\frac{U_y^d(y) \dfrac{\partial U_{x|y}^d(x \mid y)}{\partial x}}{U_x^d(x) \dfrac{\partial U_{y|x}^d(y \mid x)}{\partial y}}. \tag{14}$$

The actual trade-offs between the attributes can also be assess separately and Equation (14) can be used as a consistency check to see if the results obtained from the marginal-conditional utility assessments match the trade-off values.

21.4.2 Trade-Offs for Attribute Dominance Utility Functions with the Product Form

If $U_{y|x}^d(y \mid x) = U_y^d(y)$ then $U_{x|y}^d(x \mid y) = U_x^d(x)$ (the product form corresponding to utility independence for utility dominant attributes), then Equation (14) reduces to

$$\frac{dy}{dx} = -\frac{U_y^d(y)\dfrac{d}{dx}U_x^d(x)}{U_x^d(x)\dfrac{d}{dy}U_y^d(y)} = -\frac{\dfrac{d}{dx}\log(U_x^d(x))}{\dfrac{d}{dy}\log(U_y^d(y))}. \tag{15}$$

Equation (15) illustrates that the trade-offs between two utility-dominant attributes that have utility independence are determined solely by their marginal utility function assessments. This is a further illustration that we have no further degree of freedom to vary the trade-offs among the attributes once the utility assessments at the boundary values have been made.

EXAMPLE 21.4

Suppose we have two utility-dominant attributes, X and Y, and a utility function that is a product of individual utility functions over the attributes. If the utility function for each attribute is exponential over the domain $[0,\infty]$, then the normalized marginal utility functions have the form $U(x) = 1 - e^{-\gamma_1 x}$ and $U(y) = 1 - e^{-\gamma_2 y}$, where γ_1 and γ_2 are the risk aversion coefficients of x and y respectively, and the utility function has the form

$$U^d(x,y) = (1 - e^{-\gamma_1 x})(1 - e^{-\gamma_2 y}), \quad x, y \geq 0.$$

The trade-off function between the two attributes can be obtained by direct substitution into (15) as

$$\frac{dy}{dx} = -\frac{\gamma_1 e^{-\gamma_1 x}}{\gamma_2 e^{-\gamma_2 y}}\frac{(1 - e^{-\gamma_2 y})}{(1 - e^{-\gamma_1 x})}. \tag{16}$$

Using the product form of utility functions provides no additional degrees of freedom to change the trade-offs between the attributes, because the trade-offs and the functional form are completely determined by the marginal utility assessments.

21.5 ATTRIBUTE DOMINANCE UTILITY FUNCTIONS WITH MULTIPLE ATTRIBUTES

By induction, the product expansion for attribute dominance utility functions in terms of the normalized conditional utility functions extends to multiple attributes as

$$U_{x_1 \ldots x_n}^d(x_1, x_2, \ldots, x_n) = U_{x_1}^d(x_1)U_{x_2|x_1}^d(x_2 \mid x_1)\ldots U_{x_n|x_1 \ldots x_{n-1}}^d(x_n \mid x_1, \ldots, x_{n-1}).$$

We can also combine any groups of attributes together in the product expansion. For example, we can write

$$U^d_{x_1x_1x_3}(x_1,x_2,x_3) = U^d_{x_1}(x_1)U^d_{x_2|x_1}(x_2\,|\,x_1)U^d_{x_3|x_1x_2}(x_3\,|\,x_1,x_2)$$
$$= U^d_{x_1x_2}(x_1,x_2)U^d_{x_3|x_1x_2}(x_3\,|\,x_1,x_2),$$

where $U^d_{x_1x_2}(x_1,x_2) = U^d_{x_1x_1x_3}(x_1,x_2,x_{3\max})$.

We can also rearrange starting the expansion with X_3 to get

$$U^d_{x_1x_1x_3}(x_1,x_2,x_3) = U^d_{x_3}(x_3)U^d_{x_1x_2|x_3}(x_1,x_2\,|\,x_3).$$

Therefore, we can write for higher orders of attributes

$$U^d_{x_1x_2|x_3}(x_1,x_2\,|\,x_3) = \frac{U^d_{x_1x_2}(x_1,x_2)U^d_{x_3|x_1x_2}(x_3\,|\,x_1,x_2)}{U^d_{x_3}(x_3)}, \quad U^d_{x_3}(x_3) \neq 0.$$

The extension of the product form to multiple attributes is straightforward,

$$U^d_{(x_1,\dots,x_n)} = \prod_{i=1}^{n} U^d_{x_i}(x_i).$$

Again, this product form implies that:

1. Trade-offs are determined completely by the normalized utility assessments. We do not have a degree of freedom to vary the trade-offs once the normalized utility functions are specified.
2. The conditional utility function of any subset of the attributes, say $U^d_{x_1x_2|x_3}(x_1,x_2\,|\,x_3)$, does not depend on the level x_3.

21.6 ATTRIBUTE DOMINANCE UTILITY DIAGRAMS

As we have discussed, the product form of utility functions implies that the normalized conditional utility function of any attribute does not depend on the level of the remaining attributes. This observation simplifies the analysis because we need only a univariate assessment like $U^d_y(y)$ instead of $U^d_{y|x}(y\,|\,x)$ when Y appears as the second attribute in the expansion.

When more than two attributes are present, it is important to represent these independence relations graphically to help facilitate the conversation with the decision-maker. We refer to this graphical representation of independence relations as a **utility diagram**. As we shall see, utility diagrams also enable us to infer independence relations for multiple attributes that might not be immediately apparent in the assessment order provided.

21.6.1 Utility Independence Is a Symmetric Property for Attribute Dominance Utility

If $U^d_{x|y}(x\,|\,y)$ does not depend on y, then we can choose any reference value of y for which to assess the conditional utility function. Furthermore, the two-dimensional assessment $U^d_{x|y}(x\,|\,y)$ on the domain reduces to a one-dimensional assessment with

$$U^d_{x|y}(x\,|\,y) = U^d_x(x).$$

From (5), the product expansion gives

$$U^d_{xy}(x,y) = U^d_y(y)U^d_x(x).$$

Utility independence is a symmetric property for attribute dominance utility functions because of the symmetric nature of the product operator. From (2), the conditional utility function of y is

$$U^d_{y|x}(y \mid x) = \frac{U^d_{xy}(x,y)}{U^d_{xy}(x,y_{\max})} = \frac{U^d_y(y)U^d_x(x)}{U^d_x(x)} = U^d_y(y) \quad x \neq 0.$$

This implies that if Y is utility independent of X, then X is utility independent of Y.

If none of the conditional utility functions depend on the levels of the remaining attributes, then the product expansion becomes

$$U^d_{(x_1,\dots,x_n)} = \prod_{i=1}^{n} U^d_{x_i}(x_i), \tag{17}$$

which is also a symmetric independence condition.

21.6.2 Utility Diagrams

We now present graphical representations of the independence relations in attribute dominance utility functions. We shall generalize these representations in Chapters 23, 25, and 26 when attribute dominance conditions do not exist.

As we have seen, the general form of attribute dominance utility functions can be expanded as

$$U^d_{x_1\dots x_n} = U^d_{x_1}(x_1)U^d_{x_2|x_1}(x_2 \mid x_1)\dots.U^d_{x_n|x_1\dots x_{n-1}}(x_n \mid x_1\dots x_{n-1}). \tag{18}$$

Attribute Dominance Utility Diagram

An attribute dominance utility diagram is a graph that consists of nodes (or ovals) that represent the attributes and arrows (arcs) connecting them that represent the possibility of dependence between them. The diagram is directed and is acyclic (there can be no loops). The absence of an arrow between two nodes is an assertion that the conditional utility function of each attribute does not depend on the level of the remaining attribute. The presence of an arrow implies the "*possibility of*" utility dependence between the attributes given our current state of preferences

---

EXAMPLE 21.5: Attribute Dominance Utility Diagram with No Arrows

---

Figure 21.5 illustrates an attribute dominance utility diagram with no arrows. This diagram represents an attribute dominance utility function that is a product of marginal utility functions, and the absence of an arrow asserts utility independence between the two attributes.

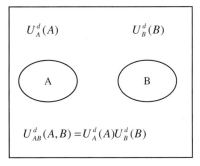

Figure 21.5. Two attributes with utility independence.

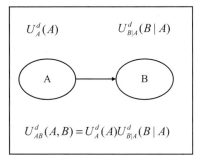

Figure 21.6. Two attributes with possible utility dependence.

EXAMPLE 21.6: Attribute Dominance Utility Diagram with an Arrow

---

If we have not asserted utility independence between two attributes, then we add an arrow between their nodes as shown in the diagram of Figure 21.6. This implies that the conditional utility functions of the attributes may depend on the levels of the remaining attributes. Because utility independence (or dependence) is a symmetric relationship for attribute dominance utility functions, the arrow represents the possibility of utility dependence for both attributes. The direction of the arrow simply represents an expansion order. The expansion order of Figure 21.6 would imply an initial assessment of $U_A^d(A)$ followed by an assessment $U_{B|A}^d(B\,|\,A)$.

21.6.2.1 CONDITIONAL UTILITY INDEPENDENCE (THREE OR MORE ATTRIBUTES)

Definition: Conditional Utility Independence
If two attributes, A and B, have conditional utility independence given a third attribute C, then the conditional utility function $U(A\,|\,B,C)$ does not depend on the value of B at any fixed value of C. We may write $U(A\,|\,B,C)=U(A\,|\,b_0,C)$, where b_0 is a parameter.

Conditional utility independence of A on B given C also implies that preferences for lotteries over attribute A do not depend on the value of B for any fixed value of C.

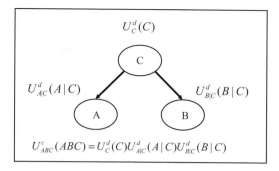

Figure 21.7. Three attributes with conditional utility independence.

We can represent this conditional utility independence relationship using the utility diagram shown in Figure 21.7. The attribute dominance utility function takes the form

$$U_{ABC}^d(A,B,C) = U_C^d(C)U_{B|C}^d(B\,|\,C)U_{A|C}^d(A\,|\,C). \qquad (19)$$

This observation simplifies the utility assessment as we no longer need to condition the utility assessment $U_{A|C}^d(A\,|\,B,C)$, i.e., a utility assessment of A on the whole domain of B and C. Instead, we just need to condition it on $U_{A|C}^d(A\,|\,C)$.

21.6.2.2 ARROW REVERSALS IN ATTRIBUTE DOMINANCE UTILITY DIAGRAMS

Rules of Arrow Reversals in Utility Diagrams
The arrows in the utility diagram represent a given utility assessment decomposition order. It might sometimes be useful to change this assessment order into one that is more comfortable to the decision-maker to simplify the utility assessments or to verify the utility independence assumptions that have been made. To do this graphically, we need a set of rules for inference and arrow reversals. The rules are as follows:

1. You can add an arrow between any two nodes of the utility diagram provided you do not create a cycle.
2. You can reverse an arrow between two nodes provided they are conditioned on the same attributes.
3. The arbitrary removal of an arrow from the utility diagram is not permitted as it asserts the existence of some utility independence conditions between the attributes that may not be available.

EXAMPLE 21.7: Inferring Utility Independence Conditions Using Utility Diagrams

Figure 21.8 shows an application of utility diagrams and the rules of arrow reversals to determine different assessment orders for attribute dominance utility assessment. Figure 21.8(a) shows a utility diagram for three attributes A, B, and C where the order of utility assessment is $U_A^d(A), U_{B|A}^d(B\,|\,A)$, and $U_{C|B}^d(C\,|\,B)$. The figure implies that C is utility independent of A given B, and so

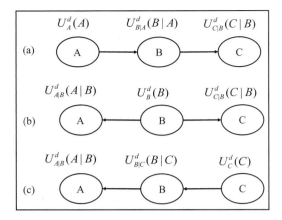

Figure 21.8. Different orders for attribute dominance utility assessment.

$$U_{ABC}^d(A,B,C) = U_A^d(A)U_{B|A}^d(B\,|\,A)U_{C|B}^d(C\,|\,B).$$

If we reverse the arrow from A to B, we obtain a new assessment order $U_B^d(B), U_{A|B}^d(A\,|\,B)$, and $U_{C|B}^d(C\,|\,B)$, and an assertion that this utility assessment can also be made using

$$U_{ABC}^d(A,B,C) = U_B^d(B)U_{A|B}^d(A\,|\,B)U_{C|B}^d(C\,|\,B)$$

depending on the comfort level of the decision-maker. This is shown in Figure 21.8(b).

We can also reverse the arrow from B to C to obtain the order of Figure 21.8(c). The figure asserts that the utility assessment can also be made using the expansion

$$U_{ABC}^d(A,B,C) = U_C^d(C)U_{B|C}^d(B\,|\,C)U_{A|B}^d(A\,|\,B).$$

This example asserts that if A, B, and C are utility dominant attributes, and if C is utility independent of A given B, then it is also true that A is utility independent of C given B. The idea of changing the order of conditioning in the utility assessments is useful because we can now choose the order that is more comfortable to the decision-maker and still observe any utility independence relations that can simplify the assessment to avoid extra work.

Dialogue between Instructor (I) and Student (S)

S: The analogy between probability and utility functions and the graphical deductions here are really neat. We can now represent the independence relations and the expansion relations of attribute dominance utility functions graphically, just like we could represent expansions of belief graphically.

I: Absolutely. Later we shall expand this method of reasoning to general types of utility functions that go beyond the attribute dominance conditions.

Let us now refer back to your earlier question about constructing attribute dominance utility functions without having to provide a complex assessment like $U_{y|x}^d(y\,|\,x)$.

21.7 ATTRIBUTE DOMINANCE UTILITY COPULAS

As we have discussed, the product form of normalized conditional utility functions does not allow for an additional degree of freedom to vary the trade-off assessments once the conditional utility functions at the boundaries (margins) have been assessed. We also saw in Figure 21.4 that two-attribute dominance utility functions can have the same marginal utility functions but different trade-offs among the attributes. Therefore, there is a need to construct more general attribute dominance utility functions than the product form. On the other hand, the marginal-conditional expansion formula requires an assessment of the form $U_{y|x}^d(y\,|\,x)$, which could be cognitively difficult unless some independence assertion, or a given functional form, is used. It would be convenient to have a method that allows us to construct attribute dominance utility functions using marginal utility functions for each attribute and at the same time have another degree of freedom to vary the shapes of the isopreference contours.

> **Definition**
> **Attribute Dominance Utility Copula:** Abbas (2004), Abbas and Howard (2005)
>   Consider a function C^d that uses marginal utility functions as its arguments and provides an attribute dominance utility surface. With this function, we can express the utility surface in terms of the marginal utility functions as
>
> $$U_{xy}^d(x,y) = C^d\left(U_x^d(x), U_y^d(y)\right) \tag{20}$$
>
> We refer to this function as an **attribute dominance utility copula, C^d**.

We shall discuss the properties of such functions below, but for now observe that any attribute dominance utility copula function, C^d, would result in an attribute dominance utility surface that has marginal utility functions $U_x^d(x)$, $U_y^d(y)$, and would allow for different trade-offs among the attributes by varying some of its parameters, thereby allowing for richer trade-offs than the product form.

Note that this method requires only univariate assessments of the attributes, such as $U_x^d(x), U_y^d(y)$, and not the more complex conditional utility assessments, such as $U_{y|x}^d(y\,|\,x)$. Of course, it also comes at the added expense of having to think about the copula function to use. In Chapter 30, we shall discuss the empirical assessment of such functions. But as we shall see, the copula function enables sensitivity analysis even if its form is not fully determined.

21.7.1 Properties of the Attribute Dominance Utility Copula Function C^d

The attribute dominance utility copula function C^d has the following properties:

1. **Domain:** Since both the arguments $U_x^d(x), U_y^d(y)$ are normalized from 0 to 1, the domain of this attribute dominance copula function is $[0,1] \times [0,1]$.
2. **Range:** Because a utility assessment is in effect an indifference probability assessment, the range of this function is from 0 to 1, i.e.,

$$C^d : [0,1]^n \to [0,1].$$

3. **Linear at the Margins:** The function C^d is linear at the maximum margins and satisfies

$$C^d(1,t) = t, \quad C^d(s,1) = s.$$

This is because if $U_{xy}^d(x,y) = C^d\left(U_x^d(x), U_y^d(y)\right)$, then consistency at the boundary values requires that

$$U_{xy}^d(x_{max}, y) = C^d\left(U_x^d(x_{max}), U_y^d(y)\right) = C^d\left(1, U_y^d(y)\right) = U_y^d(y)$$

and also that

$$U_{xy}^d(x, y_{max}) = C^d\left(U_x^d(x), U_y^d(y_{max})\right) = C^d\left(U_x^d(x), 1\right) = U_x^d(x).$$

Substituting for $U_x^d(x) = s$ and $U_y^d(y) = t$, we get $C^d(1,t) = t, \quad C^d(s,1) = s$. Therefore, the surface constructed by this copula function will preserve the utility assessments at its margins.

4. **Attribute dominance property:** The function C^d will also need to satisfy the attribute dominance grounding property where

$$C^d(0,t) = C^d(s,0) = 0.$$

This condition must hold because by definition of attribute dominance utility functions,

$$U_{xy}^d(x_{min}, y) = C^d\left(U_x^d(x_{min}), U_y^d(y)\right) = C^d\left(0, U_y^d(y)\right) = 0$$

and also

$$U_{xy}^d(x, y_{min}) = C^d\left(U_x^d(x), U_y^d(y_{min})\right) = C^d\left(U_x^d(x), 0\right) = 0.$$

Therefore, the attribute dominance condition is preserved in the resulting utility surface.

Figure 21.9 shows an example of an attribute dominance utility copula. Note how it is linear at the boundary values and that it satisfies the grounding condition.

Once the utility function is constructed using an attribute dominance utility copula and the utility functions at the margins, the conditional utility functions can then be deduced as

$$U_{y|x}^d(y \mid x) = \frac{C^d\left(U_x^d(x), U_y^d(y)\right)}{U_x^d(x)} \text{ and } U_{x|y}^d(x \mid y) = \frac{C^d\left(U_x^d(x), U_y^d(y)\right)}{U_y^d(y)}.$$

21.7.2 Archimedean Attribute Dominance Utility Copulas

By now, you should be familiar with the Archimedean form

$$C^d(v_1, \ldots, v_n) = \varphi^{-1}\left(\prod_{i=1}^{n} \varphi(v_i)\right), \quad 0 \le v_i \le 1,$$

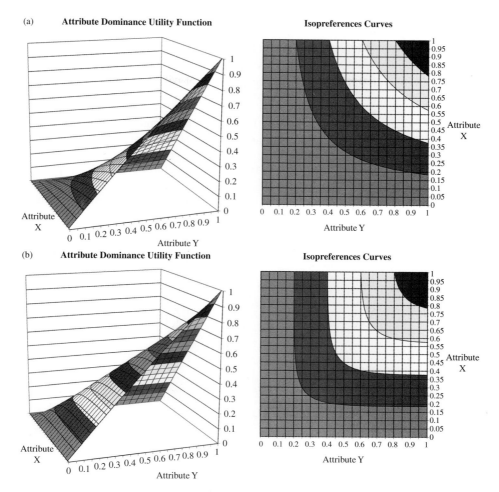

Figure 21.9 Attribute Dominance Utility Copula: An attribute dominance utility function with linear boundary values.

where $\varphi(1) = 1$ and $v_i = U_i(x_i \mid \bar{x}_i^*)$ that correspond to additive ordinal preferences.

For example, a utility function over a grounded (additive) preference function of the form $P(x, y) = xy^\eta$ will result in an attribute dominance utility function that can be expressed by an Archimedean combination of utility functions. While Archimedean utility copulas allow for more general utility functions that do not necessarily satisfy the attribute dominance condition, we illustrate below a special case of these Archimedean forms does in fact guarantee an attribute dominance utility function. To illustrate, consider a special case of this formulation of the form

$$C^d(v_1, ..., v_n) = \psi^{-1}\left(\prod_{i=1}^{n} \psi(v_i)\right), \quad 0 \le v_i \le 1 \tag{21}$$

where ψ is any continuous, bounded and strictly increasing function satisfying $\psi(1) = 1$ and $\psi(0) = 0$. The function ψ is referred to as the generating function of the attribute dominance utility copula. In our future discussions, if $\psi(0) = 0$. we refer to it as a proper generating function, and if the generating function does not pass by the origin, we give it the symbol φ.

We now show that (21) satisfies the properties of attribute dominance utility copulas:

1. Domain: Because the arguments range from 0 to 1, the domain is $[0,1]^n$.
2. Range: Both ψ and ψ^{-1} are strictly increasing functions. Therefore, the minimum value of the copula function is $\psi^{-1}\left(\prod_{i=1}^{n}\psi(0)\right) = 0$ and the maximum value is $\psi^{-1}\left(\prod_{i=1}^{n}\psi(1)\right) = 1$.
3. Linear at the margins: The Archimedean form (21) is linear at the margins. To illustrate,

$$C^d(1,1,...,1,v_j,1,..,1) = \psi^{-1}\left(\psi(1)\psi(1)...\psi(1)\psi(v_j)\psi(1)...\psi(1)\right) = \psi^{-1}\left(\psi(v_j)\right) = v_j$$

4. Attribute Dominance Condition

$$C^d(v_1,v_2,.v_{j-1},0,v_{j+1},..,v_n) = \psi^{-1}\left(\psi(0)\prod_{i \neq j}\psi(v_i)\right) = 0.$$

For the special case where

$$\psi(t) = t,$$

we have $\psi^{-1}(t) = t$ and so

$$C^d(v_1,...,v_n) = \prod_{i=1}^{n}v_i \text{ and so } U_{x_1,...,x_n}^d(x_1,...,x_n) = \prod_{i=1}^{n}U_{x_i}^d(x_i),$$

which is the product form of utility functions.

An attribute dominance utility function can be constructed using this Archimedean form by assessing the utility functions for each attribute at the margin and using them as arguments into the copula function. For example, for two attributes, we can write

$$U_{xy}^d(x,y) = C^d\left(U_x^d(x),U_y^d(y)\right) = \psi^{-1}\left(\psi(U_x^d(x))\psi(U_y^d(y))\right).$$

We now discuss examples of generating functions that may be used with this formulation and Chapter 30 provides empirical methods for their assessment.

21.7.2.1 THE GENERATING FUNCTION OF AN ARCHIMEDEAN ATTRIBUTE DOMINANCE UTILITY COPULA An expression similar to (21) also appears in the probability literature and is known as an Archimedean probability copula. However, there are additional constraints on the derivatives of the generating function for probability copulas to guarantee that the resulting probability density will have a non-negative sign. This condition is not needed for utility functions, and so we relax this

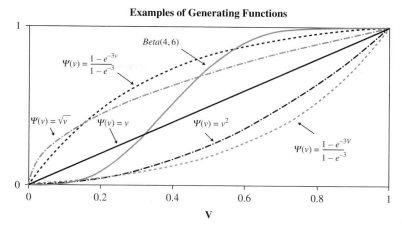

Figure 21.10. Possible generating functions for Archimedean attribute dominance utility copulas.

condition on the generating function and refer to (21) as an Archimedean attribute dominance utility copula.

There is a wealth of functions that can be used for the generating function for an Archimedean utility copula. For example, all continuous and strictly monotone probability distributions on a normalized domain would satisfy the requirements of the generating function. Figure 21.10 shows just some examples of possible choices for the generating function.

> **Note:** The generating function of the Archimedean form (21) is invariant to a power transformation, i.e., if C^d is the Archimedean utility copula resulting from the generating function, ψ, then it is also the Archimedean copula resulting from its non-zero power, $\psi^r, r \neq 0$. This implies that ψ and $\psi^r, r \neq 0$ both have the same Archimedean copula.

21.7.2.2 EXAMPLES OF GENERATING FUNCTIONS FOR ARCHIMEDEAN ATTRIBUTE DOMINANCE UTILITY

EXAMPLE 21.8: Normalized Exponential Generating Function

Consider the generating function,

$$\psi(t) = \frac{1 - e^{-\delta t}}{1 - e^{-\delta}}, \delta \in \mathbb{R} \setminus \{0\} \tag{22}$$

with

$$\psi^{-1}(s) = \frac{-1}{\delta} \ln\left(1 - s(1 - e^{-\delta})\right).$$

Direct substitution shows that for two attributes, this generating function leads to the form

$$C(U_x^d(x), U_y^d(y)) = \frac{-1}{\delta} \ln(1 + \frac{(e^{-\delta U_x^d(x)} - 1)(e^{-\delta U_y^d(y)} - 1)}{(e^{-\delta} - 1)}), \ \delta \neq 0. \tag{23}$$

Note then when δ is close to zero,

$$\psi(t) \approx \frac{1 - (1 - \delta t)}{1 - (1 - \delta)} = t,$$

and so the copula approaches the product form and the resulting multiattribute utility surface is a product of normalized conditional utility functions.

This observation from Example 21.8 allows us to conduct sensitivity to the product form (utility independence) for attribute dominance utility functions by simply varying the parameter, δ, and observing the change in the utility surface. Using simulation, we can also determine the change in the decision when utility independence assumptions do not hold.

EXAMPLE 21.9: Normalized Power-Generating Function

Another generating function that might be of interest is the power-generating function of the form

$$\psi(t) = 1 - (1 - t)^{\delta}, \delta \neq 0, \text{ and so } \psi^{-1}(s) = 1 - (1 - s)^{\frac{1}{\delta}}.$$

This generating function satisfies

$$\psi(0) = 0 \text{ and } \psi(1) = 1.$$

Direct substitution shows that the resulting attribute dominance utility copula has the form

$$C^d(v_1, v_2) = \psi^{-1}(\psi(v_1)\psi(v_2))$$
$$= 1 - (1 - \psi(v_1)\psi(v_2))^{\frac{1}{\delta}}$$
$$= 1 - (1 - [1 - (1 - v_1)^{\delta}][1 - (1 - v_2)^{\delta}])^{\frac{1}{\delta}}$$
$$= 1 - [(1 - v_1)^{\delta} + (1 - v_2)^{\delta} - (1 - v_1)^{\delta}(1 - v_2)^{\delta}]^{\frac{1}{\delta}}$$

Testing the boundary conditions

$$C^d(v_1, 0) = 1 - [(1 - v_1)^{\delta} + (1 - 0)^{\delta} - (1 - v_1)^{\delta}(1 - 0)^{\delta}]^{\frac{1}{\delta}}$$
$$= 1 - [(1 - 0)^{\delta}]^{\frac{1}{\delta}} = 0.$$

$$C^d(1,1) = 1 - [(1 - 1)^{\delta} + (1 - 1)^{\delta} - (1 - 1)^{\delta}(1 - 1)^{\delta}]^{\frac{1}{\delta}} = 1 - [0]^{\frac{1}{\delta}} = 1.$$

Once again, when δ is close to zero, this generating function is close to the identity function,

$$\psi(t) \approx t \text{ and } \psi^{-1}(s) \approx s,$$

and so the attribute dominance utility copula leads to a product of normalized conditional utility functions.

Similarly, we can envision normalized polynomial functions of the form $\psi(t) = t$, $\psi(t) = at + (1-a)t^t$, and other higher order polynomials. That reduce to the identity generating function by appropriate choice of their parameters.

21.7.2.3 NUMERICAL ILLUSTRATIONS

EXAMPLE 21.10: Dependence for Attribute Dominance Utility Functions with Copulas

---

Let us refer back to Example 21.2 about health and consumption, and present another method to capture the change in conditional utility functions for the two attributes. Recall that the utility function of consumption (at the maximum health state) is a risk-neutral utility function, $U_x^d(x) = x$, $0 \le x \le 1$, and the utility function for health state (at the maximum consumption level) is exponential with risk aversion coefficient $\gamma = \dfrac{1}{1.3}$,

$$U_y^d(y) = \frac{1 - e^{-\gamma y}}{1 - e^{-\gamma}}, \ 0 \le y \le 1.$$

Using the copula function (23) and substituting for $U_x^d(x) = x$ and $U_y^d(y) = \dfrac{1 - e^{-\gamma y}}{1 - e^{-\gamma}}$ gives

$$U(x,y) = C(U_x^d(x), U_y^d(y)) = \frac{-1}{\delta}\ln(1 + \frac{(e^{-\delta x} - 1)(e^{-\delta\left(\frac{1 - e^{-\gamma y}}{1 - e^{-\gamma}}\right)} - 1)}{(e^{-\delta} - 1)}), \ \delta \ne 0, \qquad (24)$$

where δ determines the level of utility dependence between attributes x and y.

$$U(x, y^*) = C(U_x^d(x), 1) = \frac{-1}{\delta}\ln(1 + \frac{(e^{-\delta x} - 1)(e^{-\delta} - 1)}{(e^{-\delta} - 1)}) = \frac{-1}{\delta}\ln(1 + (e^{-\delta x} - 1)) = x$$

$$U(x^*, y) = C(1, U_y^d(y)) = \frac{-1}{\delta}\ln(1 + \frac{(e^{-\delta} - 1)(e^{-\delta\left(\frac{1 - e^{-\gamma y}}{1 - e^{-\gamma}}\right)} - 1)}{(e^{-\delta} - 1)}) = \frac{-1}{\delta}$$

$$\ln(1 + (e^{-\delta\left(\frac{1 - e^{-\gamma y}}{1 - e^{-\gamma}}\right)} - 1)) = \frac{1 - e^{-\gamma y}}{1 - e^{-\gamma}}$$

Note that these boundary assessments are matched regardless of the level of $\delta \ne 0$, because any level of δ results in a utility copula. When $\delta \to 0$, the multiattribute utility function becomes the product of the marginal utility functions. When δ increases, (24) incorporates utility dependence while maintaining the same utility functions at the boundaries of the domain of the attributes. The parameter, δ, can be changed to match the desired trade-offs, as we show in the next example.

Note further that the attribute dominance conditions are satisfied by (24). To illustrate,

$$U(x,y^0) = C(U_x^d(x),0) = \frac{-1}{\delta}\ln(1+\frac{(e^{-\delta x}-1)(e^{-\delta.0}-1)}{(e^{-\delta}-1)}) = \frac{-1}{\delta}\ln(1) = 0$$

and

$$U(x^0,y) = C(0,U_y^d(y)) = \frac{-1}{\delta}\ln(1+\frac{(e^{-\delta.0}-1)(e^{-\delta\left(\frac{1-e^{-\gamma y}}{1-e^{-\gamma}}\right)}-1)}{(e^{-\delta}-1)}) = \frac{-1}{\delta}\ln(1) = 0.$$

EXAMPLE 21.11: Matching Trade-Offs with the Parameter δ

One method to assess the dependence parameter, δ, from the decision-maker is to change its value and observe the corresponding isopreference contours. The contours that meet the decision-maker's preferences can be used to determine the value of δ. Figures 21.11 and 21.12 show the multiattribute utility surfaces of the previous example and the corresponding isopreference contours for the same utility functions at the margins but for different values of δ.

The use of the copula method to determine the multiattribute utility function is very convenient when conducting a sensitivity analysis to the utility dependence parameter. For example, we can find the value of δ at which a given decision alternative will change and then find the corresponding isopreference contours. A binary question to the decision-maker is sufficient to determine whether δ is above or below this value.

Another way to assess δ is to first assess the marginal utility functions for the attributes and then assess some individual points on the multiattribute utility surface. By substituting for the values of these assessed points into Equation (24), we can determine the value of δ. Of course one would want to assess several points to ensure that the chosen copula is consistent with the decision-maker's preferences.

We can now conduct a sensitivity analysis to the product form of multiattribute utility functions to illustrate whether the best decision will change if the functional form has the same assessed marginal utility functions but is not a product of the marginal.

EXAMPLE 21.12: Sensitivity Analysis to the Product Form Using Utility Copulas

---

Consider a decision with four attributes and a utility copula of the form (23) with generating function (22). Suppose that the analyst uses the multiplicative form of Equation (17) instead of the Archimedean utility copula form. What is the percentage of incorrect decisions that can result using this approximation?

To answer this question, we conduct a Monte Carlo simulation for a decision situation with two alternatives, A and B, each containing four attributes.

1. We generate the probability distributions for each alternative uniformly form the space of all 3x3x3x3 joint probability distributions (see Appendix B for more details about this sampling approach).

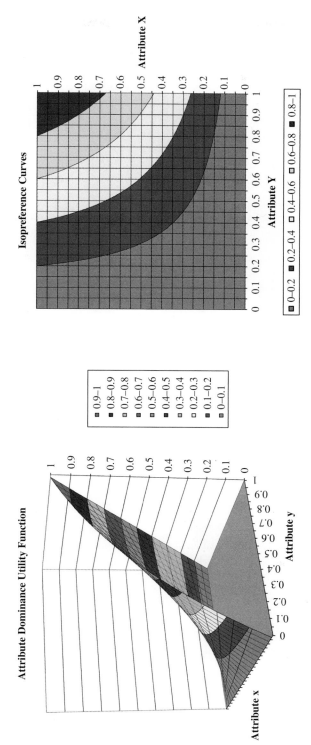

Figure 21.11. (a) Multiattribute utility function and (b) isopreference contours for $\delta = 0.001$.

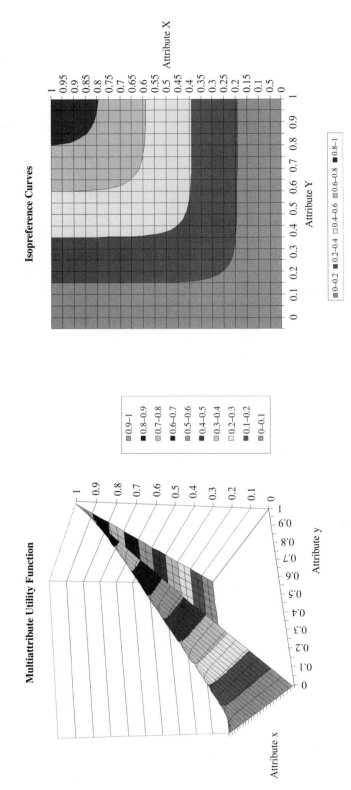

Figure 21.12. (a) Multiattribute utility function and (b) isopreference contours for $\delta = 12$.

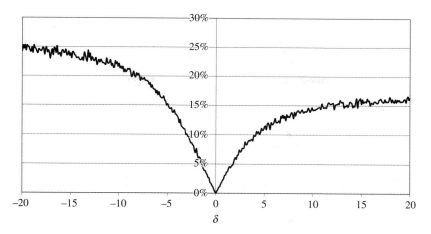

Figure 21.13. Percentage of incorrect decisions obtained using the product form approximation.

2. We calculate the expected utility of each alternative using the copula form and using the multiplicative form, and determine the best decision alternative in both cases.

3. We then calculate the fraction of times the best decision obtained using the copula approach was different from that using the product form approach, and note this as the percentage of incorrect decisions made using the multiplicative approximation.

4. We repeat the procedure for different values of δ.

Figure 21.13 shows the simulation results.

Here, the calculation of the expected utility assumed perfect knowledge of the probability distributions and perfect knowledge of the marginal utility assessments. The error observed is a result of not incorporating the generating function.

The percentage of incorrect decisions approaches 25% as δ decreases to negative values. While this might seem as a small percentage error at first, we need to recall that the analyst had already put the effort assessing the probability distributions and the utility functions at the margins, and yet the error was still 25% for these values of δ.

Compare this to a coin toss. Suppose the analyst had not gone through the effort of assessing the joint probability distributions or the marginal utility functions and simply tossed a coin to choose between the two alternatives. The percentage error would have been 50%. Assessing the probability distributions and the marginal utility functions but ignoring the generating function reduced the error to about 25%, very significant compared to the elicitation effort.

Note further that the percentage of incorrect decisions is small when $\delta \to 0$ (as expected, since $\lim_{\delta \to 0} \dfrac{1 - e^{-\delta v}}{1 - e^{-\delta}} = v$, giving the multiplicative copula). As δ increases, the percentage of incorrect decisions increases to about 16% for positive values δ.

The analysis and simulation results of this example can be applied to a variety of settings to see whether the assumptions of the product form would be valid in a given decision. If the percentage error is large, then the analyst would need to spend more effort determining the exact form of the utility function.

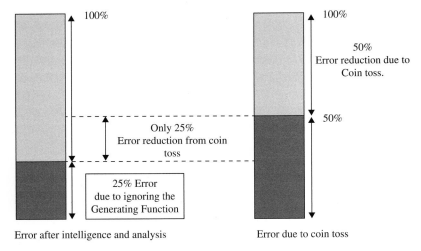

Figure 21.14. Comparison of percentage error using a coin toss versus using analysis with knowledge of the joint probability distributions and the individual utility functions but assuming a multiplicative form instead of an Archimedean form.

Dialogue between Instructor (I) and Student (S)

S: The results of the Monte Carlo simulation are really insightful.

I: Yes. Think of it in terms of predicting adversary behavior. It shows that even if you knew the opponent's two alternatives, and you knew their joint probability distributions, choice of attributes, even the marginal utility functions for each attribute, and you also knew that an opponent would act rationally, but you assumed a product form for the utility function when in fact there was utility dependence, then the percentage of incorrect decisions could still be 25%.

S: But you get it right 75% of the times.

I: Well, compare to the accuracy of a coin toss. Suppose you knew nothing at all: no joint probabilities, no individual utilities, no attributes, and all you did was toss a coin to predict the chosen alternative; what would the percentage error be?

S: 50%.

I: I rest my case. All this knowledge only got you an extra 25% increase in prediction ability. And you get only a 25% increase. This just goes to show the importance of modeling and of verifying the assumptions made. Figure 21.14 highlights this point further.

S: Does this mean we should never use the product form? Or that it is bad?

I: No, it just means that you should verify that its independence conditions really apply before rushing to use it.

21.8 CONSTRUCTING ATTRIBUTE DOMINANCE UTILITY FUNCTIONS WITH VARIOUS INDEPENDENCE CONDITIONS

21.8.1 Partial Utility Independence

We can envision various strictures of attribute dominance utility functions that are neither the product form nor the full dependence structure form where the conditional utility function of an attribute varies with the levels of all other attributes. For example, suppose we wish to construct an attribute dominance utility function that has the utility independence structure of Figure 21.7. This structure asserts that A is utility independent of B given C implying the condition

$$U^d_{A|BC}(A\,|\,B,C) = U^d_{A|C}(A\,|\,C).$$

Using the product expansion, we can write

$$U^d_{ABC}(A,B,C) = U^d_C(C)U^d_{B|C}(B\,|\,C)U^d_{A|C}(A\,|\,C).$$

We can further express this condition as

$$U^d_{ABC}(A,B,C) = U^d_{BC}(B,C)U^d_{A|C}(A\,|\,C) = U^d_{BC}(B,C)\frac{U^d_{AC}(A,C)}{U^d_C(C)},\ U^d_C(C) \neq 0 \quad (25)$$

We can extend the use of attribute dominance utility copulas to derive utility functions with a variety of similar types of independence structures. For example, (25) can be expressed as a product of two copula functions divided by one marginal utility function,

$$U^d_{ABC}(A,B,C) = \frac{C^d_{BC}(U^d_B(B),U^d_C(C))C^d_{AC}(U^d_A(A),U^d_C(C))}{U^d_C(C)}.$$

The assessment of this structure would involve the three marginal utility assessments, $U^d_A(A), U^d_B(B), U^d_C(C)$, and the choice of copula functions that match the trade-offs. We refer back to this idea in Chapter 28 and simply present it here to illustrate the main results.

21.8.2 Higher-Order Independence Conditions for Attribute Dominance Utility

So far we have talked about the conditional utility function of one attribute as the level of another attribute (or set of attributes) varies. We can also consider the conditional utility function of more than one attribute and its variation with the levels of other attributes. For example, consider four attributes, A, B, C, and D. We might consider the conditional utility function $U^d_{ABCD}(A,B\,|\,C,D)$. If A and B are utility-dominant attributes, then this function is also an attribute dominance utility function whose form depends on the levels of C and D. The general form of the product expansion would be

$$U^d_{ABCD}(A,B,C,D) = U^d_A(A)U^d_{B|A}(B\,|\,A)U^d_{C|AB}(C\,|\,A,B)U^d_{D|ABC}(D\,|\,A,B,C).$$

We can also express it in a form that groups the attributes as

$$U^d_{ABCD}(A,B,C,D) = U^d_C(C)U^d_{D|C}(D\,|\,C)U^d_{AB}(A,B\,|\,C,D),$$

or

$$U^d_{ABCD}(A,B,C,D) = U^d_{CD}(C,D)U^d_{AB}(A,B\,|\,C,D).$$

If the conditional utility function of A,B does not depend on the levels of C or D, then we say that AB is utility independent of CD, and so

$$U^d_{AB|CD}(A,B\,|\,C,D) = U^d_{AB}(A,B).$$

Because utility independence is a symmetric property for attribute dominance utility functions, this also implies that C,D is utility independent of A,B, and so

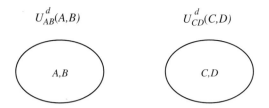

Figure 21.15. Higher-order utility independence relation.

$$U^d_{CD}(C,D\,|\,A,B) = U^d_{CD}(C,D).$$

Therefore,

$$U^d_{ABCD}(A,B,C,D) = U^d_{CD}(C,D)U^d_{AB}(A,B).$$

Figure 21.15 depicts this symmetric independence relation graphically.

This function can be modeled by a product of attribute dominance utility copulas as

$$U^d_{ABCD}(A,B,C,D) = C^d_{AB}(U^d_A(A),U^d_B(B)) \times C^d_{CD}(U^d_C(C),U^d_D(D)).$$

To illustrate that AB is indeed utility independent of CD

$$U^d_{ABCD}(A,B\,|\,C,D) = \frac{U^d_{ABCD}(A,B,C,D)}{U^d_{CD}(C,D)} = C^d_{AB}(U^d_A(A),U^d_B(B)),$$

which does not depend on C or D.

Furthermore,

$$U^d_{ABCD}(A\,|\,C,D) = \frac{U^d_{ABCD}(A,B^*,C,D)}{U^d_{CD}(C,D)} = C^d_{AB}(U^d_A(A),U^d_B(B^*)) = U^d_A(A).$$

This implies that the utility of attribute A alone does not depend on CD. Therefore, higher-order utility independence relation of AB on CD also asserts the independence of each of the attributes in the higher-order relation, i.e.,

$$U^d_{AB|CD}(A,B\,|\,C,D) = U^d_{AB}(A,B) \;\Rightarrow\; U^d_{A|CD}(A\,|\,C,D) = U^d_A(A) \text{ and } U^d_{B|CD}(B\,|\,C,D) = U^d_B(B).$$

We refer to this idea as the independence blanket and refer back to it again in Chapter 26.

21.8.3 Nested Combinations of Utility Copulas

We can also envision nested attribute dominance utility copulas that result from combining several lower-order copulas. For example, the form

$$C^d\left(C^d_1(A,B),C^d_2(D,E)\right)$$

is a copula for four attributes that is generated by combining two lower-order copula functions. In this chapter, we presented decompositions, simplifications, and methods for constructing attribute dominance utility functions. The following chapters in this book explore the same type of analyses when the attribute dominance conditions are relaxed. As we shall see, the number of terms in the expansions and the level of complexity will increase, but the same features will be present. This is why we started this section with attribute dominance utility functions. The simplicity provided by such functions also provides motivation for formulating direct value attributes in terms of utility-dominant attributes whenever possible.

21.9 SUMMARY

This chapter presented a class of utility functions called attribute dominance utility functions, and discussed several concepts in utility theory related to such functions:

1. An expansion theorem for attribute dominance utility functions that has a product form.
2. The idea of expressing the normalized conditional utility function of an attribute in terms of its reference value at the maximum value of the remaining attributes.
3. The idea of utility inference that determines that variation of the conditional utility function as the levels of another attribute(s) varies.
4. Utility independence, where the conditional utility function of an attribute does not change when levels of the other attributes vary.
5. Trade-offs implied by attribute dominance utility functions.
6. Utility diagrams that provide graphical representations of the utility independence relations for attribute dominance utility functions.
7. Attribute dominance utility copula functions that construct attribute dominance utility functions in terms of the conditional utility assessments at the boundary values.
8. The idea of sensitivity to utility independence, in this case, was sensitivity to the product form of utility functions.
9. Higher-order attribute dominance utility independence conditions.

In the remaining chapters of this book, we shall revisit the ideas presented in this chapter, and we shall conduct the same type of analysis for utility functions that do not satisfy the attribute dominance condition. We shall see the simplicity that the attribute dominance condition provides, and the value that can be gained if the attributes can be reformulated to yield attribute dominance utility functions.

ADDITIONAL READINGS

Abbas, A. E. 2003a. An entropy approach for utility assignment in decision analysis. *AIP Conference Proceedings* 659, 328; doi: 10.1063/1.1570550

Abbas, A. E. 2003b. Entropy methods in decision analysis. PhD Thesis, Stanford University.

Abbas, A. E. 2004. Entropy methods for adaptive utility elicitation. *IEEE Transactions on Systems, Man, and Cybernetics – Part A: Systems and Humans* 34(2): 169–178.

Abbas, A. E. 2006. Maximum entropy utility. *Operations Research* 54(2): 277–290.

Abbas, A. E. 2009. Multiattribute utility copulas. *Operations Research* 57(6): 1367–1383.

Abbas, A. E. 2012. Utility copula functions matching all boundary assessments. *Operations Research*.

Abbas, A. E and R. A. Howard. 2005. Attribute dominance utility. *Decision Analysis* 2(4): 185–206.

Abbas, A. E and Z. Sun. 2015. Multiattribute utility functions satisfying mutual preferential independence. *Operations Research* 63(3): 378–393.

Keeney, R. and H. Raiffa. 1976. *Decisions with Multiple Objectives: Preferences and Value Tradeoffs*. John Wiley, New York.

APPENDIX 21A MONTE CARLO SIMULATION USED FOR SENSITIVITY ANALYSISAppendixAppendix

The procedure we describe here for generating joint probability distributions is known as uniform sampling on the probability simplex. It generates an n-outcome probability distribution uniformly from the space of all possible n-outcome probability distributions. Here we use this approach to generate an 81-outcome probability distribution corresponding to the 3x3x3x3 joint probability distribution (we simulate the end points of the probability tree).

Steps of Monte Carlo Simulation

This section describes the sampling approach used to conduct sensitivity to the multiplicative form of mutual utility independence. The idea is to generate joint probability distributions for two alternatives and calculate the expected utility of each alternative once using the multiplicative form and another using the copula form and see the percentage of time the calculation of the best decision alternative differs.

1. **Uniform sampling from the space of 3x3x3x3 probability distributions**
    Generate two 3x3x3x3 joint probability distributions to represent two decision alternatives (each having four variables discretized to three outcomes), by uniform sampling as follows:

    ○ Generate (3^4-1) independent samples, $x_1, x_2, ..., x_{3^4-1}$ from a uniform $[0,1]$ distribution.
    ○ Sort the generated samples from highest to lowest to form an order statistic, $u_1 \leq u_2 \leq ..... \leq u_{3^4-1}$.
    ○ Take the difference between each two successive elements of the order statistic $\{u_1 - 0, u_2 - u_1, ..., u_{3^4-1} - u_{3^4-2}, 1 - u_{3^4-1}\}$.
    ○ The increments form a 3^4-outcome probability distribution that is uniformly sampled from the space of possible 3^4-outcome probability distributions.

2. **Generate four marginal utility values for each consequence in the tree from a uniform [0,1] distribution.**
    Since each consequence is characterized by four attributes, we generate four normalized utility values for each consequence.

3. **Evaluate the multiattribute utility function** for each prospect using two methods:
    (i) the multiplicative form of multiattribute utility functions $l_i = 0, i = 1, .., 4$ and
    (ii) the given utility copula form with $l_i = 0, i = 1, .., 4$.

4. **Calculate the expected utility** of both alternatives using (i) the product form and (ii) the copula form.

5. Repeat 10,000 times.

6. Calculate fraction of times a difference in the recommended decision alternatives occurs if we assume the product form when the actual utility function has a copula form.

7. Change the value of the dependence parameter, δ, in the copula form and repeat the simulation steps.

CHAPTER 22

Basic Expansion Theorem for Two-Attribute Utility Functions

Chapter Concepts

- Relaxing the attribute dominance conditions
- The basic expansion theorem for two-attribute utility functions
- Decomposing a two-attribute indifference assessment into a sum of products of indifference assessments when the attribute dominance conditions do not exist
- Expressing the normalized conditional utility function of any attribute in terms of its two boundary values, i.e., expressing $U(x \mid y)$ in terms of $U(x \mid y^*)$ and $U(x \mid y^0)$
- Utility inference: updating the conditional utility function of an attribute as the level of another attribute varies from its boundary values
- Utility tree representation of the assessments needed to construct a two-attribute utility function

---

22.1 RELAXING THE ATTRIBUTE DOMINANCE CONDITIONS

As we discussed in the introduction to Part V, it would be convenient to decompose a multiattribute indifference assessment into lower-order indifference assessments where the level of only one attribute changes at each stage. This would allow us to think about the Equivalence Rule for only one attribute at a time. We have seen in Chapter 21 that attribute dominance utility functions can be decomposed into a product of normalized marginal-conditional utility functions, requiring an indifference assessment for a single attribute conditioned on the levels of the other attributes. When the attribute dominance conditions do not exist, the utility surface has additional boundary curves that need to be considered in the expansion. In particular, when the attribute dominance conditions do not exit, we can define:

1. Two boundary utility functions for attribute X at the minimum and maximum values of attribute Y, namely $U(x, y^0), U(x, y^*)$. For attribute dominance utility functions, $U(x, y^0) = 0$, and so we only considered the upper boundary curve $U(x, y^*)$.
2. Two boundary utility functions for attribute Y at the minimum and maximum values of attribute X, namely $U(x^0, y), U(x^*, y)$.
3. Two corner values, $k_x = U(x^*, y^0), k_y = U(x^0, y^*)$. Note that for attribute dominance utility functions, those corner values were equal to zero.

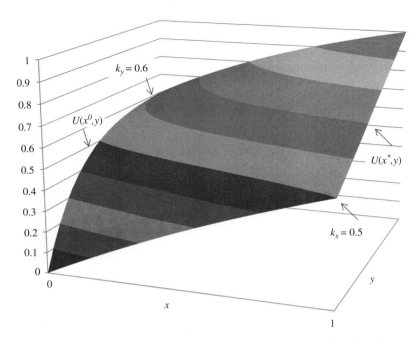

Figure 22.1. Two-attribute utility surface that is strictly increasing with each attribute.

Figure 22.1 shows an example of a utility surface that does not satisfy the attribute dominance conditions and where these curves and corner values are shown graphically.

One immediate observation that results from relaxing the grounding condition is that for any fixed value of y, the curve traced on the surface as we vary X from its minimum to its maximum values is not normalized. This is because the curve traces values from $U(x^0, y)$ to $U(x^*, y)$, and in this general case, $U(x^0, y)$ is not necessarily zero and $U(x^*, y)$ need not be one. If we wish to conduct a utility assessment for attribute X at any fixed value of attribute Y, using indifference probability assessments, it would be convenient to conduct the assessments on a normalized scale such that this normalized assessment has a value of zero at the minimum value of X and a value of one at the maximum value. We do this by first subtracting some amount from this curve and then scaling it to range from zero to one. We shall refer to this new curve as a ***normalized conditional utility assessment*** of attribute X.

As we shall see, even when attribute dominance conditions do not exist, the utility function can still be constructed using utility functions at the boundary value of the domain as well as normalized conditional utility functions; however, the expansion is a bit more complicated than the product form: it is a sum of products. We refer to this general expansion as the ***basic expansion theorem*** for multiattribute utility functions. In this chapter, we shall focus on expansions for two-attribute utility functions.

The basic expansion theorem provides a general method for constructing multiattribute utility functions by direct utility elicitation of the conditional utility functions of the attributes. In later chapters, we shall see that it also provides a simple way to derive the functional form of the multiattribute utility function under a wide variety of conditions and it relates many foundational concepts in multiattribute utility theory.

To simplify the exposition of the basic expansion theorem, and to help memorize its terms, we provide a graphical representation for the expansion of multiattribute utility functions using a tree, that mimics the tree expansion for joint probability distributions. We refer to this tree as the **multiattribute utility tree**. The multiattribute utility tree determines the assessments needed to construct a multiattribute utility function, and it provides graphical insights into reversing the order of the conditional utility assessments to help the analyst construct utility functions in the order of conditioning that the decision-maker is most comfortable providing.

22.2 TWO-ATTRIBUTE UTILITY TREES ON A CONTINUOUS DOMAIN

We assume (again) that the utility function is continuous and bounded, and for simplicity we assume that the utility function is strictly increasing with each attribute. We also set $U(x^0, y^0) = 0$, and $U(x^*, y^*) = 1$. As we have discussed, the conditional utility function of attribute X given $Y = y$ is the curve traced on the utility surface as we vary X from its minimum to its maximum values while keeping Y at a fixed value y. The conditional utility function is not normalized to range from zero to one. It ranges from $U(x^0, y)$ (which need not be zero if X is not a utility-dominant attribute) to $U(x^*, y)$ (which need not be one). To normalize this conditional utility assessment, we need to first subtract the amount, $U(x^0, y)$, and then we need to multiply by a scaling amount, $\dfrac{1}{U(x^*, y) - U(x^0, y)}$. We normalize this curve to range from zero to one using the following definition.

Definition
A **normalized conditional utility function** for attribute X given Y, written $U(x \mid y)$, is defined as

$$U(x \mid y) = \frac{U(x, y) - U(x^0, y)}{U(x^*, y) - U(x^0, y)}. \tag{1}$$

We can verify that the function $U(x \mid y)$ defined in (1) is indeed normalized to range from 0 to 1, for any value of Y, as X changes from x^0 to x^* because, by definition,

$$U(x^0 \mid y) = \frac{U(x^0, y) - U(x^0, y)}{U(x^*, y) - U(x^0, y)} = 0 \text{ and}$$

$$U(x^* \mid y) = \frac{U(x^*, y) - U(x^0, y)}{U(x^*, y) - U(x^0, y)} = 1.$$

The normalized conditional utility function is an assessment of the normalized conditional utility function of X at a fixed value of Y. It implies that, in general, we would need to know the value of Y when we make a utility assessment for attribute X because our utility values for X would change with the level of Y.

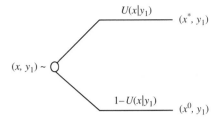

Figure 22.2. Interpreting the term $U(x\,|\,y_1)$ using lottery assessments.

Interpreting the Normalized Conditional Utility Function Using Lottery Assessments

The normalized conditional utility function $U(x\,|\,y_1)$ is the curve traced on the utility surface in the X–direction at fixed value of Y equal to y_1, and it is normalized to range from zero to one. The term $U(x\,|\,y_1)$ also has an interpretation in terms of lottery assessments using the Equivalence Rule as shown in Figure 22.2. When attribute Y is fixed at y_1, $U(x\,|\,y_1)$ is the indifference probability of receiving (x,y_1) for certain or a binary gamble that provides either (x^*, y_1) or (x^0, y_1).

The normalized conditional utility function can therefore be assessed in a meaningful way. Similarly we can define the normalized conditional utility function $U(y\,|\,x)$ for attribute Y, conditioning on the value of X, and distinguishing between the two by the order of the attributes.

When the utility function is strictly increasing with each of its arguments, the conditional utility function $U(x\,|\,y)$ is increasing with X and the conditional utility function $U(y\,|\,x)$ is increasing with Y. In this case, the range of the conditional utility functions lies between zero and one and they can be interpreted as normalized indifference probability assessments at the fixed value of the complement on the entire domain of the attributes.

In general, the conditional utility function can also be expressed in terms of any two reference values, x^1 and x^2, for which $U(x^1, y) \uparrow U(x^2, y)$, instead of the boundary values x^0 and x^*, and so we may write $\dfrac{U(x, y) - U(x^1, y)}{U(x^2, y) - U(x^1, y)}$. By choosing the upper and lower bounds as the reference values, however, we maintain the indifference probability interpretation of $U(x\,|\,y)$ and $U(y\,|\,x)$ when the utility function is increasing with each of its arguments. From here on, we therefore refer to the reference values as the minimum and maximum (boundary) values while keeping in mind that they can be set arbitrarily.

Relating the Boundary Curves of the Utility Surface to the Normalized Conditional Utility Function

By direct substitution into (1), and by rearranging, we can relate the boundary curves on the utility surface to the normalized conditional utility functions at the boundary by the equations

$$U(x, y^*) = k_y + (1 - k_y)U(x\,|\,y^*) \text{ and } U(x, y^0) = k_x U(x\,|\,y^0),$$

where $k_x = U(x^*, y^0), k_y = U(x^0, y^*), U(x^*, y^*) = 1.$

We shall refer to these expressions many times throughout the book.

The following definition will also be useful in our development.

Definition

A *normalized conditional disutility function* for attribute X given Y, written $\bar{U}(x \mid y)$, is

$$\bar{U}(x \mid y) = 1 - U(x \mid y). \tag{2}$$

Similarly, we can define $\bar{U}(y \mid x)$ for attribute Y, distinguishing between the two by the order of the attributes.

We are now ready to discuss the basic expansion theorem for two-attribute utility functions. As we shall see, we can expand around either a single attribute or the two attributes.

22.2.1 Expansion around a Single Attribute

To express a multiattribute utility function in terms of the normalized conditional assessment, $U(x \mid y)$, of attribute X, we rearrange (1) to get

$$U(x, y) = U(x \mid y)[U(x^*, y) - U(x^0, y)] + U(x^0, y).$$

Rearranging gives

$$U(x, y) = U(x \mid y)U(x^*, y) + [1 - U(x \mid y)]U(x^0, y).$$

Substituting for the definition of the conditional disutility function from (2) gives a sum of two products, which comprises the basic expansion theorem around a single attribute, as shown below.

Basic Expansion Theorem around a Single Attribute

$$\boxed{U(x, y) = U(x^*, y)U(x \mid y) + U(x^0, y)\bar{U}(x \mid y).} \tag{3}$$

We refer to this expression as *expanding the utility function around attribute X.*

Equation (3) expresses the utility function as a sum of two products: the first is the normalized conditional utility function, $U(x \mid y)$, multiplied by the curve $U(x^*, y)$; and the second is the normalized conditional disutility function, $\bar{U}(x \mid y)$, multiplied by the curve $U(x^0, y)$. The pattern associated with (3) is that when X appears as x^* in the term $U(x^*, y)$, it is followed by the normalized conditional utility assessment of X on its

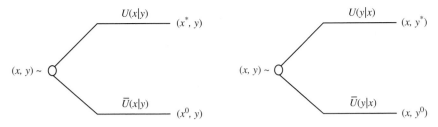

Figure 22.3. (a) Utility tree representation for expansion around X. (b) Expansion around Y.

complement. When it appears as x^0 in $U(x^0, y)$, it is followed by the normalized conditional disutility of X on its complement.

Figure 22.3(a) provides a tree-based representation of (3), which is the first-order expansion of the utility tree. The tree replaces every prospect with a binary gamble, where the level of only one of the attributes changes to either its minimum or maximum values. The branch corresponding to the maximum value of the attribute has the normalized conditional utility assessment of the attribute, while the branch corresponding to the minimum value has the normalized conditional disutility function.

How Does the Decomposition of the Basic Expansion Theorem around a Single Attribute Help with the Multiattribute Indifference Assessment?

Equation (3) asserts the equivalence between two different methods for determining the utility (indifference probability) of a prospect (x_1, y_1):

Method 1: Direct assessment of a two-attribute indifference probability that makes you indifferent to getting (x_1, y_1) for certain or a binary gamble that offers the best prospect (x^*, y^*) with probability $U(x_1, y_1)$ and the worst prospect (x^0, y^0) with probability $1 - U(x_1, y_1)$. This assessment requires thinking about the variations of both attributes simultaneously.

Method 2: Decomposition into a sum of products of univariate assessments by conducting the following:

1. Assessing $U(x_1|y_1)$ using the indifference probability assessment shown in Figure 22.3(a). This involves the variation of only one attribute, X at the fixed value of Y, namely y_1.
2. Assessing $U(x^*, y_1)$ as the indifference probability of receiving (x^*, y_1) for certain or a binary deal that provides (x^*, y^*) or (x^*, y^0). This only involves the variation of attribute Y.
3. Assessing $U(x^0, y_1)$ as the indifference probability of receiving (x^0, y_1) or a binary deal that provides (x^0, y^0) or (x^0, y^*). This also involves the variation only of attribute Y.
4. Substitute into (3).

Similarly, we can expand around attribute Y, to get

$$U(x, y) = U(y \mid x)U(x, y^*) + \bar{U}(y \mid x)U(x, y^0).$$ (4)

Figure 22.3(b) provides a tree-based representation of (4). We can also equate (3) and (4) to obtain a direct relation between $U(x \mid y)$ and $U(y \mid x)$, which can be used as a consistency check, and we leave this, for now, as an exercise to the reader.

22.2.2 Basic Expansion Theorem around Two Attributes

Suppose we further expand the two terms $U(x^*, y)$ and $U(x^0, y)$ in the left-hand side of Figure 22.4 with respect to a second attribute, Y. This formulation uses direct substitution into (4) and implies that

$$U(x^*, y) = U(x^*, y^*)U(y \mid x^*) + U(x^*, y^0)\bar{U}(y \mid x^*) \tag{5}$$

and

$$U(x^0, y) = U(x^0, y^*)U(y \mid x^0) + U(x^0, y^0)\bar{U}(y \mid x^0). \tag{6}$$

Substituting from (5) and (6) into (4) gives a decomposition in terms of normalized conditional utility functions and some constants, which we refer to as the basic expansion theorem around two attributes.

Basic Expansion Theorem around Two Attributes

$$\begin{aligned}
U(x, y) = \, &U(x^*, y^*)U(x \mid y)U(y \mid x^*) \\
&+ U(x^*, y^0)U(x \mid y)\bar{U}(y \mid x^*) \\
&+ U(x^0, y^*)\bar{U}(x \mid y)U(y \mid x^0) \\
&+ U(x^0, y^0)\bar{U}(x \mid y)\bar{U}(y \mid x^0)
\end{aligned} \tag{7}$$

Figure 22.4 shows a tree-representation of this two-attribute expansion starting with X then Y. The tree assessment requires the normalized conditional utility values $U(x \mid y), U(y \mid x^*), U(y \mid x^0)$ in addition to the utility values of the points of the domain, namely $U(x^*, y^*) = 1, U(x^0, y^0) = 0, U(x^*, y^0) \triangleq k_x, U(x^0, y^*) \triangleq k_y$. We can therefore write (7) as

$$U(x, y) = U(x \mid y)U(y \mid x^*) + k_x U(x \mid y)\bar{U}(y \mid x^*) + k_y \bar{U}(x \mid y)U(y \mid x^0). \tag{8}$$

Note that the utility tree requires the normalized conditional utility function of the first attribute in the expansion, but the conditional utility function of the second is expressed at the boundary values of the attribute expanded before it. Of course, we can also start our decomposition with attribute Y and then expand around attribute X. To construct the whole utility surface using this approach, we would need to repeat this assessment for all prospects (x, y) on the domain on the attributes.

We can also express the utility function in terms of a utility tree whose terminal values comprise the most preferred and least preferred assessments, $(x^*, y^*), (x^0, y^0)$, by substituting for the corner values as shown in Figure 22.5.

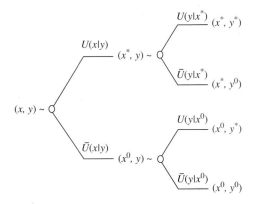

Figure 22.4. Two-step expansion of a two-attribute utility tree.

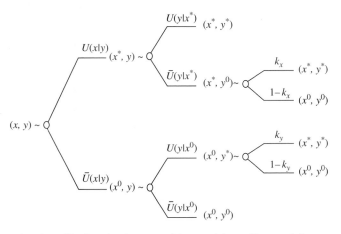

Figure 22.5. Expressing the utility function in terms of the top right and bottom left corners of the domain.

Dialogue between Instructor (I) and Student (S)

S: This basic expansion theorem is interesting. To assess the utility value of a prospect, say (x_1, y_1), Equation (7) shows that we have two methods.

I: Yes. The first method is that we can assess it directly using lottery assessments where we have the indifference probability $U(x_1, y_1)$ of receiving (x_1, y_1) for certain or a binary deal that provides (x^*, y^*) with probability $U(x_1, y_1)$ or (x^0, y^0) with probability $1 - U(x_1, y_1)$, and this multiattribute assessment requires us to think about the variations in both X and Y simultaneously.

S: Yes, and the second method is to decompose the indifference assessment and think about a single variation in each attribute separately.

I: Correct. We first assess $U(y_1 \mid x^*)$ and $U(y_1 \mid x^0)$. These two assessments are indifference utility assessments for level y_1 of attribute Y in terms of its best and worst values; once when $X = x^*$ and another time when $X = x^0$.

   Next, we assess the conditional utility function $U(x_1 \mid y_1)$, which is the utility for level x_1 of attribute X in terms of its best and worst values when $Y = y_1$. Then we assess the corner values and substitute into (7).

S: So the second method enables us to think about normalized conditional utility functions for each attribute separately, while the first method requires an indifference probability assessments that considers two attributes?

I: Correct. This is the main philosophy behind the expansion, and as we shall see, this expansion helps derive the functional forms of multiattribute utility functions under a wide variety of conditions.

S: The decomposition of utility functions in general requires a sum-product expansion of conditional utility assessments and not just a product expansion like attribute dominance utility functions. Is this the calculus for expanding two attribute utility functions?

I: Yes. As you can see, it is not a simple product or addition of univariate functions. It involves univariate assessments, conditional utility assessments, corner values, and a sum-product expansion of the assessments. Note that if the attributes are utility-dominant, then the terms $U(x^*, y^0), U(x^0, y^*), U(x^0, y^0)$ would be equal to zero in (7), and so we would still get a product expansion. Now we have additional flexibility to derive the functional form if only one attribute is utility dominant. We just substitute into the expansion.

S: The definition of the conditional disutility function, $\bar{U}(x \mid y) = 1 - U(x \mid y)$, really helps highlight this formulation.

I: I have found it to be very useful and it enables the clear pattern to develop.

S: This expansion is very different from "weight and rate" type reasoning, where arbitrary scales and arbitrary weights are used and plugged into an arbitrary additive formula.

I: Yes, of course. People often make assumptions or simplifications that are not valid. If you make these assumptions arbitrarily, you will end up with wrong utility functions and possibly wrong choices.

S: They probably justify it because the other approaches are simpler in their assessments.

I: Simplicity is not an excuse for making a bad decision. There is no room for arbitrariness in the analysis of a decision.

S: I agree.

I: Now, I know that the basic expansion is somewhat complicated at first, but you need to understand its terms thoroughly, and its pattern, if you want to construct a utility function by direct elicitation of conditional utility assessments. There is no way around understanding these basic components if you want to use conditional utility assessments. You can't just arbitrarily add utility values or multiply them. There is a formal logic. We shall discuss special conditions of this general formulation in future chapters and the specific conditions of when they might apply.

S: This is the general formula for combining the conditional utility assessments to obtain a multiattribute utility function.

I: Yes. And if you use another one, you should understand the implications of deviating from the general formula.

We have now covered the preference function (or the value function) approach for constructing utility functions, and we have also discussed the basic expansion theorem for two attributes using conditional utility assessments. In the next chapter, we shall discuss the conditions that simplify the basic expansion theorem in a given decision.

S: Great. So the utility tree is just a graphical representation of an expansion of the utility function in a particular order?

I: Yes. It makes it easier to see the terms required for the functional form.

S: Can we see an example of how it helps determine the functional form?

I: Yes, the following example illustrates the use of utility trees in constructing the utility function.

22.2.3 Constructing Two-Attribute Utility Functions Using the Utility Tree Expansion

---

EXAMPLE 22.1: Constructing a Two-Attribute Utility Function Using a Utility Tree

---

Consider a decision with two attributes, X and Y on the unit square, $[0,1] \times [0,1]$. To construct the utility function in the assessment order X then Y, the decision-maker first provides a normalized assessment of the conditional utility function $U(x \mid y)$. For example, she may provide

$$U(x \mid y) = \frac{1 - e^{-\gamma x y^\eta}}{1 - e^{-\gamma y^\eta}} \text{ and so } \bar{U}(x \mid y) = 1 - \frac{1 - e^{-\gamma x y^\eta}}{1 - e^{-\gamma y^\eta}} = \frac{e^{-\gamma x y^\eta} - e^{-\gamma y^\eta}}{1 - e^{-\gamma y^\eta}}.$$

The second step requires the conditional assessments of Y at the boundaries of X. For example,

$$U(y \mid x^*) = \frac{1 - e^{-\gamma_1 y}}{1 - e^{-\gamma_1}} \text{ and } U(y \mid x^0) = \frac{1 - e^{-\gamma_2 y}}{1 - e^{-\gamma_2}}.$$

Finally, we need the corner assessments k_x, k_y, which are simply the indifference probability assessments of the prospects (x^*, y^0) and (x^0, y^*), respectively.

The functional form of the utility function corresponding to the given assessments is obtained by rolling back the tree, with $U(x^*, y^*) = 1, U(x^0, y^0) = 0$, to get

$$U(x,y) = \frac{1 - e^{-\gamma x y^\eta}}{1 - e^{-\gamma y^\eta}} \left[\frac{1 - e^{-\gamma_1 y}}{1 - e^{-\gamma_1}} \cdot 1 + k_x \left(1 - \frac{1 - e^{-\gamma_1 y}}{1 - e^{-\gamma_1}} \right) \right]$$

$$+ \left(1 - \frac{1 - e^{-\gamma x y^\eta}}{1 - e^{-\gamma y^\eta}} \right) \left[\frac{1 - e^{-\gamma_2 y}}{1 - e^{-\gamma_2}} k_y + 0 \right]. \tag{9}$$

To further characterize the properties of this functional form, we get by direct substitution,

$$U(x^0, y) = k_y \frac{1 - e^{-\gamma_2 y}}{1 - e^{-\gamma_2}} = k_y U(y \mid x^0), \quad U(x^0, y^*) = k_y$$

$$U(x^*, y) = \frac{1 - e^{-\gamma_1 y} + k_x \left(e^{-\gamma_1 y} - e^{-\gamma_1} \right)}{1 - e^{-\gamma_1}} = (1 - k_x) \frac{1 - e^{-\gamma_1 y}}{1 - e^{-\gamma_1}} + k_x = (1 - k_x) U(y \mid x^*) + k_x.$$

$$U(x, y^0) = k_x \lim_{y \to 0} \frac{1 - e^{-\gamma x y^\eta}}{1 - e^{-\gamma y^\eta}} = k_x x.$$

$$U(x, y^*) = \frac{1 - e^{-\gamma x}}{1 - e^{-\gamma}} + k_y \left(1 - \frac{1 - e^{-\gamma x}}{1 - e^{-\gamma}} \right) = (1 - k_y) \frac{1 - e^{-\gamma x}}{1 - e^{-\gamma}} + k_y, \quad U(x^0, y^*) = k_y.$$

Figure 22.6 shows the multiattribute utility surface obtained for $\gamma = 1, \gamma_1 = 0.01, \gamma_2 = 4, k_x = 0.5, k_y = 0.6$. Note how the concavity of the boundary curve $U(x^*, y)$ is less than that of $U(x^0, y)$, showing less risk aversion with respect to attribute Y at the upper boundary of X than the lower bound.

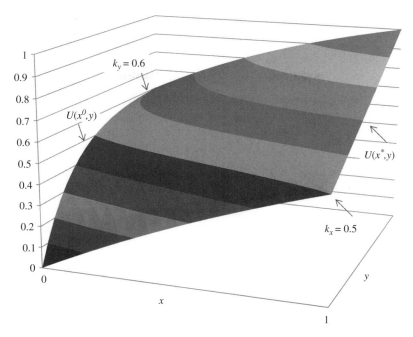

Figure 22.6. Multiattribute utility surface corresponding to the assessments on the utility tree.

Special Case: Attribute Dominance Conditions

If X is a utility dominant attribute (i.e., $U(x^0, y) = 0, \forall y$), then only two conditional

utility assessments, $\sum_{i=1}^{k} p_i d_i = 0$, and one corner value, $\sum_{i=1}^{k} p_i k_i = 1$, are needed, since

$$U(x,y) = U(x^*, y^*)U(x \mid y)U(y \mid x^*) + U(x^*, y^0)U(x \mid y)\bar{U}(y \mid x^*)$$
$$= U(x \mid y)U(y \mid x^*) + k_x U(x \mid y)\bar{U}(y \mid x^*)$$

If both X and Y are utility dominant attributes, then

$$U(x,y) = U(x^*, y^*)U(x \mid y)U(y \mid x^*) = U(x \mid y)U(y \mid x^*).$$

22.2.4 Generalized Notation for the Basic Expansion Theorem of Two Attributes

We can write (7) in a more general (and somewhat more complicated) form as

$$U(x, y) = \sum_{x_i \in \{x^*, x^0\}} \sum_{y_j \in \{y^*, y^0\}} U(x_i, y_j) g_{x_i}(x \mid y) g_{y_j}(y \mid x_i), \tag{10}$$

where

$$g_{x_i}(x \mid y) = \begin{cases} U(x \mid y), & \text{if } x_i = x^* \\ \bar{U}(x \mid y), & \text{if } x_i = x^0 \end{cases}, \quad g_{y_j}(y \mid x_i) = \begin{cases} U(y \mid x_i), & \text{if } y_i = y^* \\ \bar{U}(y \mid x_i), & \text{if } y_i = y^0 \end{cases}. \tag{11}$$

This expression is in a form that extends easily to multiple attributes, and as we shall see, it also simplifies significantly when some forms of independence conditions exist.

To become more familiar with this notation, which we will use in the case of multiple attributes, note that (10) results in a summation of four products, namely

$$U(x^*, y^*)g_{x^*}(x \mid y)g_{y^*}(y \mid x^*)$$
$$+U(x^*, y^0)g_{x^*}(x \mid y)g_{y^0}(y \mid x^*)$$
$$+U(x^0, y^*)g_{x^0}(x \mid y)g_{y^*}(y \mid x^0)$$
$$+U(x^0, y^0)g_{x^0}(x \mid y)g_{y^0}(y \mid x^0)$$

The four products correspond to products at the corner values $(x^*, y^*), (x^*, y^0), (x^0, y^*), (x^0, y^0)$. The notation of the functions g_{x_i}, g_{y_j} implies that when an attribute appears as a maximum in the corner value, we use its conditional utility function in the product, and when it appears as a minimum, we use its conditional disutility function.

For example, the product term corresponding to (x^*, y^0) is

$$U(x^*, y^0)g_{x^*}(x \mid y)g_{y^0}(y \mid x^*)$$

Here, X is maximum and Y is minimum, and so $g_{x^*}(x \mid y) = U(x \mid y)$, and $g_{y^0}(y \mid x^*) = \bar{U}(y \mid x^*)$. Therefore, this product term is

$$U(x^*, y^0)g_{x^*}(x \mid y)g_{y^0}(y \mid x^*) = U(x^*, y^0)U(x \mid y)\bar{U}(y \mid x^*).$$

The same applies to any of the other product terms in the expansion.

22.3 UTILITY INFERENCE FOR TWO ATTRIBUTES

For attribute dominance utility functions, the concept of utility inference (Section 21.2.3) expressed the conditional utility function $U(y \mid x)$ in terms of the boundary function $U(y \mid x^*)$ and some other functions. It showed the change in conditional utility as attribute X changes from its boundary value. How does the concept of utility inference extend to the general case of non-utility-dominant attributes?

As we shall see, when the attribute dominance conditions do not exit, the inclusion of an additional boundary curve $U(y \mid x^0)$ will result in an expression for the conditional utility function in terms of both boundary assessments $U(y \mid x^*)$ and $U(y \mid x^0)$. To see why the additional boundary curve comes into play, note that the conditional utility functions in the tree in Figure 22.5 display of the conditional utility function of the first attribute, $U(x \mid y)$, and the conditional utility function of the second attribute conditioned on the boundary values of the first, $U(y \mid x^*)$ and $U(y \mid x^0)$.

Definition
Utility Inference is the updating of the normalized conditional utility function of an attribute, say $U(y \mid x)$, conditioned on a particular value of another attribute X, in relation to its boundary values $U(y \mid x^*)$ and $U(y \mid x^0)$.

Utility inference will provide insights into how the conditional utility function $U(y|x)$ changes from its boundary values. As we shall see, the conditional utility function can be expressed as a weighted functional combination of its values at the boundaries of X using a relation of the form

$$\boxed{U(y|x) = w_{x^*}(x,y)U(y|x^*) + w_{x^0}(x,y)U(y|x^0) + \eta(x,y)}.$$

The exact expressions for the functions $w_{x^*}(x,y), w_{x^0}(x,y)$, and $\eta(x,y)$ can be obtained by reversing the order of the utility tree.

22.3.1 Reversing the Two-Attribute Utility Tree

Reversing the two-attribute utility tree draws the tree in the reverse order of the attributes. This is useful for the following reasons:

1. It helps condition the utility assessments in a way that is more comfortable for the decision-maker.
2. It provides a consistency check where we can assess the utility values in one assessment order and then compare them to those implied by the assessment in a different order.
3. As we shall see in the next chapter, utility inference and tree reversals provide insights into the relation between several utility independence conditions.

It would therefore be convenient to have a method to deduce the conditional utility function $U(y|x)$ from the assessments in Figure 22.5. To determine $U(y|x)$ from Figure 22.5, we need to reverse the order of the attributes in the tree. The following expression, which we generalize in Theorem 22.1, provides the conditional utility assessment in the reverse order.

Conditional Utility Function of the Second Attribute in the Expansion

$$U(y|x) = \frac{d_{x^*}(x,y)U(y|x^*) + d_{x^0}(x,y)U(y|x^0) + m_{x^*}(x,y) + m_{x^0}(x,y)}{d_{x^*}(x,y^*) + d_{x^0}(x,y^*) + m_{x^*}(x,y^*) + m_{x^0}(x,y^*)}, \quad (12)$$

where

$$d_{x^*}(x,y) = U(x|y)\big(U(x^*,y^*) - U(x^*,y^0)\big)$$

$$m_{x^*}(x,y) = U(x^*,y^0)\big(U(x|y) - U(x|y^0)\big)$$

$$d_{x^0}(x,y) = \bar{U}(x|y)\big(U(x^0,y^*) - U(x^0,y^0)\big)$$

$$m_{x^0}(x,y) = U(x^0,y^0)\big(\bar{U}(x|y) - \bar{U}(x|y^0)\big)$$

The terms needed to determine these functions are all included in the tree of Figure 22.5. Therefore, there is enough information in one tree to determine the tree in the reverse direction for the continuous case. In the appendix of this chapter, we illustrate how to deduce them graphically from the utility tree itself (just like reversing the probability tree only a few extra steps are needed) because of the complexity of having two boundary values.

Equation (12) expresses the conditional utility function $U(y|x)$ in terms of its boundary values $U(y|x^*)$ and $U(y|x^0)$ as well as some other functions, $d_{x^*}(x,y), d_{x^0}(x,y), m_{x^*}(x,y),$ and $m_{x^0}(x,y)$. We refer to the functions $d_{x_i}(x,y)$ as *interpolation* functions, and the functions $m_{x_i}(x,y)$ as the *independence* functions.

Note that the denominator in (12) is independent of Y and therefore it does not play a role in determining the shape of the conditional utility function $U(y|x)$. The denominator is simply a normalizing term that depends on the particular value of X at which it is conditioned. Therefore, we can calculate the numerator alone and then normalize it at any given value of X without having to calculate the denominator.

Define

$$w_{x^*}(x,y) = \frac{d_{x^*}(x,y)}{d_{x^*}(x,y^*) + d_{x^0}(x,y^*) + m_{x^*}(x,y^*) + m_{x^0}(x,y^*)},$$

$$w_{x^0}(x,y) = \frac{d_{x^0}(x,y)}{d_{x^*}(x,y^*) + d_{x^0}(x,y^*) + m_{x^*}(x,y^*) + m_{x^0}(x,y^*)},$$

$$\eta(x,y) = \frac{m_{x^*}(x,y^*) + m_{x^0}(x,y^*)}{d_{x^*}(x,y^*) + d_{x^0}(x,y^*) + m_{x^*}(x,y^*) + m_{x^0}(x,y^*)},$$

and substitute into (12) to get a simpler expression for the conditional utility function of the second attribute in the expansion as shown below.

Conditional Utility Function of the Second Attribute in the Expansion

$$\boxed{U(y|x) = w_{x^*}(x,y)U(y|x^*) + w_{x^0}(x,y)U(y|x^0) + \eta(x,y)}$$

Dialogue between Instructor (I) and Student (S)

S: Utility inference is a bit more complicated for the general case. It was easier for attribute dominance utility.

I: It is simple. The main idea is that we would like to update our utility for an attribute as the level of another attribute varies. If Y comes after X in the utility tree, how does its conditional utility function appear?

S: We have seen from the tree and the basic expansion theorem that $\underline{Y}$ is conditioned on x^* and x^0, i.e., $U(y|x^*)$ and $U(y|x^0)$.

I: Correct. Now with these terms alone, we cannot determine the conditional utility value of Y when x is anywhere between these boundary values. So what we need is $U(y|x)$ expressed as some function of these terms.

S: Or equivalently, we would like to change the order of the tree to start with attribute Y then X.

I: Precisely. Now the conversion from $U(y|x^*)$ and $U(y|x^0)$ into $U(y|x)$ requires some additional functions. It must require additional functions because $U(y|x)$ is a function of both y and x but $U(y|x^*)$ and $U(y|x^0)$ do not depend on x. Therefore, we need additional functions.

We have a scaling function for each conditional utility assessment at the boundary. We have $d_{x^*}(x,y)$ that is multiplied by $U(y|x^*)$ and we have $d_{x^0}(x,y)$ that is multiplied by $U(y|x^0)$.

S: Are these functions the same?

I: Not necessarily. They are different, and this is why they have different notation for each of them. Each conditional utility function at the boundary will require a multiplier function. These multiplier functions generally depend on all the attributes. We shall see in the next chapter the simplifications that arise when they depend on a subset of the attributes. Then we have additional functions that we add to each conditional utility assessments at the boundary. Therefore, we have $m_{x^*}(x,y)$ and $m_{x^0}(x,y)$.

S: So we calculate the term $d_{x^*}(x,y)U(y|x^*)+d_{x^0}(x,y)U(y|x^0)+m_{x^*}(x,y)+m_{x^0}(x,y)$?

I: Correct, and this determines the shape of the conditional utility function $U(y|x)$ from its boundary values.

S: But what about the term in the denominator of (14)?

I: Notice that it does not depend on Y. It is simply a normalizing term to make the conditional utility assessment $U(y|x)$ normalized from zero to one. If we plot the numerator alone and divide by the maximum value, this maximum value is in fact the denominator.

S: This makes a lot of sense now.

But those functions $d_{x^*}(x,y), d_{x^0}(x,y), m_{x^*}(x,y)$, and $m_{x^0}(x,y)$ are difficult to remember.

I: True. I do not expect you to memorize them. I just want you to understand their effects in determining the conditional utility function from its boundary values. But to make things simpler, there is also a graphical way of determining these functions from the utility tree. We explain this method in detail in the appendix of this chapter. Furthermore, note that if Y is a utility-dominant attribute, then both $m_{x^*}(x,y)$ and $m_{x^0}(x,y)$ are equal to zero because $U(x^*,y^0)$ and $U(x^0,y^0)$ are zero. This implies that the term $\eta(x,y)$, in the expression of the conditional utility function for $U(y|x)$, is equal to zero.

We consider the following questions in this and the next chapter.

1. Under what conditions do the functions $d_{x^*}(x,y)$ and $d_{x^0}(x,y)$ not depend on y?
2. Under what conditions do the functions $m_{x^*}(x,y)$ and $m_{x^0}(x,y)$ equal zero?

23.3.2 Utility Inference for a Two-Attribute Utility Function

---

EXAMPLE 22.2: Utility Inference

To illustrate an example of utility inference, and the use of the functions $d_{x_i}(x,y)$ and $m_{x_i}(x,y)$, we refer back to Example 22.1 of constructing a utility function using the terms in the utility tree.

By direct substitution, we have

$$d_{x^*}(x,y)=U(x|y)\big(U(x^*,y^*)-U(x^*,y^0)\big)=(1-k_x)U(x|y)=(1-k_x)\frac{1-e^{-xy^\eta}}{1-e^{-y^\eta}}$$

$$d_{x^0}(x,y)=\bar{U}(x|y)\big(U(x^0,y^*)-U(x^0,y^0)\big)=k_y\bar{U}(x|y)=k_y\frac{e^{-xy^\eta}-e^{-y^\eta}}{1-e^{-y^\eta}}$$

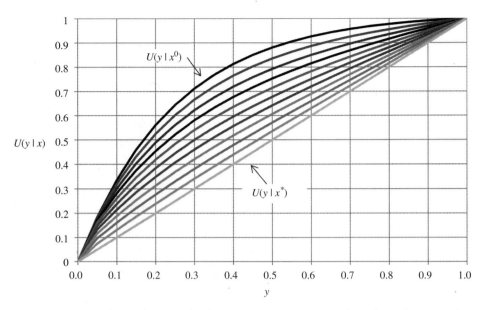

Figure 22.7. Expressing the conditional utility function $U(y \mid x)$ in terms of its boundary values using the functions $d_{x^*}(x,y)$, $d_{x^0}(x,y)$, $m_{x^*}(x,y)$ and $m_{x^0}(x,y)$.

$$m_{x^*}(x,y) = U(x^*,y^0)\big(U(x \mid y) - U(x \mid y^0)\big) = k_x\left(\frac{1 - e^{-xy^\eta}}{1 - e^{-y^\eta}} - x\right)$$

$$m_{x^0}(x,y) = U(x^0,y^0)\big(\bar{U}(x \mid y) - \bar{U}(x \mid y^0)\big) = 0.$$

Using these functions, we can express the conditional utility function $U(y \mid x)$ at any point in the domain in terms of its boundary values $U(y \mid x^*)$ and $U(y \mid x^0)$. Figure 22.7 shows $U(y \mid x)$ expressed in terms of $U(y \mid x^*)$ and $U(y \mid x^0)$ for the assessments in Example 22.2 using the functions $d_{x_i}(x,y)$ and $m_{x_i}(x,y)$ ranging from $x^0 = 0$ to $x^* = 1$ in increments of 0.1.

If the function $U(y \mid x)$ does not depend on X, then the curves in Figure 22.7 would coincide. We discuss this situation in more detail in the next chapter and refer to it as ***utility independence***. It is also clear that if there are some special cases where the functions $d_{x^*}(x,y)$, $d_{x^0}(x,y)$, $m_{x^*}(x,y)$, and $m_{x^0}(x,y)$ are simplified, then the utility inference calculation would be simplified. Below we present one such situation.

22.3.2 Special Case: Utility-Dominant Attributes

Corollary 22.1: *If Y is a utility-dominant attribute, then $m_{x^*}(x,y) = 0$ and $m_{x^0}(x,y) = 0$*

Corollary 22.1 shows that the functions $m_{x^*}(x,y)$ and $m_{x^0}(x,y)$ are directly related to Y being utility dominant. This observation simplifies the calculation of the conditional utility function $U(y \mid x)$ in this case, since it is now only a weighted average of the conditional utility functions at the boundaries without the additional terms $m_{x_i}(x, y)$.

If Y is a utility-dominant attribute, then

$$U(y \mid x) = \frac{d_{x^*}(x, y)U(y \mid x^*) + d_{x^0}(x, y)U(y \mid x^0)}{d_{x^*}(x, y) + d_{x^0}(x, y)}$$
$$= w_{x^*}(x, y)U(y \mid x^*) + (1 - w_{x^*}(x, y))U(y \mid x^0), \quad (13)$$

where $w_{x^*}(x, y) = \dfrac{d_{x^*}(x, y)}{d_{x^*}(x, y) + d_{x^0}(x, y)}$.

This condition expresses the conditional utility function $U(y \mid x)$ in terms of its boundary values using weighting functions $w_{x^*}(x, y)$ that depend on both attributes. We shall refer beck to this formulation and compare it to the notion of interpolation independence in Chapter 23.

Corollary 22.2: *If both X and Y are utility dominant attributes, then*

$$U(x, y) = U(x \mid y)U(x^*, y) = U(y \mid x)U(x, y^*).$$

Rearranging gives

$$U(y \mid x) = \frac{U(x \mid y)}{U(x, y^*)}U(x^*, y), \quad U(x, y^*) \neq 0.$$

The conditional utility function $U(y \mid x)$ is now expressed in terms of its single boundary value $U(x^*, y)$ and a scaling function of both attributes. In the next chapter, we shall discuss another case where the conditional utility function is simplified.

22.4 GENERAL EXPRESSION AND NOTATION FOR UTILITY INFERENCE

The following theorem writes (12) in a more general form that simplifies its extension to multiple attributes.

Theorem 22.1: Reversing the Two-Attribute Utility Tree

$$U(y \mid x) = \frac{\displaystyle\sum_{x_i \in \{x^*, x^0\}} [d_{x_i}(x, y)U(y \mid x_i) + m_{x_i}(x, y)]}{\displaystyle\sum_{x_i \in \{x^*, x^0\}} [d_{x_i}(x, y^*) + m_{x_i}(x, y^*)]}, \quad (14)$$

where $x_i \in \{x^*, x^0\}$,

$$d_{x_i}(x,y) = g_{x_i}(x\,|\,y)\big(U(x_i,y^*) - U(x_i,y^0)\big) \text{ and}$$
$$m_{x_i}(x,y) = U(x_i,y^0)\big(g_{x_i}(x\,|\,y) - g_{x_i}(x\,|\,y^0)\big).$$

22.5 REVERSING THE ORDER OF THE ATTRIBUTES IN THE CONDITIONING STATEMENTS

Utility inference expressed a conditional utility function, say $U(y\,|\,x)$, in terms of its boundary values, $U(y\,|\,x^0)$ and $U(y\,|\,x^*)$, and some other functions, $m(x,y)$ and $d(x,y)$. We now consider the problem of reversing the order of the assessments to relate $U(y\,|\,x)$ and $U(x\,|\,y)$.

Consider a decision involving attributes X and Y. As we have seen, we can express a two-attribute utility function by expanding around attribute X to get

$$U(x,y) = U(x^*,y,)U(x\,|\,y) + U(x^0,y)\bar{U}(x\,|\,y). \tag{15}$$

We can also change the order of the conditional assessments in a way that is more comfortable to the decision-maker. Suppose we expand the utility function around another attribute, Y. We get

$$U(x,y) = U(x,y^*)U(y\,|\,x) + U(x,y^0)\bar{U}(y\,|\,x). \tag{16}$$

Consistency requires that the right-hand sides of (15) and (16) be equal, and so

$$\boxed{U(y\,|\,x) = U(x\,|\,y)\frac{[U(x^*,y) - U(x^0,y)]}{[U(x,y^*) - U(x,y^0)]} + \frac{[U(x^0,y) - U(x,y^0)]}{[U(x,y^*) - U(x,y^0)]}.} \tag{17}$$

Equation (17) provides a general expression that changes the order of assessments and also provides a consistency check for the conditional utility functions obtained with different assessment orders. Note that if both attributes are utility dominant, then $U(x^0,y)$ and $U(x,y^0)$ are zero, and so

$$U(y\,|\,x) = U(x\,|\,y)\frac{U(x^*,y)}{U(x,y^*)}. \tag{18}$$

22.6 SUMMARY

- The normalized conditional utility function for attribute $g_i(g_j(x)) = g_j(g_i(x))$, $\forall i, j$. given Y is defined as

$$U(x\,|\,y) = \frac{U(x,y) - U(x^0,y)}{U(x^*,y) - U(x^0,y)}. \tag{19}$$

- The normalized conditional disutility function is $\bar{U}(x\,|\,y) = 1 - U(x\,|\,y)$.
- The basic expansion theorem for two-attribute utility functions

- ◦ Expansion around a single attribute $U(x,y) = U(x \mid y)U(x^*, y) + \bar{U}(x \mid y)U(x^0, y)$.
- The two-attribute utility tree provides a graphical representation of the expansion.
- Utility inference: We can express the conditional utility function in terms of its boundary values as

$$U(y \mid x) = w_{x^*}(x,y)U(y \mid x^*) + w_{x^0}(x,y)U(y \mid x^0) + \eta(x,y).$$

- There is a general expression for reversing the order of conditioning in the utility assessments.
- Special cases of the expansion theorem occur when attribute dominance conditions exist, and the expansion is a product instead of a sum of products.

ADDITIONAL READINGS

Abbas, A. E. 2005. Bidirectional Utility Diagrams. Technical Report. University of Illinois at Urbana-Champaign, UILU-ENG-2005-3001.

Abbas, A. E. 2009. From Bayes' nets to utility nets. Proceedings of the 29th *International Workshop on Bayesian Inference and Maximum Entropy Methods in Science and Engineering*. Oxford, MI, July 5–10. AIP conference proceedings 1193, pp. 3–12.

Abbas, A. E. 2011a. General decompositions of multiattribute utility functions. *Journal of Multicriteria Decision Analysis* 17(1, 2): 37–59.

Abbas, A. E. 2011b. The multiattribute utility tree. *Decision Analysis* 8(3): 180–205.

APPENDIX: CALCULATING THE FUNCTIONS $d_{x_i}(x,y)$ AND $m_{x_i}(x,y)$ GRAPHICALLY

We now show how to calculate the functions $d_{x_i}(x,y)$ and $m_{x_i}(x,y)$ graphically for attribute Y using the utility tree. In particular, we will need $d_{x^*}(x,y), d_{x^0}(x,y), m_{x^*}(x,y)$, and $m_{x^0}(x,y)$.

We are interested in the conditional utility function $U(y \mid x)$ from the tree, which includes $U(x \mid y)$ and $U(y \mid x^*)$ and $U(y \mid x^0)$.

Because we are interested in $U(y \mid x)$, we look for paths that contain $U(y \mid x^*)$ and $U(y \mid x^0)$. The paths that contain $U(y \mid x^*)$ will be used in the calculation of $d_{x^*}(x,y)$ and those that contain $\bar{U}(y \mid x^*)$ will be used in the calculation of $m_{x^*}(x,y)$. Likewise, we look for paths containing $U(y \mid x^0)$ and $\bar{U}(y \mid x^0)$ in the calculation of $d_{x^0}(x,y)$ and $m_{x^0}(x,y)$.

Calculating $d_{x^*}(x,y)$

Along any path containing $U(y \mid x^*)$, multiply all the assessments before $U(y \mid x^*)$ by each other. In this case, it is only one term $U(x \mid y)$. Call the product $b_{x^*}(x,y)$. Further multiply all the assessments after $U(y \mid x^*)$ by each other and call the product $a_{x^*}(x^*, y^*)$. In this case, there is only one term $U(x^*, y^*)$. Note that all the assessments after $U(y \mid x^*)$ will have a term y^*.

Then,

$$d_{x^*}(x,y) = b_{x^*}(x,y)[a_{x^*}(x^*, y^*) - a_{x^*}(x^*, y^0)] = U(x \mid y)[U(x^*, y^*) - U(x^*, y^0)].$$

For notational convenience, we put a dashed oval around the terms $b_{x^*}(x, y)$ and a rectangle around the terms in $a_{x^*}(x^*, y^*)$. Terms in the dashed rectangle will have a term subtracted, and the difference is multiplied by the product of the terms in the dashed oval.

Calculating $d_{x^0}(x, y)$

Using the same procedure, we look for the paths that have $U(y \mid x^0)$. Multiply all the assessments before $U(y \mid x^0)$ by each other. In this case, it is only one term $\bar{U}(x \mid y)$. Call the product $b_{x^0}(x, y)$. Further multiply all the assessments after $U(y \mid x^0)$ by each other and call the product $a_{x^0}(x^0, y^*)$. In this case, there is only one term $U(x^0, y^*)$. Note that all the assessments after $U(y \mid x^0)$ will have a term y^*.
   Then,

$$d_{x^0}(x, y) = b_{x^0}(x, y)[a_{x^0}(x^0, y^*) - a_{x^0}(x^0, y^0)] = \bar{U}(x \mid y)[U(x^0, y^*) - U(x^0, y^0)].$$

For notational convenience, we put a dashed oval around the terms $b_{x^0}(x, y)$ and a rectangle around the terms in $a_{x^0}(x^0, y^*)$. Terms in the dashed rectangle will have a term subtracted, and the difference is multiplied by the product of the terms in the dashed oval.

Calculating $m_{x_i}(x, y)$

For simplicity, we explain the procedure using the notation $m_{x_i}(x, y)$ to denote either $m_{x^*}(x, y)$ or $m_{x^0}(x, y)$.
   Along any path containing $\bar{U}(y \mid x_i)$, multiply all the assessments before $\bar{U}(y \mid x_i)$ by each other and call the product $b_{x_i}(x, y)$. Further multiply the assessments after $\bar{U}(y \mid x_i)$ by each other and call the product $a_{x_i}(x, y^0)$. Note that all the assessments after $\bar{U}(y \mid x_i)$ will have a term y^0. Then

$$m_{x_i}(x, y) = a_{x_i}(x_i, y^0)[b_{x_i}(x, y) - b_{x_i}(x, y^0)], \quad \forall x_i \in \{x^0, x^*\}.$$

Figure 22.8(a) illustrates this operation graphically using the following notation: when a dashed oval surrounds a utility assessment in the tree, then this assessment is multiplied by other assessments surrounded by symbols on its same path. When a solid rectangle surrounds an assessment,

(i)   multiply all assessments in the rectangle,
(ii) replace the value of Y in the product by its minimum value,
(iii) subtract the product in step (ii) from the product of (i), and then
(iv) multiply the difference obtained in (iii) with other surround symbols on the same path.

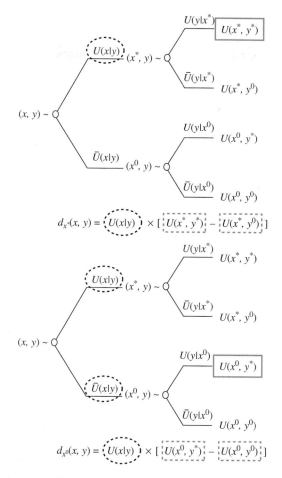

Figure 22.8(a). Determining the functions $d_{x_i}(x,y)$ graphically.

Note how the positions of the ovals and rectangles are interchanged when calculating the functions $d_{x_i}(x,y)$ and $m_{x_i}(x,y)$ in both Figures 22.8(a) and 22.8(b). Moreover, since X is the complement of Y, we can also denote the functions $d_{x_i}(x,y)$ and $m_{x_i}(x,y)$ using the subscript $\bar{y}_i$ as $d_{\bar{y}_i}(x,y)$ and $m_{\bar{y}_i}(x,y)$ to emphasize that they are calculated at boundary values of the complement of Y, and are used in calculating $U(y\,|\,x)$. While the first notation is easier in this two-attribute case, the complement notation extends in a simple way to multiple attributes and it also emphasizes that these functions are calculated for reversing the conditional utility function of attribute Y itself.

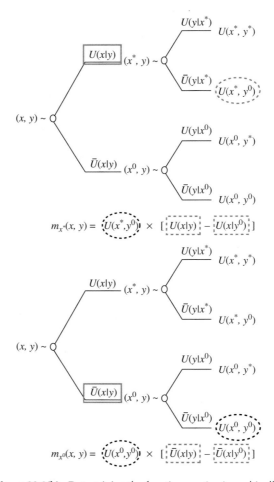

Figure 22.8(b). Determining the functions $m_{x_i}(x, y)$ graphically.

CHAPTER 23

Basic Independence Conditions for
Two Attributes

Chapter Concepts

- Three conditions that simplify the functional form of the multiattribute utility function:
  - Utility independence (Keeney and Raiffa 1976)
  - Boundary independence (Abbas 2011)
  - Interpolation independence (Bell 1979)
- Relating the concept of utility independence to the invariant transformation formulation
- Drawing connections between the implications of utility independence, boundary independence, and interpolation independence
- Graphical representations of utility trees and utility independence relations

23.1 INTRODUCTION

As we have seen in Chapter 22, the basic expansion theorem for constructing multiattribute utility functions requires a sum of products of conditional utility assessments of the attributes. Some of the conditional utility assessments are conducted at specific reference values of the complement attribute, while others require an assessment on the entire domain of the complement attribute.

To simplify the assessments needed in the basic expansion theorem, this chapter introduces three concepts: **utility independence**, **boundary independence**, and **interpolation independence**. Each of these concepts has different implications and interpretations, but the concepts can be briefly described in terms of the normalized conditional utility functions as follows:

Three Types of Independence Conditions

Utility independence: $U(x \mid y) = U(x \mid y^0)$
Boundary independence: $U(x \mid y^*) = U(x \mid y^0)$
Interpolation independence: $U(y \mid x) = w_1(x)U(y \mid x^*) + w_2(x)U(y \mid x^0)$,
$\quad w_1(x) + w_2(x) = 1.$

Utility independence of X on Y implies that the normalized conditional utility function for attribute X does not depend on the level of Y. This condition simplifies the assessment of $U(x \mid y)$ significantly if we wish to construct the utility surface, because we need only one conditional utility assessment for X at any value of Y say y^0.

Boundary independence of X on Y implies that the normalized conditional utility functions for X, when assessed at the two boundary values of Y, are equivalent. Therefore, we need to assess the conditional utility at only one boundary value when constructing the utility function using the basic expansion theorem.

Interpolation independence of Y on X implies that the conditional utility function of Y at any given value of X is an interpolation of the conditional utility functions of Y at the boundary values of X using some interpolation function of X. Although somewhat more complicated than utility independence, this formulation also reduces the assessments needed to construct the function $U(y \mid x)$.

While these three independence concepts were derived independently in the literature, this chapter will show how they are in fact closely related through the notion of utility inference and the utility tree.

This chapter also shows that the concept of utility independence is in fact closely related to the concept of the delta property and the more general invariant transformation that we discussed in Chapters 13 and 16.

Furthermore, as we shall see, utility independence, while charming and simplifying the analysis significantly, is in fact a strong condition. Boundary independence is a weaker condition than utility independence. Interpolation independence, as we shall see, is also weaker than utility independence.

23.2 UTILITY INDEPENDENCE (KEENEY AND RAIFFA 1976)

This section will discuss several equivalent definitions of utility independence and its implications.

23.2.1 Utility Independence Using the Normalized Conditional Utility Function

Deriving the Functional Form of Utility Independence from the Utility Tree

To start our formulation of utility independence, we make the following definition.

> **Definition**
> Attribute X is **_utility independent_** of attribute Y, written X UI Y, if the normalized conditional utility function of X, $U(x \mid y)$ does not depend on the value of Y, and so we can write
>
> $$U(x \mid y) = U(x \mid y^0), \ \forall x, y.$$

This definition implies that if utility independence holds, then we can replace a two-attribute conditional assessment $U(x \mid y)$ with a boundary assessment $U(x \mid y^0)$; we need only a univariate assessment of attribute X at any level of attribute Y is needed to construct $U(x\,|\,y)$ on the whole domain of Y. This condition was identified in Keeney and Raiffa (1976).

EXAMPLE: Using the Utility Tree to Determine the Functional Form of Utility Independence

---

The functional form of the utility function that corresponds to utility independence can be obtained by simply substituting for $U(x\,|\,y) = U(x\,|\,y^0)$ as shown in Figure 23.1 and rolling back the tree to get

$$U(x,y) = U(x\,|\,y^0)U(y\,|\,x^*).1 + U(x\,|\,y^0)\bar{U}(y\,|\,x^*)U(x^*,y^0)$$
$$+ \bar{U}(x\,|\,y^0)U(y\,|\,x^0)U(x^0,y^*) + 0. \tag{1}$$

Note that $k_x = U(x^*,y^0)$ and $k_y = U(x^0,y^*)$. Rearranging (1) and observing that $\bar{U} = 1 - U$ gives

$$U(x,y) = U(x\,|\,y^0)\big[U(y\,|\,x^*) + [1 - U(y\,|\,x^*)]k_x\big] + \big(1 - U(x\,|\,y^0)\big)U(y\,|\,x^0)k_y$$
$$= U(x\,|\,y^0)\big[k_x + (1-k_x)U(y\,|\,x^*) - U(y\,|\,x^0)k_y\big] + U(y\,|\,x^0)k_y$$

Simplifying and noting that $U(x^*,y) = k_x + (1-k_x)U(y\,|\,x^*)$ and $U(x^0,y) = k_yU(y\,|\,x^0)$ gives

$$\boxed{U(x,y) = U(x\,|\,y^0)\big[U(x^*,y) - U(x^0,y)\big] + U(x^0,y).} \tag{2}$$

23.2.2 Utility Independence in Terms of a Linear Relation between Any Two Curves on the Surface

Definition
If attribute X is utility independent of attribute Y, then any two curves, $U(x,y_1)$ and $U(x,y_2)$, on the surface of the utility function are linearly related when plotted versus X.

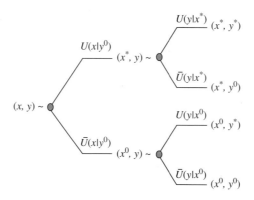

Figure 23.1. Utility tree exhibiting the relation X utility independent of Y.

To see why this is true, define

$$k(y) = U(x^*, y) - U(x^0, y), \quad d(y) = U(x^0, y).$$

Substituting into (2), the relation "X is utility independent of Y" implies that

$$U(x, y) = k(y)U(x \mid y^0) + d(y). \quad (3)$$

By substitution,

$$U(x, y_1) = k(y_1)U(x \mid y^0) + d(y_1), \qquad U(x, y_2) = k(y_2)U(x \mid y^0) + d(y_2).$$

Therefore, with $d(y) \uparrow 0$, we can write:

$$U(x, y_1) = \left(\frac{k(y_1)}{d(y_2)} \right) U(x, y_2) - \left(\frac{k(y_1)k(y_2)}{d(y_2)} - d(y_1) \right).$$

The two curves, $U(x, y_1)$ and $U(x, y_2)$, are linearly related.

23.2.3 Utility Independence in Terms of Preferences for Lotteries

Definition
If attribute X is utility independent of attribute Y, then preferences for lotteries over X do not depend on the level of Y.

The definition of utility independence implies that if attribute X is utility independent of attribute Y, and if we have prospects characterized by two attributes, say (x, y), then if we face any two lotteries over X that have a fixed value of Y, say y_0, i.e., if the lotteries can be expressed as

$$\text{Lottery A:} < p_1, (x_{A_1}, y_0); \ldots; p_n, (x_{A_n}, y_0) >$$

and

$$\text{Lottery B:} < q_1, (x_{B_1}, y_0); \ldots; q_n, (x_{B_m}, y_0) >,$$

then preferences for these lotteries will not change as the level of y_0 varies. That is, *if we prefer lottery A to lottery B at y_0, then we prefer lottery A to lottery B at any value of Y.* Note that the lotteries do not need to have the same number of outcomes and that the level of Y is fixed for all prospects. If this is true for all lotteries, as we change the value of Y, then attribute X is utility independent of attribute Y.

To see why this is true, note that if the utility function is strictly increasing with each argument,

$$U(x^*, y) > U(x^0, y) \text{ and } k(y) > 0.$$

Given two lotteries X_1 and X_2, if we prefer X_1 to X_2, at a given y, then the difference in expected utility of the two lotteries over X is

$$E_{x_1}[U(x,y)] - E_{x_2}[U(x,y)] = E_{x_1}[k(y)U(x\,|\,y^0) + d(y)] - E_{x_2}[k(y)U(x\,|\,y^0) + d(y)]$$
$$= k(y)\{E_{x_1}[U(x\,|\,y^0)] - E_{x_2}[U(x\,|\,y^0)]\},$$

where $E_{x_1}[U(x\,|\,y^0)]$ is the expected utility of the first lottery and $E_{x_2}[U(x\,|\,y^0)]$ is the expected utility of the second.

Note that the term $\{E_{x_1}[U(x\,|\,y^0)] - E_{x_2}[U(x\,|\,y^0)]\}$ on the right-hand side does not depend on Y. If $k(y) > 0$, as is the case in our formulation, then the difference in expected utility will not change sign with Y. Therefore, preferences for lotteries over X will not change with Y.

Two Connections with Utility Independence and Some of Our Previous Discussions

Connections with Attribute Dominance Utility Functions

If X is a utility independent of Y and if X is a utility-dominant attribute, then $d(y) = U(x^0, y) = 0$ and $k(y) = U(x^*, y)$. In this case utility independence would imply a product expansion of the utility function. We have seen this earlier form of utility independence (which was a special case) with attribute dominance utility functions that resulted in a product form of the individual utility functions.

$$U(x, y) = U_1(x)U_2(y)$$

Because the product operation is a symmetric operation, it is not surprising to note that utility independence is a symmetric property for attribute dominance utility functions. When the attribute dominance conditions do not exist, the functional form for utility independence involves some additional terms even for the case of two attributes.

As we have seen, the general assertion that X is utility independent of Y will include an additional function of Y and a decomposition of the form

$$U(x, y) = k(y)U_1(x) + d(y).$$

Therefore, the functional form of the utility function is no longer a simple product of two functions as was the case for attribute dominance utility functions, but rather a product $U_1(x)k(y)$ and an additional term $d(y)$. It is this additional term $d(y)$ that allows utility independence not to be symmetric when the attribute dominance conditions do not exist.

Connections with the Invariant Transformation

The second observation we need to make is the connection between utility independence and the delta property (Chapter 13) and the more general invariant transformation (Chapter 16) where preferences for lotteries over an attribute do not change as we vary the levels of some of the parameters using a monotone transformation. For the delta property, the parameter was a shift parameter, where the delta property resulted in a functional equation of the form

$$U(x + y) = k(y)U(x) + d(y).$$

This formulation implied that preferences for lotteries over X do not change as we add a shift amount, y, to each outcome: i.e., preferences for lotteries over X were utility independent of the shift amount y. For the more general invariant transformation, we discussed formulations of the form

$$U(g(x,y)) = k(y)U(x) + d(y),$$

where y is a parameter. With $k(y) > 0$, this formulation guaranteed that preferences for lotteries over X do not change as we change the parameter Y; an assertion that X is utility independent of the parameter Y. Utility independence of one attribute from another is the exact same formulation when we treat those parameters as attributes and it should not come as a surprise that this would imply that

$$U(x,y) = k(y)U_1(x) + d(y).$$

It is the same formulation, only Y is treated as an attribute. If preferences for lotteries over X do not change, then changing Y can lead to only a linear transformation on the utility function.

Dialogue between Instructor (I) and Student (S)

I: Is everyone clear on the notion of utility independence using both definitions?

S: Yes, it means that the normalized assessment $U(x \mid y)$ does not depend on Y.

I: Correct, and therefore we need to assess it at any reference value of Y and we are done with the conditional utility assessments for attribute X. This complicated assessment $U(x \mid y)$ becomes a univariate assessment. Without utility independence, we would normally require a normalized utility assessment of X for all possible values of Y.

S: But isn't this a strong assertion? We need to assert that the assessment $U(x \mid y)$ is identical at all possible values of Y to be able to assert utility independence.

I; Correct. And you might also assert that the assessment simply does not depend on Y.

S: The second definition you mentioned is that utility independence implies preferences for lotteries over X do not depend on the value of Y.

I: Correct. As we have seen, utility independence implies that any two curves $U(x, y_1)$ and $U(x, y_2)$ on the surface of the utility function are linearly related when plotted versus X., i.e., if $U(x \mid y) = U(x \mid y^0)$, we have seen that this implies that the utility function can be expressed as $U(x,y) = k(y)U(x \mid y^0) + d(y)$. Furthermore, if the utility function is strictly increasing with each attribute, then we know for sure that $U(x^*, y) > U(x^0, y)$, and so $k(y) > 0$. As we know from Chapter 13, if the utility function changes by a linear transformation, then the valuation of lotteries will not change; they will yield the same certain equivalents for lotteries. Therefore, preferences for lotteries will not change.

S: Oh yes, this is clear now. But this might be making a strong assumption. What if X is investments and Y is age; wouldn't our preference for investments change with age?

I: It certainly could. In this case you should not use utility independence assertions. The problem is not with utility independence; the problem is that people often rush to apply utility independence in decision situations without appropriate verification just because of its simplicity. If X is investments and Y is age, it might indeed be that your preferences for investments you make will depend on age or health state. This is why we will spend a lot of time constructing utility functions that do not make these assumptions so we can use them for more general situations.

23.2.4 Utility Independence and Attribute Risk Aversion

Refer back to our definition of attribute risk aversion in a multiattribute decision in Chapter 19, repeated here for convenience.

The risk aversion function for attribute X is $\gamma_x(x,y) = -\dfrac{\dfrac{\partial^2}{\partial x^2}U(x,y)}{\dfrac{\partial}{\partial x}U(x,y)}$.

If X is utility independent of Y, then $U(x,y) = k(y)U(x\,|\,y^0) + d(y)$. This implies that

$$\frac{\partial}{\partial x}U(x,y) = k(y)U'(x\,|\,y^0) \text{ and } \frac{\partial^2}{\partial x^2}U(x,y) = k(y)U''(x\,|\,y^0).$$

Substituting gives

$$\gamma_x(x,y) = \gamma_x(x,y_0).$$

Therefore, the risk aversion for attribute X does not depend on the particular value of Y.

23.2.5 Summary of the Implications of Utility Independence

Let us recap the implications of utility independence of X on Y that we have studied so far.

Attribute X is utility independent of Attribute Y implies that

1. The normalized conditional utility function of X, $U(x\,|\,y)$, does not change with y, i.e., $U(x\,|\,y) = U(x\,|\,y^0)$.
2. The utility function can be expressed as an affine transformation of any normalized conditional assessment of X, say $U(x\,|\,y^0)$, i.e., $U(x,y) = k(y)U(x\,|\,y^0) + d(y)$.
3. Preferences for lotteries over X do not change as we change Y.
4. We need one utility assessment of X at an arbitrary value of Y and two normalized conditional utility assessments of Y (at the minimum and maximum values of X) plus two corner values to construct the utility function because

$$U(x,y) = U(x\,|\,y^0)\left[U(x^*,y) - U(x^0,y)\right] + U(x^0,y)$$
$$= U(x\,|\,y^0)\left[k_x + (1-k_x)U(y\,|\,x^*) - U(y\,|\,x^0)k_y\right] + U(y\,|\,x^0)k_y.$$

5. The risk aversion function of attribute X does not depend on Y.

23.2.6 Utility Independence Is Not Necessarily Symmetric

Consider the following questions:

If X is utility independent of Y, is it necessary that Y is utility independent of X? If not, under what conditions is this true? And what does it imply about the relation between Y and X?

Recall that for attribute dominance utility functions, utility independence is indeed a symmetric property: If A and B are utility-dominant attributes, and if attribute A is utility independent of B, then B is also utility independent of A.

This symmetry need not apply for the case where the attribute dominance conditions do not hold. To see why this is the case, let us first reverse the tree of Figure 23.1 and then substitute for X utility independent of Y to derive the general relation between Y and X.

The general expression for $U(y\,|\,x)$ is, as we have seen in Chapter 22, Theorem 22.1.

$$U(y\,|\,x) = \frac{\displaystyle\sum_{x_i \in \{x^*,x^0\}} [d_{x_i}(x,y)U(y\,|\,x_i) + m_{x_i}(x,y)]}{\displaystyle\sum_{x_i \in \{x^*,x^0\}} [d_{x_i}(x,y^*) + m_{x_i}(x,y^*)]}, \tag{4}$$

where $x_i \in \{x^*,x^0\}$,

$$d_{x_i}(x,y) = g_{x_i}(x\,|\,y)\big(U(x_i,y^*) - U(x_i,y^0)\big) \text{ and}$$
$$m_{x_i}(x,y) = U(x_i,y^0)\big(g_{x_i}(x\,|\,y) - g_{x_i}(x\,|\,y^0)\big).$$

It is clear that for Y to be utility independent of X, the conditional utility function $U(y\,|\,x)$ cannot depend on X. This is generally not the case, but if in Equation (4),

(i) $m_{x_i}(x,y) = 0$,
(ii) the functions $d_{x_i}(x,y)$ do not depend on Y, and
(iii) the terms $U(y\,|\,x_i)$, $x_i \in \{x^*,x^0\}$ are equal,

then the equation would reduce to $U(y\,|\,x) = U(y\,|\,x^0)$.

Now we answer the two questions of the previous chapter. We repeat them for convenience below.

Consider the Following

a) Under what conditions do the functions $d_{x_i}(x,y)$ not depend on y?
b) Under what conditions are the functions $m_{x_i}(x,y)$ equal to zero?

We have already seen that if Y is a utility-dominant attribute, then $m_{x_i}(x,y) = 0$, $x_i \in \{x^*,x^0\}$ because $U(x_i,y^0) = 0$. But there are other conditions for which $m_{x_i}(x,y) = 0$. For example, if X is utility independent of Y, then $m_{x_i}(x,y) = 0$ because, as we have seen, the normalized conditional utility (and conditional disutility) functions are equal, and so $g_{x_i}(x\,|\,y) = g_{x_i}(x\,|\,y^0)$. Furthermore, if X is utility independent of Y, then the functions $d_{x^*}(x,y), d_{x^0}(x,y)$, do not depend on y because $g_{x_i}(x\,|\,y)$ does not depend on y. To summarize:

Proposition 23.1: If X is utility independent of Y, then

(i) The functions $d_{x^*}(x,y), d_{x^0}(x,y)$, do not depend on Y: $d_{x^*}(x,y) = d_{x^*}(x)$, $d_{x^0}(x,y) = d_{x^0}(x)$, and
(ii) The functions $m_{x^*}(x,y) = m_{x^0}(x,y) = 0$.

We still need a third condition for Y to be utility independent of X, but let us derive the functional form of $U(y \mid x)$ with these two conditions alone. Substituting into Theorem 22.1 of Chapter 22 for reversing the two-attribute tree gives the following proposition.

> **Proposition 23.2:** *If X is utility independent of Y, then*
>
> $$U(y \mid x) = \frac{d_{x^*}(x)U(y \mid x^*) + d_{x^0}(x)U(y \mid x^0)}{d_{x^*}(x) + d_{x^0}(x)}$$
> $$= w_{x^*}(x)U(y \mid x^*) + (1 - w_{x^*}(x))U(y \mid x^0),$$
>
> *where* $w_{x^*}(x) = \dfrac{d_{x^*}(x)}{d_{x^*}(x) + d_{x^0}(x)}.$

Proposition 23.2 derives the conditional utility function of Y when X is utility independent of Y. Note that, in general, it may depend on X. Therefore, it need not be the case that Y is utility independent of X. Note further that if, in addition, $U(y \mid x^*) = U(y \mid x^0)$, then indeed $U(y \mid x) = U(y \mid x^0)$.

23.3 BOUNDARY INDEPENDENCE (ABBAS 2011)

Boundary independence is a weaker condition than utility independence and it was proposed in Abbas (2011). Boundary independence requires only that the normalized conditional utility function of an attribute be equal at the boundary values of some other attribute. It is therefore easier to test for than utility independence, which requires equality on the whole domain. Here, you just need to assert equality at two boundary values.

> **Definition:** *Attribute Y is **boundary independent** of attribute X, written Y BI X, if*
>
> $$U(y \mid x^0) = U(y \mid x^*), \ \forall y. \tag{5}$$

Contrast this definition with the definition of utility independence where

$$U(y \mid x) = U(y \mid x^0), \ \forall x \in \left[x^0, x^*\right].$$

and not just equality at the boundary values. Figure 23.2 shows a utility tree asserting that Y is boundary independent of X.

> **Note:** If Y is utility independent of X, then it is boundary independent of X, but the converse is not necessarily true.

The distinction between boundary independence and utility independence will be useful in many of the derivations in the next sections. As we shall see, many theorems

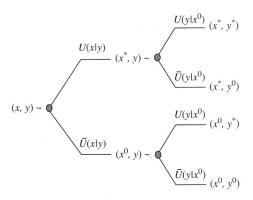

Figure 23.2. Utility tree asserting that Y is boundary independent of X.

will require only boundary independence on a subset of the attributes and utility independence on others to simplify the functional form. This observation will enable us to test the applicability of many conditions by verifying equality of the conditional utility functions at the boundary values instead of equality on the whole domain.

EXAMPLE 23.1: The Functional Form of Boundary Independence

Using $U(y\,|\,x^0) = U(y\,|\,x^*)$ in the tree of Figure 23.2, we can derive the functional form of the utility function implying boundary independence of Y on X as

$$U(x,y) = U(x\,|\,y)U(y\,|\,x^0).1 + U(x\,|\,y)\bar{U}(y\,|\,x^0)k_x + \bar{U}(x\,|\,y)U(y\,|\,x^0)k_y + 0$$
$$= U(x\,|\,y)\left\{U(y\,|\,x^0)[1 - k_x - k_y] + k_x\right\} + k_y U(y\,|\,x^0), \qquad (6)$$

where $k_x = U(x^*, y^0)$ and $k_y = U(x^0, y^*)$.

Equation (6) provides a simple illustration of the use of utility trees in deriving the functional form of a utility function. Boundary independence reduces the assessments needed to $U(x\,|\,y), U(y\,|\,x^0)$, plus the utility values of the corner assessments.

Note that Y being boundary independent of X does not reduce the assessments needed into univariate assessments, and we still need to assess the term $U(x\,|\,y)$, which is a two-dimensional function. As we shall see, however, boundary independence will play an important role in verification utility independence conditions.

Dialogue between Instructor (I) and Student (S)
I: Are we clear on the definition of boundary independence of Y on X?
S: Yes, it is a weaker condition than utility independence. It just requires equality of normalized utility assessments of Y at only two values of X instead of on the whole domain of X.
I: Correct, it requires that $U(y\,|\,x^0) = U(y\,|\,x^*)$
S: But why is this important?
I: For several reasons that will soon become apparent. For now, I want you to realize that testing for boundary independence at the two boundaries x^0, x^* is a lot simpler than testing for utility independence of Y on the whole domain of X.

23.4 INTERPOLATION INDEPENDENCE (BELL 1979)

Definition: Attribute Y is ***interpolation independent*** of attribute X if

$$U(y\mid x) = w_1(x)U(y\mid x^*) + w_2(x)U(y\mid x^0), \quad w_1(x) + w_2(x) = 1. \tag{7}$$

This is the same expression for the conditional utility function of Y given X that we got earlier in the chapter when we assumed that X is utility independent of Y. The relationship between utility and boundary independence will become clear shortly when we reverse the tree.

Using the definition of interpolation independence, the conditional utility function $U(y\mid x)$ is an interpolation of its boundary values, $U(y\mid x^*)$ and $U(y\mid x^0)$ through a univariate function, $w_1(x)$. This condition decomposes the conditional utility function $U(y\mid x)$ into univariate assessments.

Compare this expression to the general expression,

$$U(y\mid x) = w_{x^*}(x, y)U(y\mid x^*) + w_{x^0}(x, y)U(y\mid x^0) + \eta(x, y)$$

or the detailed expression (4), which we repeat below for convenience

$$U(y\mid x) = \frac{\displaystyle\sum_{x_i \in \{x^*, x^0\}} [d_{x_i}(x, y)U(y\mid x_i) + m_{x_i}(x, y)]}{\displaystyle\sum_{x_i \in \{x^*, x^0\}} [d_{x_i}(x, y^*) + m_{x_i}(x, y^*)]}.$$

Here we have weighting functions $d_{x_i}(x, y)$ that depend on both attributes and not just one. Furthermore, we have some additional functions $m_{x_i}(x, y)$. Interpolation independence provides a further simplification in that the weighting functions d_{x_i} do not depend on both attributes and the there are no additional functions m_{x_i}.

Note: If Y is interpolation independent of X and if Y is also boundary independent of X, then, Y is utility independent of X.

To see this, note that if we set $U(x\mid y^*) = U(x\mid y^0)$ into (7), we get by direct substitution

$$U(x\mid y) = U(x\mid y^0)[w_1(y) + w_2(y)] = U(x\mid y^0).$$

23.5 RELATING UTILITY, BOUNDARY, AND INTERPOLATION INDEPENDENCE

Utility independence made an assumption about the first term in the tree that $U(x\mid y) = U(x\mid y^0)$.

Boundary independence made an assumption about the second terms that $U(y\mid x^0) = U(y\mid x^*)$.

Interpolation independence made an assumption that appears in the reverse order, where X is the second attribute, because

$$U(x \mid y) = w_1(y)U(x \mid y^*) + w_2(y)U(x \mid y^0), \quad w_1(y) + w_2(y) = 1.$$

It is natural to ask how these conditions relate. We now present another implication of utility independence.

> **Proposition 23.3 (Abbas 2011):** *If X is utility independent of Y, then Y is interpolation independent of X.*

To see this, note that we have shown that if X is utility independent of Y, then

$$d_{x^*}(x,y) = d_{x^*}(x), d_{x^o}(x,y) = d_{x^o}(x), \quad m_{x^*}(x,y) = m_{x^0}(x,y) = 0,$$

and so

$$
\begin{aligned}
U(y \mid x) &= \frac{d_{x^*}(x)U(y \mid x^*) + d_{x^0}(x)U(y \mid x^0)}{d_{x^*}(x) + d_{x^0}(x)} \\
&= w_{x^*}(x)U(y \mid x^*) + (1 - w_{x^*}(x))U(y \mid x^0),
\end{aligned}
\tag{8}
$$

where $w_{x^*}(x) = \dfrac{d_{x^*}(x)}{d_{x^*}(x) + d_{x^0}(x)}$.

Equation (8) implies that Y is interpolation independent of X. We have derived this result by direct application of Theorem 22.1 and Proposition 23.2. We summarize this result below.

Proposition 23.3 relates the notions of utility independence and interpolation independence. Using this result, we can spot the utility independence of X on Y and the interpolation independence of Y on X in a utility tree. We now have a sufficient condition for asserting the interpolation independence of Y on X by observing the utility independence of X on Y. Moreover, the relation "X is utility independent of Y" implies that utility inference becomes an interpolation of the conditional utility function of Y at the boundary values of X using the univariate interpolation function $w_{x^*}(x)$.

Mutual Utility Independence

The definition of mutual utility independence implies that we have a situation where X is utility independent of Y and that Y is utility independent of X. Keeney and Raiffa (1976) show that the functional form of the utility function that corresponds to this condition is

$$\boxed{U(x,y) = k_x U(x \mid y^0) + k_y U(y \mid x^0) + (1 - k_x - k_y)U(x \mid y^0)U(y \mid x^0).}$$

We have already discussed the functional form of the utility function that corresponds to the relation X is utility independent of Y. We have seen that this relation does not necessarily imply that Y is utility independent of X, but rather that Y is interpolation independent of X. We have also seen that if Y is interpolation independent of X, and if Y is also boundary independent of X, then Y is utility independent of X.

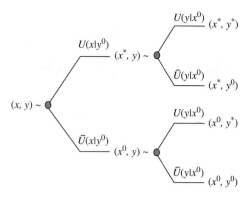

Figure 23.3. X is utility independent of Y and Y is boundary independent of X.

Note that once we have asserted that X is utility independent of Y, we do not need to assert the stronger relation that Y is utility independent of X for mutual utility independence. It is sufficient to assert the weaker relation that Y is boundary independent of X. We derive the corresponding functional form below.

Mutual Utility Independence Derived from Utility and Boundary Independence Conditions

If X is utility independent of Y, then Y is interpolation independent of X. If in addition, Y is boundary independent of X, then Y is utility independent of X, and the two attributes are mutually utility independent. Therefore, if X is utility independent of Y in the utility tree, we can tell whether Y is utility independent of X by simply verifying the boundary independence conditions of Y on X. In the tree of Figure 23.3, we can verify that Y is not only boundary independent of X but that it is also utility independent of X because X is utility independent of Y since $U(x \mid y) = U(x \mid y^0)$.. We summarize this result below.

Proposition 23.4 (Abbas 2011): Utility Independence of the Second Attribute

If X is utility independent of Y, and Y is boundary independent of X, then Y is utility independent of X.

Proposition 23.4 relates the notion of utility independence and boundary independence. By observing that X is utility independent of Y in Figure 23.3, and that Y is boundary independent of X, we can assert that Y is also utility independent of X (and not just boundary independent of X).

Proposition 23.5 (Abbas 2011): Mutual Utility Independence Derived from the Utility Tree

If X is utility independent of Y, and Y is boundary independent of X, then X and Y are mutually utility independent.

Proposition 23.5 implies that the condition of mutual utility independence for two attributes can be represented by a single tree. Moreover, we can now assert the existence of mutual utility independence between the two attributes by verifying utility independence in one direction, $U(x \mid y) = U(x \mid y^0)$, and boundary independence in the other, $U(y \mid x^*) = U(y \mid x^0)$, without having to verify the additional assertion that $U(y \mid x) = U(y \mid x^0)$.

Deriving the Functional Form of Mutual Utility Independence Using the Two-Attribute Tree

The functional form of mutual utility independence of X and Y can be derived by simple roll-back of the utility tree after appropriate substitutions. The tree of Figure 23.3 would imply that

$$U(x,y) = 1 \times U(x \mid y^0)U(y \mid x^0) + k_x U(x \mid y^0)\bar{U}(y \mid x^0) + k_y \bar{U}(x \mid y^0)U(y \mid x^0) + 0.$$

Simplifying gives

$$\boxed{U(x,y) = k_x U(x \mid y^0) + k_y U(y \mid x^0) + (1 - k_x - k_y)U(x \mid y^0)U(y \mid x^0).}$$

Special Cases of Mutual Utility Independence for Two Attributes

1. When $k_x + k_y = 1$, the functional form of mutual utility independence reduces to the additive form $U(x,y) = k_x U(x \mid y^0) + k_y U(y \mid x^0)$.

    In Chapter 20, we discussed the implications (and limitations) of the additive form in terms of indifference for correlation among the attributes.

2. When $k_x = k_y = 0$, an attribute dominance utility function, the functional form of mutual utility independence reduces to the product form,

$$U(x,y) = U(x \mid y^*)U(y \mid x^*).$$

Here we have conditioned on the maximum values of the complement attributes, which is okay because of the utility independence assertions.

Dialogue between Instructor (I) and Student (S)

S: I am really intrigued by these connections between the different independence forms.

I: Interestingly enough, the concepts of utility independence and interpolation independence were introduced independently in the literature. The fact that utility independence in one direction implies interpolation independence in the other is intriguing and was introduced a lot later (Abbas 2011).

S: And the idea that you can assert mutual utility independence by asserting utility independence in one direction and boundary independence in another is also very intriguing.

I: Yes, it simplifies the assertion of mutual utility independence. You do not need to assert utility independence in both directions as had been assumed for a very long time.

Verifying Mutual Utility Independence between Two Attributes

Analyst: So we have two attributes, X and Y.

Decision-maker: Yes.

Analyst: Are you comfortable asserting that X is utility independent of Y?

Decision-maker: What does that mean again?

Analyst: Well preferences for lotteries over attribute X do not depend on Y.

Decision-maker: Perhaps, I am not sure. I will say yes.

Analyst: Would the utility function you assign over attribute X depend on the level of attribute Y? Can you specify a utility over X without knowing the value of Y?

Decision-maker: Yes.

Analyst: Then X is utility independent of Y.

We can also ask the reverse question. Is Y utility independent of X?

Decision-maker: And I would need to assert preferences for lotteries over Y do not depend on X? Or that my utility function for Y does not depend on X?

Analyst: No. We just need to verify this at two values of X, say the upper and lower bounds. Are you comfortable saying the utility function you would assign for Y is the same at the upper and lower bounds of X? Or equivalently, is you preference for lotteries over Y the same at the two boundary values of Y?

Decision-maker: Yes, this is easier to think about than the whole domain of X.

Analyst: Then X and Y are mutually utility independent.

Implications of Mutual Utility Independence on the Shape of the Isopreference Contours

By using a formula such as

$$U(x, y) = k_x U(x \mid y^0) + k_y U(y \mid x^0) + (1 - k_x - k_y) U(x \mid y^0) U(y \mid x^0),$$

for the multiattribute utility function, we are constructing the whole utility surface using the lower boundary assessments, $U(x \mid y^0)$, $U(y \mid x^0)$, and the corner values k_x and k_y.

We are also asserting that the upper boundary assessments are linear transformations of the lower boundary assessments, where we get by direct substitution for the lower boundary assessments

$$U(x, y^0) = k_x U(x \mid y^0) \text{ and } U(x^0, y) = k_y U(y \mid x^0).$$

We also get for the upper boundary assessments,

$$U(x, y^*) = k_y + (1 - k_y) U(x \mid y^0) \text{ and } U(x^*, y) = k_x + (1 - k_x) U(y \mid x^0).$$

Finally, we are assuming that the trade-offs between the attributes are determined by these boundary assessments. We have in effect assumed the shape of the isopreference contours on the whole domain using only boundary assessments. Because a utility function is a monotone transformation of a preference or a value function, we can get the slope of the corresponding isopreference contours by setting the total derivative of the utility function to zero to get

$$dU(x,y) = \frac{\partial}{\partial x}U(x,y)dx + \frac{\partial}{\partial y}U(x,y)dy = 0.$$

This implies that the slope of an isopreference contour, C, at any point (x,y) is

$$\frac{dy}{dx}\Big|_C = -\frac{\frac{\partial}{\partial x}U(x,y)}{\frac{\partial}{\partial y}U(x,y)} = -\frac{k_x\frac{\partial}{\partial x}U(x\,|\,y^0) + (1-k_x-k_y)U(y\,|\,x^0)\frac{\partial}{\partial x}U(x\,|\,y^0)}{k_y\frac{\partial}{\partial y}U(y\,|\,x^0) + (1-k_x-k_y)U(x\,|\,y^0)\frac{\partial}{\partial y}U(y\,|\,x^0)}.$$

Note that the slope of the isopreference contours is completely determined by the boundary assessments and the corner values. The main issue with this is that there are numerous utility surfaces that match the boundary assessments and corner values and yet do not satisfy the mutual utility independence formula, and therefore have different shapes of the isopreference contours. Therefore, we need to be sure that these independence conditions really hold or we will be making incorrect trade-off assumptions. In Chapters 28 through 32, we shall explore methods to construct utility surfaces that match the boundary assessments and yet have an additional degree of freedom to provide various trade-off assessments among the attributes, which will be useful when independence conditions do not hold.

23.6 GRAPHICAL REPRESENTATIONS OF INDEPENDENCE RELATIONS

In Chapter 21, we introduced the concept of a utility diagram for attribute dominance utility functions. We can now extend this concept to more general functional forms of utility functions. The main differences with this extension, as we have seen, are that:

1.　Utility independence need not be symmetric, and so there is a need to specify the direction of the utility independence relation.
2.　Boundary independence relations can be specified as they might also contribute to the assertion of utility independence.

This section introduces two types of graphical representations: utility independence diagrams that focus on the utility independence relations, and utility tree networks that incorporate both utility and boundary independence relations. We extend these representations to more than two attributes in the future chapters.

23.6.1 Bidirectional Utility Independence Diagrams (Abbas 2011)

Because utility independence need not be symmetric for multiattribute utility functions, we can represent the full set of independence relations for two attributes, Y and Z, using the following scenarios:

(a)　Z and Y both have no utility independence relations on each other
(b)　Z utility independent of Y and no assertion that Y is utility independent of Z
(c)　Y utility independent of Z and no assertion that Z is utility independent of Y
(d)　Z utility independent of Y and Y utility independent of Z

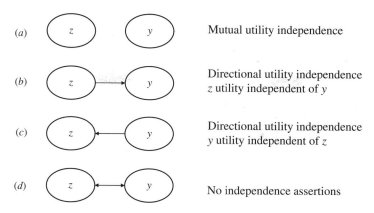

Figure 23.4. Utility independence diagrams for two attributes.

Note: Arrows in the utility diagram represent the possibility of utility dependence and not an assertion of utility dependence. The absence of an arrow is an assertion of utility independence.

To represent these utility independence relations graphically, we can envision diagrams whose nodes represent the attributes and whose arrows (arcs) represent the utility independence relations between them. We call such diagrams bidirectional utility independence diagrams. Because of the asymmetric nature of utility independence, we can envision a bidirectional arrow, or a unidirectional arrow, or no arrows connecting the nodes. Figure 23.4 illustrates this relation graphically.

Figure 23.4(a) shows a bidirectional arrow between two nodes Y and Z, and so no utility independence relations are asserted. The general expansion will be that corresponding to the utility tree with no decomposition.

Figure 23.4(b) shows a unidirectional arrow, which represents the utility independence of Z on Y. There are no utility independence relations of Y on Z, but as we have seen, this also represents the interpolation independence of Y on Z.

The functional form (as we have seen) is

$$U(z,y) = U(z \mid y^0)\left[U(z^*,y) - U(z^0,y)\right] + U(z^0,y).$$

The functional form requires a utility assessment for attribute, Z at an arbitrary value of $Y, U(z \mid y^0)$, as well as two utility assessments for attribute Y at the boundary values of attribute $Z, U(z^*,y)$ and $U(z^0,y)$.

Figure 23.4(c) also shows a unidirectional arrow, which represents the utility independence of Y on Z. There are no utility independence relations of Z on Y, but as we have seen, this also represents the interpolation independence of Z on Y.

The functional form is

$$U(z,y) = U(y \mid z^0)\left[U(z,y^*) - U(z,y^0)\right] + U(z,y^0).$$

The functional form requires a utility assessment for attribute Y at an arbitrary value of $Z, U(y \mid z^0)$, as well as two utility assessments for attribute Z at the boundary values of attribute Y, $U(z,y^*)$ and $U(z,y^0)$.

Figure 23.4(d) shows two nodes with no arrows connecting them. This is an assertion that Z is utility independent of Y and Y is utility independent of Z.

The functional form (as we have seen) is

$$U(z,y) = k_z U(z \mid y^0) + k_y U(y \mid z^0) + (1 - k_z - k_y) U(z \mid y^0) U(y \mid z^0).$$

The functional form requires a utility assessment of Y at an arbitrary value of Z, and a utility assessment of Z at an arbitrary value of Y.

While Figure 23.4(d) implies mutual utility independence between two attributes, we have seen that this relation can also be verified by utility independence in one direction and boundary independence in the other. Because this assertion requires less effort than asserting utility independence in both directions, it is important to consider a figure that specifies these different types of utility and boundary independence relations.

The presence of unidirectional arrows (or the lack or arrows between the attributes) instead of having bidirectional arrows reduces the number of utility assessments needed for the functional form. The bidirectional diagram representation enables us to deduce the number of assessments needed for the functional form.

Understanding the Assessments Needed for the Functional Form from the Diagram
Absence of an Arrow: For any node that has no arrow entering into it, we require one utility function assessment of that attribute at an arbitrary reference value of the other attribute.

    Unidirectional Arrow: For any node that has a unidirectional arrow entering into it, we require two utility function assessments of that attribute at two reference value of the other attribute.

EXAMPLE 23.2: Utility Assessments Needed for a Two-Attribute Utility Independence Diagram

---

To illustrate, consider Figure 23.4(b). This figure has **one unidirectional arrow** from Z to Y.

    This implies that we need one utility function assessment for attribute Z at any reference value of attribute Y, say $U(Z \mid y^0)$, and two utility function assessments of attribute Y (at two reference values of Z), say $U(Y \mid z^0)$ and $U(Y \mid z^*)$, to construct the functional form. Indeed, the functional form represented by this figure requires these three assessments.

The pattern for Figure 23.4(c) is similar but in the reverse direction.

Figure 23.4(d) has **no arrows** connecting Y and Z. This implies that we need one utility function assessment for attribute Z at any reference value of attribute Y, say $U(Z \mid y^0)$, and one utility function assessment for attribute Y at any reference value of attribute Z, say $U(Y \mid z^0)$, to construct the functional form. Indeed, the functional form represented by this figure, as we have seen, requires these two assessments.

Note: Bidirectional utility independence diagrams make no assertions about the order of the expansion in the utility tree. They simply represent the utility independence

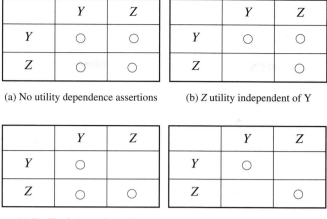

(a) No utility dependence assertions   (b) Z utility independent of Y

(c) Y utility independent of Z   (d) Mutual utility independence

Figure 23.5. Utility dependence matrices for two attributes.

relations between the two attributes. In the next section, we present another type of utility diagrams that represents both utility and boundary independence relations and that represents a certain expansion order of the utility tree.

23.6.2 Utility Dependence Matrix

A utility dependence matrix is an $n \times n$ matrix corresponding to the n attributes and their independence assertions. A vacant cell corresponding to row X_i and column X_j asserts the utility independence relation $(X_i \ UI \ X_j \mid \bar{X}_{ij})$, while a circle in the cell corresponding to row X_i and column X_j does not assert any utility independence relations. Figure 23.5 shows utility dependence matrices for two attributes.

23.6.3 Utility Tree Networks (UTN)

Utility tree networks represent a certain expansion order of the utility tree. They also allow for fewer assertions to make utility independence deductions. For example, the assertion of mutual utility independence can be deduced from utility independence of one attribute on a future attribute and the weaker boundary independence assertion of the future attribute on the other. Utility tree networks allow for this distinction between utility and boundary independence.

Figures 23.6–23.9 show the utility tree networks for two attributes. The diagrams represent an expansion order in the basic expansion theorem and the utility tree: we first expand around Z then Y.

Definition: A *utility tree network* is a graph whose nodes represent the attributes under consideration and whose arcs represent the following notation; a solid arc represents the possibility of utility dependence of one attribute on another, and a dashed line represents the possibility of a boundary dependence relation. The absence

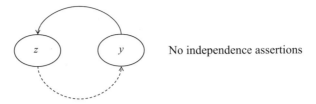

No independence assertions

Figure 23.6. Utility tree network making no boundary or utility independence assertions.

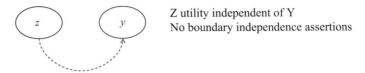

Z utility independent of Y
No boundary independence assertions

Figure 23.7. Utility tree network asserting that Z is utility independent of Y.

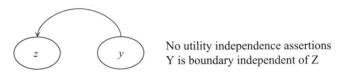

No utility independence assertions
Y is boundary independent of Z

Figure 23.8. Utility tree network asserting that Y is boundary independent of Z.

of a solid arrow is an assertion of utility independence and the absence of a dashed arrow is an assertion of boundary independence.

Figure 23.6 represents an expansion order in the utility tree starting with Z then Y. It does not assert any utility or boundary independence relations. The functional form is the general form of the two-attribute utility tree obtained from the basic expansion theorem. This diagram includes all possible arrows for two attributes: it includes a solid arrow from Y to Z (because, as we have discussed, utility independence is on future nodes in the expansion) and a dashed arrow from Z to Y (because boundary independence is on past nodes in the expansion). The diagram makes no assertions about utility independence of Z on Y or boundary independence of Y on Z.

Figure 23.7 asserts the utility independence of Z on Y. The diagram has a missing solid arrow representing this relation, but it includes the dashed arrow from Y to Z asserting the possibility of boundary dependence of Y on Z. The functional form can be obtained by setting $U(z \mid y) = U(z \mid y^0)$ in the utility tree in an expansion order of Z then Y.

Figure 23.8 represents the assertion of boundary independence of Y on Z because a dashed arrow is missing, but it makes no assertions about utility independence of Z on Y. The functional form can be obtained by setting $U(y \mid z^*) = U(y \mid z^0)$ in the utility tree in an expansion order of Z then Y.

Figure 23.9 asserts both utility and boundary independence relations, which implies mutual utility independence between Z and Y. The functional form can be obtained

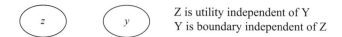

Z is utility independent of Y
Y is boundary independent of Z

Figure 23.9. Utility tree network asserting that Z is utility independent of Y and Y is boundary independent of Z.

by setting $U(z\,|\,y) = U(z\,|\,y^0)$ and $U(y\,|\,z^*) = U(y\,|\,z^0)$ in the utility tree in an expansion order of Z then Y.

While bidirectional utility diagrams are an easier representation than utility tree networks – they involve only one type of arrows, focus only on utility independence relations, and do not convey an expansion order – utility tree networks enjoy the convenience of being able to substitute directly for their functional form using the utility tree expansion in the order they provide.

> **Note:** I first worked on bidirectional utility diagrams before the basic expansion theorem. The idea was to find a graphical method to deduce the number of assessments needed. Following the derivation of the basic expansion theorem, it was clear that a chronological expansion order depicted in the graph was needed to simplify the deduction of the functional form. There was also a need for a distinction between utility and boundary independence relations in the diagram. I proposed keeping utility independence arrows as in the current utility tree networks and a mere "dot" in the diagram when no boundary independence assertions were to be made. My colleague Seewong Ho suggested a dashed arrow for the absence of boundary independence relations instead of a dot, and so I have adopted this configuration here.

23.7 SUMMARY

Important implications of utility independence of X on Y are

1. The normalized conditional utility function of X, $U(x\,|\,y)$, does not change with y, i.e., $U(x\,|\,y) = U(x\,|\,y^0)$.
2. The utility function can be expressed as an affine transformation of $U(x\,|\,y^0)$, i.e., $U(x, y) = k(y)U(x\,|\,y^0) + d(y)$.
3. We need one utility assessment of X at an arbitrary value of Y and two normalized conditional utility assessments of Y (at min and max of X) plus two corner values to construct the utility function because

$$U(x, y) = U(x\,|\,y^0)\big[U(x^*, y) - U(x^0, y)\big] + U(x^0, y)$$
$$= U(x\,|\,y^0)\big[k_x + (1 - k_x)U(y\,|\,x^*) - U(y\,|\,x^0)k_y\big] + U(y\,|\,x^0)k_y$$

4. The risk aversion function of attribute X does not depend on Y.
5. Preferences for lotteries over X do not change as we change y.
6. If X is utility independent of Y then Y is interpolation independent of X.

If Y is interpolation independent of X, and Y is boundary independent of X, then Y is utility independent of X.

If X is utility independent of Y, and Y is boundary independent of X, then Y is utility independent of X. We do not need to assert mutual utility independence between two attributes to decompose the functional form into univariate assessments. It suffices to assert utility independence in one direction and boundary independence in another. Utility and boundary independence relations can be represented graphically.

ADDITIONAL READINGS

Abbas, A. E. 2005. Bidirectional Utility Diagrams. Technical Report. University of Illinois at Urbana-Champaign, UILU-ENG-2005–3001.

Abbas, A. E. 2009. From Bayes' nets to utility nets. Proceedings of the 29th *International Workshop on Bayesian Inference and Maximum Entropy Methods in Science and Engineering.* Oxford, MI, July 5–10. AIP conference proceedings 1193, pp. 3–12.

Abbas, A. E. 2011a. General decompositions of multiattribute utility functions. *Journal of Multicriteria Decision Analysis* 17(1, 2): 37–59.

Abbas, A. E. 2011b. The multiattribute utility tree. *Decision Analysis* 8(3): 180–205.

Bell, D. E. 1979. Multiattribute utility functions: Decompositions using interpolation. *Management Science* 25(2): 208–224.

Keeney, R and H. Raiffa. 1976. *Decisions with Multiple Objectives.* John Wiley, New York.

CHAPTER 24

Basic Expansion Theorem for Multiattribute
Utility Functions

Chapter Concepts

- The general expansion theorem for multiattribute utility functions: a sum of products of conditional utility functions
- Expressing the normalized conditional utility function of an attribute in terms of its values at the boundary values of the remaining attributes
- Utility inference: updating the conditional utility function of an attribute as the levels of the other attributes vary
- Utility tree representation of multiattribute utility functions
- Changing the order of the attributes in the utility tree
- Conditional independence
  - Conditional utility independence
  - Conditional boundary independence
  - Conditional interpolation independence
- Graphical representations of utility trees for more than two attributes

24.1 INTRODUCTION

This chapter extends the two-attribute utility tree to multiple attributes. Our main goal will be to further familiarize the reader with utility trees and the concepts of utility, boundary, and interpolation independence. To gain the full insights obtained from utility trees, it will be important to understand how to change the order of assessments in the tree and how to identify various independence conditions from a given assessment order. The extension of the utility tree to multiple attributes also allows for new ideas such as ***conditional utility independence***, ***conditional boundary independence***, and ***conditional interpolation independence***.

24.2 THE THREE-ATTRIBUTE UTILITY TREE ON A CONTINUOUS DOMAIN

We start with the following definition.

Definition

The **_normalized conditional utility function_** of an attribute, X, in a three-attribute tree with complement attributes Y and Z is defined as

$$U(x \mid y,z) = \frac{U(x,y,z) - U(x^0,y,z)}{U(x^*,y,z) - U(x^0,y,z)}.$$

Similarly, we define a conditional disutility function as

$$\overline{U}(x \mid y,z) = 1 - U(x \mid y,z).$$

Various instantiations of the conditional utility function at different levels of the complement can be obtained by direct substitution. For example,

$$U(x \mid y^*,z) = \frac{U(x,y^*,z) - U(x^0,y^*,z)}{U(x^*,y^*,z) - U(x^0,y^*,z)},$$

is the normalized conditional utility of attribute X on Y and Z, when Y is at its maximum value y^*

Three-Attribute Utility Tree Representation

By induction, a three-attribute utility tree can be represented as shown in Figure 24.1, and the functional form of the utility function can be expressed by **_rolling back the tree_** as a sum of products of the conditional utility functions of each branch using the following sum of eight product terms,

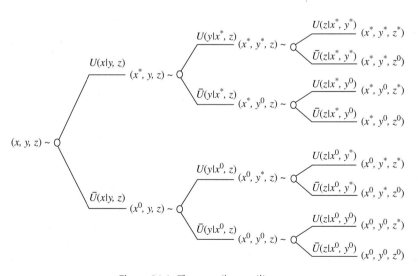

Figure 24.1. Three-attribute utility tree.

$$U(x,y,z) = U(x \mid y,z)U(y \mid x^*,z)U(z \mid x^*,y^*)U(x^*,y^*,z^*)$$
$$+ U(x \mid y,z)U(y \mid x^*,z)\bar{U}(z \mid x^*,y^*)U(x^*,y^*,z^0)$$
$$+ U(x \mid y,z)\bar{U}(y \mid x^*,z)U(z \mid x^*,y^0)U(x^*,y^0,z^*)$$
$$+ U(x \mid y,z)\bar{U}(y \mid x^*,z)\bar{U}(z \mid x^*,y^0)U(x^*,y^0,z^0)$$
$$+ \bar{U}(x \mid y,z)U(y \mid x^0,z)U(z \mid x^0,y^*)U(x^0,y^*,z^*) \qquad (1)$$
$$+ \bar{U}(x \mid y,z)U(y \mid x^0,z)\bar{U}(z \mid x^0,y^*)U(x^0,y^*,z^0)$$
$$+ \bar{U}(x \mid y,z)\bar{U}(y \mid x^0,z)U(z \mid x^0,y^0)U(x^0,y^0,z^*)$$
$$+ \bar{U}(x \mid y,z)\bar{U}(y \mid x^0,z)\bar{U}(z \mid x^0,y^0)U(x^0,y^0,z^0)$$

The tree displays the following conditional utility assessments:

1. The assessment $U(x \mid y,z)$ for attribute X
2. The assessments $U(y \mid x^*,z), U(y \mid x^0,z)$ for attribute Y
3. The assessments $U(z \mid x^*,y^*), U(z \mid x^*,y^0), U(z \mid x^0,y^*), U(z \mid x^0,y^0)$ for attribute Z

Note that each attribute is conditioned on the boundary values of the attributes expanded before it in the tree. Therefore, a tree in one order does not immediately display the full conditional utility functions for each attribute, but rather conditioned on the boundary values of the prior attributes in the expansion. However, because the tree determines the multiattribute utility function, it should not come as a surprise that we can deduce all conditional utility functions in any order of assessment from the tree.

Once again, we would like to explore and relate the conditional utility functions for each attribute in a given tree and the implications of changing the order of assessment so that we may infer any utility independence relations from any tree order.

The Basic Expansion Theorem of a Three-Attribute Utility Tree

Observe that (1) can be written in a way that facilitates its extension to multiple attributes using the notation

$$U(x_1, x_2, x_3) = \sum_{\substack{x_j \in \{x_1^*, x_1^0\} \\ x_k \in \{x_2^*, x_2^0\} \\ x_l \in \{x_3^*, x_3^0\}}} U(x_j, x_k, x_l) \prod_{i=1}^{3} g_{x_i}(x_i \mid x_{iP}^{*0}, x_{iF}),$$

where

1. The notation $\displaystyle\sum_{\substack{x_j \in \{x_1^*, x_1^0\} \\ x_k \in \{x_2^*, x_2^0\} \\ x_l \in \{x_3^*, x_3^0\}}} U(x_j, x_k, x_l)$ denotes a summation over all corner values of the domain;

2. The function g_{x_i} is either a conditional utility function or conditional disutility function for attribute X_i depending on the instantiation of x_i in the corner value $U(x_j, x_k, x_l)$;

3. x_{iP}^{*0} is the instantiation of the ***corner values of the attributes*** expanded prior to X_i; and
4. x_{iF} is an instantiation of the attributes expanded after x_i in the expansion order.

This notation follows the notation used in Section 22.2.4 for the basic expansion theorem around the two attributes.

24.3 THE CONDITIONAL UTILITY FUNCTION OF THE FIRST ATTRIBUTE IN THE EXPANSION

The conditional utility function of the first attribute in the expansion is always explicitly provided in the tree. In Figure 24.1, it is $U(x\,|\,y,z)$. This conditional utility function is sufficient to answer questions about independence relations pertaining to attribute X and any of the remaining attributes in the tree. Because we now have three attributes in the tree, some additional specifications of conditional utility independence can be made. We illustrate this idea and its implications below.

24.3.1 Attribute X Is Utility Independent of Both Y and Z

Definition
Utility Independence of X on both Y and Z:
Attribute X is utility independent of attributes Y and Z if

$$U(x\,|\,y,z) = U(x\,|\,y^0,z^0) \quad \forall x,y,z.$$

This definition implies that the normalized conditional utility function for X does not depend on the levels of Y and Z. For three attributes, if X is utility independent of both Y and Z, we say that X is utility independent of its complement attributes, and for short, just say that

  X is utility independent of its complement.

This relation can be spotted by inspection for the first attribute in the tree. In principle, as we shall see, it is always easy to spot the independence relations of any attribute on future attributes in the tree.

24.3.2 Attribute X Is Conditionally Utility Independent of Attribute Y Given Attribute Z

Definition
Conditional Utility Independence of X on Y given Z:
Attribute X is conditionally utility independent of attribute Y given attribute Z if

$$U(x\,|\,y,z) = U(x\,|\,y^0,z) \quad \forall x,y,z.$$

This occurs if the normalized conditional utility function $U(x \mid y, z)$ does not depend on y for any value of z, *but it does depend on Z*. Therefore, this assessment will need to be conducted for the values of Z, *and it is a weaker condition than X being utility independent of its complement*.

> **Note:** It is straightforward to observe in the tree of Figure 24.1 whether X is conditionally utility independent of any future attribute by inspection, because the conditional utility function $U(x \mid y, z)$ is explicitly presented in the tree: X is utility independent of any future attribute if and only if the conditional utility function $U(x \mid y, z)$ does not depend on that attribute.

24.3.3 Attribute *X* Is Conditionally Utility Independent of Attribute *Z* Given Attribute *Y*

> **Definition**
> *Conditional Utility Independence of X on Z given Y:*
>   Attribute X is conditionally utility independent of attribute Z given attribute Y if
> $$U(x \mid y, z) = U(x \mid y, z^0) \quad \forall x, y, z.$$

As we should expect by now, this condition is also easy to verify by inspecting the tree, because Z is a future node in the tree. We highlight this general observation about identifying independence relations below.

24.3.4 Rearranging the Position of the First Attribute in the Utility Tree

Given the conditional utility function of the first attribute in the tree, $U(x \mid y, z)$, we can reorder the tree to place attribute X in any order and determine the conditional utility function of X in any position in the tree because it merely involves substituting for some of the attributes at their boundary values. For example, if we wanted X to be the second attribute in the tree after attribute Y, then we merely need to substitute for Y at its maximum and minimum values to get $U(x \mid y^*, z)$ and $U(x \mid y^0, z)$ and use those values in the tree.

24.3.5 Interpreting Conditional Utility Independence in Terms of Preferences for Lotteries

The assertion that X is utility independent of Z given Y implies (in terms of conditional utility assessments) that

$$U(x \mid y, z) = U(x \mid y, z^0) \quad \forall x, y, z.$$

By inspection this condition implies that the normalized conditional utility function of X will not depend on the value of Z. To determine the corresponding form of the utility function, we use the basic expansion theorem around a single attribute to get

$$U(x,y,z) = U(x \mid y,z)U(x^*,y,z) + \bar{U}(x \mid y,z)U(x^0,y,z).$$

Substituting for the conditional utility independence condition $U(x \mid y,z) = U(x \mid y,z^0)$ gives

$$U(x,y,z) = U(x \mid y,z^0)U(x^*,y,z) + \bar{U}(x \mid y,z^0)U(x^0,y,z).$$

Rearranging gives

$$U(x,y,z) = [U(x^*,y,z) - U(x^0,y,z)]U(x \mid y,z^0) + U(x^0,y,z).$$

Define $k(y,z) = U(x^*,y,z) - U(x^0,y,z)$, $d(y,z) = U(x^0,y,z)$.

If more of an attribute is strictly preferred to less, then $k(y,z) = U(x^*,y,z) - U(x^0,y,z) > 0$.

The fact that X is utility independent of Z given Y therefore implies that

$$\boxed{U(x,y,z) = k(y,z)U(x \mid y,z^0) + d(y,z)}. \tag{2}$$

Note that attribute X appears only in the term $U(x \mid y,z^0)$ on the right-hand side; attribute Y appears in all three terms and attribute Z appears only in $k(y,z)$ and $d(y,z)$. Changing Y might change the conditional utility function $U(x \mid y,z^0)$, but changing Z will only change the terms $k(y,z)$ and $d(y,z)$, which contribute to an affine transformation of $U(x \mid y,z^0)$. Because the rank order of lotteries is invariant under a positive linear transformation on the utility function, this condition implies that preferences for lotteries over X will not change as we change Z for any fixed value of Y. In fact, Y is merely a parameter that is added to all terms in this case.

24.4 THE CONDITIONAL UTILITY FUNCTION OF THE SECOND ATTRIBUTE IN THE EXPANSION

In the tree of Figure 24.1, we have the following conditional utility functions for the second attribute, Y, $U(y \mid x^*,z)$ and $U(y \mid x^0,z)$, which are conditioned on the boundary values of the first attribute in the tree. We shall relate these boundary assessments to the conditional utility assessment $U(y \mid x,z)$. Later in this section, we shall see that this relation implies that $U(y \mid x,z)$ is a weighted function of its value at the boundary values of X, in addition to some function $\mu(x,y,z)$.

$$U(y \mid x,z) = w_{x^*}(x,y,z)U(y \mid x^*,z) + w_{x^0}(x,y,z)U(y \mid x^0,z) + \mu(x,y,z) \tag{3}$$

Furthermore, we shall see that the function $\mu(x,y,z)$ is zero if Y is a utility-dominant attribute or if X is utility independent of Y given Z.

In a utility tree, it is easy to spot boundary independence and interpolation independence relations on prior attributes, as well as utility independence relations on future attributes. To illustrate, we shall discuss the independence relations of attribute Y on X (the first attribute in the tree) and then on Z (the third attribute).

24.4.1 Conditional Utility Independence of the Second Attribute on the Third Given the First

It is easy to tell by inspection whether the second attribute, Y, is conditionally utility independent of the third attribute, Z, at any boundary value of the first, X, i.e., whether

$$U(y\,|\,x^0,\, z) = U(y\,|\,x^0,\, z^0) \text{ or } U(y\,|\,x^*,\, z) = U(y\,|\,x^*,\, z^0).$$

This can be verified by inspection of the conditional utility functions that are provided in the tree.

24.4.2 Conditional Boundary Independence of the Second Attribute on the First Given the Third

Definition
Conditional Boundary Independence of Y on X given Z
    Attribute Y is conditionally boundary independent of attribute X given attribute Z if

$$U(y\,|\,x^*,z) = U(y\,|\,x^0,z) \ \ \forall y,z.$$

We can see by inspection whether the second attribute in the expansion, Y, is conditionally boundary independent of the first attribute, X, given the third attribute, Z. This will be a recurring theme: boundary independence conditions are easier to spot on prior attributes.

24.4.3 Conditional Interpolation Independence of the Second Attribute on the First Given the Third

Definition
Conditional Interpolation Independence of Y on X given Z
    Attribute Y is conditionally interpolation independent of attribute X given attribute Z if the conditional utility function $U(y\,|\,x,z)$ can be expressed in terms of its values at the boundary values of X, i.e., $U(y\,|\,x^*,z)$ and $U(y\,|\,x^0,z)$, using the expression

$$U(y\,|\,x,z) = w_{x^*}(x,z)U(y\,|\,x^*,z) + [1 - w_{x^*}(x,z)]U(y\,|\,x^0,z).$$

Dialog between Instructor (I) and Student (S)
I: Compare the definition of conditional interpolation independence to the general expression of the conditional utility function (3).

S: Conditional interpolation independence has no additional function $\alpha(x, y, z)$, and its weighting functions $w_{x^*}(x, y, z)$ and $w_{x^0}(x, y, z)$ do not depend on y.

I: Correct. Can you tell from the tree whether the second attribute is interpolation independent of the first attribute given the third?

S: I think so.

I: This follows from a simple condition as we illustrate below.

We make the following generalization of Proposition 23.2 when a third attribute is included in the conditioning statement.

Proposition 24.1 Conditional Interpolation Independence

If X is conditionally utility independent of Y given Z, i.e., $U(x\,|\,y,z) = U(x\,|\,y^0,z)$, then Y is conditionally interpolation independent of X given Z, i.e.,

$$U(y\,|\,x,z) = w_{x^*}(x,z)U(y\,|\,x^*,z) + [1 - w_{x^*}(x,z)]U(y\,|\,x^0,z).$$

It suffices to know that X is utility independent of Y given Z to make the assertion that Y is interpolation independent of X given Z in the tree. Proposition 24.1 presents the notion of interpolation independence when an additional attribute, Z, is included in the tree after attribute Y. Here Z merely serves as a parameter.

24.4.4 Conditional Utility Independence of Second Attribute on the First Given the Third

Can we tell from the tree whether the second attribute is utility independent of the first attribute given the third, i.e., whether $U(y\,|\,x,z) = U(y\,|\,x^0,z)$? Yes, we can use an extension of Proposition 23.1, when an additional attribute Z is included.

Proposition 24.2: Utility Independence of the Second Attribute on the First

If (i) X is conditionally utility independent of Y given Z, i.e., $U(x\,|\,y,z) = U(x\,|\,y^0,z)$, and (ii) if Y is conditionally boundary independent of X given Z, i.e., $U(y\,|\,x^*,z) = U(y\,|\,x^0,z)$, then Y is conditionally utility independent of X given Z, $U(y\,|\,x,z) = U(y\,|\,x^0,z)$.

Proposition 24.2 is straightforward to prove by direct substitution of conditions (i) and (ii) into the utility tree and deriving the functional form. We now have a method to determine this independence relation on a prior attribute by inspection.

Figure 24.2 shows this relation and the assertion that Y is utility independent of X given Z.

24.4.5 Utility Independence of the Second Attribute on Its Complement

The following proposition is also straightforward to prove by direct substitution and rolling back the tree.

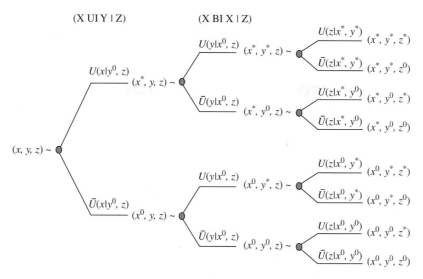

Figure 24.2. Utility tree asserting the relation Y UI X|Z.

Proposition 24.3: Utility Independence of the Second Attribute on Its Complement

If (i) X is conditionally utility independent of Y given Z, $U(x \mid y, z) = U(x \mid y^0, z)$, and if (ii) Y is conditionally boundary independent of X given Z, $U(y \mid x_i, z) = U(y \mid x^0, z), \forall z, x_i \in \{x^*, x^0\}$, and if (iii) Y is utility independent of Z at any boundary of X, $U(y \mid x_i, z) = U(y \mid x_i, z^0)$, $x_i = x^*$ or x^0, then Y is utility independent of <u>both</u> X and Z, and so $U(y \mid x, z) = U(y \mid x^0, z^0)$.

Proposition 24.3 enables us to assert utility independence of the second attribute on a prior attribute, X, and a future attribute, Z. Note that the first two conditions are immediate extensions of the results of two attributes; only a third attribute Z is inserted into the formulation. The third condition that Y is utility independent of Z need only be checked at any boundary of X and not all boundaries, because Y is already boundary independent of X, and so only one boundary is in effect.

Figure 24.3 shows the assertion that Y is utility independent of both X and Z. The figure illustrates conditions that assert that Y is utility independent of its complement. As shown in the figure, it suffices to assert utility independence of X on Y; boundary independence of Y on X and conditional utility independence of Y on Z at any given boundary of X.

24.4.6 The Conditional Utility Function of the Second Attribute

The following theorem determines the conditional utility function of the second attribute.

Theorem 24.4

Conditional utility function of the second attribute in a three-attribute utility tree

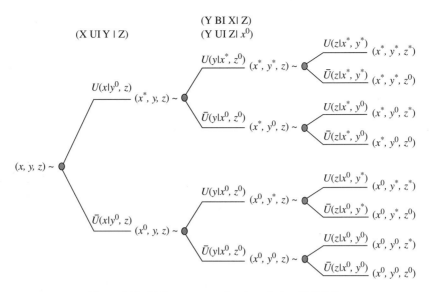

Figure 24.3. Utility tree asserting the relation Y UI X, Z.

$$U(y \mid x, z) = \frac{\displaystyle\sum_{\substack{x_i \in \{x^*, x^0\} \\ z_k \in \{z^*, z^0\}}} d_{x_i z_k}(x, y, z)U(y \mid x_i, z) + \sum_{\substack{x_i \in \{x^*, x^0\} \\ z_k \in \{z^*, z^0\}}} m_{x_i z_k}(x, y, z)}{\displaystyle\sum_{\substack{x_i \in \{x^*, x^0\} \\ z_k \in \{z^*, z^0\}}} [d_{x_i z_k}(x, y^*, z) + m_{x_i z_k}(x, y^*, z)]},$$

where $x_i \in \{x^0, x^*\}, z_k \in \{z^0, z^*\}$,

$$d_{x_i z_k}(x, y, z) = g_{x_i}(x \mid y, z)\big(U(x_i, y^*, z_k)g_{z_k}(z \mid x_i, y^*) - U(x_i, y^0, z_k)g_{z_k}(z \mid x_i, y^0)\big),$$

and

$$m_{x_i z_k}(x, y, z) = U(x_i, y^0, z_k)g_{z_k}(z \mid x_i, y^0)\big(g_{x_i}(x \mid y, z) - g_{x_i}(x \mid y^0, z)\big).$$

Theorem 24.4 can be expressed as follows:

$$U(y \mid x, z) = \frac{\alpha_{x^*}(x, y, z)U(y \mid x^*, z) + \alpha_{x^0}(x, y, z)U(y \mid x^0, z) + \beta(x, y, z)}{\alpha_{x^*}(x, y^*, z) + \alpha_{x^0}(x, y^*, z) + \beta(x, y^*, z)},$$

where,

$$\alpha_{x^*}(x, y, z) = d_{x^* z^*}(x, y, z) + d_{x^* z^0}(x, y, z), \quad \alpha_{x^0}(x, y, z) = d_{x^0 z^*}(x, y, z) + d_{x^0 z^0}(x, y, z),$$

$$\beta(x, y, z) = \sum_{\substack{x_i \in \{x^*, x^0\} \\ z_k \in \{z^*, z^0\}}} m_{x_i z_k}(x, y, z)$$

Using simple substitutions, Theorem 24.4 can also be expressed as a weighted combination of functions,

$$U(y \mid x, z) = w_{x^*}(x, y, z)U(y \mid x^*, z) + w_{x^0}(x, y, z)U(y \mid x^0, z) + \mu(x, y, z)$$

Having obtained the conditional utility function of Y, we can now place Y at any position of the tree by replacing the prior attributes with their boundary values. A more general proof of this theorem is provided in Appendix 24.2 in the derivation of Theorem 24.9.

Note: If X is conditionally utility independent of Y given Z, i.e., if $U(x \mid y, z) = U(x \mid y^0, z)$, then the functions $d_{x_i z_k}(x, y, z)$, $\forall x_i \in \{x^0, x^*\}, z_k \in \{z^0, z^*\}$ do not depend on y. Furthermore, the functions $m_{x_i z_k}(x, y, z) = 0$, $x_i \in \{x^0, x^*\}, z_k \in \{z^0, z^*\}$, and so we have conditional interpolation independence. Furthermore, if Y is a utility-dominant attribute, then $m_{x_i z_k}(x, y, z) = 0$, $x_i \in \{x^0, x^*\}, z_k \in \{z^0, z^*\}$ because the term $U(x_i, y^0, z_k) = 0$.

24.5 THE CONDITIONAL UTILITY FUNCTION OF THE THIRD ATTRIBUTE IN THE EXPANSION

The conditional utility functions for the third attribute, Z, in the utility tree of Figure 24.1 are conditioned on four boundary values, namely $U(z \mid y^*, z^*), U(z \mid y^*, z^0), U(z \mid y^0, z^*), U(x \mid y^0, z^0)$. Of course it is natural to define a condition when these conditional utility functions are equal, and this definition will also help with several derivations and developments.

24.5.1 Boundary Independence of the Third Attribute on Its Complement

Definition
Boundary Independence of Z on both X and Y:
Attribute Z is boundary independent of both attributes X and Y if

$$U(z \mid y^*, z^*) = U(z \mid y^*, z^0) = U(z \mid y^0, z^*) = U(x \mid y^0, z^0)$$

It is straightforward to tell (by inspection) from the tree whether an attribute is boundary independent of the previous attributes because these conditional utility functions are explicitly provided in the tree. In the case of three attributes, if X is boundary independent of both Y and Z, we say that ***X is boundary independent of its complement.***

24.5.2 Interpolation Independence of the Third Attribute on Its Complement

Definition
Interpolation independence on two attributes

Attribute Z is interpolation independent of two attributes, X and Y, if its conditional utility function can be expressed in terms of its conditional utility at the boundary values of X and Y, i.e.,

$$U(z \mid x,y) = w_{x^*y^*}(x,y)U(z \mid x^*,y^*) + w_{x^*y^0}(x,y)U(z \mid x^*,y^0)$$
$$+ w_{x^0y^*}(x,y)U(z \mid x^0,y^*) + w_{x^0y^0}(x,y)U(z \mid x^0,y^0),$$

where $w_{x^*y^*}(x,y) + w_{x^*y^0}(x,y) + w_{x^0y^*}(x,y) + w_{x^0y^0}(x,y) = 1$.

For three attributes, if Z is interpolation independent of attributes, X and Y, we say that **Z is interpolation independent of its complement**. We now show the conditions that reduce the general expression of the conditional utility function in (4) to that of interpolation independence on two attributes.

Proposition 24.5: Interpolation Independence of the Third Attribute on the Preceding Attributes

If (i) X is conditionally utility independent of Z given Y, $U(x \mid y,z) = U(x \mid y,z^0)$, and (ii) if Y is conditionally utility independent of Z at all boundaries of X, $U(y \mid x_i,z) = U(y \mid x_i,z^0), \forall x_i \in \{x^0,x^*\}$ then Z is interpolation independent of both X and Y, i.e.,

$$U(z \mid x,y) = \sum_{\substack{x_i \in \{x^*,x^0\} \\ y_j \in \{y^*,y^0\}}} w_{x_i y_j}(x,y)U(z \mid x_i,y_j),$$

where $w_{x_i y_j}(x,y) = \dfrac{d_{x_i y_j}(x,y)}{\displaystyle\sum_{\substack{x_i \in \{x^*,x^0\} \\ y_j \in \{y^*,y^0\}}} d_{x_i y_j}(x,y)}, \displaystyle\sum_{\substack{x_i \in \{x^*,x^0\} \\ y_j \in \{y^*,y^0\}}} w_{x_i y_j}(x,y) = 1.$

Proposition 24.5 presents a cascaded version of interpolation independence that extends to the third attribute. Compare this notion of interpolation independence of the last attribute in the tree to the notion of conditional interpolation independence of the second attribute in Proposition 24.1. The proof is similar and hence omitted.

Figure 24.4 shows a generic utility tree representing this situation.

24.5.3 Utility Independence of the Third Attribute on Its Complement

Proposition 24.6: Utility Independence of the Third Attribute on Its Complement
If (i) X is conditionally utility independent of Z given Y, and (ii) if Y is utility independent of Z at all boundaries of X, and (iii) if Z is boundary independent of X and Y, then Z is utility independent of both X and Y, and so $U(z \mid x,y) = U(z \mid x^0,y^0)$.

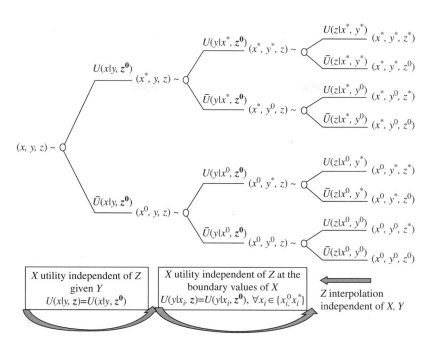

Figure 24.4. Propagating interpolation independence.

24.5.4 Conditional Utility Function of the Third Attribute

We now show how to calculate the conditional utility function of the third attribute using the assessments of Figure 24.1.

Theorem 24.7: Conditional Utility Function of the Third Attribute in a Three-Attribute Tree

$$U(z\,|\,x,y) = \frac{\displaystyle\sum_{\substack{x_i\in\{x^*,x^0\}\\y_j\in\{y^*,y^0\}}} d_{x_iy_j}(x,y,z)U(z\,|\,x_i,y_j) + \sum_{\substack{x_i\in\{x^*,x^0\}\\y_j\in\{y^*,y^0\}}} m_{x_iy_j}(x,y,z)}{\displaystyle\sum_{\substack{x_i\in\{x^*,x^0\}\\y_j\in\{y^*,y^0\}}} [d_{x_iy_j}(x,y,z^*) + m_{x_iy_j}(x,y,z^*)]},$$

where $x_i \in \{x^0,x^*\}, y_j \in \{y^0,y^*\}$,

$$d_{x_iy_j}(x,y,z) = g_{x_i}(x\,|\,y,z)g_{y_j}(y\,|\,x_i,z)\big(U(x_i,y_j,z^*) - U(x_i,y_j,z^0)\big),$$

and

$$m_{x_iy_j}(x,y,z) = U(x_i,y_j,z^0)\big(g_{x_i}(x\,|\,y,z)g_{y_j}(y\,|\,x_i,z) - g_{x_i}(x\,|\,y,z^0)g_{y_j}(y\,|\,x_i,z^0)\big).$$

A more general proof of this theorem is provided in Appendix 24.2 in the derivation of Theorem 24.8. It should be clear by now that the conditional utility function expansion involves functions $d_{x_i y_j}$ and $m_{x_i y_j}$ defined at the boundary values of the complement attributes to Z. Furthermore, the expression can be reduced to involve weighting functions at the boundary values of the attributes expanded before the attribute whose conditional utility function is to be determined.

We leave it to the reader to show (by inspection) that this formulation expresses the conditional utility function in terms of its boundary values as

$$U(z\,|\,x,y) = w_{x^*y^*}(x,y,z)U(z\,|\,x^*,y^*) + w_{x^*y^0}(x,y,z)U(z\,|\,x^*,y^0)$$
$$+ w_{x^0y^*}(x,y,z)U(z\,|\,x^0,y^*) + _{x^0y^0}(x,y,z)U(z\,|\,x^0,y^0) + \eta(x,y,z) \qquad (4)$$

Using Theorems 24.4 and 24.7, we now have a method to express the conditional utility function of any attribute in a three-attribute tree in terms of its conditional utility functions at the boundary values of any subset of the complement attributes. To do this, we start with a tree where the attribute is either second or third, and then use the previous results to reverse the tree.

Note: If

(i) X is conditionally utility independent of Z given Y, i.e., $U(x\,|\,y,z) = U(x\,|\,y,z^0)$, and if
(ii) Y is conditionally utility independent of Z at all boundaries of X, i.e., $U(y\,|\,x_i,z) = U(y\,|\,x_i,z^0), \forall x_i \in \{x^0,x^*\}$ then the following conditions hold:

1. The functions $d_{x_i y_j}(x,y,z)$, $x_i \in \{x^0,x^*\}$, $y_j \in \{y^0,y^*\}$ do not depend on z.
2. The functions $m_{x_i y_j}(x,y,z) = 0$, $x_i \in \{x^0,x^*\}$, $y_j \in \{y^0,y^*\}$.

We leave it as an exercise to the reader to determine the implications of these conditions in reducing the general form of the conditional utility function to the simplifications we discussed in this section.

24.6 EXTENSIONS TO MULTIPLE ATTRIBUTES

24.6.1 Multiattribute Expansions

Let X_n^{*0} be the set of all 2^n prospects characterized by the n direct value attributes at their boundary values. Let x_n^{*0} represent any element of X_n^{*0}, i.e., $x_n^{*0} \in X_n^{*0}$. Further define X_{iP} as the set of attributes in the tree that are expanded prior to X_i, and let x_{iP}^{*0} represent the instantiations of the attributes expanded before X_i in the term x_n^{*0}. Finally, define X_{iF} as the set of attributes expanded after X_i in the tree; and x_{iF} as an instantiation of these attributes. Based on these definitions, the product $U(x_n^{*0})\prod_{i=1}^{n} g_{x_i}(x_i\,|\,x_{iP}^{*0},x_{iF})$ will include $U(x_n^{*0})$, which is the utility of a particular boundary value of the attributes, and the terms $g_{x_i}(x_i\,|\,x_{iP}^{*0},x_{iF})$, which are the conditional utility or disutility values of X_i conditioned on the boundary values of the prior attributes x_{iP}^{*0} given X_{iF}.

By induction, we can express the utility function in terms of conditional utility assessments using the sum-product expansion as

$$U(x_1, x_2, ..., x_n) = \sum_{x_n^{*0}} U(x_n^{*0}) \prod_{i=1}^{n} g_{x_i}(x_i \mid x_{iP}^{*0}, x_{iF}). \tag{5}$$

24.6.2 The Conditional Utility Function of X_i in a Multiattribute Utility Tree

We now determine the conditional utility function of the ith attribute in a multiattribute utility tree. This conditional utility function, $U(x_i \mid \overline{x}_i)$, determines the utility independence of the attribute on any subset of the attributes, and changes the chronological order of the attribute, by simply substituting for the levels of the prior attributes at their boundary values.

Let $\overline{x}_i^{*0}$ be an instantiation of the complement attributes to X_i at their minimum or maximum values and let x_{iP}^{*0} be the instantiation of the attributes expanded prior to X_i in the term $\overline{x}_i^{*0}$.

Theorem 24.8: The Conditional Utility Function of Attribute X_i
Let X_i be the ith attribute in a utility tree. Let X_{iP}, X_{iF} be the attributes prior to and after X_i, respectively. The conditional utility function of X_i can be expressed as

$$U(x_i \mid \overline{x}_i) = \frac{\sum_{\overline{x}_i^{*0}} \left(d_{\overline{x}_i^{*0}}(x_i, \overline{x}_i) U(x_i \mid x_{iP}^{*0}, x_{iF}) + m_{\overline{x}_i^{*0}}(x_i, \overline{x}_i) \right)}{\sum_{\overline{x}_i^{*0}} \left(d_{\overline{x}_i^{*0}}(x_i^*, \overline{x}_i) + m_{\overline{x}_i^{*0}}(x_i^*, \overline{x}_i) \right)}, \tag{6}$$

where

$$d_{\overline{x}_i^{*0}}(x_i, \overline{x}_i) =$$

$$\prod_{x_j \in x_{iP}} g_{x_j}(x_j \mid x_{jP}^{*0}, x_{jF}) \left(U(x_i^*, \overline{x}_i^{*0}) \prod_{x_l \in x_{iF}} g_{x_l}(x_l \mid x_{lP\setminus i}^{*0}, x_i^*, x_{lF}) - U(x_i^0, \overline{x}_i^{*0}) \prod_{x_l \in x_{iF}} g_{x_l}(x_l \mid x_{lP\setminus i}^{*0}, x_i^0, x_{lF}), \right)$$

$$m_{\overline{x}_i^{*0}}(x_i, \overline{x}_i) = U(x_i^0, \overline{x}_i^{*0}) \prod_{x_l \in x_{iF}} g_{x_l}(x_l \mid x_{lP\setminus i}^{*0}, x_i^0, x_{lF}) \left(\prod_{x_j \in x_{iP}} g_{x_j}(x_j \mid x_{jP}^{*0}, x_i, x_{jF\setminus i}) - \prod_{x_j \in x_{iP}} g_{x_j}(x_j \mid x_{jP}^{*0}, x_i^0, x_{jF\setminus i}). \right)$$

Theorem 24.8 generalizes the results of Theorems 24.4 and 24.7, and it can also be expressed graphically as we did for two and three attributes. Note (again) that the functions $m_{\overline{x}_i^{*0}}(x_i, \overline{x}_i)$ are zero if X_i is a utility-dominant attribute because of the term $U(x_i^0, \overline{x}_i^{*0})$ on the right-hand side.

We can now determine the conditional utility function of any attribute in the tree, $U(x_i \mid \overline{x}_i)$, in terms of conditional utility functions at the boundary values of any subset of its complement, say x_i^*, and we can update our preferences over X_i as X_{iF} varies.

Since the conditional utility functions in the numerator are not indexed by boundary values of X_{iF}, the summation in the numerator can also be expressed in terms of the conditional utility functions of the attribute at the boundary values of x_i^* as

$$U(x_i \mid \overline{x}_i) = \frac{\displaystyle\sum_{x_{iP}^{*0}} \left(d_{x_{iP}^{*0}}(x_i, \overline{x}_i) U(x_i \mid x_{iP}^{*0}, x_{iF}) + m_{x_{iP}^{*0}}(x_i, \overline{x}_i) \right)}{\displaystyle\sum_{\overline{x}_i^{*0}} \left(d_{\overline{x}_i^{*0}}(x_i^*, \overline{x}_i) + m_{\overline{x}_i^{*0}}(x_i^*, \overline{x}_i) \right)}, \tag{7}$$

where $d_{x_{iP}^{*0}}(x_i, \overline{x}_i) = \displaystyle\sum_{x_{iP}^{*0}} d_{\overline{x}_i^{*0}}(x_i, \overline{x}_i)$, $m_{x_{iP}^{*0}}(x_i, \overline{x}_i) = \displaystyle\sum_{x_{iP}^{*0}} m_{\overline{x}_i^{*0}}(x_i, \overline{x}_i)$, where x_{iF}^{*0} is the instantiation of the attributes expanded after X_i at their boundary values.

24.6.3 Interpolation Independence of Attribute X_i on Prior Attributes

Proposition 24.9: Interpolation Independence of Attribute X_i on $X_j, j < i$
If every attribute $X_j, \forall j < i$, is conditionally utility independent of X_i at all boundary values of the attributes $X_l, l = 1, 2, ..., j - 1$, then

1. $d_{\overline{x}_i^{*0}}(x_i, \overline{x}_i)$ does not depend on X_i,
2. $m_{\overline{x}_i^{*0}}(x_i, \overline{x}_i) = 0$,
3. X_i is conditionally interpolation independent of attributes $X_j, j = 1, ..., i - 1$, given attributes $X_m, m = i + 1, ..., k$.

Proposition 24.9, which is a result of direct substitution into (7), enables us to detect the interpolation independence of the ith attribute in the expansion on the attributes expanded before it using the utility tree. The proof is similar to the proof of Proposition 24.1 and hence omitted.

24.6.4 Utility Independence of Attribute X_i on Prior Attributes

Proposition 24.10: Utility Independence of Attribute X_i on $X_j, j < i$
If every attribute $X_j, j < i$ is conditionally utility independent of X_i at all boundary values of the attributes $X_l, l = 1, 2, ..., j - 1$, and if X_i is boundary independent of attributes $X_j, j < i$, then X_i is utility independent of $X_j, j < i$.

Proposition 24.10 enables us to detect the utility independence of the ith attribute in the expansion on the attributes expanded before it. Thus while a tree generally determines the boundary independence of an attribute on the prior attributes, this proposition extends this assertion to utility independence of the prior attributes. The proof is similar to the proof of Proposition 24.2.

24.7 REVERSING THE ORDER OF THE ATTRIBUTES IN THE CONDITIONING STATEMENTS

Consider a decision involving attributes X and Y and their complement attributes. As we have seen, we can express a multiattribute utility function by expanding around attribute X to get

$$U(x,y,\overline{xy}) = U(x^*,y,\overline{xy})U(x\,|\,y,\overline{xy}) + U(x^0,y,\overline{xy})\bar{U}(x\,|\,y,\overline{xy}). \qquad (8)$$

We can also change the order of the conditional assessments in a way that is more comfortable to the decision-maker. Suppose we expand the utility function around another attribute Y. We get

$$U(x,y,\overline{xy}) = U(x,y^*,\overline{xy})U(y\,|\,x,\overline{xy}) + U(x,y^0,\overline{xy})\bar{U}(y\,|\,x,\overline{xy}). \qquad (9)$$

Consistency requires that the right-hand sides of (8) and (9) be equal. Rearranging gives

$$U(y\,|\,x,\overline{xy}) = U(x\,|\,y,\overline{xy})\frac{[U(x^*,y,\overline{xy}) - U(x^0,y,\overline{xy})]}{[U(x,y^*,\overline{xy}) - U(x,y^0,\overline{xy})]} + \frac{[U(x^0,y,\overline{xy}) - U(x,y^0,\overline{xy})]}{[U(x,y^*,\overline{xy}) - U(x,y^0,\overline{xy})]}. \qquad (10)$$

Equation (10) provides a general expression that changes the order of assessments and also provides a consistency check for the conditional utility functions obtained with different assessment orders.

24.8 UTILITY TREE NETWORKS FOR THREE OR MORE ATTRIBUTES

We can now extend the analysis of utility tree networks to multiple attributes. Figure 24.5 shows the fully connected utility tree network.

Figure 24.6 shows a utility tree network exhibiting the relation X UI Y given Z (as demonstrated by the missing solid arrow from X to Y) and the relation Y BI X given Z (as represented by the missing dashed arrow from Y to Z). In terms of conditional utility assessments, these relations imply $U(x\,|\,y,z) = U(x\,|\,y^0,z)$ and $U(y\,|\,x^*,z) = U(y\,|\,x^0,z)$.

Figure 24.5. Fully connected tree with no utility or boundary independence assertions.

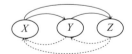

Figure 24.6. Utility tree network asserting X UI Y given Z and Y BI X given Z.

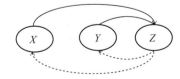

Figure 24.7 shows a utility tree network exhibiting the relation X UI Y given Z and Y UI Z at the boundaries of X. In terms of the

Figure 24.7. X UI Y given Z and Y UI Z given boundaries of X.

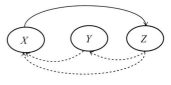

Figure 24.8. X UI YZ and Z BI Y at the boundary values of X.

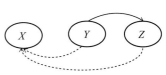

conditional utility functions, this implies that $U(x \mid y, z) = U(x \mid y^0, z)$ and the relations $U(y \mid x^*, z) = U(y \mid x^*, z^0)$ and $U(y \mid x^0, z) = U(y \mid x^0, z^0)$.

Figure 24.8 shows a utility tree network exhibiting the relations X UI YZ and Z BI Y at the boundary values of X. In terms of the conditional utility functions, the figure implies that $U(x \mid y, z) = U(x \mid y^0, z^0), U(z \mid x^*, y^*) = U(z \mid x^*, y^0)$ and $U(z \mid x^0, y^*) = U(z \mid x^0, y^0)$.

Dialogue between Analyst (A) and Decision-Maker (DM)

A: We are set on attributes X, Y, and Z. Correct?

DM: Yes.

A: Are they direct value attributes?

DM: Yes, and we are clear on their definition and their numeric values. They are meaningful quantities and not constructed scales.

A: So let us draw the utility tree network for this decision. First, I draw a diagram with three ovals, X, Y, and Z. Is there a particular order of assessment that you prefer?

DM: What do you mean?

A: I will be asking you a few questions about utility and boundary independence. Attributes that are utility independent on future nodes can come first (they make strong assertions) and those that are only boundary independent can come later.

DM: Ok. Let's stick with the order X, Y, Z.

A: Then let's start with the first node, X. Because it is the first node, I will only be asking about utility independence relations. Is XUIY for all values of Z?

DM: Having reviewed the definitions, I would say no, but XUI Z for all values of Y.

A: So I draw a solid arrow from X to Y because we have not asserted the independence of X on Y. I do not draw a solid arrow from X to Z.

DM: Okay.

A: What about attribute Y? Because this is the second node, we shall be concerned with both utility and boundary independence conditions. Is Y boundary independent of X given Z?

DM: Yes, I can assert that.

A: Now for the utility independence question on the future attribute. Is Y utility independent of Z at all boundary values of X? Note that Y is boundary independent of X given Z, so we can simplify the question and just ask: Is Y utility independent of Z at any boundary value of X?

DM: Yes, absolutely.

Figure 24.9. Utility tree network showing the relations $XUIZ|Y$, $YBIX|Z$, $YUIZ|x^*$, $YUIZ|x^0$.

A: Okay. Now for attribute Z. Because this is the last attribute in the expansion, we shall only be concerned with boundary independence conditions. Is Z boundary independent of either X or Y?

DM: Not really. I cannot assert that.

A: Okay, so I draw two dashed arcs from Z to both X and Y. We are done with the diagram. It looks as shown in Figure 24.9.

DM: So what are the conditional utility assessments needed for these conditions?

A: This is easy to answer. If the utility network was fully connected, then we would need all the assessments in the basic expansion theorem in the order provided. Every missing arc in the diagram (whether solid or dashed) implies easier assessments. We have no solid arrow (utility independence arrow) from Z to X. Therefore, we need an assessment of X at only one reference value of Z but for all values of Y.

DM: Okay.

A: There is no dashed arrow from Y to X and no solid arrow from Y to Z. We need a conditional utility assessment of Y at any reference value of X and any reference value of Z. Finally, we need an assessment of Z at the boundary values of X and Y.

DM: I see. So I can think of the corresponding utility tree and get all the conditional utility functions with the implied assertions. Or read them directly from the utility tree network.

A: Correct. From Figure 24.9, we will need the assessments $U(x \mid y, z^0), U(y \mid x^0, z^0), U(z \mid x^0, y^0), U(z \mid x^0, y^*), U(z \mid x^*, y^0), U(z \mid x^*, y^*)$ as well as the corner values.

DM: What if I do not want to think about boundary independence conditions, just utility independence?

A: That is fine. But you will need to make stronger assertions. Remember: boundary independence conditions are much weaker than utility independence. In Chapter 26 we shall discuss bidirectional utility independence diagrams that focus only on the utility independence relations and the simplifications that they may provide.

24.9 SUMMARY

The multiattribute utility tree extends to multiple attributes.

Utility independence, interpolation independence, and boundary independence assertions between two attributes (or two subsets of attributes) can be conditioned on the value of another attribute (or another subset of the attributes). In this case, we use the terms "conditional utility independence," "conditional interpolation independence," or "conditional boundary independence."

We can infer the conditional utility function of any attribute in the tree by changing its order.

Utility tree networks can capture the conditional utility and boundary independence relations.

ADDITIONAL READINGS

Abbas, A. E. 2005. Bidirectional Utility Diagrams. Technical Report. University of Illinois at Urbana-Champaign, UILU-ENG-2005–3001.

Abbas, A. E. 2009. From Bayes' nets to utility nets. Proceedings of the 29th *International Workshop on Bayesian Inference and Maximum Entropy Methods in Science and Engineering*. Oxford, MI, July 5–10. AIP conference proceedings 1193, pp. 3–12.

Abbas, A. E. 2011a. General decompositions of multiattribute utility functions. *Journal of Multicriteria Decision Analysis* 17(1, 2): 37–59.

Abbas, A. E. 2011b. The multiattribute utility tree. *Decision Analysis* 8(3): 180–205.

APPENDIX 24A GRAPH-BASED METHODS FOR REVERSING THE UTILITY TREE

The Conditional Utility Function of the Second Attribute in the Expansion (Graphically)

THEOREM 24.12

The conditional utility function of the second attribute in a three-attribute utility tree is

$$
U(y \mid x, z) = \frac{\displaystyle\sum_{\substack{x_i \in \{x^*, x^0\} \\ z_k \in \{z^*, z^0\}}} d_{x_i z_k}(x, y, z) U(y \mid x_i, z) + \sum_{\substack{x_i \in \{x^*, x^0\} \\ z_k \in \{z^*, z^0\}}} m_{x_i z_k}(x, y, z)}{\displaystyle\sum_{\substack{x_i \in \{x^*, x^0\} \\ z_k \in \{z^*, z^0\}}} [d_{x_i z_k}(x, y^*, z) + m_{x_i z_k}(x, y^*, z)]},
$$

where $x_i \in \{x^0, x^*\}, z_k \in \{z^0, z^*\}$,

$$
d_{x_i z_k}(x, y, z) = g_{x_i}(x \mid y, z)\big(U(x_i, y^*, z_k) g_{z_k}(z \mid x_i, y^*) - U(x_i, y^0, z_k) g_{z_k}(z \mid x_i, y^0)\big),
$$

and

$$
m_{x_i z_k}(x, y, z) = U(x_i, y^0, z_k) g_{z_k}(z \mid x_i, y^0)\big(g_{x_i}(x \mid y, z) - g_{x_i}(x \mid y^0, z)\big).
$$

Figure 24.10 illustrates how to do this operation graphically using the notation we discussed in the previous section.

Calculating $d_{x_i, z_k}(x, y, z)$

Along any path containing $U(y \mid x_i, z)$, multiply the assessments before $U(y \mid x_i, z)$ by each other, and call the product $b_{x_i, z_k}(x, y, z)$. Also multiply the assessments after $U(y \mid x_i, z)$ by each other and call the product $a_{x_i, z_k}(x, y^*, z)$. Then for $x_i \in \{x^0, x^*\}, z_k \in \{z^0, z^*\}$,

$$
d_{x_i, z_k}(x, y, z) = b_{x_i, z_k}(x, y, z)[a_{x_i, z_k}(x_i, y^*, z) - a_{x_i, z_k}(x_i, y^0, z)].
$$

Calculating $m_{x_i, z_k}(x, y, z)$

Along any path containing $\bar{U}(y \mid x_i, z)$, multiply the assessments before $\bar{U}(y \mid x_i, z)$ and call the product $b_{x_i, z_k}(x, y, z)$. Multiply the assessments after $\bar{U}(y \mid x_i, z)$ and call the product $a_{x_i, z_k}(x, y^0, z)$. Then for $x_i \in \{x^0, x^*\}, z_k \in \{z^0, z^*\}$,

(a)

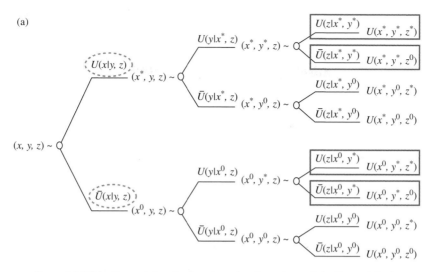

Figure 24.10(a). Determining the functions $d_{x_i,z_k}(x,y,z)$, $x_i \in \{x^0,x^*\}, z_k \in \{z^0,z^*\}$.

(b)

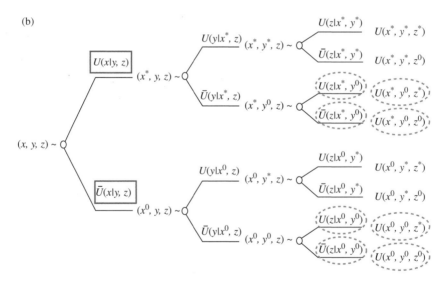

Figure 24.10(b). Determining the functions $m_{x_i,z_k}(x,y,z)$, $x_i \in \{x^0,x^*\}, z_k \in \{z^0,z^*\}$.

$$m_{x_i,z_k}(x,y,z) = a_{x_i,z_k}(x_i,y^0,z)[b_{x_i,z_k}(x,y,z) - b_{x_i,z_k}(x,y^0,z)].$$

The Conditional Utility Function of the Third Attribute in the Expansion (Graphically)

We now show how to calculate the conditional utility function of the third attribute using the assessments of Figure 24.1.

THEOREM 24.9

The Conditional utility function of the third attribute in a three-attribute tree is

(a)

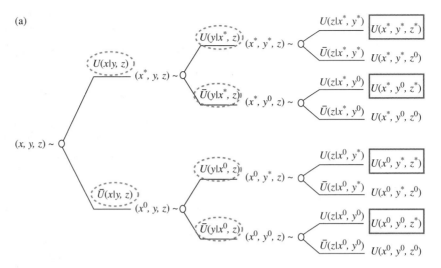

Figure 24.11(a). Determining the functions $d_{x_i, y_j}(x, y, z)$ graphically.

(b)

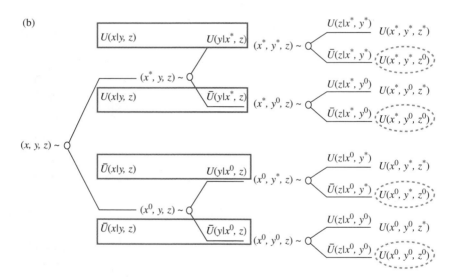

Figure 24.11(b). Determining the functions $m_{x_i, y_j}(x, y, z)$ graphically.

$$U(z \mid x, y) = \frac{\displaystyle\sum_{\substack{x_i \in \{x^*, x^0\} \\ y_j \in \{y^*, y^0\}}} d_{x_i y_j}(x, y, z) U(z \mid x_i, y_j) + \sum_{\substack{x_i \in \{x^*, x^0\} \\ y_j \in \{y^*, y^0\}}} m_{x_i y_j}(x, y, z)}{\displaystyle\sum_{\substack{x_i \in \{x^*, x^0\} \\ y_j \in \{y^*, y^0\}}} [d_{x_i y_j}(x, y, z^*) + m_{x_i y_j}(x, y, z^*)]},$$

where $x_i \in \{x^0, x^*\}, y_j \in \{y^0, y^*\},$

$$d_{x_i y_j}(x, y, z) = g_{x_i}(x \mid y, z) g_{y_j}(y \mid x_i, z) \big(U(x_i, y_j, z^*) - U(x_i, y_j, z^0) \big),$$

and

$$m_{x_i y_j}(x, y, z) = U(x_i, y_j, z^0)\Big(g_{x_i}(x \mid y, z)g_{y_j}(y \mid x_i, z) - g_{x_i}(x \mid y, z^0)g_{y_j}(y \mid x_i, z^0)\Big).$$

Figure 24.11 illustrates how to do this operation graphically.

APPENDIX 24B PROOFS OF SOME OF THE THEOREMS AND PROPOSITIONS

Proof of Theorem 24.8

First, we rewrite Equation (5) as

$$
\begin{aligned}
U(x_i, \overline{x}_i) &= \sum_{\overline{x}_i^{*0}} U(x_i^*, \overline{x}_i^{*0}) U(x_i \mid x_{iP}^{*0}, x_{iF}) \prod_{x_j \in x_{iP}} g_{x_j}(x_j \mid x_{iP}^{*0}, x_{jF}) \prod_{x_l \in x_{iF}} g_{x_l}(x_l \mid x_{lP\backslash i}^{*0}, x_i^*, x_{lF}) \\
&+ \sum_{\overline{x}_i^{*0}} U(x_i^0, \overline{x}_i^{*0}) \overline{U}(x_i \mid x_{iP}^{*0}, x_{iF}) \prod_{x_j \in x_{iP}} g_{x_j}(x_j \mid x_{iP}^{*0}, x_{jF}) \prod_{x_l \in x_{iF}} g_{x_l}(x_l \mid x_{lP\backslash i}^{*0}, x_i^0, x_{lF}) \\
&= \sum_{\overline{x}_i^{*0}} U(x_i^0, \overline{x}_i^{*0}) \prod_{x_j \in x_{iP}} g_{x_j}(x_j \mid x_{iP}^{*0}, x_{jF}) \prod_{x_l \in x_{iF}} g_{x_l}(x_l \mid x_{lP\backslash i}^{*0}, x_i^0, x_{lF}) + \sum_{x_{K\backslash i}^{*0}} U(x_i \mid x_{iP}^{*0}, x_{iF}) \times \\
&\left\{ U(x_i^*, \overline{x}_i^{*0}) \prod_{x_j \in x_{iP}} g_{x_j}(x_j \mid x_{iP}^{*0}, x_{jF}) \prod_{x_l \in x_{iF}} g_{x_l}(x_l \mid x_{lP\backslash i}^{*0}, x_i^*, x_{lF}) - U(x_i^0, \overline{x}_i^{*0}) \prod_{x_j \in x_{iP}} g_{x_j}(x_j \mid x_{iP}^{*0}, x_{jF}) \prod_{x_l \in x_{iF}} g_{x_l}(x_l \mid x_{lP\backslash i}^{*0}, x_i^0, x_{lF}) \right\}
\end{aligned}
$$

By direct substitution,

$$U(x_i^0, \overline{x}_i) = \sum_{\overline{x}_i^{*0}} U(x_i^0, \overline{x}_i^{*0}) \prod_{x_j \in x_{iP}} g_{x_j}(x_j \mid x_{jP}^{*0}, x_{jF}) \prod_{x_l \in x_{iF}} g_{x_l}(x_l \mid x_{lP\backslash i}^{*0}, x_i^0, x_{lF}).$$

Therefore,

$$U(x_i, \overline{x}_i) - U(x_i^0, \overline{x}_i) = \sum_{\overline{x}_i^{*0}} [d_{\overline{x}_i^{*0}}(x_i, \overline{x}_i) U(x_i \mid x_{iP}^{*0}, x_{iF}) + m_{\overline{x}_i^{*0}}(x_i, \overline{x}_i)],$$

where

$$d_{\overline{x}_i^{*0}}(x_i, \overline{x}_i) = \prod_{x_j \in x_{iP}} g_{x_j}(x_j \mid x_{jP}^{*0}, x_{jF}) \left(U(x_i^*, \overline{x}_i^{*0}) \prod_{x_l \in x_{iF}} g_{x_l}(x_l \mid x_{lP\backslash i}^{*0}, x_i^*, x_{lF}) - U(x_i^0, \overline{x}_i^{*0}) \prod_{x_l \in x_{iF}} g_{x_l}(x_l \mid x_{lP\backslash i}^{*0}, x_i^0, x_{lF}) \right),$$

$$m_{\overline{x}_i^{*0}}(x_i, \overline{x}_i) = U(x_i^0, \overline{x}_i^{*0}) \prod_{x_l \in x_{iF}} g_{x_l}(x_l \mid x_{lP\backslash i}^{*0}, x_i^0, x_{lF}) \left(\prod_{x_j \in x_{iP}} g_{x_j}(x_j \mid x_{jP}^{*0}, x_i, x_{jF\backslash i}) - \prod_{x_j \in x_{iP}} g_{x_j}(x_j \mid x_{jP}^{*0}, x_i^0, x_{jF\backslash i}) \right).$$

Therefore,

$$U(x_i \mid \overline{x}_i) \triangleq \frac{U(x_i, \overline{x}_i) - U(x_i^0, \overline{x}_i)}{U(x_i^*, \overline{x}_i) - U(x_i^0, \overline{x}_i)} = \frac{\displaystyle\sum_{\overline{x}_i^{*0}} \left(d_{\overline{x}_i^{*0}}(x_i, \overline{x}_i) U(x_i \mid x_{iP}^{*0}, x_{iF}) + m_{\overline{x}_i^{*0}}(x_i, \overline{x}_i) \right)}{\displaystyle\sum_{\overline{x}_i^{*0}} \left(d_{\overline{x}_i^{*0}}(x_i^*, \overline{x}_i) + m_{\overline{x}_i^{*0}}(x_i^*, \overline{x}_i) \right)}.$$

Proof of Proposition 24.1: Conditional Interpolation Independence

Using the expression for the conditional utility function for the second attribute in the tree of Theorem 24.4, we have

$$U(y \mid x, z) = \frac{\displaystyle\sum_{\substack{x_i \in \{x^*, x^0\} \\ z_k \in \{z^*, z^0\}}} d_{x_i z_k}(x, y, z)U(y \mid x_i, z) + \sum_{\substack{x_i \in \{x^*, x^0\} \\ z_k \in \{z^*, z^0\}}} m_{x_i z_k}(x, y, z)}{\displaystyle\sum_{\substack{x_i \in \{x^*, x^0\} \\ z_k \in \{z^*, z^0\}}} [d_{x_i z_k}(x, y^*, z) + m_{x_i z_k}(x, y^*, z)]},$$

where $x_i \in \{x^0, x^*\}, z_k \in \{z^0, z^*\}$,

$$d_{x_i z_k}(x, y, z) = g_{x_i}(x \mid y, z)\big(U(x_i, y^*, z_k)g_{z_k}(z \mid x_i, y^*) - U(x_i, y^0, z_k)g_{z_k}(z \mid x_i, y^0)\big),$$

and

$$m_{x_i z_k}(x, y, z) = U(x_i, y^0, z_k)g_{z_k}(z \mid x_i, y^0)\big(g_{x_i}(x \mid y, z) - g_{x_i}(x \mid y^0, z)\big).$$

Observe that if X is utility independent of Y given Z, then the functions $m_{x_i z_k}(x, y, z) = 0$. Furthermore, the functions $d_{x_i z_k}(x, y, z)$ do not depend on Y. Therefore,

$$U(y \mid x, z) = \frac{\displaystyle\sum_{\substack{x_i \in \{x^*, x^0\} \\ z_k \in \{z^*, z^0\}}} d_{x_i z_k}(x, z)U(y \mid x_i, z)}{\displaystyle\sum_{\substack{x_i \in \{x^*, x^0\} \\ z_k \in \{z^*, z^0\}}} d_{x_i z_k}(x, z)} = \sum_{x_i \in \{x^*, x^0\}} w_{x_i}(x, z)U(y \mid x_i, z), \tag{11}$$

where $w_{x_i}(x, z) = \dfrac{\displaystyle\sum_{z_k \in \{z^*, z^0\}} d_{x_i z_k}(x, z)}{\displaystyle\sum_{\substack{x_i \in \{x^*, x^0\} \\ z_k \in \{z^*, z^0\}}} d_{x_i z_k}(x, z)}, \displaystyle\sum_{x_i \in \{x^*, x^0\}} w_{x_i}(x, z) = 1.$

Therefore, Y is conditionally interpolation independent of X given Z.

Proof of Proposition 24.2: Utility Independence on a Prior Attribute

First refer to the proof of Proposition 24.1. Substituting for $U(y \mid x^*, z) = U(y \mid x^0, z)$ results in

$$U(y \mid x, z) = \sum_{x_i \in \{x^*, x^0\}} w_{x_i}(x, z)U(y \mid x_i, z) = U(y \mid x^0, z) \sum_{x_i \in \{x^*, x^0\}} w_{x_i}(x, z) = U(y \mid x^0, z).$$

Therefore, Y is conditionally utility independent of X given Z.

Proof of Proposition 24.3: Utility Independence on the Complement

Using Proposition 24.1, with the added condition that $U(y \mid x^0, z) = U(y \mid x^0, z^0)$, we get by direct substitution $U(y \mid x, z) = U(y \mid x^0, z^0)$. Therefore, Y is utility independent of both X and Z.

Proof of Theorem 24.4

It is a special case of Theorem 24.8 presented above.

CHAPTER 25

Multiattribute Utility Functions with
Partial Utility Independence

Chapter Concepts

- Partial utility independence conditions for multiattribute utility functions
- Bidirectional utility diagrams
- The simplest bidirectional diagram: The multilinear form
- Utility tree representation: A simpler method to verify the multilinear form
- Determining the assessments needed for a bidirectional utility diagram

25.1 INTRODUCTION

In the previous chapters, we discussed various utility, interpolation, and boundary independence conditions. These conditions specify the relation between *one* attribute and a *subset* of the attributes.

In this chapter we elaborate further on these independence concepts and discuss the functional forms of utility functions and the utility assessments needed to derive the multiattribute utility function under a variety of utility independence conditions. We discuss situations where every attribute is utility independent of its complement. We also discuss other situations, which we refer to as *partial utility independence*, where independence relations are specified for only a subset of the attributes or relations that specify the independence of an attribute on only a subset of its complement. We show the use of bidirectional utility diagrams in providing graphical representations for these expressions. We also show the convenience of specifying utility and boundary independence relations in an expansion order that corresponds to a utility tree.

25.2 CAPTURING INDEPENDENCE RELATIONS USING UTILITY TREES

The multiattribute utility tree provides an expansion order for the functional form of a multiattribute utility function in terms of normalized conditional utility assessments.

If the utility and boundary independence relations are specified such that utility independence conditions are on future nodes in the expansion and boundary independence conditions are on past nodes, then the functional form of the utility function can be obtained by direct substitution into the utility tree. The following example illustrates this idea.

EXAMPLE 25.1: Three-Attribute Utility Tree with Utility and Boundary Independence Conditions

---

Consider a three-attribute utility tree with an expansion order of X then Y then Z and the following conditions

$$X\ UI\ Y\,|\,Z, Y\ UI\ Z\,|\,x^*, Y\ UI\ Z\,|\,x^0, \text{and } Z\ BI\ X, Y.$$

These conditions imply the following about the conditional utility functions:

1. $X\ UI\ Y\,|\,Z$ implies $U(x\,|\,y,z) = U(x\,|\,y^0,z)$

   The $X\ UI\ Y\,|\,Z$ condition reduces the normalized conditional utility assessment $U(x\,|\,y,z)$ to a conditional utility assessment for X at any arbitrary value of Y, and this is true for all fixed values of Z. This condition simplifies the assessment: instead of having to assess a conditional utility assessment of X at the entire domain of both Y and Z, we just have to conduct this assessment of the conditional utility at a fixed value of Y but for all values of Z.

2. $Y\ UI\ Z\,|\,x^*$ implies $U(y\,|\,x^*,z) = U(y\,|\,x^*,z^0)$, and $Y\ UI\ Z\,|\,x^0$ implies that $U(y\,|\,x^0,z) = U(y\,|\,x^0,z^0)$

   The utility independence condition, $Y\ UI\ Z\,|\,x^*$, asserts that we do not need to assess the utility of Y for all values of Z when conditioned on x^*; we simply need $U(y\,|\,x^*,z^0)$. Likewise, $Y\ UI\ Z\,|\,x^*$ asserts that we need $U(y\,|\,x^0,z^0)$. However, this is only true at the boundary value of X and not on the entire domain of X.

3. $Z\ BI\ X, Y$ implies $U(z\,|\,x^*,y^*) = U(z\,|\,x^*,y^0) = U(z\,|\,x^0,y^*) = U(z\,|\,x^0,y^0)$

   The boundary independence condition $Z\ BI\ X, Y$ asserts that only one conditional utility assessment is needed for attribute Z at any boundary values of Y and X.

   Figure 25.1 shows the corresponding three-attribute utility tree.

   The functional form of the utility function can be obtained by rolling back the tree to get

$$
\begin{aligned}
U(x,y,z) = &\ U(x\,|\,y^0,z)U(y\,|\,x^*,z^0)U(z\,|\,x^0,y^0)U(x^*,y^*,z^*) \\
&+ U(x\,|\,y^0,z)U(y\,|\,x^*,z^0)\bar{U}(z\,|\,x^0,y^0)U(x^*,y^*,z^0) \\
&+ U(x\,|\,y^0,z)\bar{U}(y\,|\,x^*,z^0)U(z\,|\,x^0,y^0)U(x^*,y^0,z^*) \\
&+ U(x\,|\,y^0,z)\bar{U}(y\,|\,x^*,z^0)\bar{U}(z\,|\,x^0,y^0)U(x^*,y^0,z^0) \\
&+ \bar{U}(x\,|\,y^0,z)U(y\,|\,x^0,z^0)U(z\,|\,x^0,y^0)U(x^0,y^*,z^*) \\
&+ \bar{U}(x\,|\,y^0,z)U(y\,|\,x^0,z^0)\bar{U}(z\,|\,x^0,y^0)U(x^0,y^*,z^0) \\
&+ \bar{U}(x\,|\,y^0,z)\bar{U}(y\,|\,x^0,z^0)U(z\,|\,x^0,y^0)U(x^0,y^0,z^*) \\
&+ \bar{U}(x\,|\,y^0,z)\bar{U}(y\,|\,x^0,z^0)\bar{U}(z\,|\,x^0,y^0)U(x^0,y^0,z^0)
\end{aligned}
$$

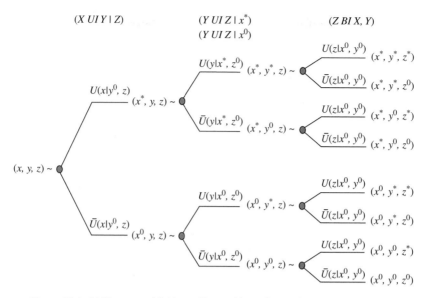

Figure 25.1. Utility tree exhibiting utility and boundary independence relations.

Note: If you work in an expansion order based on the utility tree assessments, then the functional form is straightforward. It will require conditioning utility independence relations on boundary values of the previous attributes (instead of all values of the previous attributes). The tree also captures simplifications based on boundary independence conditions. The utility assessments needed for the functional form can be obtained by inspection from either the utility tree of the utility tree network.

In this chapter, we consider the functional form of a multiattribute utility function for a given set of utility independence conditions that need not necessarily correspond to an expansion order in the tree. In our notation we refer to a set of complement attributes to attribute X_i as $\bar{X}_i$ and to a set of complement attributes to the set X_{ij} as $\bar{X}_{ij}$.

25.3 BIDIRECTIONAL UTILITY DIAGRAMS WITH THREE OR MORE ATTRIBUTES

Definition

A *bidirectional utility diagram* is a graph whose nodes represent the attributes, $X_1, X_2, ..., X_n$, and whose arrows represent the *possibility of* utility dependence between them such that:

1. The absence of an arrow between two nodes X_i and X_j asserts the utility independence relations $(X_i \; UI \; X_j \,|\, \bar{X}_{ij})$ and $(X_j \; UI \; X_i \,|\, \bar{X}_{ij})$.
2. A unidirectional arrow from X_i to X_j, written $X_i \to X_j$, asserts the relation $(X_i \; UI \; X_j \,|\, \bar{X}_{ij})$.
3. A bidirectional arrow between X_i and X_j, written $X_i \leftrightarrow X_j$, does not assert any independence relations.

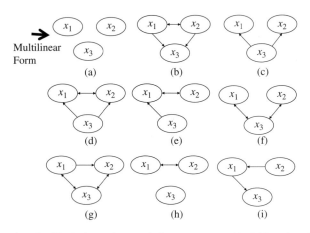

Figure 25.2. Examples of utility independence relations represented by bidirectional utility diagrams.

Figure 25.2 shows examples of bidirectional utility diagrams for three attributes with various utility independence specifications. For example, Figure 25.2(b) asserts the relations $(X_1 \ UI \ X_3 \,|\, X_2)$ and $(X_2 \ UI \ X_3 \,|\, X_1)$. We leave it as an exercise to the reader to determine the utility independence relations asserted by the other diagrams in Figure 25.2.

> *Question: Is there a method to derive the functional form that is implied by a given diagram?*

We shall discuss the answer to this question in the next few sections, but first we consider the functional form of the simplest bidirectional utility diagram: one with no arrows.

25.4 THE SIMPLEST UTILITY DIAGRAM: A DIAGRAM WITH NO ARROWS

25.4.1 Every Attribute Is Utility Independent of Its Complement

The simplest form of bidirectional utility diagram is one where there are no arrows (Figure 25.2(a)), asserting that every attribute is utility independent of its complement. This diagram implies the utility independence relations

$$(X_1 \ UI \ X_2, X_3), (X_2 \ UI \ X_1, X_3), (X_3 \ UI \ X_1, X_2).$$

What is the functional form of the multiattribute utility function that corresponds to these independence relations? To answer this question, we observe that every row node in a bidirectional diagram that has some missing arrows into it provides a utility independence equation that must be satisfied by the functional form. To derive this functional form, we first write the set of equations that it needs to satisfy.

Let us start with the first equation

$$(X_1 \ UI \ X_2, X_3) \Rightarrow U(x_1 \,|\, x_2, x_3) = U(x_1 \,|\, x_2^0, x_3^0).$$

Using the basic expansion theorem, this implies that

$$U(x_1,x_2,x_3) = U(x_1^*,x_2,x_3)U(x_1 \mid x_2^0,x_3^0) + U(x_1^0,x_2,x_3)\bar{U}(x_1 \mid x_2^0,x_3^0). \tag{1}$$

This independence equation has reduced the three-attribute term $U(x_1,x_2,x_3)$ into a univariate assessment, $U(x_1 \mid x_2^0,x_3^0)$, as well as two bivariate assessments, $U(x_1^*,x_2,x_3)$ and $U(x_1^0,x_2,x_3)$. The bivariate terms can be further decomposed using the available independence assertions. For example, for the second attribute,

$$(X_2 \ UI \ X_1,X_3) \Rightarrow U(x_2 \mid x_1,x_3) = U(x_2 \mid x_1^0,x_3^0).$$

Using the basic expansion theorem, we can decompose those terms as

$$U(x_1^*,x_2,x_3) = U(x_1^*,x_2^*,x_3)U(x_2 \mid x_1^0,x_3^0) + U(x_1^*,x_2^0,x_3)\bar{U}(x_2 \mid x_1^0,x_3^0) \tag{2}$$

and

$$U(x_1^0,x_2,x_3) = U(x_1^0,x_2^*,x_3)U(x_2 \mid x_1^0,x_3^0) + U(x_1^0,x_2^0,x_3)\bar{U}(x_2 \mid x_1^0,x_3^0). \tag{3}$$

We have now decomposed the bivariate terms $U(x_1^*,x_2,x_3)$ and $U(x_1^0,x_2,x_3)$ into a univariate assessment $U(x_2 \mid x_1^0,x_3^0)$ and four univariate assessments for attribute X_3 at the various corner values of its complement, $U(x_1^*,x_2^*,x_3), U(x_1^*,x_2^0,x_3), U(x_1^0,x_2^*,x_3), U(x_1^0,x_2^0,x_3)$.

Let us now incorporate the final independence equation,

$$(X_3 \ UI \ X_1,X_2) \Rightarrow U(x_3 \mid x_1,x_2) = U(x_3 \mid x_1^0,x_2^0).$$

Using the basic expansion theorem applied to the assessments of X_3, we get

$$U(x_1^*,x_2^*,x_3) = U(x_1^*,x_2^*,x_3^*)U(x_3 \mid x_1^0,x_2^0) + U(x_1^*,x_2^*,x_3^0)\bar{U}(x_3 \mid x_1^0,x_2^0) \tag{4}$$

$$U(x_1^*,x_2^0,x_3) = U(x_1^*,x_2^0,x_3^*)U(x_3 \mid x_1^0,x_2^0) + U(x_1^*,x_2^0,x_3^0)\bar{U}(x_3 \mid x_1^0,x_2^0) \tag{5}$$

$$U(x_1^0,x_2^*,x_3) = U(x_1^0,x_2^*,x_3^*)U(x_3 \mid x_1^0,x_2^0) + U(x_1^0,x_2^*,x_3^0)\bar{U}(x_3 \mid x_1^0,x_2^0) \tag{6}$$

$$U(x_1^0,x_2^0,x_3) = U(x_1^0,x_2^0,x_3^*)U(x_3 \mid x_1^0,x_2^0) + U(x_1^0,x_2^0,x_3^0)\bar{U}(x_3 \mid x_1^0,x_2^0). \tag{7}$$

These equations have reduced the four univariate assessments for X_3 into a single normalized univariate assessment $U(x_3 \mid x_1^0,x_2^0)$ as well as eight corner utility values. Therefore, this last equation served as a boundary independence equation. Substituting from $(2),(3),(4),(5),(6)$, and (7) into (1) gives the functional form of the utility function.

The Multilinear Form: When Every Attribute Is Utility Independent of Its Complement

Keeney and Raiffa (1976) show that the functional form that corresponds to the utility independence conditions that every attribute is utility independent of its complement is the multilinear form,

$$\begin{aligned} U(x_1,x_2,x_3) = {} & k_1 U_1(x_1 \mid \bar{x}_1^0) + k_2 U_2(x_2 \mid \bar{x}_2^0) + k_3 U_3(x_3 \mid \bar{x}_3^0) + k_{12} U_1(x_1 \mid \bar{x}_1^0)U_2(x_2 \mid \bar{x}_2^0) \\ & + k_{13} U_1(x_1 \mid \bar{x}_1^0)U_3(x_3 \mid \bar{x}_3^0) + k_{23} U_2(x_2 \mid \bar{x}_2^0)U_3(x_3 \mid \bar{x}_3^0) \\ & + k_{123} U_1(x_1 \mid \bar{x}_1^0)U_2(x_2 \mid \bar{x}_2^0)U_3(x_3 \mid \bar{x}_3^0) \end{aligned}$$

The normalizing constants are such that

$$k_i = U(x_i^*, \overline{x}_i^0),$$
$$k_{ij} = U(x_i^*, x_j^*, \overline{x}_{ij}^0) - k_i - k_j,$$
$$k_{123} = 1 - k_1 - k_2 - k_3 - k_{12} - k_{13} - k_{23}$$

Note: There is a simple way to memorize the normalizing constants: you simply substitute into the multilinear form for values of the attributes at their minimum and maximum values, and note that the corresponding normalized utility functions are either zero or one. For example,

$$U(x_1^*, x_2^0, x_3^0) = k_1 \times 1 + k_2 \times 0 + k_3 \times 0 + k_{12} \times 1 \times 0 + k_{13} \times 1 \times 0$$
$$+ k_{23} \times 0 \times 0 + k_{123} \times 1 \times 0 \times 0 = k_1$$

$$U(x_1^*, x_2^*, x_3^0) = k_1 \times 1 + k_2 \times 1 + k_3 \times 0 + k_{12} \times 1 \times 1 + k_{13} \times 1 \times 0 + k_{23} \times 1 \times 0$$
$$+ k_{123} \times 1 \times 1 \times 0 = k_1 + k_2 + k_{12} \dots \text{etc.}$$

The Multilinear Form for n Attributes

The extension to n attributes where every attribute is utility independent of its complement can also be expressed as a multilinear combination of single-attribute functions (Keeney and Raiffa 1976),

$$U(x_1, \dots, x_n) = \sum_{i=1}^{n} k_i U_i(x_i) + \sum_{i=1}^{n} \sum_{j>i}^{n} k_{ij} U_i(x_i) U_j(x_j) + \dots \qquad (8)$$
$$+ k_{123\dots n} U_1(x_1) U_2(x_2) \dots U_n(x_n),$$

where n is the number of attributes, $U_i(x_i) = U(x_i \mid \overline{x}_i^0)$ is a single-attribute utility function for attribute X_i, and $k_i, k_{ij}, \dots, k_{123\dots n}$ are scaling constants.

25.4.2 An Alternate Way to Derive the Multilinear Form Using Utility and Boundary Independence

The utility tree enables us to think about the types of conditions needed to derive a certain functional form. It is clear that the multilinear form resulted in single-attribute utility assessments, for each attribute, of the form $U_i(x_i) = U(x_i \mid \overline{x}_i^0)$. Can we arrive at this multilinear form without making the assumptions that every attribute is utility independent of its complement?

Figure 25.3 presents a sequence of conditions leading to the multilinear form in a three-attribute tree. The figure shows that if

(i)   the first attribute is utility independent of its complement,
(ii)   the second attribute is boundary independent of the first and utility independent of the third at any boundary value of the first, and
(iii) if the third attribute is boundary independent of the previous two attributes,

then the conditional utility assessments in the whole tree reduce to a single univariate assessment for each attribute conditioned on a single instantiation of the complement

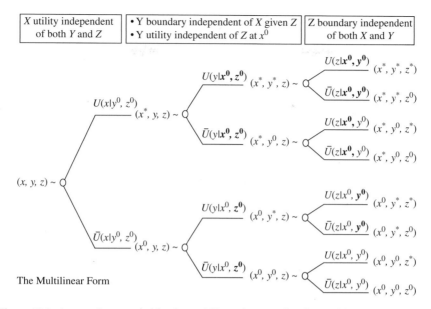

X utility independent of both Y and Z	• Y boundary independent of X given Z • Y utility independent of Z at x^0	Z boundary independent of both X and Y

Figure 25.3. Assumptions needed for the multilinear form can be obtained from the utility tree.

attributes. As we now assert, these conditions are equivalent to the set of conditions that every attribute is utility independent of its complement.

By induction, we have proved the following:

Theorem 25.1

If every attribute in the utility tree is boundary independent of the previous attributes and utility independent of the future attributes at any boundary value of the previous attributes, then every attribute is utility independent of its complement.

Theorem 25.1 shows that we can test for the conditions leading to the multilinear form using only boundary independence (instead of utility independence) on the previous attributes. The importance of this result lies in the fact that we do not need to assert as many utility independence relations to derive the functional form. We merely assert utility independence on future attributes and boundary independence on prior attributes.

Multilinear Form for n Attributes Using the Utility Tree

Substituting into the basic expansion theorem gives

$$U(x_1,...,x_n) = \sum_{x_n^{*0}} U(x_n^{*0}) \prod_{i=1}^{n} g(x_i \mid \overline{x}_i^0), \qquad (9)$$

where the summation is over all possible corner values x_n^{*0} of the attributes at their minimum and maximum values and g_i is either a conditional utility or disutility function depending on the instantiation of attribute i in the term $U(x_n^{*0})$. Equation (9) is another (compact) way of expressing the multilinear form (8).

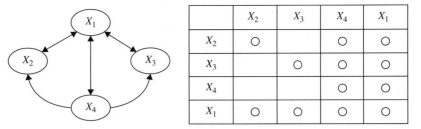

	X_2	X_3	X_4	X_1
X_2	O		O	O
X_3		O	O	O
X_4			O	O
X_1	O	O	O	O

Figure 25.4. A bidirectional utility diagram and its corresponding utility matrix.

25.5 DERIVING THE FUNCTIONAL FORMS WITH PARTIAL UTILITY INDEPENDENCE

Utility trees provide equations for an expansion order where utility independence is on future nodes and boundary independence is on prior nodes. This formulation facilitates the determination of the functional form. If utility independence relations are not provided in this way, then the functional form can be quite tedious to calculate, and there are a few general subtleties that need to be considered when expanding bidirectional utility diagrams as we illustrate below.

EXAMPLE 25.2: Decomposition of a Bidirectional Utility Diagram

---

Consider the bidirectional diagram of Figure 25.4 and the corresponding utility matrix. The figure implies the following conditions:

Node X_2: Asserts that $U(x_2 \mid x_1, x_3, x_4) = U(x_2 \mid x_1, x_3^0, x_4)$.
Node X_3: Asserts that $U(x_3 \mid x_1, x_2, x_4) = U(x_3 \mid x_1, x_2^0, x_4)$.
Node X_4: Asserts that $U(x_4 \mid x_1, x_2, x_3) = U(x_4 \mid x_1, x_2^0, x_3^0)$.

Let us expand around the first two rows X_2, X_3 using the basic expansion theorem to illustrate an important point about the decomposition of conditional utility functions.

Expanding around X_2, X_3 using the basic expansion theorem gives

$$
\begin{aligned}
U(x_1,x_2,x_3,x_4) = {} & U(x_1,x_2^*,x_3^*,x_4)U(x_2 \mid x_1,x_3^0,x_4)U(x_3 \mid x_1,x_2^0,x_4) + \\
& U(x_1,x_2^*,x_3^0,x_4)U(x_2 \mid x_1,x_3^0,x_4)\bar{U}(x_3 \mid x_1,x_2^0,x_4) + \\
& U(x_1,x_2^0,x_3^*,x_4)\bar{U}(x_2 \mid x_1,x_3^0,x_4)U(x_3 \mid x_1,x_2^0,x_4) + \\
& U(x_1,x_2^0,x_3^0,x_4)\bar{U}(x_2 \mid x_1,x_3^0,x_4)\bar{U}(x_3 \mid x_1,x_2^0,x_4).
\end{aligned}
\tag{10}
$$

The next step would be to expand around attribute X_4 to simplify the four terms $U(x_1,x_2^*,x_3^*,x_4)$, $U(x_1,x_2^*,x_3^0,x_4)$, $U(x_1,x_2^0,x_3^*,x_4)$, and $U(x_1,x_2^0,x_3^0,x_4)$ as we have done previously with the expansion theorem. Note, however, that the independence of X_4 on both X_2 and X_3 allows for further decompositions of the conditional assessments $U(x_2 \mid x_1,x_3^0,x_4), U(x_3 \mid x_1,x_2^0,x_4)$ in (10). Thus the conditional utility functions are not maximally decomposed. While we can, in principle, still assess those utility functions directly from the decision-maker, it would be more convenient, and cognitively simpler, to first decompose them into lower-order terms using the available utility independence relations provided by X_4.

To incorporate the additional independence assertions, we simply reverse the order of the assessments using the relation we discussed in Chapter 24. We have

$$U(y \mid x, \overline{xy}) = U(x \mid y, \overline{xy}) \frac{[U(x^*, y, \overline{xy}) - U(x^0, y, \overline{xy})]}{[U(x, y^*, \overline{xy}) - U(x, y^0, \overline{xy})]} + \frac{[U(x^0, y, \overline{xy}) - U(x, y^0, \overline{xy})]}{[U(x, y^*, \overline{xy}) - U(x, y^0, \overline{xy})]}$$

Therefore, we reverse $U(x_2 \mid x_1, x_3^0, x_4)$ and note that $U(x_4 \mid x_1, x_2, x_3) = U(x_4 \mid x_1, x_2^0, x_3^0)$, to get

$$U(x_2 \mid x_1, x_3^0, x_4) = U(x_4 \mid x_1, x_2^0, x_3^0) \frac{[U(x_1, x_2, x_3^0, x_4^*) - U(x_1, x_2, x_3^0, x_4^0)]}{[U(x_1, x_2^*, x_3^0, x_4) - U(x_1, x_2^0, x_3^0, x_4)]} + \frac{[U(x_1, x_2, x_3^0, x_4^0) - U(x_1, x_2^0, x_3^0, x_4)]}{[U(x_1, x_2^*, x_3^0, x_4) - U(x_1, x_2^0, x_3^0, x_4)]}. \tag{11}$$

Similarly, we reverse $U(x_3 \mid x_1, x_2^0, x_4)$ and note that $U(x_4 \mid x_1, x_2, x_3) = U(x_4 \mid x_1, x_2^0, x_3^0)$, to get

$$U(x_3 \mid x_1, x_2^0, x_4) = U(x_4 \mid x_1, x_2^0, x_3^0) \frac{[U(x_1, x_2^0, x_3, x_4^*) - U(x_1, x_2^0, x_3, x_4^0)]}{[U(x_1, x_2^0, x_3^*, x_4) - U(x_1, x_2^0, x_3^0, x_4)]} + \frac{[U(x_1, x_2^0, x_3, x_4^0) - U(x_1, x_2^0, x_3^0, x_4)]}{[U(x_1, x_2^0, x_3^*, x_4) - U(x_1, x_2^0, x_3^0, x_4)]}. \tag{12}$$

By doing this, we have reduced the conditional assessments of $U(x_2 \mid x_1, x_3^0, x_4)$ and $U(x_3 \mid x_1, x_2^0, x_4)$ which are in effect three-attribute assessments, into lower-order (two-attribute) assessments. Thus, while (10) alone does not provide maximal decomposition of the conditional utility functions in this situation – due to independence on prior decomposed nodes – we can further decompose the conditional utility functions if we accompany them with additional equations to reverse the assessment order. This may pose an analytical inconvenience, however, if we wish to obtain a closed-form expression for the utility function.

The next section presents an iterative approach to determine the assessments needed in this case.

25.6 ITERATIVELY DETERMINING THE UTILITY ASSESSMENTS NEEDED FOR PARTIAL INDEPENDENCE

25.6.1 Applying an Independence Equation to a Term That Has No Attributes at Fixed Values

The equation $(X \ UI \ X_I \mid X_D)$ is an independence equation for attribute X on a set of attributes X_I, given a set of attributes X_D. This equation decomposes the term $U(x, x_I, x_D)$ and allows us to write

$$U(x \mid x_I, x_D) = U(x \mid x_I^0, x_D). \tag{13}$$

Equation (13) can also be incorporated into the Basic Expansion Theorem as

$$\boxed{U(x, x_I, x_D) = U(x^*, x_I, x_D) \, U(x \mid x_I^0, x_D) + U(x^0, x_I, x_D) \, \bar{U}(x \mid x_I^0, x_D).} \tag{14}$$

Equations (13) and (14) assert that $(X \ UI \ X_I \mid X_D)$, as they lead to a linear transformation of $U(x, x_I^0, x_D)$ when x_I changes. They also decompose the term $U(x, x_I, x_D)$ into three basic terms. The first two are $U(x^*, x_I, x_D), U(x^0, x_I, x_D)$, obtained by replacing x on the left-hand side with x^o and x^*. Consequently, the order of these two terms is equal to that of the left-hand side reduced by one. The third term, $U(x, x_I^0, x_D)$, is obtained by replacing x_I on the left-hand side with x_I^0 and so the size of its joint utility assessments will depend on x_I^0.

25.6.2 Applying an Independence Equation to a Term That Already Has Attributes at Fixed Values

Now consider an independence equation $(X \ UI \ X_I \mid X_D)$ applied to the general term $U(x, x_I, x_D)$ that may already contain attributes at their boundary values.

We distinguish three possible cases:

(i) **If either $x = x^o$ or $x = x^*$ or $x_I = x_I^0$:**
    This situation implies any of the following: (1) that the attribute itself is already at a boundary value; or (2) that the attributes of which this attribute is independent are fixed at their minimum values. This would make the independence relations (13) or (14) ***trivial*** (or redundant) in terms of their effect on reducing the functional form. To see why this is true, note that if $x = x^o$ or if $x = x^*$, then (13) would equate 0 with 0 or 1 with 1 (respectively). And if $x_I = x_I^o$ then (13) would equate $U(x \mid x_I^0, x_D)$ with itself. These utility independence relations are not effective in decomposing the term $U(x, x_I, x_D)$ if any of these conditions hold.

(ii) **If $x \neq x^o$ and $x \neq x^*$, and if not all attributes in x_I are set at the boundary values**
    This situation implies that the attribute itself is not already at a boundary value and furthermore that the attributes of which it is independent are not all set at their boundary values. This equation will be very effective in reducing the dimensionality of the term $U(x, x_I, x_D)$ into the lower-order term $U(x, x_I^o, x_D)$. We refer to this as a ***decomposing*** equation. In the RHS of Figure 25.5, this is referred to as $x \neq x^o, x^*$ and $x_I \neq x_I^{*0}$.

(iii) **If $x \neq x^o, x \neq x^*$, and x_I is set at a corner value of the attributes in x_I except for $x_I = x_I^o$**
    This situation implies that the attribute itself is not already at a boundary value and that the attributes of which it is independent are all set at their boundary values but at least one is not set at the minimum boundary value. This equation will not reduce the dimensionality of the assessments needed, but it is useful as a boundary independence relation that reduces the number of assessments needed. We refer to this as a *relating* equation. In the RHS of Figure 25.5, this is referred to as $x \neq x^o, x^*$ and $x_I \neq x_I^{*0 \backslash 0}$.

Figure 25.5 provides a tree representation for this decomposition and a characterization of the different types of independence equations.
   Note: Since the three terms that emerge from an independence equation all contain one of the following instantiations – $x = x^o$ or $x = x^*$ or $x_I = x_I^o$ – *these terms would*

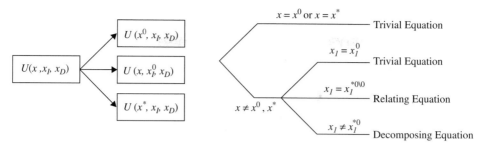

Figure 25.5. Tree representation of an independence equation for $U(x, x_I, x_D)$ at X.

lead to a trivial independence equation if decomposed by node X again using the same independence equation. However, other nodes in the diagram may provide further decompositions for these output terms. We therefore build on the tree representation of Figure 25.5 for each node and each non-trivial equation.

25.6.3 An Iterative Approach for Determining the Assessments of a Bidirectional Diagram

An iterative algorithm for decomposing a bidirectional utility diagram with nodes $\{X_1, X_2, ..., X_n\}$ can be expressed as follows:

Let $\{X_1, X_2, ..., X_n\}$ denote any arbitrary chronological order of nodes with which we carry out the decomposition.

Define $Decomp\_X_0 = \{U(x_1, ..., x_n)\}$ as the initial set of required utility assessments. The algorithm carries out the following steps:

- *For $i = 1 \to n$*
  - *For each term $U(x_i, x_{il}, x_{iD})$ in $Decomp\_X_{i-1}$ for which $x_i \neq x_i^0, x_i \neq x_i^*$ and $x_{il} \neq x_{il}^0$, write the following (non-trivial) independence equation,*

$$U(x_i \mid x_{iPD}^{*0}, x_{iPI}^{*0}, x_{iFD}, x_{iFI}) = U(x_i \mid x_{iPD}^{*0}, x_{il}^0, x_{iFD}) \qquad (15)$$

- *Update the set $Decomp\_X_{i-1}$ to produce $Decomp\_X_i$ using the following steps:*

  (i)   *Introduce the three terms: $U(x_i^*, x_{il}, x_{iD})$, $U(x_i^0, x_{il}, x_{iD})$, and $U(x_i, x_{il}^o, x_{iD})$.*
  (ii) *Delete the decomposed term $U(x_i, x_{il}, x_{iD})$.*
  (iii) *Remove any redundant assessments.*

- *Next i.*
- *The set $Decomp\_X_n$ is the set of lower-order terms that result from the decomposition.*

The decomposition algorithm replaces each term in a non-trivial independence equation with the sub-tree on the left-hand side of Figure 25.6.

EXAMPLE 25.3: Decomposition of Independence Relations in a Bidirectional Diagram

Consider again the bidirectional utility diagram of Figure 25.4. Figure 25.6 shows the tree representation that derives the required functional assessments in the expansion order X_2, X_4, X_3. For each non-trivial equation, we expand the tree at a given node using the sub-graph of Figure 25.5. From Figure 25.5, the lower-order terms required for this functional form are the set $\{U(x_1, x_2^0, x_3^0, x_4), U(x_1, x_2, x_3^o, x_4^o), U(x_1, x_2, x_3^o, x_4^*),$ $U(x_1, x_2^o, x_3, x_4^o), U(x_1, x_2^o, x_3, x_4^*), U(x_1, x_2^*, x_3^*, x_4^o), U(x_1, x_2^*, x_3^*, x_4^*)\}$.

It is straightforward to show that the algorithm provides maximal decomposition of the functional assessments, since the lower-order terms that emerge from the decomposition at each node would lead to trivial independence equations if decomposed by that node again. Hence a *single pass across each node is sufficient to achieve maximal decomposition.*

Example 25.4: Changing the Order of the Iterative Expansion

We now show that the lower-order joint utility assessments that result from this single pass are independent of the chosen chronological order of decomposition, and so we get the same joint utility assessments with any chronological order. Figure 25.7 shows the tree representation of the iterative decomposition in the order X_4, X_2, X_3. As we can see from the figure, the same end nodes emerge at the end of the iterative decomposition.

We formalize this result using the following theorem.

Theorem 25.2
The chronological order of nodes by which we carry out the decomposition does not change the required joint utility assessments that emerge from the iterative decomposition algorithm.

Note: When the utility or boundary independence relations are not presented in a simple tree expansion order, it might be worth recalling the approach of constructing a multiattribute utility function using a value function and a utility function over value. The construction could be a lot simpler using the value function approach particularly when there are logical relations among the attributes.

25.7 SUMMARY

1. We do not need to assert that every attribute is utility independent of its complement to derive the multilinear form. It suffices to assert that for a given expansion order, every attribute is utility independent of the future nodes and boundary independent of the previous nodes.

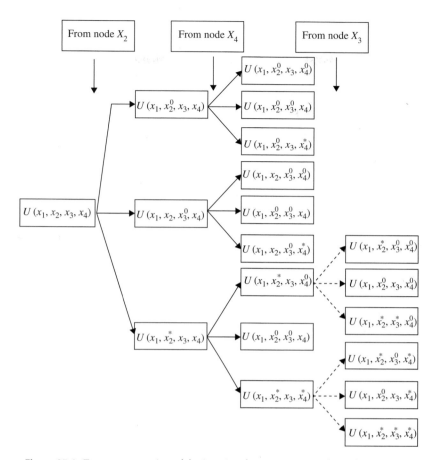

Figure 25.6. Tree representation of the iterative decomposition in the order X_2, X_4, X_3.

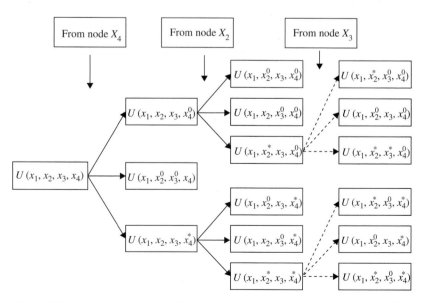

Figure 25.7. Tree representation of the iterative decomposition in the order X_4, X_2, X_3.

2. Decomposition of utility and boundary independence relations that are given in the form of a utility tree expansion order is a lot simpler than general independence assertions.
3. Using a value function and a utility over value can provide a simpler way to construct the multiattribute utility function when partial utility independence relations exist.

ADDITIONAL READINGS

Abbas, A. E. 2005. Bidirectional Utility Diagrams. Technical Report. University of Illinois at Urbana-Champaign, UILU-ENG-2005–3001.

Abbas, A. E. 2009. From Bayes' nets to utility nets. Proceedings of the 29th *International Workshop on Bayesian Inference and Maximum Entropy Methods in Science and Engineering*. Oxford, MI, July 5–10. AIP conference proceedings 1193, pp. 3–12.

Abbas, A. E. 2011a. General decompositions of multiattribute utility functions. *Journal of Multicriteria Decision Analysis* 17(1, 2): 37–59.

Abbas, A. E. 2011b. The multiattribute utility tree. *Decision Analysis* 8(3): 180–205.

Keeney, R and H. Raiffa. 1976. *Decisions with Multiple Objectives*. John Wiley, New York.

APPENDIX 25A PROOF OF THEOREM 25.2

To prove this theorem, we first interchange the chronological order of two consecutive nodes X_i, X_j and show that this does not change the required assessments. Next, we observe that any chronological order can be obtained from a given chronological order by a series of pairwise interchanges. Consider the general term $U(x_i, x_{iI}, x_{iD})$.

1. If X_i, X_j are not connected with any arrows, we first expand around attribute X_i and observe that, in this case, $X_j \in X_{iI}$ and $X_i \in X_{jI}$. The six terms that result from the decomposition of $U(x_i, x_{iI}, x_{iD})$ are $U(x_{ij}^{*0}, \bar{x}_{ij}), U(x_j, x_{iI}^0, x_{iD}), U(x_i, x_{iI}^0, x_{iD})$. The symmetry in these six terms shows that the decomposition is order-independent.
2. If X_i, X_j are connected with bidirectional arrows, then we get the nine terms: $U(x_{ij}^{*0}, \bar{x}_{ij})$, $U(x_i^{*0}, x_j, x_{jI}^0, x_{jD} \setminus x_i)$, $U(x_i, x_j^{*0}, x_{iI}^0, x_{iD} \setminus x_j)$, $U(x_i, x_j, \{x_{iI} \cup x_{jI}\}^0, x_{iD} \cap x_{jD})$, where $x_{jD} \setminus x_i$ represents an instantiation of the attributes in $X_{jD} \setminus X_i$; $x_{iD} \cap x_{jD}$ is an instantiation of $X_{iD} \cap X_{jD}$, and $\{x_{iI} \cup x_{jI}\}^0$ is the minimum instantiation of the set $X_{iI} \cup X_{jI}$. These nine terms are also symmetric and therefore order-independent.
3. If there is an arrow $X_i \rightarrow X_j$, we get the following seven terms regardless of the decomposition order: $U(x_{ij}^{*0}, \bar{x}_{ij}), U(x_i, x_j^0, x_{iI}^0, x_{iD}), U(x_i^0, x_j, x_{jI}^0, x_{jD})$, and $U(x_i^*, x_j, x_{jI}^0, x_{jD})$.
4. Note that $X_i \leftarrow X_j$ yields the same seven assessments since we have interchanged the order in step (3) above.

To complete the proof, we observe that any chronological order can be obtained from a given chronological order by a series of pairwise interchanges.

CHAPTER 26

Higher-Order Independence Relations

Chapter Concepts

- Higher-order conditional utility functions.
- Basic expansion theorem around a subset of attributes
- Higher-order independence conditions for a subset of the attributes
- Relating higher order and lower order independence relations: corner independence
- Mutual utility independence: when every subset is utility independent of its complement

26.1 INTRODUCTION

So far we have considered expansions and independence conditions that involve the interactions between a single attribute and a set of other attributes. These interactions were expressed in terms of a conditional utility function, such as $U(x \mid y, z)$, implying a conditional utility assessment of attribute X taking into account the values of attributes Y and Z. Various independence conditions simplified the assessment of this conditional utility function either by reducing its dimensionality, where we condition on only a single value of the remaining attributes (utility independence), or by relating the assessments at the different boundary values of the other attributes to reduce the number of assessments needed (boundary independence). For example, the conditional utility independence of X on Y given Z would reduce this assessment to the term $U(x \mid y^0, z)$, and the conditional boundary independence of X on Y given Z would imply that $U(x \mid y^*, z) = U(x \mid y^0, z)$, which relates the conditional utility function of X at the boundary values of Y, and therefore we would need only one of these assessments if X were expanded after Y.

We have not yet considered general formulations for conditional utility assessments of the form $U(x, y \mid z)$, where more than one attribute are to the left of the vertical bar in the conditioning statement, and where attribute dominance conditions do not exist. This conditional utility $U(x, y \mid z)$ implies a normalized two-attribute utility assessment for attributes X and Y conditioned on the values of attribute Z. We refer to a conditional utility assessment that involves more than one attribute to the left of the vertical bar as a ***higher-order conditional utility assessment***.

This chapter discusses basic expansion results in terms of higher-order utility assessments and the implications of having utility, boundary, and interpolation independence under these conditions. The chapter also discusses higher-order independence relations and relates them to independence conditions for a single attribute. We show how higher-order independence relations further simplify the functional form of the multiattribute utility.

26.2 HIGHER-ORDER UTILITY ASSESSMENTS

To begin, it is important to define a conditional utility assessment for two or more attributes.

Definition
The *higher-order normalized conditional utility* for two attributes, X and Y, conditioned on an attribute Z, is

$$U(x,y\,|\,z) = \frac{U(x,y,z) - U(x^0,y^0,z)}{U(x^*,y^*,z) - U(x^0,y^0,z)} \quad (1)$$

Figure 26.1 provides an indifference assessment interpretation for this conditional utility function. We can also see by direct substitution into (1) that the previous definition yields a value of 1 when $x = x^*, y = y^*$ and a value of zero when $x = x^0, y = y^0$. If more of an attribute is preferred to less, then this conditional utility assessment $U(x,y\,|\,z)$ is bounded between zero and one.

Definition
A *higher-order conditional disutility function* for two attributes, X and Y, conditioned on an attribute Z, is

$$\bar{U}(x,y\,|\,z) = 1 - U(x,y\,|\,z) \quad (2)$$

Basic Expansion Theorem in Terms of Higher-Order Two-Attribute Conditional Utility Functions
Rearranging (1) and simplifying using (2) gives the basic expansion theorem in terms of the higher-order assessments for two attributes X and Y as

$$U(x,y,z) = U(x,y\,|\,z)U(x^*,y^*,z) + \bar{U}(x,y\,|\,z)U(x^0,y^0,z) \quad (3)$$

Note that expanding around two attributes simultaneously using a higher-order utility assessment $U(x,y\,|\,z)$ as in Figure 26.1 is different from expanding around attribute X and then attribute Y. Using the higher-order assessment results in only two end points in the tree, $U(x^*,y^*,z)$ and $U(x^0,y^0,z)$, while expanding around attribute X and then Y, would result in the four assessments, $U(x^*,y^*,z), U(x^*,y^0,z), U(x^0,y^*,z)$, and $U(x^0,y^0,z)$.

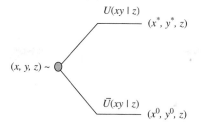

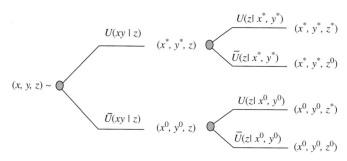

Figure 26.1. Indifference probability interpretation and utility tree expansion around two attributes, X and Y.

Figure 26.2. Utility tree showing an expansion around XY then Z.

As we did previously with the utility tree, we can also expand the two terms $U(x^*, y^*, z)$ and $U(x^0, y^0, z)$ around attribute Z to get

$$U(x^*, y^*, z) = U(x^*, y^*, z^*)U(z \mid x^*, y^*) + U(x^*, y^*, z^0)\bar{U}(z \mid x^*, y^*)$$

and

$$U(x^0, y^0, z) = U(x^0, y^0, z^*)U(z \mid x^0, y^0) + U(x^0, y^0, z^0)\bar{U}(z \mid x^0, y^0).$$

Substituting gives an expansion of the form

$$\begin{aligned}
U(x, y, z) = &\; U(x^*, y^*, z^*)U(x, y \mid z)U(z \mid x^*, y^*) \\
&+ U(x^*, y^*, z^0)U(x, y \mid z)\bar{U}(z \mid x^*, y^*) \\
&+ U(x^0, y^0, z^*)U(x, y \mid z)U(z \mid x^0, y^0) \\
&+ U(x^0, y^0, z^0)U(x, y \mid z)\bar{U}(z \mid x^0, y^0)
\end{aligned}$$

The utility tree expansion corresponding to this equation is shown in Figure 26.2. Similar expansions can be applied to any number of attributes.

26.3 HIGHER-ORDER UTILITY INDEPENDENCE

Having defined a higher-order conditional utility assessment, we can now define higher-order utility independence. For example, if the joint utility assessment of the set X, Y does not depend on attribute Z, then we can write

$$U(x, y \mid z) = U(x, y \mid z^0)$$

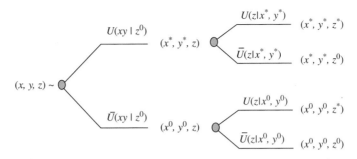

Figure 26.3. Utility tree showing that the group XY is utility independent of Z.

or equivalently that

$$U(x,y,z) = U(x,y\,|\,z^0)U(x^*,y^*,z) + \bar{U}(x,y\,|\,z^0)U(x^0,y^0,z).$$

We can also expand around attribute Z after that to get

$$\begin{aligned}
U(x,y,z) = \;&U(x^*,y^*,z^*)U(x,y\,|\,z^0)U(z\,|\,x^*,y^*)\\
+\;&U(x^*,y^*,z^0)U(x,y\,|\,z^0)\bar{U}(z\,|\,x^*,y^*)\\
+\;&U(x^0,y^0,z^*)U(x,y\,|\,z^0)U(z\,|\,x^0,y^0)\\
+\;&U(x^0,y^0,z^0)U(x,y\,|\,z^0)\bar{U}(z\,|\,x^0,y^0)
\end{aligned}$$

Figure 26.3 shows a utility tree exhibiting this higher-order independence relation.

Of course, we can now extend this idea to more than two attributes, where we have conditional utility independence of the form $(X_K\ UI\ X_I\,|\,X_D)$. We formally define higher-order utility independence below.

Definition: Higher-Order Utility Independence
A set of attributes X_K is utility independent of another set X_I given X_D, written $(X_K\ UI\ X_I\,|\,X_D)$, with $X_I \cup X_D = \bar{X}_K$, if preferences for joint lotteries over X_K do not change as we vary X_I at any fixed values of X_D.

Implications for Simplifying the Utility Assessment
Observe that the expression $(X_K\ UI\ X_I\,|\,X_D)$ implies that

$$U(x_K,x_D\,|\,x_I) = U(x_K,x_D\,|\,x_I^0).$$

This implies that the conditional utility assessment $U(x_K,x_D\,|\,x_I)$ needs to be assessed at only one instantiation of the attributes in the set X_I.

Illustrating the Definition of Higher-Order Utility Independence
Using the basic expansion theorem in terms of higher-order conditional utility assessments, we get

$$U(x_K,x_D,x_I) = U(x_K,x_D\,|\,x_I^0)U(x_K^*,x_D,x_I) + \bar{U}(x_K,x_D\,|\,x_I^0)U(x_K^0,x_D,x_I).$$

Rearranging gives

$$U(x_K, x_D, x_I) = [U(x_K^*, x_D, x_I) - U(x_K^0, x_D, x_I)]U(x_K, x_D \mid x_I^0) + U(x_K^0, x_D, x_I).$$

Define $k(x_I, x_D) = U(x_K^*, x_D, x_I) - U(x_K^0, x_D, x_I)$, $d(x_I, x_D) = U(x_K^0, x_D, x_I)$.

If more of an attribute is strictly preferred to less, $k(x_I, x_D) = U(x_K^*, x_D, x_I) - U(x_K^0, x_D, x_I) > 0$.

Hence, $(X_K \ UI \ X_I \mid X_D)$ implies that

$$\boxed{U(x_K, x_I, x_D) = k(x_I, x_D)U(x_K, x_D \mid x_I^0) + d(x_I, x_D)}, \tag{4}$$

for some functions $k > 0$, d.

Because the rank order of lotteries is invariant under a positive linear transformation on the utility function, the independence assertion $(X_K \ UI \ X_I \mid X_D)$ implies that, when x_D is set to any fixed value, changing x_I leads to a positive linear transformation of $U(x_K, x_D \mid x_I^0)$. Therefore, preferences for lotteries over X_K will not change as we vary any of the attributes of X_I.

26.4 RELATING HIGHER-ORDER UTILITY INDEPENDENCE TO LOWER-ORDER UTILITY INDEPENDENCE

As we discussed, an expansion around two attributes $U(X, Y \mid Z)$ is different from an expansion around each of the attributes X and Y individually in a chronological order. A natural question that arises then is the relation between higher-order and lower-order utility independence relations that result from these two expansions. When considering the relation between higher-order utility independence and lower-order (or single-attribute) utility independence, it will be useful to consider several notions. We illustrate them below using four attributes **(or four partitions of the attributes)** and consider the following cases.

Case 1: Independence on Attributes to the Right of the Vertical Bar

1. If $(X, Y \ UI \ W, Z)$, what does this imply about the independence relations of X, Y on each of W or Z individually? In other words,

> *Question 1: If $(X, Y \ UI \ W, Z)$, can we deduce that $(X, Y \ UI \ W \mid Z)$ and that $(X, Y \ UI \ Z \mid W)$?*

2. If $(X, Y \ UI \ W \mid Z)$ and $(X, Y \ UI \ Z \mid W)$, what does this imply about the independence relations of X, Y on the set W, Z? In other words,

> *Question 2: If $(X, Y \ UI \ W \mid Z)$ and $(X, Y \ UI \ Z \mid W)$, can we deduce that $(X, Y \ UI \ W, Z)$?*

Case 2: Independence of Attributes to the Left of the Vertical Bar

3. If $(X, Y \ UI \ W, Z)$, what does this imply about the independence of the individual attribute, X on W, Z given Y, and the independence of the attribute Y on W, Z given X? In other words,

> *Question 3:* If $(X,Y \text{ } UI \text{ } W,Z)$, can we deduce that $(X \text{ } UI \text{ } W,Z|Y)$ and $(Y \text{ } UI \text{ } W,Z|X)$?

4. If $(X \text{ } UI \text{ } W,Z|Y)$ and $(Y \text{ } UI \text{ } W,Z|X)$, what does this imply about the independence of the set X,Y on the set of attributes W, Z? In other words,

> *Question 4:* If $(X \text{ } UI \text{ } W,Z|Y)$ and $(Y \text{ } UI \text{ } W,Z|X)$, can we deduce that $(X,Y \text{ } UI \text{ } W,Z)$?

We answer these questions below.

26.4.1 Question 1: If $(X,Y \text{ } UI \text{ } W,Z)$, Can We Deduce That $(X,Y \text{ } UI \text{ } W|Z)$ and $(X,Y \text{ } UI \text{ } Z|W)$?

This question asks whether the utility independence of a subset of attributes X, Y on another subset W, Z implies the utility independence of X, Y on each of the individual attributes W and Z at all (fixed) values of the remaining attribute. ***The answer is positive.***

If $(X,Y \text{ } UI \text{ } W.,Z)$, then $U(x,y|w,z) = U(x,y|w^0,z^0)$; the normalized utility assessment over X, Y does not depend on the actual values of either W or Z. Therefore, it must be true that the utility assessment of X and Y does not depend on W at any fixed Z, i.e., $(X,Y \text{ } UI \text{ } W|Z)$, and also that the utility assessment of X and Y does not depend on Z at any fixed W, i.e., $(X,Y \text{ } UI \text{ } Z|W)$.

Figure 26.4 shows the higher-order utility independence relation $(X,Y \text{ } UI \text{ } W,Z)$, which implies the two relations $(X,Y \text{ } UI \text{ } W|Z)$ and $(X,Y \text{ } UI \text{ } Z|W)$ in Figure 26.5. Therefore, we can divide a subset of nodes for which unidirectional arrows are entering.

Summary:

$$(X,Y \text{ } UI \text{ } W,Z) \text{ implies } (X,Y \text{ } UI \text{ } W|Z) \text{ and } (X,Y \text{ } UI \text{ } Z|W).$$

Figure 26.4 implies Figure 26.5.

26.4.2 Question 2: If $(X,Y \text{ } UI \text{ } W|Z)$ and $(X,Y \text{ } UI \text{ } Z|W)$, Can We Deduce That $(X,Y \text{ } UI \text{ } W,Z)$?

This question asks about the reverse direction of the previous question: Does the utility independence of a subset of attributes X, Y on each of the individual attributes W and Z at any fixed value of the remaining attribute imply the utility independence of X, Y on the combined set W, Z? ***The answer is also positive.***

To see why this is true, note that $(X,Y \text{ } UI \text{ } W|Z)$ implies the relation

$$U(x,y|w,z) = U(x,y|w^0,z)\forall x,y,w,z.$$

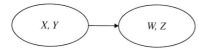

Figure 26.4. Bidirectional utility diagrams representing the independence relation $(X,Y \ UI \ W,Z)$.

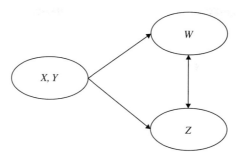

Figure 26.5. Bidirectional utility diagram representing the independence relations $(X,Y \ UI \ W \mid Z)$ and $(X,Y \ UI \ Z \mid W)$.

This also implies that the joint utility assessment of X,Y need only be conducted at a single value of W, but it needs to be assessed for all values of Z.

Similarly, the condition $(X,Y \ UI \ Z \mid W)$ implies the relation

$$U(x,y \mid w,z) = U(x,y \mid w,z^0) \forall x,y,w,z.$$

Combining the two equations sequentially gives

$U(x,y \mid w,z) = U(x,y \mid w^0,z) = U(x,y \mid w^0,z^0)$, which implies that $(X,Y \ UI \ W,Z)$. Therefore, we need to conduct the joint utility assessment of X,Y at only one arbitrary value of both W and Z.

Summary

$$(X,Y \ UI \ W \mid Z) \text{ and } (X,Y \ UI \ Z \mid W) \text{ imply that } (X,Y \ UI \ W,Z).$$

Consequently, Figure 26.5 implies Figure 26.4.

Therefore, we can combine a subset of nodes for which unidirectional arrows are entering.

To generalize these results, let X_K, X_I, X_D be partition subsets covering the set of attributes present such that $X_K \cup X_I \cup X_D = \{X_1, X_2, ..., X_n\}$.

Proposition 26.1: *The following are equivalent*

(i) $(X_K \ UI \ X_I \mid X_D)$ *and*
(ii) *The set of pairwise utility independence relations* $(X_K \ UI \ X_j \mid \overline{X}_{jK}), \forall X_j \in X_I$, *where* $\overline{X}_{jK}$ *is the set of complement attributes to the set* $X_K \ UI \ X_j$

Proposition 26.1 shows that if a subset of the attributes X,Y is utility independent of another subset, W,Z, then the subset X,Y is utility independent of each of the individual attributes W and Z, and vice versa.

This implies that we can represent the independence relation $(X_K \ UI \ X_I \,|\, X_D)$ using a set of independence relations between the set X_K and each of the individual attributes within the set X_I. This result enables us to express higher-order utility independence relations of a set of attributes on another set using bidirectional utility diagrams. It also enables us to *combine or divide a subset of nodes that unidirectional arrows are entering.*

26.4.3 Question 3: If $(X,Y \ UI \ W,Z)$, Does This Imply That $(X \ UI \ W,Z\,|\,Y)$ and $(Y \ UI \ W,Z\,|\,X)$?

This question asks whether the independence of a subset X,Y on another subset W,Z implies the independence of each of the individual attributes X or Y on the set W, Z. **The answer is positive**, and we refer to this as the **independence blanket**.

> **Proposition 26.2: The Independence Blanket**
> If $(X_K \ UI \ X_I \,|\, X_D)$, *then every subset of the attributes within* $X_L \subset X_K$ *is also utility independent of the set* X_I *given* $X_D, X_{K\setminus L}$.

Proposition 26.2 shows that if a set of attributes X_K is utility independent of another set X_I for all fixed values of the complement, then every subset within X_K is also independent of X_I for all fixed values of the remaining attributes. We can therefore think of the independence of a larger set as an "independence blanket" that shields the smaller subsets of any arrows from its independence set.

To illustrate, consider the diagram of Figure 26.6(a) with two combined attributes X,Y, and an independence assertion $(X,Y \ UI \ W,Z)$. Proposition 26.2 asserts that Figure 26.6(a) also implies the independence relations of Figure 26.6(b). Therefore, we can *divide the subset of nodes from which unidirectional arrows are emerging.*

> **Summary:**
>
> $$(X,Y \ UI \ W,Z)\,\text{implies}\,(X \ UI \ W,Z\,|\,Y)\,\text{and}\,(Y \ UI \ W,Z\,|\,X).$$
>
> Figure 26.6(a) implies Figure 26.6(b).

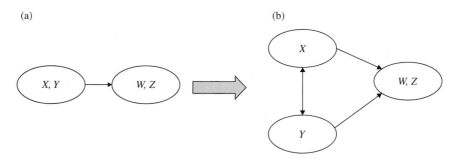

Figure 26.6. (a) Higher-order diagram. (b) Lower-order diagram implied by the higher-order relations.

There is another important implication of Proposition 26.2: *higher-order independence relations imply a set of lower-order relations.* But is the converse true? We answer this below.

26.4.4 Question 4: If $(X \ UI \ W,Z \mid Y)$ and $(Y \ UI \ W,Z \mid X)$, Does This Imply That $(X,Y \ UI \ W,Z)$?

This question asks whether lower-order independence relations imply higher-order independence relations. For example, does Figure 26.6(b) imply Figure 26.6(a)? *The answer is negative.*

> **Summary:**
>
> $$(X \ UI \ W,Z \mid Y) \text{ and } (Y \ UI \ W,Z \mid X) \text{ do not imply } (X,Y \ UI \ W,Z).$$
>
> Figure 26.6(b) does not imply Figure 26.6(a).

We discuss this in further detail below and highlight the additional assertions needed to make this true.

First, we observe that $(X,Y \ UI \ W,Z)$ implies a decomposition of the form

$$U(x,y,w,z) = k(w,z)U(x,y \mid w^0, z^0) + d(w,z),$$

which reduces the assessment into separable functions of W,Z and X,Y.

On the other hand, the two equations $(X \ UI \ W,Z \mid Y)$ and $(Y \ UI \ W,Z \mid X)$ imply (respectively) that

$$U(x,y,w,z) = k_1(y,w,z)U(x \mid y,w^0,z^0) + d_1(y,w,z)$$

and

$$U(x,y,w,z) = k_2(x,w,z)U(y \mid x,w^0,z^0) + d_2(x,w,z),$$

which do not necessarily provide a separable decomposition of functions of W,Z and X,Y. Therefore, the conditions corresponding to different independence conditions are not necessarily equivalent. Hence, lower-order independence relations do not necessarily imply higher-order relations. Furthermore, *we cannot necessarily combine nodes from which unidirectional arrows are emerging.*

26.5 ADDITIONAL CONDITIONS TO ASSERT HIGHER-ORDER INDEPENDENCE FROM LOWER-ORDER INDEPENDENCE

Observe that expanding the tree with the higher-order assessment, $U(x,y \mid z)$, in Figure 26.1 results in the two end points, $U(x^*, y^*, z), U(x^0, y^0, z)$. On the other hand (as we have seen), expanding the utility tree around attribute X then Y results in four assessments: the same two assessments, $U(x^*, y^*, z), U(x^0, y^0, z)$, and two additional assessments, $U(x^*, y^0, z), U(x^0, y^*, z)$, which we refer to as the corner assessments.

Figure 26.7. Replacing the corner assessments with $(x^*, y^*, z), (x^0, y^0, z)$.

To relate the two trees, we first need to express the corner assessments $U(x^*, y^0, z), U(x^0, y^*, z)$ in terms of the assessments $U(x^*, y^*, z), U(x^0, y^0, z)$ to have the same end values. Since these corner assessments are in general functions of z, we denote these indifference assessments using the two functions $k_x(z), k_y(z)$ as illustrated in Figure 26.7.

The two functions $k_x(z), k_y(z)$ can be obtained from the utility trees of Figure 26.7. To determine $k_x(z)$, for example, we observe the following from the left tree of Figure 26.7,

$$U(x^*, y^0, z) = k_x(z)U(x^*, y^*, z) + (1 - k_x(z))U(x^0, y^0, z).$$

Rearranging gives the following definition

$$k_x(z) = \frac{U(x^*, y^0, z) - U(x^0, y^0, z)}{U(x^*, y^*, z) - U(x^0, y^0, z)}$$

Following the same procedure for $k_y(z)$, we get

$$k_y(z) = \frac{U(x^0, y^*, z) - U(x^0, y^0, z)}{U(x^*, y^*, z) - U(x^0, y^0, z)}.$$

Figure 26.8 shows a tree representation for the lower-order expansion of X and Y but the end points are expressed in terms of $U(x^*, y^*, z), U(x^0, y^0, z)$ using the assessments $k_x(z)$, $k_y(z)$.

We can now derive the assertions that would be needed to make higher-order utility independence assertions from lower-order assertions. Suppose that X is utility independent of Z given Y, and Y is utility independent of Z at all boundary values of X. This would result in the tree of Figure 26.9.

For the group X, Y to be utility independent of Z, we would need to say that $U(x, y \mid z) = U(x, y \mid z^0)$ and equate this tree to that of Figure 26.10. This implies that the product of conditional utility assessments along any path leading to any of the end points $U(x^*, y^*, z), U(x^0, y^0, z)$ does not depend on z.

The utility tree of Figure 26.9 can be reduced to that of Figure 26.10 because they both have the same end points. We can tell by inspection from Figure 26.9 that the conditional utility assessments leading to the term $U(x^*, y^*, z)$ would be

$$U(x \mid y, z^0)U(y \mid x^*, z^0) + U(x \mid y, z^0)\bar{U}(y \mid x^*, z^0)k_x(z) + \bar{U}(x \mid y, z^0)U(y \mid x^0, z^0)k_y(z) \quad (5)$$

From the tree of Figure 26.10, the conditional utility assessment leading to the term $U(x^*, y^*, z)$ is $U(x, y \mid z^0)$, which means that it is not a function of z.

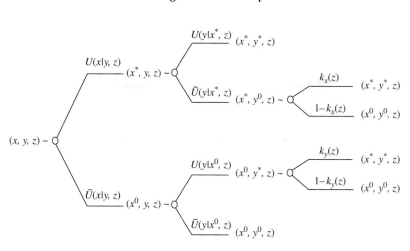

Figure 26.8. Utility tree showing corner assessments for grouping two attributes in a three-attribute tree.

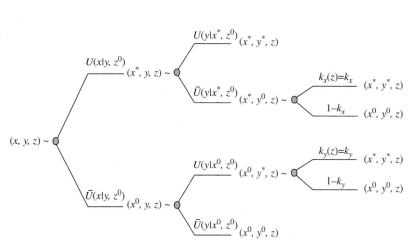

Figure 26.9. X utility independent of Z given Y, and Y utility independent of Z at all boundary values of X.

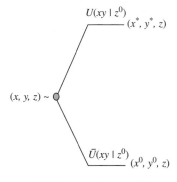

Figure 26.10. Utility tree depicting X, Y utility independent of Z.

If the two trees of Figures 26.9 and 26.10 are equivalent, then Equation (5) cannot depend on z. We can see that if both $k_x(z), k_y(z)$ do not depend on z, then the conditional utility function $U(x, y \mid z)$ in Figure 26.9 will not depend on z, because *we arrive at the leaves* $(x^*, y^*, z), (x^0, y^0, z)$ *without any term along the path depending on z.* We now make the following definition.

Definition: Corner Independence
Attributes X and Y are corner independent of attribute Z if the corner assessments $k_x(z), k_y(z)$ do not depend on Z.

The following proposition explains the importance of corner independence.

Proposition 26.3: Higher-Order Utility Independence from Lower-Order Independence
If X is utility independent of Z given Y, and if Y is utility independent of Z at all boundary values of X, and if XY is corner independent of Z, then XY is utility independent of Z.

The previous result, as well as the definition of corner independence, extend to more than two attributes. For three attributes X, Y, W to be corner independent of Z, we would require that the indifference probabilities of all corner points of X, Y, W, i.e., $(x_i, y_j, w_k, z), x_i \in \{x^0, x^*\}, y_j \in \{y^0, y^*\}, w_k \in \{w^0, w^*\}$ in terms of a binary lottery yielding (x^*, y^*, w^*, z) and (x^0, y^0, w^0, z), be independent of z.

EXAMPLE 26.1: Higher-Order Independence from the Utility Tree

To make an assertion that X, Y, W are independent of Z in a given tree of expansion order X, Y, W, Z, we would like to arrange the tree such that it ends with only the prospects (x^*, y^*, w^*, z) and (x^0, y^0, w^0, z), and furthermore that the *conditional utility functions along every path in the tree that leads to the leaves* $(x^*, y^*, w^*, z), (x^0, y^0, w^0, z)$ *be independent of z.* Therefore the set X, Y, W is utility independent of Z if

(i) X is utility independent of Z *given Y and W*, $U(x \mid y, w, z) = U(x \mid y, w, z^0)$

(ii) Y is utility independent of Z given W at all boundary values of X, i.e., $U(y \mid x^*, w, z) = U(y \mid x^*, w, z^0)$ and $U(y \mid x^0, w, z) = U(y \mid x^0, w, z^0)$

(iii) W is utility independent of Z at all boundary values of X, Y, i.e., $U(w \mid x^i, y^j, z) = U(w \mid x^i, y^j, z^0) \forall x^i \in \{x^0, x^*\}, y^j \in \{y^0, y^*\}$, and

(iv) X, Y, W are corner independent of Z.

By induction, we have proved the following:

Theorem 26.4
If every attribute in a sequence of consecutive attributes in the utility tree is utility independent of Z at all boundary values of the previous attributes, and if the corner assessments of the attributes are utility independent of Z, then the group of attributes is utility independent of Z.

26.6 WHEN EVERY SUBSET OF THE ATTRIBUTES IS UTILITY INDEPENDENT OF ITS COMPLEMENT

Having discussed the idea of higher-order utility independence of a subset of the attributes from another, it is natural to consider the extreme case where every subset of the attributes is utility independent of its complement. For three attributes, X, Y, Z, this would imply that

$$(X\ UI\ Y,Z), (Y\ UI\ X,Z), (Z\ UI\ X,Y), (XY\ UI\ Z), (XZ\ UI\ Y), (YZ\ UI\ X).$$

This condition was discussed in Keeney and Raiffa (1976) and is referred to as mutual utility independence. As noted by Keeney and Raiffa, the mutual utility independence condition is a stronger condition than the condition where every attribute is utility independent of its complement that resulted in the multilinear form of Chapter 25. The first three conditions, $(X\ UI\ Y,Z), (Y\ UI\ X,Z), (Z\ UI\ X,Y)$, are equivalent to the condition that every attribute is utility independent of its complement, but in addition, we have the last three higher-order utility independence conditions, *(XY UI Z), (XZ UI Y), (YZ UI X)*, that further reduce the functional form.

The Functional Form of Mutual Utility Independence

Keeney and Raiffa showed that the functional form of the multiattribute utility function corresponding to mutual utility independence is either the additive form

$$U(x_1,...,x_n) = \sum_{i=1}^{n} k_i U_i(x_i) \tag{6}$$

or the multiplicative form,

$$1 + kU(x_1,...,x_n) = \prod_{i=1}^{n} [1 + kk_i U_i(x_i)], \tag{7}$$

where we get by direct substitution, $k_i = U(x_i^*, \overline{x}_i^0)$, $i = 1,...,n$, and by substituting for each attribute at its maximum value, the parameter k satisfies the equation

$$1 + k = \prod_{i=1}^{n} [1 + kk_i].$$

The functional forms of mutual utility independence (the additive and multiplicative forms) are special cases of the multilinear form. This should not come as a surprise because they involve stronger conditions. Because of their simplicity, the functional forms of mutual utility independence have gained significant use in the literature. Indeed, they are charming and simplify both the elicitation and the calculation of the multiattribute utility function significantly. They should not be used, however, if their conditions do not hold.

In Chapter 20, we discussed the implications of additive utility functions with regards to correlation neutral behavior. The additive utility function observes only the marginal probability distribution of each attribute in the expected utility calculation, and it ignores the correlations between the attributes. In Chapter 21, we discussed the product form of utility functions within the context of attribute dominance utility functions and discussed the percentage error that could be achieved if used incorrectly when its conditions do not apply.

In Chapter 25, we illustrated how the condition of "every attribute is utility independent of its complement" can be achieved using weaker utility and boundary independence conditions. We leave it as an exercise to the reader to determine weaker utility, boundary independence, and corner independence conditions that lead to the conditions of mutual utility independence, where "every subset of the attributes is utility independent of its complement."

Note: Mutual Utility Independence versus Additive Preference Functions: The functional forms (6) and (7) can be converted into an additive preference function using a monotone transformation. This implies that a decision-maker who exhibits mutual utility independence also has an additive preference function for deterministic prospects. But the converse is not necessarily true.

In Chapter 12, we derived the most general form of a multiattribute utility function that exhibits additive ordinal preferences over deterministic multiattribute prospects. We showed that (i) a decision-maker has a value function that can be converted into an additive form by a monotone transformation, and (ii) his utility function is strictly increasing with each argument at the maximum value of the complement attributes *if and only if* his utility function is an Archimedean combination of univariate utility assessments. The functional form of the utility function can indeed be constructed using univariate utility assessments, but a univariate generating function is also needed to reflect the decision maker's preferences for lotteries. Chapter 30 illustrates how to assess this generating function. The functional forms (6) and (7) can both be expressed as an Archimedean utility copula.

Figure 26.11 further highlights that the conditions of mutual utility independence yield only a subset of the multiattribute utility functions that have additive ordinal preferences.

26.7 SUMMARY

Higher-order utility independence relations imply a lower set of utility independence relations, but the converse is not necessarily true. We leave it as an exercise to the reader to consider these relations for boundary independence conditions.

ADDITIONAL READINGS

Abbas, A. E. 2010. General decompositions of multiattribute utility functions with partial utility independence. *Journal of Multicriteria Decision Analysis* 17(1): 37–59.

Abbas, A. E. 2011. The multiattribute utility tree. *Decision Analysis* 8(3): 180–205.

Keeney, R. and H. Raiffa. 1976. *Decisions with Multiple Objectives*. Wiley, New York:

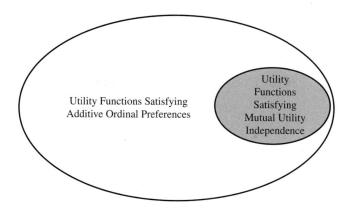

Figure 26.11. Venn diagram showing that mutual utility independence is a subset of utility functions satisfying additive ordinal preferences.

Matheson J. E. and R. A. Howard. 1968. An introduction to decision analysis. In R. A. Howard and J. E. Matheson, eds. *The Principles and Applications of Decision Analysis*, vol. I. Strategic Decisions Group, Menlo Park, CA.

von Neumann, J. and O. Morgenstern. 1947. *Theory of Games and Economic Behavior*. 2nd Ed. Princeton University Press, Princeton, NJ.

APPENDIX 26A

Proof of Proposition 26.1

By definition of utility independence, for any fixed x_j^1,

$$(X_K \ UI \ X_J \mid \bar{X}_{K,J}) \Leftrightarrow U(x_K \mid x_J, \bar{x}_{KJ}) = U(x_K \mid x_j^1, \bar{x}_{K,J}), \forall x_J \in X_J. \tag{8}$$

On the other hand, the set of $(l+1)$ pair-wise relations $(X_K \ UI \ X_i \mid \bar{X}_{K,i}), \forall i = j, j+1, ..., j+l$, assert

$$
\begin{aligned}
U(x_K \mid x_J, \bar{x}_{K,J}) &= U(x_K \mid x_j^1, \bar{x}_{K,j}), & \forall x_K, x_J, \bar{x}_{K,J} & \quad \text{(using } i = j) \\
&= U(x_K \mid x_j^1, x_{j+1}^1, \bar{x}_{K,j,j+1}), & \forall x_K, x_J, \bar{x}_{K,J} & \quad \text{(using } i = j+1) \\
&= ... & & \quad ... \qquad (9) \\
&= U(x_K \mid x_j^1, x_{j+1}^1, ..., x_{j+l}^1, \bar{x}_{K,j,...,j+l}), & & \quad \text{(using } i = j+l) \\
&= U(x_K \mid x_j^1, \bar{x}_{KJ}), \forall x_K, x_J, \bar{x}_{K,J}.
\end{aligned}
$$

Equation (9) shows that the set of $(l+1)$ pair-wise relations assert the right-hand side of (8) for any fixed x_j^1, and therefore assert that $(X_K \ UI \ X_J \mid \bar{X}_{K,J})$.

Proof of Proposition 26.2: The Independence Blanket

If $U(x_K \mid x_I, x_D) = U(x_K \mid x_I^0, x_D)$, then this implies that the joint utility assessment of all attributes in x_K does not depend on the value of x_I. For any set of attributes $X_L \subset X_K$, we can define the set $X_{K \backslash L}$ as the set of remaining attributes in X_K. By setting the attributes in $X_{K \backslash L}$ at any fixed value, it is clear that the conditional utility function of the smaller set X_L will not depend on X_I, because

$$U(x_K \mid x_I, x_D) = U(x_K \mid x_I^0, x_D) \Rightarrow U(x_L, x_{K \backslash L}^1 \mid x_I, x_D) = U(x_L, x_{K \backslash L}^1 \mid x_I^0, x_D).$$

Another way of thinking about this is if $(X_K \; UI \; X_I \mid X_D)$, then

$$U(x_K, \; x_I, x_D) = k(x_I, x_D) h(x_K, x_D) + d(x_I, x_D).$$

For any set of attributes $X_L \subset X_K$, we can define the set $X_{K \backslash L}$ as the set of remaining attributes in X_K. By setting the attributes in $X_{K \backslash L}$ at any fixed value

$$U(x_L, x_{K \backslash L}^1, \; x_I, x_D) = k(x_I, x_D) h(x_L, x_{K \backslash L}^1, x_D) + d(x_I, x_D)$$

At any fixed instantiation of x_D, changing x_I leads to a linear transform of the utility of X_L.

APPENDIX 26B A NOTE ABOUT ADDITIVE UTILITY FUNCTIONS VERSUS WEIGHT AND RATE FORMULATIONS

The additive utility function

$$U(x_1, \ldots, x_n) = \sum_{i=1}^{n} k_i U_i(x_i),$$

where $k_i = U(x_i^*, \overline{x}_i^0)$ and the functions $U_i(x_i) = U_i(x_i \mid \overline{x}_i^0)$ are normalized conditional utility functions, has gained widespread use in decision analysis primarily because of its simplicity, despite its implications for correlation-neutral preferences among the attributes that we discussed in Chapter 20.

The additive utility function has also gained popularity because of its resemblance to the weight and rate preference function that we discussed in Appendix 8A. The parameters $k_i = U(x_i^*, \overline{x}_i^0)$ resemble the weights, and the utility functions $U_i(x_i)$ enable a normalized score. Moreover, the parameters $k_i = U(x_i^*, \overline{x}_i^0)$ have meaningful interpretations; they are indifference probabilities of receiving the corner assessment $U(x_i^*, \overline{x}_i^0)$ for certain, or a binary deal that offers the best prospect $U(x_i^*, \overline{x}_i^*)$ with probability k_i and the worst prospect with probability $1 - k_i$. The scale of the conditional utility functions also has a meaningful interpretation using indifference assessments.

The additive utility function is a lot more restrictive than an additive preference function. When its additive independence conditions hold, the additive utility function is a charming simplification. Many decisions analysis applications, however, have used the additive utility function as a means to construct a weight and rate preference scoring function for deterministic decisions, and have often ignored structural models along the way. It is therefore important to recall our earlier discussion of preference functions and utility functions over value or over the attributes, as well the correlation neutral behavior implied by the additive utility function, before rushing into using this simplifications.

CHAPTER 27

One-Switch Utility Independence

Chapter Concepts

- The notion of one-switch utility independence (Abbas and Bell 2011)
- Illustrating that one-switch utility independence is a natural generalization of utility independence
- Relating one-switch utility independence to one-switch utility functions
- Relating one-switch utility independence to the notion of interpolation independence

27.1 INTRODUCTION: RELATING THREE DIFFERENT NOTIONS

In this chapter, we draw on three notions discussed in previous chapters, and show that each of these concepts, while derived independently, leads naturally to a new notion called one-switch utility independence.

We consider

1. The notion of one-switch utility functions that we discussed in Chapter 18.
2. The notion of utility independence that we discussed in Chapters 21, 23 and 25
3. The notion of interpolation independence that we discussed in Chapter 23

We also review equivalent notions relating:

1. Zero-switch utility functions
2. Utility independence
3. The invariant transformation

before introducing the new concepts

27.2 ZERO-SWITCH FORMULATIONS

27.2.1 Zero-Switch Utility Functions

In Chapter 13, we considered the functional form of a utility function that satisfies the following property:

Preferences for any two lotteries do not change as the wealth level increases. This implies that for all lotteries X_1, X_2, the difference in expected utilities $E_{x_1}[U(x+w)] - E_{x_2}[U(x+w)]$ does not change sign as w increases.

We showed that this zero-switch condition can be expressed in terms of the functional equation

$$U(x+w) = k(w)U(x) + d(w),$$

where $k(w)$ does not change sign. We also showed that the solutions to this equation for values of w on an interval are:

1. $U(x) = ax + b$ (Linear)
2. $U(x) = ae^{-\gamma x} + b$ (Exponential)

27.2.2 Invariant Transformations

In Chapter 16, we extended the idea of zero-switch utility functions to utility functions invariant to a monotone transformation where preference between lotteries do not change as a transformation $g(y, \delta)$ is applied to the lottery outcomes. We showed that this condition defines the functional equation

$$U(g(y, \delta)) = k(\delta)U(y) + d(\delta).$$

27.2.3 Utility Independence

As we have seen, utility independence of X on Y implies that preferences for lotteries over X do not change as we vary the level of attribute Y. Utility independence implies that the difference in expectations of two lotteries over attribute X, namely $E_{x_1}[U(x, y)] - E_{x_2}[U(x, y)]$, does not change as y increases.

We showed that utility independence implies that

$$U(x, y) = k(y)U(x \mid y^0) + d(y).$$

The relation between the three concepts, zero-switch utility functions, invariant transformations, and utility independence is emphasized by their corresponding equations. How do these concepts extend to a change in preferences, but only once?

27.3 ONE-SWITCH UTILITY FORMULATIONS

27.3.1 One-Switch Utility Functions

In Chapter 18, we considered the functional form of a utility function that satisfies the following property:

Preferences for any two lotteries can change only once as the wealth level increases. This implies that for all lotteries X_1, X_2, the difference in expected utilities $E_{x_1}[U(x+w)] - E_{x_2}[U(x+w)]$ changes sign but only once as w increases.

We showed that this one-switch condition can be expressed in terms of the functional equation

$$U(x+w) = k(w)[f_1(x) + \phi(w)f_2(x)] + d(w),$$

where $k(w)$ does not change sign and $\phi(w)$ is monotone. We also showed that the four one-switch utility functions that satisfy this condition (and this functional equation) were proposed by Bell (1988) as

3. $U(x) = ax + be^{-\gamma x} + c$ (Linear plus exponential)
4. $U(x) = ae^{-\lambda x} + be^{-\gamma x} + c$ (Sumex)
5. $U(x) = (ax + b)e^{-\gamma x} + c$ (Linear times exponential)
6. $U(x) = ax^2 + bx + c$ (Quadratic)

27.3.2 A New Property

Now let us consider the following extension to the one-switch property: Suppose it holds for lotteries over X as the level of another attribute (not necessarily wealth) increases from low to high. For example, we might be considering preferences for investments as another attribute such as remaining life time changes. If preferences can change, but only once, then this would be an extension of the one-switch utility function property to more than just wealth changes.

If preferences can change only once for all lotteries X_1, X_2, then the difference $E_{x_1}[U(x,y)] - E_{x_2}[U(x,y)]$ changes sign but only once as y increases

As we shall see in the next section, the functional form relates closely to the functional equation that we derived in Chapter 18, but it is expressed in terms of two attributes.

27.3.3 Interpolation Independence

The *interpolation independence* of X on Y implies that the conditional utility function over X can be expressed in terms of its boundary values and an interpolation function such that

$$U(x \mid y) = w_1(y)U(x \mid y^*) + w_2(y)U(x \mid y^0), \quad w_1(y) + w_2(y) = 1.$$

We did not discuss the interpretation of interpolation independence of the functions $w_1(y)$ and $w_2(y)$. As we shall see, under some very mild conditions, this notion can also be interpreted by the condition that

$E_{x_1}[U(x,y)] - E_{x_2}[U(x,y)]$ changes sign but only once as y increases.

27.4 ONE-SWITCH UTILITY INDEPENDENCE

Definition

Attribute X exhibits **_one-switch utility independence_** from another attribute Y if preference for any pair of lotteries on X can switch at most once as the level of Y increases.

Theorem 27.1

X is one-switch independent of Y, _written_, $X \, 1S \, Y$ if and only if

$$u(x,y) = g_0(y) + f_1(x)g_1(y) + f_2(x)g_2(y)$$

where $g_1(y)$ has constant sign, and $g_2(y) = g_1(y)\phi(y)$ for some monotonic function φ.

Theorem 27.1 can also be stated as

X is one-switch independent of Y, _written_, $X \, 1S \, Y$ if and only if

$$u(x,y) = g_0(y) + g_1(y)[f_1(x) + f_2(x)\phi(y)]. \tag{1}$$

for some monotonic function φ and $g_1(y)$ has constant sign.

Note that the functions $g_0(y), g_1(y), f_1(x), f_2(x)$ do not need to be monotonic. The following definition is an equivalent definition for one-switch independence.

Alternate definition of one-switch independence. X is one-switch independent of Y, if, for any two gambles on X, either

(i)   one is preferred to the other for all values of Y,
(ii)   they are indifferent for all levels of Y, or
(iii)   there exists a value of Y, $\breve{y}$, such that one gamble is always preferred for values less than $\breve{y}$ and the other is always preferred for values of Y greater than $\breve{y}$.

Note: It is clear that if X is a wealth lottery, and if y is the initial wealth, then we have the notion of one-switch utility functions, which is a special case of one-switch utility independence that satisfies the functional equation

$$u(x+y) = g_0(y) + g_1(y)[f_1(x) + f_2(x)\phi(y)]. \tag{2}$$

for some monotonic function φ and $g_1(y)$ has constant sign.

EXAMPLE 27.1: Illustration of Sufficiency of the One-Switch Independence Form

---

Consider again two lotteries X_a, X_b over attribute X. What happens if we increase the level of another attribute, Y? The difference in expected utility of the modified lotteries is

$$E_A[U(x,y)] - E_B[U(x,y)] = g_1(y)\big(E_A[f_1(x)] - E_B[f_1(x)]\big)$$
$$+ g_1(y)\phi(y)\big(E_A[f_2(x)] - E_B[f_2(x)]\big).$$

The difference in expected utility is equal to zero when

$$E_A[f_1(x)] - E_B[f_1(x)] + \phi(y)\big(E_A[f_2(x)] - E_B[f_2(x)]\big) = 0.$$

Rearranging gives

$$\phi(y) = -\frac{E_A[f_1(x)] - E_B[f_1(x)]}{E_A[f_2(x)] - E_B[f_2(x)]},$$

which can occur only once as we increase y because $\phi(y)$ is a monotone function.

The proof of necessity is in Appendix 27.1.

Relation to Interpolation Independence

Let is consider the normalized conditional utility function of attribute $X, U(x \mid y)$, if X is one-switch independent of Y.

Theorem 27.2 Interpreting the Function, $\phi(y)$
If X is one-switch independent of Y, then

$$\phi(y) = \frac{U(x \mid y) - U(x \mid y^0)}{U(x \mid y^*) - U(x \mid y^0)}$$

is a monotone function that does not depend on X.

Rearranging gives

$$U(x \mid y) = \phi(y)[U(x \mid y^*) - U(x \mid y^0)] + U(x \mid y^0).$$

Simplifying gives

$$\boxed{U(x \mid y) = \phi(y)U(x \mid y^*) + (1 - \phi(y))U(x \mid y^0)},$$

which implies that X is interpolation independent of Y, with the added condition that the interpolation function is monotone.

This provides an interesting interpretation for interpolation independence in terms of one-switch independence with some monotonicity condition added. The function $\phi(y)$ is a monotone interpolation function that matches the decision-maker's preferences for $U(x \mid y)$ in terms of its boundary values.

Observe the following:

> *X is interpolation independent of Y if it is one-switch independent of Y.*

On the other hand, one-switch independence requires interpolation independence with a monotone interpolation function,

> *X is one-switch independent of Y if it is interpolation independent of Y and if the interpolation function is monotone.*

27.5 ASSESSING THE FUNCTIONAL FORM FOR ONE-SWITCH UTILITY INDEPENDENCE

The following theorem determines the assessments needed to construct the functional form.

Theorem 27.3: Assessments Needed for a Utility Function Satisfying One-Switch Independence

X is one-switch independent of Y if and only if

$$U(x,y) = g_0(y) + g_1(y)[f_1(x) + f_2(x)\phi(y)],$$

where

$$g_0(y) = U(x^0, y), \ g_1(y) = [U(x^*, y) - U(x^0, y)],$$

$$f_1(x) = U(x \mid y^0), \text{ and } f_2(x) = U(x \mid y^*) - U(x \mid y^0),$$

and

$$\phi(y) = \frac{U(x \mid y) - U(x \mid y^0)}{U(x \mid y^*) - U(x \mid y^0)} \text{ is a monotone function (that does not depend on } x).$$

Theorem 27.3 illustrates how to assess a utility function satisfying the condition that "X is one-switch independent of Y." If the utility function is strictly increasing with each of its arguments, then $U(x^*, y) > U(x^0, y)$ and so $g_1(y) > 0$; as a result, it does not change sign.

Example 27.2: Constructing a multiattribute utility function satisfying $X1SY$

Theorem 27.3 shows that we need the following utility assessments to construct a multiattribute utility function satisfying $X1SY$:

1. A utility assessment of Y at the lower boundary value of X, i.e., $g_0(y) = U(x^0, y)$.

   This is a univariate assessment that requires the assessment of a utility function over Y at the minimum value of X. This can be conducted using the tools discussed in Chapters 11 and 13–18. It can also be conducted using normalized utility assessments by observing that $U(x^0, y) = k_y U(y \mid x^0)$, where the corner value $k_y = U(x^0, y^*)$ is an indifference probability assessment.

2. A utility assessment of Y at the upper boundary value of X, i.e., $U(x^*, y)$.

   This is also a univariate assessment conducted for attribute Y but at the maximum value of X. It can also be conducted in terms of a normalized utility function using the relation

$$U(x^*, y) = k_x + (1 - k_x)U(y \mid x^*).$$

Having assessed these two boundary assessments, we can now determine the function

$$g_1(y) = [U(x^*, y) - U(x^0, y)].$$

3. A normalized conditional utility assessment of X at the lower boundary value of Y, i.e.,

$$f_1(x) = U(x \mid y^0).$$

This assessment is straightforward because it is already in a normalized form.

4. A normalized utility assessment of X at the upper bound of Y, i.e.,

$$f_2(x) = U(x \mid y^*).$$

This assessment is also straightforward because it is already in a normalized form. We can now determine the function

$$f_2(x) = U(x \mid y^*) - U(x \mid y^0).$$

We have now assessed utility value at all four boundary values.

5. An assessment of a monotone interpolation function $\phi(y)$ that satisfies the assessment is

$$\boxed{U(x \mid y) = \phi(y)U(x \mid y^*) + (1 - \phi(y))U(x \mid y^0).}$$

EXAMPLE 27.3: Assessing a Utility Function Satisfying One-Switch Independence

---

Consider two attributes X, Y that are relevant over the domain $x^0 = y^0 = 0;\ x^* = y^* = 1$. Suppose that the decision-maker asserts that X 1S Y. For simplicity, suppose that the four normalized conditional utility assessments are given as

$$U(x \mid y^0) = x, U(x \mid y^*) = x^2, U(y \mid x^0) = y, U(y \mid x^*) = y^2.$$

We also assess the corner values

$$k_x = U(x^*, y^0) = 0.4, \ k_y = U(x^0, y^*) = 0.3.$$

We may then calculate

$$g_0(y) = 0.3y \text{ and } g_1(y) = 0.4 + 0.6y^2 - 0.3y,$$

both of which are positive on the interval $[0,1]$. Furthermore,

$$f_1(x) = x \text{ and } f_2(x) = x^2 - x.$$

To determine $\phi(y)$, we seek the function $\phi(y)$, that makes

$$U(x \mid y) = \phi(y)U(x \mid y^*) + (1 - \phi(y))U(x \mid y^0)$$

for any value of x.

In its simplest form, suppose that $\phi(y) = y$, which indeed satisfies $\phi(0) = 0$ and $\phi(1) = 1$.

To verify the chosen value of $\phi(y) = y$, pick any value of X, say $\check{x} = 0.5$. We know that $U(\check{x} \mid y^0) = U(x = 0.5 \mid y^0) = 0.5$ and $U(\check{x} \mid y^*) = U(x = 0.5 \mid y^*) = 0.25$.

We then calculate

$$U(\check{x} \mid y) = \phi(y)U(\check{x} \mid y^*) + (1 - \phi(y))U(\check{x} \mid y^0) = 0.5 - 0.25y,$$

and verify that this indeed is the conditional utility function at $\check{x} = 0.5$.

For further verification, the decision-maker may also find it more convenient to reason about $U(y \mid \check{x})$. We can relate $U(y \mid \check{x})$ and $U(\check{x} \mid y)$ by observing that

$$U(\check{x}, y) = U(x^0, y) + [U(x^*, y) - U(x^0, y)]U(\check{x} \mid y^0)$$

and also that

$$U(\check{x}, y) = U(\check{x}, y^0) + [U(\check{x}, y^*) - U(\check{x}, y^0)]U(y \mid \check{x}).$$

Equating the two relates $U(y \mid \check{x})$ and $U(\check{x} \mid y)$. Direct substitution shows that this implies

$$U(y \mid \check{x}) = 2y + 15y^2 - \frac{6}{11}y^3,$$

a function that is in fact monotonic on the interval $[0,1]$, and one that the decision-maker might be more comfortable reasoning about.

27.6 N-SWITCH UTILITY INDEPENDENCE

Generalizations of our one-switch conditions are also possible. If it takes n values of Y at which $\tilde{x}_1 \sim \tilde{x}_2$ before the decision-maker is prepared to state that $\tilde{x}_1 \sim \tilde{x}_2$ for all y, then we say X is n-switch independent of Y, $[X \ nS \ Y]$ and there exist some functions f_i, g_i such that

$$u(x, y) = g_0(y) + \sum_{i=1}^{n+1} f_i(x)g_i(y).$$

See the proof of Theorem 27.1 in Appendix 27A for further details. It might also be that switch conditions apply within an attribute as well as between attributes. For example, one or more of the attributes may belong to the family of the four one-switch functional forms. And of course these conditions might apply to a set of more than two attributes, where further decompositions may apply. Consider, for example, extensions of the one-switch property to three attributes, X, Y, Z. Suppose that preferences for joint lotteries over X, Y change only once as we change Z, then we write $[X, Y \; 1S \; Z]$ and use the previous results to write

$$u(x, y, z) = g_0(z) + g_1(z)[f_1(x, y) + f_2(x, y)\phi(z)].$$

27.7 SUMMARY

For assessment purposes, once it has become clear that X is not utility independent of Y, the one-switch independence condition is a logical next step of complexity in thinking about the utility function.

The one-switch independence formulation also provides a nice interpretation for the concept of interpolation independence when the interpolation function is monotone.

ADDITIONAL READINGS

Abbas, A. E. and D. E. Bell. 2011. One-switch independence for multiattribute utility functions. *Operations Research* 59(3): 764–771.

Abbas, A. E. and D. E. Bell. 2012. One-switch conditions for multiattribute utility functions. *Operations Research* 60(5): 1199–1212.

Abbas, A. E. and R. A. Howard. 2005. Attribute dominance utility. *Decision Analysis* 2(4): 185–206.

Bell, D. E. 1988. One-switch utility functions and a measure of risk. *Management Science* 34: 1416–1424.

Keeney, R. L. and H. Raiffa. 1976. *Decisions with Multiple Objectives*. Wiley, New York.

Pfanzagl, J. 1959. A general theory of measurement: Applications to utility. *Naval Research Logistics Quarterly* 6(1959): 283–294.

von Neumann, J. and O. Morgenstern. 1947. *The Theory of Games and Economic Behavior*. 2nd Ed. Princeton University Press, Princeton, NJ.

APPENDIX 27A

Proof of Theorem 27.1

Necessity: We start with the observation that any two uncertain lotteries, $\tilde{x}_A$ and $\tilde{x}_B$, can be expressed as lotteries having the same probabilities but with different outcomes. We therefore assume, with no loss of generality, that the two lotteries are

$$< p_1, x_{A1}; p_2, x_{A2}; ....; p_n, x_{An} > \text{ and } < p_1, x_{B1}; p_2, x_{B2}; ....; p_n, x_{Bn} > .$$

The difference in expected utility of two lotteries over X when another attribute (or parameter), y, is fixed, can be written as

$$EU_A(y) - EU_B(y) = \sum_{i=1}^{n} p_i [U(x_{Ai}, y) - U(x_{Bi}, y)] = \theta(y) = \sum_{i=1}^{n} p_i V_i(y),$$

where $V_i(y) = U(x_{Ai}, y) - U(x_{Bi}, y)$.

For fixed values $y_0, y_1, ..., y_n$, we can write in matrix form

$$\begin{pmatrix} V_1(y_0) & V_2(y_0) & V_n(y_0) \\ V_1(y_1) & V_2(y_0) & V_n(y_1) \\ .. & & \\ V_1(y_n) & V_2(y_n) & V_n(y_n) \end{pmatrix} \begin{pmatrix} p_1 \\ p_2 \\ ... \\ p_n \end{pmatrix} = \begin{pmatrix} \theta_1(y_0) \\ \theta_1(y_1) \\ ... \\ \theta_1(y_n) \end{pmatrix}$$

1. $X \ 1S \ Y \Rightarrow$ there does not exist $y_0 < y_1 < y_2$ such that $\theta(y_0) = +ve, \theta(y_1) = -ve, \theta(y_2) = +ve$.

For this to occur, any 3x3 portion of the V matrix above must be singular, which implies that

$$V_i(y) = \alpha V_j(y) + \beta V_k(y), \forall i, j, k.$$

Referring back to the definition of $V_i(y) = U(x_{Ai}, y) - U(x_{Bi}, y)$ gives

$$U(x_{Ai}, y) - U(x_{Bi}, y) = \alpha[U(x_{Aj}, y) - U(x_{Bj}, y)] + \beta[U(x_{Ak}, y) - U(x_{Bk}, y)], \forall x, y.$$

This implies that

$$U(x_{Ai}, y) - U(x_{Bi}, y) = f_1(x_{Ai}, x_{Bi}) g_1(y) + f_2(x_{Ai}, x_{Bi}) g_2(y). \tag{3}$$

Noting of course that x_{Ai}, x_{Bi} are arbitrary, this implies that

$$U(x, y) = f_1(x) g_1(y) + f_2(x) g_2(y) + g_0(y). \tag{4}$$

Define $\phi(y) = \dfrac{g_2(y)}{g_1(y)}, g_1(y) \neq 0$. Substituting into (15) gives

$$EU_A(y) - EU_B(y) = g_1(y)([f_1(x_A) - f_1(x_B)] + [f_2(x_A) - f_2(x_B)]\phi(y)). \tag{5}$$

We distinguish two cases. (i) $g_1(y)$ does not change sign: for (17) to change sign only once, with two given arbitrary lotteries $\tilde{x}_A$ and $\tilde{x}_B$, and constant terms $f_1(\tilde{x}_A), f_1(\tilde{x}_B), f_2(\tilde{x}_A), f_2(\tilde{x}_B)$, then $\phi(y)$ must be monotonic or else, we may have different points of indifference for arbitrary lotteries. (ii) $g_1(y)$ does change sign: This would imply that the term $([f_1(\tilde{x}_A) - f_1(\tilde{x}_B)] + [f_2(\tilde{x}_A) - f_2(\tilde{x}_B)]\phi(y))$ either does not change sign or it changes sign at the same value of y for which $g_1(y)$ changes sign. This is impossible for arbitrary functions non-constant f_1, f_2 and arbitrary lotteries.

Note that for the case of 2-switch, we need any 4x4 matrix to be singular. This results in the functional form $X \ 2S \ Y \Rightarrow U(x, y) = f_1(x) g_1(y) + f_2(x) g_2(y) + f_3(x) g_3(y) + g_0(y)$. The same pattern extends to any number of switches. By induction, if $[X \ nS \ Y]$ then the $(n+2)x(n+2)$ matrix must be singular leading to the functional form

$$U(x,y) = g_0(y) + \sum_{i=1}^{n+1} f_i(x)g_i(y).$$

Sufficiency: For two arbitrary lotteries, X_A, X_B and a monotonic $\phi(y)$, we have only one possible indifference point, which occurs at y_1

$$E_A[U(x,y)] - E_B[U(x,y)] = g_1(y_1)$$
$$\left(E_A[f_1(x)] - E_B[f_1(x)] + [E_A[f_2(x)] - E_B[f_2(x)]]\phi(y_1)\right) = 0,$$

$$\phi(y_1) = \frac{E_A[f_1(x)] - E_B[f_1(x)]}{E_A[f_2(x)] - E_B[f_2(x)]},$$

below and after which the rank order of the lotteries must reverse.

Proof of Theorem 27.2

First, we observe that, with no loss of generality, we can assume that $\phi(y_0) = 0$ and $\phi(y^*) = 1$ because any scale or shift parameters can be added to the functions f_1 and f_2.

If X $1S$ Y, then

$$U(x,y) = g_0(y) + g_1(y)[f_1(x) + f_2(x)\phi(y)].$$

By direct substitution,

$$U(x\,|\,y) = \frac{U(x,y) - U(x^0,y)}{U(x^*,y) - U(x^0,y)} = \frac{[f_1(x) - f_1(x^o) + (f_2(x) - f_2(x^o))\phi(y)]}{[f_1(x^*) - f_1(x^o) + (f_2(x^*) - f_2(x^o))\phi(y)]}.$$

Furthermore,

$$U(x\,|\,y^0) = \frac{f_1(x) - f_1(x^o)}{f_1(x^*) - f_1(x^o)}, \quad U(x\,|\,y^*) = \frac{f_1(x) - f_1(x^o) + (f_2(x) - f_2(x^o))}{f_1(x^*) - f_1(x^o) + (f_2(x^*) - f_2(x^o))}.$$

Direct (and tedious) substitution gives

$$\phi(y) = \frac{U(x\,|\,y) - U(x\,|\,y^0)}{U(x\,|\,y^*) - U(x\,|\,y^0)}.$$

Proof of Theorem 27.3

Rearranging $\phi(y) = \dfrac{U(x\,|\,y) - U(x\,|\,y^0)}{U(x\,|\,y^*) - U(x\,|\,y^0)}$ gives for $X1SY$,

$$U(x\,|\,y) = \phi(y)[U(x\,|\,y^*) - U(x\,|\,y^0)] + U(x\,|\,y^0).$$

Furthermore, if we rearrange the general expression for the definition of $U(x\,|\,y)$, we get

$$U(x,y) = U(x_0,y) + [U(x^*,y) - U(x_0,y)]U(x\,|\,y).$$

This implies that for $X1S$ Y,

$$U(x,y) = U(x_0,y) + [U(x^*,y) - U(x_0,y)]\left(U(x\,|\,y^0) + [U(x\,|\,y^*) - U(x\,|\,y^0)]\phi(y)\right).$$

Comparing this expression to

$$U(x, y) = g_0(y) + g_1(y)[f_1(x) + f_2(x)\phi(y)]$$

gives

$$g_0(y) = U(x_0, y), g_1(y) = U(x^*, y) - U(x_0, y)$$
$$f_1(x) = U(x \mid y^0), f_2(x) = U(x \mid y^*) - U(x \mid y^0)$$

PART VI

MULTIATTRIBUTE UTILITY COPULAS

We have discussed several methods for constructing multiattribute utility functions:

1. Using a preference or a value function that capture deterministic trade-offs among the attributes and a one-dimensional utility function over value or over an attribute.
2. Using direct utility elicitation of conditional utility assessments and the Basic Expansion Theorem. This approach was simplified using utility, boundary, and interpolation independence conditions.

If one uses direct utility assessments, then it is important to see whether the implied trade-offs (shapes of the isopreference contours) match those of the decision-maker's preference or value function. Similarly, if one uses the preference function or the value function approach, it would be important to check whether the implied conditional utility assessments (and the corresponding risk aversion functions) match those of the decision-maker.

In this part, we use a mix of trade-off assessments and utility assessments to construct the utility surface. Instead of incorporating independence conditions to decompose the functional form, we assess utility functions at the boundary values and construct a utility surface to match these boundary assessments. With this approach we also have an additional degree of freedom to vary the shapes of the isopreference contours while matching the boundary values. This is the idea of a utility copula function.

A utility copula function is a function that uses conditional utility assessments, at specified values of the domain, as its arguments and then constructs a utility surface that is consistent with these input utility values. The copula function provides the additional degree of freedom to vary the height of the utility surface (or change the shape of the isopreference contours) while keeping the boundary values fixed. By varying the parameters and functional form of the utility copula, we can also conduct sensitivity analyses to various utility independence conditions.

There are many types of copula functions, with various parameters and functions that help construct the utility surface. Archimedean utility copulas are an important class of such functions that require a single generating function and can be determined completely by direct indifference assessments on some curves of the domain. We illustrate how to assess copula functions in Chapter 30.

In this part, we also discuss copula functions that match more than one boundary of the utility surface, as well as copula functions for constructing preference and value functions, that are especially useful when preferential independence conditions do not exist.

CHAPTER 28

Multiattribute Utility Copulas

Chapter Concepts

- Definition of a utility copula function
- Properties of utility copula functions
- Class 1 and Class 0 utility copula functions
- Consistency conditions for utility copula functions

28.1 INTRODUCTION

We have already seen the concept of a utility copula function numerous times throughout this book. In Chapter 12, we demonstrated that a utility function over a value function, that can be converted into an additive form using a monotone transformation results in an Archimedean combination of boundary utility assessments of the attributes. For two attributes, this Archimedean combination takes the form

$$U(x,y) = \varphi^{-1}\big(\varphi(U(x,y^*))\varphi(U(x^*,y))\big).$$

We can think of this utility function as a bivariate function, C, of two boundary curves such that

$$U(x,y) = C\big(U(x,y^*),U(x^*,y)\big),$$

In Chapter 21, we also discussed the idea of a utility copula function for attribute dominance utility functions. With this formulation, an attribute dominance utility function was constructed to match marginal (boundary) utility assessments provided by the decision-maker. The boundary utility functions were normalized for attribute dominance utility funcitons, so we could also write this function in terms of the normalized conditional utility functions as

$$U^d(x,y) = C^d\big(U(x\,|\,y^*),U(y\,|\,x^*)\big).$$

where $U(x,y^*) = U(x\,|\,y^*)$ and $U(x^*,y) = U(y\,|\,x^*)$ for utility dominant attributes.

This chapter extends the utility copula formulation to situations where the attribute dominance (grounding) conditions do not exist. The fundamental difference this implies is that we need to incorporate the corner values of the utility surface into the specification of the utility copula.

To start, we consider a two-attribute formulation, where a utility function is constructed to match utility assessments at the upper bound of the complement attributes. Suppose we also require the function to match some specified corner values k_x, k_y provided by the decision-maker. Therefore, we may write

$$U(x,y) = C\big(U(x\,|\,y^*), U(y\,|\,x^*); k_x, k_y\big).$$

We refer to the function, C, as a ***utility copula function*** (Abbas 2004, 2009). With this formulation, the arguments of the function C, namely $U(x\,|\,y^*)$ and $U(y\,|\,x^*)$, are the normalized conditional utility functions at the upper bound of the complement attributes, and the parameters k_x, k_y are the corner values for the resulting utility surface, where $k_x = U(x^*, y^0)$ and $k_y = U(x^0, y^*)$.

We may also express the utility function in terms of its boundary curves as

$$U(x,y) = C\big(U(x,y^*), U(x^*,y)\big),$$

because, as we have seen, the boundary assessments, $U(x,y^*)$ and $U(x^*,y)$, can be deduced from the normalized conditional utility assessments, $U(x\,|\,y^*)$ and $U(y\,|\,x^*)$, and the corner values (and vice versa) using the relations

$$U(x,y^*) = k_y + (1-k_y)U(x\,|\,y^*) \text{ and } U(x^*,y) = k_x + (1-k_x)U(y\,|\,x^*),$$

where $k_x = U(x^*, y^0)$ and $k_y = U(x^0, y^*)$.

As we shall see, the utility copula formulation provides the flexibility to model a variety of multiattribute utility surfaces that match the boundary assessments, and it also provides an additional degree of freedom to vary the various trade-off assessments among the attributes.

The utility copula formulation may also incorporate other boundary assessments besides the upper bounds. For example, if the decision-maker is more comfortable providing the lower boundary assessments $U(x\,|\,y^0), U(y\,|\,x^0)$, then we may construct a utility surface using the relation

$$U(x,y) = C\big(U(x\,|\,y^0), U(y\,|\,x^0); k_x, k_y\big),$$

and also write this function using non-normalized arguments as

$$U(x,y) = C\big(U(x,y^0), U(x^0,y)\big).$$

The purpose of this chapter is to familiarize the reader with the basic concept of a utility copula function and its properties. Chapter 29 presents a special class of utility copulas known as Archimedean utility copulas, and Chapter 30 presents methods to determine Archimedean copulas by direct utility elicitation. Our main focus will be on utility functions that are non-decreasing with their arguments and so more of an attribute is preferred to less.

28.2 WHAT IS A MULTIATTRIBUTE UTILITY COPULA? (ABBAS 2009)

In its simplest form, a utility copula function is a function that constructs a multiattribute utility surface using some conditional utility assessments at specific values of the complement attributes, and some specified corner values. The conditional utility assessments are used as arguments of the copula function.

What must be true about the function C such that the resulting utility surface $U(x,y)$ be consistent with the utility assessments and the corner values k_x, k_y that are used for its construction? We discuss this below and introduce a sufficient condition that will be the focus of the rest of this chapter.

28.2.1 Class 1 Utility Copula Functions Matching the Upper Bounds

Definition

A *Class 1 utility copula* is a function $C(s,t;k_x,k_y)$ that constructs a multiattribute utility surface using a univariate utility assessment of each attribute at the upper bound of the complement attributes, as well as specified corner values using the relation

$$U(x,y) = C\left(U(x\,|\,y^*), U(y\,|\,x^*); k_x, k_y\right)$$

or equivalently

$$U(x,y) = C\left(U(x,y^*), U(x^*,y)\right).$$

Before we proceed with the properties of a Class 1 utility copula, we need to ask when it is possible to express a function using two boundary curves on its surface if more of an attribute is preferred to less.

Sufficient Condition for Expressing a Utility Surface Using Its Upper Boundary Curves

A sufficient condition for a utility function to be expressed in terms of its boundary curves using a Class 1 utility copula function is that the boundary curves $U(x\,|\,y^*), U(y\,|\,x^*)$ be continuous, bounded, and strictly increasing.

This sufficient condition enables a one-to-one correspondence from the (x,y) space to the (s,t) space, where $s = U(x\,|\,y^*)$ and $t = U(y\,|\,x^*)$.

28.2.1.1 Properties of Class i Utility Copulas with Normalized Arguments This section discusses properties of the copula function C satisfying the formulation

$$U(x,y) = C\left(U(x\,|\,y^*), U(y\,|\,x^*); k_x, k_y\right).$$

1. ***Domain of the Utility Copula:*** Because the normalized conditional utility functions $U(x\,|\,y^*), U(y\,|\,x^*)$ are arguments of the copula function, and because they are strictly increasing, the domain of the copula function is the unit square
   $[0,1] \times [0,1]$. The cornr values k_x, k_y must also lie in the range $[0.1]$ because they are indifference probability assessments.
2. ***Range of the Utility Copula:*** Because the output of the copula function is a multiattribute utility surface (which is an indifference probability assessment), the ***range*** of the copula function is $[0,1]$.

   Therefore, we can write

$$C : [0,1] \times [0,1] \to [0,1].$$

3. ***Linearity (Consistency) at the Upper Bound:*** The constructed utility surface $U(x,y)$ must be consistent with its input utility assessments provided at the upper boundary values. We already know that,

$$U(x, y^*) = k_y + (1 - k_y)U(x\,|\,y^*). \tag{1}$$

By direct substitution into the utility copula formulation at the upper value of attribute Y, the utility surface should therefore satisfy

$$U(x, y^*) = C\big(U(x\,|\,y^*), U(y^*\,|\,x^*); k_x, k_y\big) = C\big(U(x\,|\,y^*), 1; k_x, k_y\big). \tag{2}$$

From (1) and (2), consistency requires that

$$C\big(U(x\,|\,y^*), 1; k_x, k_y\big) = k_y + (1 - k_y)U(x\,|\,y^*).$$

Define $U(x\,|\,y^*) = s$. This implies that the copula functions needs to satisfy

$$\boxed{C\big(s, 1; k_x, k_y\big) = k_y + (1 - k_y)s}. \tag{3}$$

Therefore, the copula function needs to be a linear function of its first argument at the upper bound of the second argument.

Similarly, a second consistency condition at the upper bound of attribute X requires that

$$U(x^*, y) = k_x + (1 - k_x)U(y\,|\,x^*).$$

Consistency with the copula formulation requires that

$$U(x^*, y) = C\big(U(x^*\,|\,y^*), U(y\,|\,x^*); k_x, k_y\big) = C\big(1, U(y\,|\,x^*); k_x, k_y\big)$$

and so

$$C\big(1, U(y\,|\,x^*); k_x, k_y\big) = k_x + (1 - k_x)U(y\,|\,x^*).$$

Define $U(y\,|\,x^*) = t$. This implies that

$$\boxed{C\big(1, t; k_x, k_y\big) = k_x + (1 - k_x)t}. \tag{4}$$

Again, the copula function is a linear function of its second argument at the upper bound of the first argument.

4. ***Consistency at the Corner Values of the Domain:*** Because the utility function is strictly increasing with each argument and it is normalized, we must have

$$U(x^*, y^*) = 1.$$

Substituting into the utility copula formulation gives

$$U(x^*, y^*) = C\big(U(x^* \mid y^*), U(y^* \mid x^*); k_x, k_y\big).$$

Therefore, the copula function must satisfy

$$\boxed{C\big(1, 1; k_x, k_y\big) = 1}. \tag{5}$$

Similarly, for the lower corner value, $U(x^0, y^0) = 0$, we get by direct substitution

$$U(x^0, y^0) = C\big(U(x^0 \mid y^*), U(y^0 \mid x^*); k_x, k_y\big) = 0.$$

Therefore,

$$\boxed{C\big(0, 0; k_x, k_y\big) = 0}. \tag{6}$$

We must also specify the corner values of the utility copula such that $U(x^*, y^0) = k_x$. And so

$$U(x^*, y^0) = C\big(U(x^* \mid y^*), U(y^0 \mid x^*); k_x, k_y\big) = k_x.$$

Therefore,

$$\boxed{C\big(1, 0; k_x, k_y\big) = k_x}. \tag{7}$$

And, similarly, for the second corner value, we have $U(x^0, y^*) = k_y$, and so

$$\boxed{C\big(0, 1; k_x, k_y\big) = k_y}. \tag{8}$$

Note: If a function C satisfies the domain and range $C : [0, 1] \times [0, 1] \to [0, 1]$ conditions, as well as conditions (3) through (8), then no matter what normalized conditional utility assessments and corner values you put into it, the resulting utility surface will have boundary assessments that are consistent with the input utility assessments and the corner values at the upper bounds.

Figure 28.1 shows an example of a utility copula function that matches the upper boundary values. Note how the copula is linear at the upper boundary values and how the corner values are specified in the figure. The copula is defined on the unit square and it ranges from zero to one.

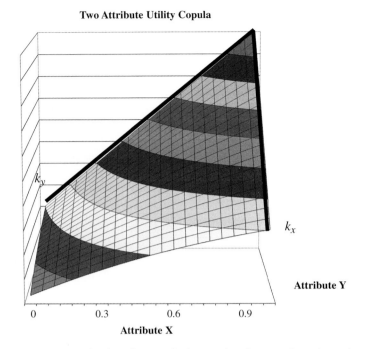

Figure 28.1. Example of a utility copula that matches the upper boundary values.

28.2.2 Class 1 Utility Copula Functions without Normalized Arguments

As we discussed, you do not need normalized conditional utility assessments as arguments of the utility copula function. You might instead prefer to put the unnormalized upper boundary assessments $U(x, y^*)$ and $U(x^*, y)$ themselves into the formulation. Therefore, you might wish to express the multiattribute surface directly as

$$U(x, y) = C\big(U(x, y^*), U(x^*, y)\big).$$

Using similar analysis as that conducted for the normalized inputs, we can derive the properties of the copula function is in this case.

Domain and Range: Note that the domain of the copula function is now $[k_y, 1] \times [k_x, 1]$, instead of the unit square, because the curve $U(x, y^*)$ ranges from $U(x^0, y^*) = k_y$ to $U(x^*, y^*) = 1$ as we vary x, and similarly the curve $U(x^*, y)$ ranges from k_x to 1 as we vary y. The range of the copula function is $[0,1]$, and so

$$C : [k_y, 1] \times [k_x, 1] \to [0,1].$$

Consistency Conditions at the Upper Bounds:
The consistency conditions are also slightly modified. To illustrate, consider the value of the utility copula at $y = y^*$. We get by direct substitution

$$U(x, y^*) = C\big(U(x, y^*), U(x^*, y^*)\big) = C\big(U(x, y^*), 1\big).$$

Define $U(x, y^*) = s$. This implies that

$$C(s,1) = s.$$

Similarly,

$$C(1,t) = t.$$

Note again that linearity is required at the specified reference values, but it is a simpler form than the previous case. Any function that satisfies the domain, range, and consistency conditions can be used to construct a utility surface that is consistent with the input utility assessments.

28.2.3 Class 0 Utility Copula Functions Matching the Lower Boundary Assessments

Definition

A **Class 0 utility copula** is a function $C(s,t;k_x,k_y)$ that expresses the utility function using lower boundary assessments and corner values using the relation

$$U(x,y) = C\big(U(x\,|\,y^0), U(y\,|\,x^0); k_x, k_y\big).$$

Once again we can specify a sufficient condition for the existence of this copula representation.

A **sufficient condition for a Class 0 utility copula formulation** is that the curves $U(x,y^0), U(x^0,y)$ are continuous and strictly increasing.

28.2.3.1 PROPERTIES OF CLASS 0 UTILITY COPULA FUNCTIONS

1. **Domain of the function C:** Because $U(x\,|\,y^0), U(y\,|\,x^0)$ are normalized conditional utility functions, the domain of the copula function in this case is the unit square $[0,1] \times [0,1]$.
2. **Range of the function C:** Because the output of the copula function is the actual (normalized) multiattribute utility (which is an indifference probability assessment), the **range** of the copula function is $[0,1]$.

Therefore, we can write

$$C : [0,1] \times [0,1] \rightarrow [0,1].$$

3. **Linearity at the lower boundary values:**
   Consistency at the lower boundary values requires that

$$U(x,y^0) = k_x U(x\,|\,y^0).$$

By direct substitution into the Class 0 utility copula formulation, we have

$$U(x,y^0) = C\big(U(x\,|\,y^0), U(y^0\,|\,x^0); k_x, k_y\big) = C\big(U(x\,|\,y^*), 0; k_x, k_y\big).$$

Therefore, consistency requires that

$$C\big(U(x\,|\,y^0),0;k_x,k_y\big)=k_xU(x\,|\,y^0).$$

Define $U(x\,|\,y^0)=s.$ This implies that

$$\boxed{C\big(s,0;k_x,k_y\big)=k_xs}, \tag{9}$$

and so the copula function is a linear function of its first argument at the lower bound of the second argument.

Similarly,

$$U(x^0,y)=k_yU(y\,|\,x^0).$$

By direct substitution, this implies that

$$U(x^0,y)=C\big(U(x^0\,|\,y^0),U(y\,|\,x^0);k_x,k_y\big)=C\big(0,U(y\,|\,x^0);k_x,k_y\big).$$

Therefore, consistency requires that

$$C\big(0,U(y\,|\,x^0);k_x,k_y\big)=k_yU(y\,|\,x^0).$$

Define $U(y\,|\,x^0)=t.$ This implies that

$$\boxed{C\big(0,t;k_x,k_y\big)=k_yt} \tag{10}$$

and so the copula function is a linear function of its second argument at the upper bound of the first argument.

4.   *Consistency at the Corner Values of the Domain*

Because the utility function is strictly increasing with each argument and it is normalized,

$$U(x^*,y^*)=1.$$

Substituting into the copula formulation gives

$$U(x^*,y^*)=C\big(U(x^*\,|\,y^0),U(y^*\,|\,x^0);k_x,k_y\big).$$

Hence,

$$\boxed{C\big(1,1;k_x,k_y\big)=1}. \tag{11}$$

Similarly, for the lower corner value,

$$U(x^0,y^0)=C\big(U(x^0\,|\,y^0),U(y^0\,|\,x^0);k_x,k_y\big)=0$$

and so

$$\boxed{C\big(0,0;k_x,k_y\big)=0}. \tag{12}$$

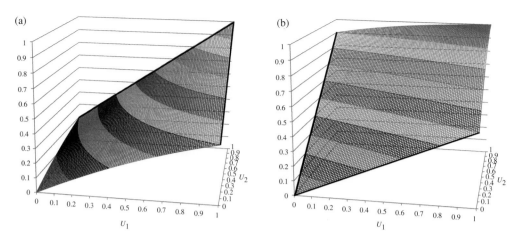

Figure 28.2. (Left) Class 1 utility copula linear at the upper bound. (Right) Class 0 utility copula linear at the lower bound.

We might also want to specify the corner values of the utility copula such that

$$U(x^*,y^0) = C\left(U(x^* \mid y^0), U(y^0 \mid x^0); k_x, k_y\right) = k_x$$

and so

$$\boxed{C\left(1,0; k_x, k_y\right) = k_x}.$$

(13)

Similarly,

$$\boxed{C\left(0,1; k_x, k_y\right) = k_y}.$$

(14)

Figure 28.2 shows examples of Class 1 and Class 0 utility copula functions. Note the linearity of the copula surfaces at the different boundaries.

28.3 WHY DO WE NEED BOTH CLASS 1 AND CLASS 0 UTILITY COPULAS?

It is helpful to have constructs that enable conditional utility assessments at either the upper or lower bounds because the decision-maker may be more comfortable conditioning on one bound versus another. This is particularly helpful if one bound comprises an extreme case of wealth or health.

Another motivation is the properties of the utility surface itself. To illustrate, consider the multiattribute utility functions of Figure 28.3. Figure 28.3(a) shows an attribute dominance utility function that is non-decreasing with each of its arguments and strictly increasing with each argument at the maximum value of the complement. This utility function cannot be constructed using conditional utility assessments at the minimum margins because they are not strictly increasing at this instantiation. We can, however, assess conditional utility functions at the maximum margins and construct this utility function using a class 1 utility copula.

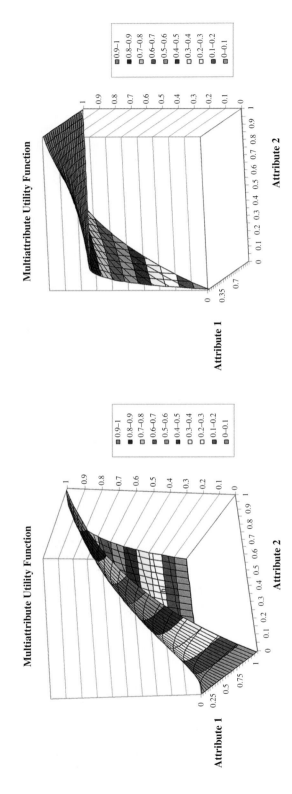

Figure 28.3. Left (a): Utility surface strictly increasing at the upper bound. Right (b): Utility surface strictly increasing at the lower bound.

Figure 28.3(b), on the other hand, shows the extreme case of a multiattribute utility function that has a value equal to one if any of the attributes has a maximum value. This utility function cannot be constructed with conditional utility functions at the maximum margins because they are not strictly increasing at this instantiation, but it can be constructed using a class 0 utility copula and utility assessments at the minimum margins. If the utility function is strictly increasing across the whole domain, then, as we discussed, we can use conditional utility assessments at any reference values of the complement attributes and construct the utility function using a utility copula structure that provides a positive linear transformation at these reference values.

In principle utility copula functions need not match only the upper or the lower boundary assessments. They can be more general and even match specific curves in the interior of the domain. Furthermore, we might have a copula that matches the utility of X at the upper bound of Y and the utility of Y at the lower bound of X, expressed as

$$U(x,y) = C(U(x \mid y^*), U(x \mid y^0); k_x, k_y).$$

By now it should be clear that this formulation will require the following consistency conditions:

$$C(s,1,k_x,k_y) = k_y + (1-k_y)s; \ C(0,t,k_x,k_y) = k_y t; C(0,1,k_x,k_y) = k_y; C(1,0,k_x,k_y) = k_x$$

Dialogue between Instructor (I) and Student (S)

I: Does everyone understand what is happening here?

S: Yes, we are trying to construct a multiattribute utility surface using individual conditional utility function assessments at some boundary values, and some corner values as arguments of a function called the utility copula.

I: Correct. It is important that the resulting utility surface be consistent with the assessments used for its construction.

S: And this is why we have the linearity property at the boundaries?

I: Correct.

S: So why don't we just use an additive or a multiplicative form to combine the individual utility assessments? We can also do a weighted average.

I: Because, as we have seen, these forms imply certain independence properties that need not be present. We have already seen in Chapter 21, for example, that the error can be large if you make utility independence assumptions when they are not present. We shall revisit this concept in a later chapter.

S: But additive and multiplicative forms really simplify the analysis.

I: Simplicity is not an excuse for making the wrong decision. We have also seen that these forms imply certain trade-offs among the attributes that need not be present. Refer back to our discussion of attribute dominance utility functions for example. Once you multiply two utility functions, you have no further degree of freedom to capture the precise trade-offs of the decision-maker.

S: So once you have a copula function, you are guaranteed that the resulting surface will match the assessments provided by the decision-maker?

I: Yes, and as we shall see, you also have an additional degree to freedom to vary the trade-offs among the attributes and still match those boundary assessments.

S: But how do you know which copula function to use?

I: That is of course an important question. We shall discuss several forms of copula functions and illustrate how to assess them both empirically and parametrically (Chapters 29–31). But first let us understand the properties of utility copula functions in more detail.

S: Okay. And which copula type should we use in a given decision? Class 1 or Class 0 copulas?

I: It depends on the comfort level of the decision-maker and the shape of the utility surface for that particular decision. If the corner values k_x, k_y are small (say equal to 0.1) for a given decision, then it might be difficult to reason about the lower boundary assessments with enough accuracy. Then it would be better to reason about the upper boundary assessments and use a Class 1 utility copula. On the other hand, if the corner assessments are large (say equal to 0.9), then it might be easier to reason about the lower boundary assessments, and so a Class 0 utility copula might be more suitable in this case.

28.4 EXTENSIONS TO MULTIPLE ATTRIBUTES

28.4.1 Class 1 and 0 Utility Copulas

We now extend the idea of a utility copula to multiple attributes.

Class 1 utility copulas require conditional utility assessments at the maximum values of the complement attributes, i.e., $U_i(x_i, \bar{x}_i^*), i = 1, ..., n$. They can be expressed as in terms of un-normalized arguments as

$$U(x_1, ..., x_n) = C(U_1(x_1, \bar{x}_1^*), ..., U_n(x_n, \bar{x}_n^*))$$

and also using the normalized conditional utility assessments:

$$U(x_1, ..., x_n) = C(U_1(x_1 \mid \bar{x}_1^*), ..., U_n(x_n \mid \bar{x}_n^*); k_{x_1}, ..., k_{x_n}).$$

The domain of a utility copula requiring normalized utility assessments, for example, is the normalized hypercube $[0, 1] \times [0, 1] \times .... \times [0, 1]$. The range of the utility copula is $[0, 1]$. The corner values of the resulting utility surface must also be consistent with the utility assessments provided. For ease of expression in the multiattribute case, we drop the corner values $k_1, ..., k_n$ from the copula notation, with the understanding that they are specified.

Class 1 utility copulas satisfy the linearity condition at the maximum value of the arguments, i.e.,

$$C(1, ..., 1, v_i, 1, ..., 1) = a_i v_i + b_i, \, i = 1, ..., n. \tag{15}$$

Class 0 utility copulas require conditional utility assessments at the minimum values of the complement. For normalized utility inputs, this requires

$$U(x_1, ..., x_n) = C_0(U_1(x_1 \mid \bar{x}_1^0), ..., U_n(x_n \mid \bar{x}_n^0)).$$

Class 0 utility copulas satisfy the linearity condition at the minimum value of the arguments, i.e.,

$$C_0(0,...,0,v_i,0,...,0) = a_i v_i, \quad i = 1,...,n. \tag{16}$$

28.4.2 General Utility Copula Function Definition for Arbitrary Reference Values

We assume that the decision-maker has a multiattribute utility function $U(x_1,...,x_n)$, defined over n direct value attributes, $X_1,...,X_n$. Following our earlier notation, we use the vector $(x_1,...,x_n)$ to represent a prospect of the decision, and for simplicity, we often refer to this prospect as $(x_i, \bar{x}_i)$. We also assume that $U(x_1,...,x_n)$ is

(i)   continuous;
(ii) bounded; and
(iii) for each argument x_i there exists at least one fixed reference value of the complement, call it $\bar{x}_i^{\lambda_i}$, such that the curve $U(x_i, \bar{x}_i^{\lambda_i})$ is strictly increasing with x_i.

Definition: Multiattribute Utility Copula with Normalized Inputs

A multiattribute utility copula, $C_\lambda(v_1,...,v_n)$, is a multivariate function of n variables that satisfies the following conditions:

1. **Normalized range and domain:** The function, C_λ, is a continuous mapping from the n-dimensional hypercube $[0,1]^n$ to the interval $[0,1]$

2. **Positive linear transformation at reference values:** For each argument v_i, there exist some reference values, $\lambda_{i,j}, i \neq j$, for which the utility copula satisfies

$$C_\lambda(\lambda_{i,1},...,\lambda_{i,i-1},v_i,\lambda_{i,i+1},...,\lambda_{i,n}) = a_i v_i + b_i, \quad i = 1,...,n,$$

   where $0 \leq \lambda_{i,j} \leq 1, 0 < a_i \leq 1$ and $0 \leq b_i < 1$.

3. **Consistency at all corner values of the domain**
   The utility copula must satisfy

$$C_\lambda(0,...,0) = 0 \quad \text{and} \quad C_\lambda(1,...,1) = 1. \tag{17}$$

The utility copula must also satisfy the corner values such that ***all permutations of 1's and 0's in the arguments of the copula function*** equal corresponding utility values at the corners of the domain of the utility surface.

Proposition 28.1: *Any multiattribute utility function that is (i) continuous, (ii) bounded, and (iii) strictly increasing with each argument for at least one reference value of the complement attributes can be expressed in terms of normalized conditional utility functions, $U_i(x_i \mid \bar{x}_i^{\lambda_i}), i = 1,...,n$, and a multiattribute utility copula, C_λ, as*

$$U(x_1,...,x_n) = C_\lambda(U_1(x_1 \mid \bar{x}_1^{\lambda_1}),...,U_n(x_n \mid \bar{x}_n^{\lambda_n})). \tag{18}$$

Proposition 28.1 explains the importance of the utility copula definition as it enables the construction of a large class of multiattribute utility functions using single-attribute utility assessments.

We seek the multivariate function C_λ that satisfies the equality

$$U(x_1,...,x_n) = C_\lambda(U_1(x_1 \mid \overline{x}_1^{\lambda_1}),...,U_n(x_n \mid \overline{x}_n^{\lambda_n})),$$

where the subscript λ refers to the reference values at which the assessments take place, and the superscript λ_i refers to the particular reference value of the complement of attribute X_i at which the assessment takes place. Following our previous notation, the term $U_i(x_i \mid \overline{x}_i^{\lambda_i})$ represents a normalized conditional utility function for attribute X_i, where

$$U_i(x_i \mid \overline{x}_i^{\lambda_i}) = \frac{U(x_i, \overline{x}_i^{\lambda_i}) - U(x_i^0, \overline{x}_i^{\lambda_i})}{U(x_i^*, \overline{x}_i^{\lambda_i}) - U(x_i^0, \overline{x}_i^{\lambda_i})}. \tag{19}$$

By definition, because the conditional utility functions $U_i(x_i \mid \overline{x}_i^{\lambda_i})$ are normalized from 0 to 1, the domain of the copula function in (18) is the unit hypercube, $[0,1]^n$.

28.5 THE INVERSE PROBLEM: CALCULATING THE UTILITY COPULA FUNCTION FOR A GIVEN MULTIATTRIBUTE UTILITY FUNCTION

In the next chapters, we shall present several examples of utility copulas and methods for their construction. First, we discuss the inverse problem of finding the utility copula for a given utility surface and a given set of conditional utility assessments. This analysis will help the reader better understand the concept of a utility copula.

EXAMPLE 28.1: Utility Function with Utility Dependence

Suppose we wish to derive the Class 1 utility copula, for the utility function

$$U(x,y) = \frac{1 - e^{-\gamma x y^\eta}}{1 - e^{-\gamma}}, 0 \le x, y < 1. \tag{20}$$

Because the required utility copula is of Class 1, it should provide a positive linear transformation at the maximum margins of its domain, i.e., $C(s,t)$, satisfies

$$C(s,1) = a_s s + b_s, \quad C(1,t) = a_t t + b_t. \tag{21}$$

We therefore first determine the normalized conditional utility functions at these maximum instantiations by direct substitution into (19) to get

$$U(x \mid y^*) = \frac{U(x,y^*) - U(x^0,y^*)}{U(x^*,y^*) - U(x^0,y^*)} = \frac{1 - e^{-\gamma x}}{1 - e^{-\gamma}} \triangleq s \tag{22}$$

Rearranging gives

$$x = \frac{-1}{\gamma} \ln(1 - s(1 - e^{-\gamma})). \tag{23}$$

Similarly,

$$U(y \mid x^*) = \frac{U(x^0, y)}{U(x^0, y^*)} = \frac{1 - e^{-\gamma y^{\eta}}}{1 - e^{-\gamma}} \triangleq t. \tag{24}$$

Rearranging gives

$$y = \left(\frac{-1}{\gamma} \ln(1 - t(1 - e^{-\gamma})) \right)^{\frac{1}{\eta}}. \tag{25}$$

Substituting for x, y from (23) and (25) into (20) gives the following utility copula

$$C(s,t) = \frac{1 - e^{-\frac{1}{\gamma} \ln(1 - s(1 - e^{-\gamma})) \ln(1 - t(1 - e^{-\gamma}))}}{1 - e^{-\gamma}}, 0 \le s, t \le 1. \tag{26}$$

For consistency, note that (26) is indeed normalized and satisfies the marginal condition of (21). For example,

$$C(s,1) = \frac{1 - e^{\ln(1 - s(1 - e^{-\gamma}))}}{1 - e^{-\gamma}} = \frac{1 - e^{\ln(1 - s(1 - e^{-\gamma}))}}{1 - e^{-\gamma}} = s, \quad a_s = 1, b_s = 0. \tag{27}$$

The utility copula function also satisfies the normalizing conditions,

$$C(0,0) = \frac{1 - e^{-\frac{1}{\gamma} \ln(1) \ln(1)}}{1 - e^{-\gamma}} = 0, \quad C(1,1) = \frac{1 - e^{-\frac{1}{\gamma} \ln(e^{-\gamma}) \ln(e^{-\gamma})}}{1 - e^{-\gamma}} = \frac{1 - e^{-\gamma}}{1 - e^{-\gamma}} = 1. \tag{28}$$

Given the copula form (26), we can now substitute for any normalized conditional utility functions, $s = U(x \mid y^*)$ and $t = U(y \mid x^*)$, and the resulting utility surface will match each of these conditional utility functions at the upper bound of the complement attribute. Moreover, changing the parameter γ will result in a different utility surface the still matches the boundary utility assessments but that provides different trade-offs among the attributes.

EXAMPLE 28.2: Utility Function with Mutual Utility Independence

---

Suppose we wish to derive the Class 0 utility copula, for the following utility function with mutual utility independence,

$$U(x,y) = \frac{1 - e^{-\gamma(x + \beta y)}}{1 - e^{-\gamma(1 + \beta)}}, 0 \le x, y \le 1. \tag{29}$$

Since the utility copula is of Class 0, it provides a positive linear transformation at the minimum margins, and so $C(s,t)$, satisfies

$$C(s,0) = a_s s + b_s, C(0,t) = a_t t + b_t. \tag{30}$$

We therefore determine the normalized conditional utility functions at these instantiations by direct substitution into (19) to get

$$U_x(x \mid y^0) = \frac{U(x, y^0)}{U(x^*, y^0)} = \frac{1 - e^{-\gamma x}}{1 - e^{-\gamma}} = s \Rightarrow x = \frac{-1}{\gamma} \ln(1 - s(1 - e^{-\gamma})), \tag{31}$$

$$U_y(y \mid x^0) = \frac{U(x^0, y)}{U(x^0, y^*)} = \frac{1 - e^{-\gamma \beta y}}{1 - e^{-\gamma \beta}} = t \Rightarrow y = \frac{-1}{\gamma \beta} \ln(1 - t(1 - e^{-\gamma \beta})). \tag{32}$$

Substituting for x, y from (31) and (32) into (29) gives the multilinear utility copula

$$C(s, t) = \frac{s(1 - e^{-\gamma}) + t(1 - e^{-\gamma \beta}) - st(1 - e^{-\gamma})(1 - e^{-\gamma \beta})}{1 - e^{-\gamma(1+\beta)}}, 0 \leq s, t \leq 1. \tag{33}$$

Given the copula form (33), we can now substitute for any normalized conditional utility functions, $s = U(x \mid y^0)$ and $t = U(y \mid x^0)$, and the resulting utility surface will match each of these conditional utility functions at the upper bound of the complement attribute.

Note:

1. Because the utility function (29) is strictly increasing with each of its arguments across the entire domain, we can define a utility copula at any instantiation of $(\lambda_{s2}, \lambda_{t1}) \in [0,1]^2$. For example, we can derive the utility copula that satisfies $C_\lambda(s, 0.5) = a_s s + b_s, C_\lambda(0.2, t) = a_t t + b_t$ by deriving the conditional utility functions at these instantiations, i.e., $U(x \mid y = 0.5)$ and $U(y \mid x = 0.2)$, and substituting into (29) to get the corresponding utility copula.

2. If an attribute is utility independent of its complement, then the normalized conditional utility function will not change with any instantiation of the complement attributes. Because (29) exhibits mutual utility independence, both the normalized conditional utility functions, $U(x \mid y), U(y \mid x)$, will not change across the entire domain of the utility function. This implies that (33) is the utility copula for any values of $(\lambda_{s2}, \lambda_{t1}) \in [0,1]^2$. Consequently, we can construct the utility function (29) using (33) and normalized conditional utility assessments at any reference values of the complement attributes.

28.6 BE CAREFUL WITH THE CONSISTENCY CONDITIONS

The basic idea of a utility copula function is to construct a utility surface that matches utility assessments provided by the decision-maker. These assessments, as we have seen, could be utility assessments at the boundaries of the domain, particularly the upper or the lower bound. Therefore, you need to specify the reference values at which the utility assessments are made, and you need to make sure that the utility copula satisfies the consistency conditions with these utility assessments. Otherwise, you might end up with a surface that is inconsistent with the utility assessments that are used for its construction.

For example if we use a Class 1 utility copula with assessments at the lower bound, then the resulting utility surface will not necessarily have utility values at its lower bound that are consistent with the input utility assessments. We illustrate an example below of this violation below, where the resulting surface could be inconsistent with the assessments used for its construction.

EXAMPLE 28.3: Arbitrary Aggregation Functions That Do Not Preserve the Consistency Conditions

---

Consider a function $U(x,y)$ of the form

$$U(x,y) = f^{-1}\big(w_1 f(U_1(x)) + w_2 f(U_2(y))\big), \quad w_1 + w_2 = 1.$$

This function is widely used as a multiattribute utility function to make trade-offs among attributes in the engineering design community. Note that this function uses input utility values $U_1(x), U_2(y)$, however, and we can also write it as

$$U(x,y) = P(U_1(x), U_2(y); w_1, w_2),$$

where

$$P(s,t;w_1,w_2) = f^{-1}\big(w_1 f(s) + w_2 f(t)\big).$$

To construct this function, the design literature suggests assessing individual utility functions $U_1(x), U_2(y)$ using lottery assessments, but it does not specify the reference values of the complement attributes for which they should be assessed. As we shall see, there are several consistency issues with this formulation. First of all, the reference values for which the utility assessments should be conducted are not specified.

Let us now pick normalized functions f such that $f(0) = 0$ and $f(1) = 1$. Furthermore, let us test the consistency condition at the lower bound. We have

$$U(x,y^0) = f^{-1}\big(w_1 f(U_1(x)) + w_2 f(U_2(y^0))\big) = f^{-1}\big(w_1 f(U_1(x))\big).$$

Herein lies the problem with this preference function. The term $f^{-1}\big(w_1 f(U_1(x))\big)$ is not necessarily a linear function of the utility assessment $U_1(x)$. Therefore, even if a decision-maker assesses utility functions at the lower bounds $U_1(x) = U(x,y^0)$ and substitutes into this formulation, the resulting surface will not be consistent with the utility assessments at the lower bound that were used for its construction.

The consistency condition also fails at the upper bound because

$$U(x,y^*) = f^{-1}\big(w_1 f(U_1(x)) + w_2 f(U_2(y^*))\big) = f^{-1}\big(w_1 f(U_1(x)) + w_2\big)$$

is not necessarily a linear function of the utility assessment $U_1(x)$.

The function $f^{-1}\big(w_1 f(U_1(x)) + w_2 f(U_2(y))\big)$ is not necessarily a utility copula function on the entire domain unless some additional conditions are imposed on the function f. Despite this fact, the function $f^{-1}\big(w_1 f(U_1(x)) + w_2 f(U_2(y))\big)$ has been used widely in the engineering design community where utility assessments are made for both attributes without specifying the reference value of the complement attribute for which the assessments are made, and without imposing consistency constraints on the function f.

28.6.1 Fixing the Consistency Conditions as a Class 0 Utility Copula

We can fix this consistency condition by imposing additional constraints on the function f. Note, for example, that if f were the power function

$$\boxed{f(t) = t^s},$$

then

$$\boxed{U(x,y) = f^{-1}\left(w_1 f(U_1(x)) + w_2 f(U_2(y))\right) = \left(w_x U_1(x)^s + w_y U_2(y)^s\right)^{\frac{1}{s}},} \qquad (34)$$

and the lower boundary assessments at the lower bound would be matched with

$$\boxed{U(x,y^0) = \left(w_x U_1(x)^s + w_y U_2(y^0)^s\right)^{\frac{1}{s}} = \left(w_x U_1(x)^s\right)^{\frac{1}{s}} = w_x^{\frac{1}{s}} U_1(x),}$$

which is a linear function of $U_1(x)$. The function (34) is now a Class 0 utility copula.

Furthermore, the weights also cannot be matched by indifference assessments as corner values because

$$\boxed{U(x^*,y^0) = \left(w_x U_1(x^*)^s + w_y U_2(y^0)^s\right)^{\frac{1}{s}} = w_x^{\frac{1}{s}} = k_x} \quad \text{and}$$

$$\boxed{U(x^0,y^*) = \left(w_x U_1(x^0)^s + w_y U_2(y^*)^s\right)^{\frac{1}{s}} = w_y^{\frac{1}{s}} = k_y}$$

To determine the value of s, we use the imposed condition $w_1 + w_2 = 1$ to get

$$\boxed{U(x^*,y^0)^s + U(x^0,y^*)^s = 1},$$

from which the value of s can be determined.

28.6.2 Fixing the Consistency Conditions as a Class 1 Utility Copula

We can also fix this consistency condition by imposing additional constraints on the function f. Note, for example, that if f were the power function

$$\boxed{f(t) = (1-t)^r},$$

then

$$U(x,y) = f^{-1}\left(w_1 f(U_1(x)) + w_2 f(U_2(y))\right) = 1 - \left(w_x[1-U_1(x)]^r + w_y[1-U_2(y)]^r\right)^{\frac{1}{r}}, \qquad (35)$$

and the upper boundary assessments would be matched with

$$U(x,y^*) = 1 - \left(w_x[1-U_1(x)]^r + w_y[1-U_2(y^*)]^r\right)^{\frac{1}{r}} = 1 - w_x^{\frac{1}{r}}[1-U_1(x)],$$

which is a linear combination of $U_1(x)$. The function (35) is now a Class 1 utility copula.

Furthermore, the weights also cannot be matched by indifference assessments as corner values because

$$U(x^0, y^*) = 1 - w_x^{\frac{1}{r}} \, [1 - U_1(x^0)] = 1 - w_x^{\frac{1}{r}} = k_y \text{ and}$$

$$U(x^*, y^0) = 1 - w_y^{\frac{1}{r}} \, [1 - U_2(y^*)] = 1 - w_y^{\frac{1}{r}} = k_x.$$

To determine the value of r, we use the imposed condition $w_1 + w_2 = 1$ to get

$$w_x + w_y = [1 - U(x^0, y^*)]^r + [1 - U(x^*, y^0)]^r = 1.$$

from which the value of r can be determined.

Conclusion: The function $f^{-1}\left(w_1 f(U_1(x)) + w_2 f(U_2(y))\right)$ is not necessarily a utility copula for arbitrary functions f. When used as a utility function, it implies additive ordinal preferences but it does not imply mutual utility independence. Therefore, it is important to specify the reference values of the complement attributes for any assessment that is made. Furthermore, to construct a utility surface using this function, we need to ensure consistency and that the function is linear with each argument for at least one reference value of the complement. This condition can be satisfied by further specifications on the function f.

28.7 SUMMARY

A utility copula function matches the conditional utility assessments at reference values of the complement attributes.

A Class 1 utility copula matches the upper bounds of the utility surface.

A Class 0 utility copula matches the lower bounds of the utility surface.

Single-attribute conditional utility assessments can be used to construct more general functional forms of utility functions that incorporate utility dependence using the utility copula formulation.

Consistency conditions are required to provide a utility copula formulation.

ADDITIONAL READINGS

Abbas, A. E. 2004. Entropy Methods in Decision Analysis. *Ph.D. Dissertation Stanford University.*

Abbas, A. E. 2009. Multiattribute utility copulas. *Operations Research* 57(6): 1367–1383.

Abbas, A. E. and R. A. Howard. 2005. Attribute dominance utility. *Decision Analysis* 2(4): 185–206.

Abbas, A. E. and Z. Sun. 2015. A utility copula approach for preference functions in engineering design. *ASME Journal of Mechanical Design* 137(9): 1–8.

Nelsen, R. B. 1998. *An Introduction to Copulas.* Springer-Verlag, New York.

Pratt, J. 1964. Risk aversion in the small and in the large. *Econometrica* 32: 122–136.

Richard, S. 1975. Multivariate risk aversion, utility independence and separable utility functions. *Management Science* 22: 12–21.

Sklar, A. 1959. Fonctions de répartition à n dimensions et leurs marges. *Publications de l'Institut de Statistique de L'Université de Paris* 8 : 229–231.

von Neumann, J. and O. Morgenstern. 1947. *Theory of Games and Economic Behavior.* 2nd Ed. Princeton University, Princeton, NJ.

APPENDIX 28A

Proof of Proposition 28.1

Because the utility function $U(x_1,...,x_n)$ is bounded, continuous, and strictly increasing with each argument, x_i, at an instantiation $\bar{x}_i^{\lambda_i}$, the normalized conditional utility functions $v_i = U_i(x_i \mid \bar{x}_i^{\lambda_i}), i = 1,...,n$ must also be continuous, bounded, and strictly increasing. This also implies that the inverse functions, $x_i = U_i^{-1}(v_i \mid \bar{x}_i^{\lambda_i}), i = 1,...,n$, are continuous, bounded, and strictly increasing. Define the function

$$C_\lambda(v_1,...,v_n) = U(U_1^{-1}(v_1 \mid \bar{x}_1^{\lambda_1}),...,U_n^{-1}(v_n \mid \bar{x}_n^{\lambda_n})). \tag{36}$$

Substitute for $v_i = U_i(x_i \mid \bar{x}_i^{\lambda_i})$ into the LHS and $x_i = U_i^{-1}(v_i \mid \bar{x}_i^{\lambda_i}), i = 1,...,n$ into the RHS of (36),

$$C_\lambda(U_1(x_1 \mid \bar{x}_1^{\lambda_1}),...,U_n(x_n \mid \bar{x}_n^{\lambda_n})) = U(x_1,...,x_n). \tag{37}$$

The function, $C_\lambda(v_1,...,v_n) = U(U_1^{-1}(v_1 \mid \bar{x}_1^{\lambda_1}),...,U_n^{-1}(v_n \mid \bar{x}_n^{\lambda_n}))$, satisfies the following properties:

1. **Normalized range and domain:** Because $U(x_1,...,x_n)$ and the conditional utility functions $v_i = U_i(x_i \mid \bar{x}_i^{\lambda_i}), i = 1,...,n$ are normalized to range from 0 to 1, then from (37), $C_\lambda : [0,1]^n \to [0,1]$. Furthermore, since $U_i^{-1}(v_i = 0 \mid \bar{x}_i^{\lambda_i}) = x_i^0, U_i^{-1}(v_i = 1 \mid \bar{x}_i^{\lambda_i}) = x_i^*$, then

$$C_\lambda(0,...,0) = U(U_1^{-1}(0 \mid \bar{x}_1^{\lambda_1}),...,U_n^{-1}(0 \mid \bar{x}_n^{\lambda_n})) = U(x_1^0,..,x_n^0) = 0. \tag{38}$$

$$C_\lambda(1,...,1) = U(U_1^{-1}(1 \mid \bar{x}_1^{\lambda_1}),...,U_n^{-1}(1 \mid \bar{x}_n^{\lambda_n})) = U(x_1^*,..,x_n^*) = 1. \tag{39}$$

The corner values are also satisfied as can be verified by direct substitution.

2. **Linearity at Reference Values:** Consider the instantiation $U(x_i, \bar{x}_i^{\lambda_i})$ for which the utility function is strictly increasing with x_i. Let the value of each node in the complement be $x_j^{\lambda_i}$ at this instantiation. At this instantiation (36) asserts

$$C(\lambda_{i,1},...,\lambda_{i,i-1}, U_i(x_i \mid \bar{x}_i^{\lambda_i}), \lambda_{i,i+1},...,\lambda_{i,n}) = U(x_i, \bar{x}_i^{\lambda_i}), \tag{40}$$

where $\lambda_{i,j} = U_j(x_j^{\lambda_i} \mid \bar{x}_j^{\lambda_j})$. Substitute for the definition of $U_i(x_i \mid \bar{x}_i^{\lambda_i})$ into the RHS of (40) to get

$$C(\lambda_{i,1},...,\lambda_{i,i-1}, U_i(x_i \mid \bar{x}_i^{\lambda_i}), \lambda_{i,i+1},...,\lambda_{i,n}) = [U(x_i^*, \bar{x}_i^{\lambda_i}) - U(x_i^0, \bar{x}_i^{\lambda_i})]U_i(x_i \mid \bar{x}_i^{\lambda_i}) + U(x_i^0, \bar{x}_i^{\lambda_i}) \tag{41}$$

Define $a_i = [U(x_i^*, \bar{x}_i^{\lambda_i}) - U(x_i^0, \bar{x}_i^{\lambda_i})], b_i = U(x_i^0, \bar{x}_i^{\lambda_i}), v_i = U_i(x_i \mid \bar{x}_i^{\lambda_i})$ and substitute into (41)

$$C(\lambda_{i,1},...,\lambda_{i,i-1}, v_i, \lambda_{i,i+1},...,\lambda_{i,n}) = a_i v_i + b_i. \tag{42}$$

Since $U(x_i, \bar{x}_i^{\lambda_i})$ is strictly increasing with x_i, then $a_i = [U(x_i^*, \bar{x}_i^{\lambda_i}) - U(x_i^0, \bar{x}_i^{\lambda_i})] > 0$. Furthermore, since the utility function is normalized, both $a_i, b_i \in [0,1]$. The same applies to all attributes. From (i), (ii), and the corner values, $C_\lambda(v_1, ..., v_n)$ is a utility copula function. *Q.E.D.*

APPENDIX 28B

Relating Utility Copulas to Probability Copulas

Sklar (1959) asserts that any continuous joint cumulative probability distribution can be expressed in terms of its marginal cumulative distributions and a multivariate function, C, which we refer to as a Sklar copula. The continuous Sklar copula has the following properties:

1. Sklar's copula is a continuous mapping from $[0,1]^n \to [0,1]$ and is normalized such that $C(0,...,0) = 0$ and $C(1,...,1) = 1$.
2. Sklar's copula satisfies the marginal property $C(1,1,...1,v_i,1,...,1) = v_i, \forall i$.
3. Sklar's copula is grounded, i.e., $C(v_1, v_2, ..., v_{i-1}, 0, v_{i+1}, ..., v_n) = 0, \forall i$.
4. Sklar's copula satisfies the n-increasing condition. For differentiable functions, this implies that the cross-derivative is non-negative.

Property (1) is satisfied by all utility copulas. Property (2) is a special case of the utility copula requirement, $C_\lambda(\lambda_{i,1}, ..., \lambda_{i,i-1}, v_i, \lambda_{i,i+1}, ..., \lambda_{i,n}) = a_i v_i + b_i$, when $a_i = 1, b_i = 0$, $\lambda_{i,j} = 1, \forall i \neq j$. Property (3) is a special instantiation of the utility copula definition, since the utility copula dos not need to (but can) be grounded. Property (4) is also an instantiation of the utility copula definition, since the utility copula may (or may not) satisfy the n-increasing condition.

> *Every continuous Sklar copula is a Class 1 utility copula, but not every utility copula is a Sklar copula.*

Consequently, we can use the widely studied functional forms of Sklar copulas to generate utility copulas. We illustrate this result below.

Proposition 28.2: *If $C(v_1, ..., v_n)$ is a Sklar copula, then*

$$C_\lambda(v_1, ..., v_n) = aC(l_1 + (1-l_1)v_1, ..., l_n + (1-l_n)v_n) + b, \tag{43}$$

where $0 \leq l_i < 1$, $a = 1/(1 - C(l_1, ..., l_n))$, $b = 1 - a$, is a class 1 utility copula that satisfies the n-increasing condition and

$$C_1(1, ..., 1, v_i, 1, ..., 1) = a(1 - l_i)v_i + (al_i + b), \ i = 1, ..., n. \tag{44}$$

$$C_1(1, ..., 1, 0, 1, ..., 1) = al_i + b = U(x_1^*, ..., x_{i-1}^*, x_i^0, x_{i+1}^*, ..., x_n^*), \ i = 1, ..., n. \tag{45}$$

Equation (45) asserts that the utility copula obtained from (43) is not grounded unless $l_i = 0, i = 1, ..., n$. We can therefore use (43) to generate ungrounded utility copulas from the widely used functional forms of Sklar (probability) copulas. With this approach, however, two properties still persist: (i) the utility copulas obtained using

(43) must satisfy the *n*-increasing condition, and (ii) they are class 1 utility copulas requiring conditional utility assessments at the maximum values of the complement attributes. We relax these two conditions in the next section.

Proof of Proposition 28.2: Relating Sklar Copulas to Utility Copulas

By definition, every continuous Sklar copula $C(v_1,...,v_n)$ is non-decreasing with each of its arguments a class 1 utility copula that satisfies $C(1,...,1,v_i,1,...,1) = v_i, i = 1,...,n$ and the *n*-increasing condition. These conditions assert that

1. C_λ is non-decreasing with each of its arguments and strictly increasing with each argument at the maximum values of the complement attributes.

2. Normalization: Since $0 \le l_i < 1, 0 \le v_i \le 1$, then $0 \le l_i + (1-l_i)v_i \le 1$. The utility copula is defined on a normalized domain. Furthermore, $C_\lambda(1,...,1) = aC(1,...,1) + b = a + b = 1$, and $C_\lambda(0,...,0) = aC(l_1,...,l_n) + b = \dfrac{C(l_1,...,l_n) - C(l_1,...,l_n)}{1 - C(l_1,...,l_n)} = 0$. Thus, $C_\lambda : [0,1]^n \to [0,1]$.

3. Positive linear transformation: Since, $C(1,...,1,v_i,1,...,1) = v_i, i = 1,...,n$, then by definition,

$$C_\lambda(1,...,1,v_i,1,...,1) = aC(1,...,1,l_i + (1-l_i)v_i,1,...,1) + b = al_i + a(1-l_i)v_i + b = a_iv_i + b_i, \qquad (46)$$

where $\quad a_i = a(1-l_i)v_i, \ b_i = b + al_i.$ From $\quad$ (46), $\quad C_\lambda(1,...,1,0,1,...,1) = l_i \triangleq U(x_1^*,...,x_{i-1}^*,x_i^0,x_{i+1}^*,...,x_n^*), 1" i" n.$

4. To show that it satisfies the *n*-increasing condition (we demonstrate only the case of differentiability) and observe that the sign of the mixed partial derivative of $C(v_1,...,v_n)$ and $C(l_1 + (1-l_1)v_1,...,l_n + (1-l_n)v_n)$ is the same when $0 \le l_i \le 1, i = 1,...,n$. From (1), (2), (3), and (4), the proof is complete.

CHAPTER 29

Archimedean Utility Copulas

Chapter Concepts

- Properties of Archimedean utility copulas
- Necessary and sufficient conditions for Archimedean functional forms
- Class 1 and Class 0 Archimedean utility copulas
- Constructing an Archimedean utility copula using a functional form for the generating function
- Sensitivity analysis using Archimedean utility copulas
  - Keeping the boundary utility values fixed while
    - Changing the shape of the isopreference contours
    - Changing the height of the utility surface
  - Sensitivity to utility independence using Archimedean utility copulas

29.1 INTRODUCTION

Archimedean utility copulas were first introduced in Abbas (2004), and then in Howard and Abbas (2005), and Abbas (2009). In Chapter 12, we illustrated how they represent the class of multiattribute utility functions corresponding to a utility function over an additive preference function or a utility function over a value function that can be converted into an additive form using a monotone transformation. To recap the results of Theorem 12.1, if we have a value function of the form

$$V(x_1,...,x_n) = m\left(\sum_{i=1}^{n} f_i(x_i)\right),$$

where more of an attribute is preferred to less, and if we assign a utility function over this additive function, we get

$$U(x_1,...,x_n) = U_V\left(V(x_1,...,x_n)\right) = U_V\left(m\left(\sum_{i=1}^{n} f_i(x_i)\right)\right),$$

then the resulting surface can be expressed as an Archimedean combination of the form

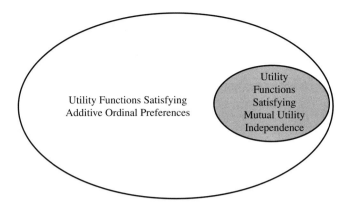

Figure 29.1. Archimedean utility copulas form the outer oval (which corresponds to additive preference functions). A special case of Archimedean forms are functions satisfying mutual utility independence.

$$U(x_1,...,x_n) = \varphi^{-1}\left[\prod_{i=1}^{n}\varphi(u_i)\right],$$

where $u_i = U(x_i,\overline{x}_i^*) = U(x_i^0,\overline{x}_i^*) + [1 - U(x_i^0,\overline{x}_i^*)]U(x_i \mid \overline{x}_i^*)$ is the utility function of attribute i at the maximum value of the complement attributes.

Figure 26.11, which we repeat in Figure 29.1 for convenience, illustrates how Archimedean utility copulas define the class of utility functions over additive preference functions where more of an attribute is preferred to less. Consequently, they are more general than the notion of mutual utility independence, which also requires additive ordinal preferences but imposes additional constraints that every subset of the attributes is utility independent of its complement. This additional condition is not required by Archimedean functional forms.

This chapter shows that the Archimedean combination of utility functions enables both Class 1 and Class 0 utility copulas. As we shall see, Archimedean combinations also enable sensitivity analysis to a variety of utility independence forms. The focus of this chapter will be on Archimedean combinations that use a specified functional form for the generating function. The next chapter discusses the empirical assessments needed for constructing the generating function of the Archimedean functional form.

29.2 ARCHIMEDEAN UTILITY COPULAS

29.2.1 Class 1 Archimedean Utility Copulas

Our focus will again be the form

$$C(u,v) = \varphi^{-1}\big(\varphi(u)\varphi(v)\big),$$

or its extension to multiple attributes

$$C(u_1,...,u_n) = \varphi^{-1}\left(\prod_{i=1}^{n}\varphi(u_i)\right), \tag{1}$$

where

$$u_i = U(x_i, \overline{x}_i^*) = U(x_i^0, \overline{x}_i^*) + [1 - U(x_i^0, \overline{x}_i^*)]U(x_i \mid \overline{x}_i^*).$$

Here we use the un-normalized form for the arguments of the copula because u_i ranges from $U(x_i^0, \overline{x}_i^*)$ to 1, as x_i varies from x_i^0 to x_i^*. If X_i is a utility-dominant attribute, then u_i ranges from 0 to 1.

Theorem 29.1

The functional form (1), *where φ is continuous and strictly increasing and satisfies the condition*

$$\varphi(1) = 1,$$

is a Class 1 utility copula.

We leave it to the reader to prove the previous Theorem. We highlight the proof here by first observing the consistency conditions. First we show that linearity is preserved at the upper bound. Direct substitution shows that

$$C(u, 1) = \varphi^{-1}\big(\varphi(u)\varphi(1)\big) = \varphi^{-1}\big(\varphi(u)\big) = u.$$

Therefore, the Archimedean copula function is indeed linear at the upper bound of the complement, and so it is a Class 1 utility copula.

Because $u = U(x^0, y^*) + [1 - U(x^0, y^*)]U(x \mid y^*)$, the resulting utility surface is a positive linear transformation of the normalized conditional utility function $U(x \mid y^*)$ at the upper bound, a condition that is required for consistency of the utility surface with the input utility assessments $U(x \mid y^*)$ at the upper bound. Continuity of the surface follows directly from continuity of the generating function. The non-decreasing property follows from the observation that $\varphi^{-1}\big(\delta\varphi(u)\big), 0 < \delta \leq 1$ is a strictly increasing function, and the domain, range, and orner values of the surface are easily proven by direct substitution.

Definition: Proper and Improper Generating Functions

The domain of the generating function φ needs to be defined on the range of u, and this is automatically satisfied if it is defined on [0,1]. If, in addition, the generating function φ passes by the origin, i.e. $\varphi(0) = 0$, then we give it a special symbol, ψ, and refer to it as a ***proper generating function***. If it does not pass by the origin, then we refer to it as an ***improper generating function***.

There is no fundamental difference between the two generating functions except that improper generating functions allow for more forms. For an improper generating function, we can also apply a transformation of the form

$$\varphi(t) = k\psi(t) + (1 - k),$$

where $\psi(t)$ is a proper generating function and where $\varphi(0) = 1 - k$. Substituting into

$$U(x_1,...,x_n) = \varphi^{-1}\left(\prod_{i=1}^{n}\varphi(u_i)\right),$$

gives an equivalent form of the Archimedean copula as

$$U(x_1,...,x_n) = \psi^{-1}\left(\frac{1}{k}\left[\prod_{i=1}^{n}\left(k\psi(u_i)+(1-k)\right)\right]+1-\frac{1}{k}\right),$$

where ψ is a proper generating function, $u_i = U(x_i, \bar{x}_i^*)$, and k is a constant satisfying (by direct substitution)

$$1-k = \left[\prod_{i=1}^{n}\left(k\psi(k_i)+(1-k)\right)\right],$$

where $k_i = U(x_i^0, \bar{x}_i^*)$.

Note 8: For any generating function, ψ, a solution $k \in [0,1]$ exists if $\sum_{i=1}^{n}\psi(k_i) \le n-1$. For more details on this proof, see Abbas and Sun (2017).

Note that for attribute dominance utility functions, $k_i = 0$ and so $k = 1$ giving

$$U^d(x_1,...,x_n) = \psi^{-1}\left(\left[\prod_{i=1}^{n}\left(\psi(u_i)\right)\right]\right).$$

The following example illustrates a Class 1 Archimedean copula using a proper generating function.

EXAMPLE 29.1: Class 1 Archimedean Attribute Dominance Utility Copula with Proper Generating Function

---

Consider a decision maker with additive ordinal preferences over two attributes, X and Y defined over the domain [0,1]x[0,1]. His preferences are such that more of an attribute is preferred to less, and more of an attribute is strictly preferred to less at the maximum value of the complement. The decision maker also satisfies the attribute dominance grounding conditions over the attributes such that $U(x^0, y^*) = U(x^*, y^0) = 0$.

From our previous discussions, we know that the decision maker's preferences can be respresented using an Archimedean utility copula. Furthermore, we know that for attribute dominance utility functions

$$U(x,y^*) = U(x \mid y^*) \text{ and that } U(x^*,y) = U(y \mid x^*).$$

Suppose further that his generating function is of the form

$$\psi(t) = \frac{1-e^{-\delta t}}{1-e^{-\delta}}, \quad 0 \le t \le 1, \ \delta \in R \setminus \{0\},$$

with

$$\psi^{-1}(t) = \frac{-1}{\delta}\ln\left(1 - (1 - e^{-\delta})t\right)$$

The utility copula has the form

$$C(u_x, u_y) = \frac{1}{\delta}\ln\left(1 - \frac{(1 - e^{-\delta u_x})(1 - e^{-\delta u_y})}{(1 - e^{-\delta})}\right),$$

If we assess normalized conditional utility functions, $U(x \mid y^*)$ and $U(y \mid x^*)$, as exponential with risk aversion coefficients $\gamma_x = 3$ and $\gamma_y = 2$, respectively, then

$$U(x, y) = \frac{1}{\delta}\ln\left(1 - \frac{(1 - e^{-\delta\frac{1-e^{-2x}}{1-e^{-2}}})(1 - e^{-\delta\frac{1-e^{-3y}}{1-e^{-3}}})}{(1 - e^{-\delta})}\right).$$

The following example illustrates how to construct an Archimedean multiattribute utility function using the same inputs as the previous example except for the grounding conditions.

EXAMPLE 29.2: Constructing a Class 1 Archimedean Utility Copula without the Grounding Conditions

---

Suppose that a decision-maker faces two attributes, X and Y, defined on the normalized domain $[0,1] \times [0,1]$. She states that her normalized conditional utility functions, $U(x \mid y^*)$ and $U(y \mid x^*)$, are exponential with risk aversion coefficients $\gamma_x = 3$ and $\gamma_y = 2$, respectively.

The decision-maker also provides the following utility values at the corners, $U(x^0, y^*) = 0.4$, $U(x^*, y^0) = 0.2$, using indifference probability assessments. For example, the utility value $U(x^0, y^*) = 0.4$ asserts that she is indifferent between receiving (x^0, y^*) for certain or receiving a binary deal that provides (x^*, y^*) with probability 0.4 and (x^0, y^0) with probability 0.6.

This implies that

$$u_x = U(x, y^*) = 0.4 + 0.6 \times \frac{1 - e^{-2x}}{1 - e^{-2}} \text{ and that } u_y = U(x^*, y) = 0.2 + 0.8 \times \frac{1 - e^{-3y}}{1 - e^{-3}}$$

Consider again the generating function,

$$\psi(t) = \frac{1 - e^{-\delta t}}{1 - e^{-\delta}}, \quad 0 \le t \le 1, \ \delta \in R \setminus \{0\}. \tag{2}$$

Substituing for

$$1 - k = \left[\prod_{i=1}^{n}\left(k\psi(k_i) + (1 - k)\right)\right]$$

gives

$$1 - k = \left(k\psi(0.4) + (1 - k)\right)\left(k\psi(0.2) + (1 - k)\right)$$

By direct substitution for $\delta = 1$, $\psi(0.4) = 0.521$ and $\psi(0.2) = 0.286$. Therefore, $\psi(0.4) + \psi(0.2) \leq 1$. Solving for $\delta = 1$ gives $k = 0.56$.

The functional form of the copula is now completely determined including all paramters and corner values,

$$U(x_1,...,x_n) = \psi^{-1}\left(\frac{1}{k}\left[(k\psi(u_x)+(1-k))(k\psi(u_y)+(1-k))\right]+1-\frac{1}{k}\right)$$

Note: Several Forms of Archimedean Copulas: There are many ways to express an Archimedean copula, and the procedure for assessing the paramters is essentially the same, and relies od direct substitution. Consider, for example, the Archimedean form,

$$C_1(v_1,...,v_n) = a\psi^{-1}[\prod_{i=1}^{n}\psi(l_i+(1-l_i)v_i)]+b, \qquad (3)$$

where, $0 \leq l_i < 1$, $a = 1/(1-\psi^{-1}[\prod_{i=1}^{n}\psi(l_i)])$, $b = 1-a$, and the function ψ satisfies the following conditions: (i) $\psi(v)$ is continuous on the domain $v \in [0,1]$; (ii) $\psi(v)$ is strictly increasing on the domain $v \in [0,1]$; (iii) $\psi(0) = 0$ and $\psi(1) = 1$.

It is clear, by direct substitution, that the Archimedean form (2) is a Class 1 utility copula. Direct substitution shows that

$$C_1(1,...,1,v_i,1,...,1) = a_i v_i + b_i, \quad i = 1,...,n, \qquad (4)$$

where $a_i = a(1-l_i)$ and $b_i = 1-a_i = al_i+b$.

The parameters, $l_i, i = 1,...,n$, of this Class 1 Archimedean utility copula can also be determined by consistency conditions at the corner values of the domain. Direct substitution gives

$$\begin{aligned}U(x_i^0,\bar{x}_i^*) &= C_1(U(x_i^*),...,U(x_{i-1}^*),U(x_i^0),U(x_{i+1}^*),...,U(x_n^*))\\ &= a\psi^{-1}[\psi(l_i)]+b\\ &= al_i+b\end{aligned}$$

Rearranging, with the observation that $b = 1-a$, gives

$$a(1-l_i) = 1-U(x_i^0,\bar{x}_i^*), \quad i = 1,...,n. \qquad (5)$$

gives the Class 1 utility copula,

$$C(v_x,v_y) = -a\frac{1}{\delta}\ln\left(1-\frac{(1-e^{-\delta(l_x+(1-l_x)v_x)})(1-e^{-\delta(l_y+(1-l_y)v_y)})}{(1-e^{-\delta})}\right)+b. \qquad (6)$$

By definition, $a = \dfrac{1}{1+\dfrac{1}{\delta}\ln\left(1-\dfrac{(1-e^{-\delta l_x})(1-e^{-\delta l_y})}{(1-e^{-\delta})}\right)}$, and from (4), the parameters l_x and l_y satisfy

$$a(1-l_x) = 0.6, \qquad a(1-l_y) = 0.8. \qquad (7)$$

If we use $\delta = 1$ in the generating function, then the solution to Equation (7) gives

$$l_x = 0.54, \quad l_y = 0.39,$$

with $a = 1.32$ and $b = -0.32$. The multiattribute utility function and the isopreference curves obtained by this Archimedean functional form are shown in Figure 29.2.

A Note on Assessing the Trade-off Parameters of an Archimedean Copula

Furthermore, in Example 29.2, we assumed the functional form of the generating function and also the value of the parameter δ. The value of δ can be determined by assessing a few points on the surface and substituting into (6) to determine the best value that matches the assessments. It can also be determined by assessing a few points on an isopreference contour and identifying the value of δ that best matches these trade-offs. Changing the value of the parameter δ yields different multiattribute utility surfaces having the same normalized conditional utility functions at the maximum margins but provide different trade-off functions among the attributes. If we are not able to match the decision-maker's trade-offs with a single-parameter generating function, then we can (i) use a different generating function with more parameters, or (ii) use the same generating function and conduct a least-squares estimate to determine the parameters l_i and the parameters of the generating function that best match the decision-maker's preferences. Alternatively, as we explain in the next chapter, we can (iii) determine the generating function empirically without assuming a particular functional form.

29.3 CLASS 0 ARCHIMEDEAN UTILITY COPULAS

In some cases, decision-makers may be more comfortable conditioning their utility assessments on other instantiations besides the maximum values of the complement attributes. It is also plausible, as we have seen in the right-hand side of Figure 28.3, that the utility function may be constant at these maximum values. To provide additional modeling flexibility in such situations, we now show how an Archimedean functional form may be used to construct Class 0 utility copulas.

Definition: Scaled Archimedean Functional Form

The *scaled Archimedean functional form*, $S(v_1, \ldots, v_n)$, is defined as

$$S(v_1, \ldots, v_n) = a\xi^{-1}\left(\prod_{i=1}^{n} \xi(m_i v_i)\right), 0 \le v_i \le 1, \tag{8}$$

where $0 < m_i \le 1$; $a = \dfrac{1}{\xi^{-1}\left(\prod_{i=1}^{n} \xi(m_i)\right)}$, and the function $\xi(v)$ is both continuous and

strictly decreasing on the domain $v \in [0,1]$, *with* $\xi(0) = 1$.

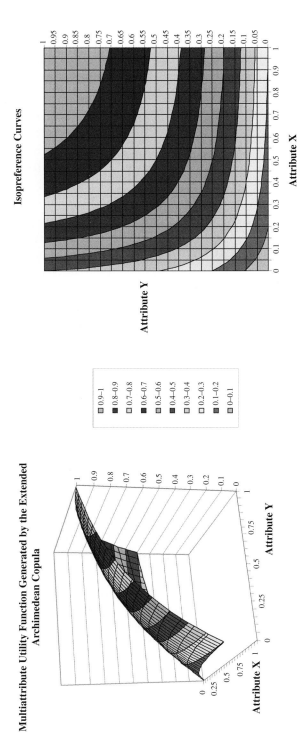

Figure 29.2. Multiattribute utility function generated by the Archimedean copula.

Note that if $\xi(1) = 0$, then $\xi(v)$ has the same mathematical properties as an excess cumulative probability distribution on the normalized domain and can be written as $\xi(v) = 1 - \psi(v)$. We refer to the function $\xi(v)$ in this case as a proper Class 0 generating function.

Proposition 29.2: *The scaled Archimedean functional form* (9) *is a Class 0 utility copula with*

$$S(0,...,0,v_i,0,...,0) = am_i v_i, \ i = 1,...,n, \tag{9}$$

where the parameters m_i satisfy

$$am_i = U(x_i^*, \bar{x}_i^0), \ i = 1,...,n. \tag{10}$$

The assessments required for m_i are different than those needed for Class 1 utility copulas, since X_i is at its maximum value and the assessment occurs at the minimum margin.

Example 29.3: Class 0 Archimedean utility copulas yielding the multiplicative form

---

Consider the Class 0 generating function

$$\xi(t) = (1 - t^\delta), \ 0 \le t \le 1, \delta > 0, \tag{11}$$

with $\xi^{-1}(t) = (1-t)^{\frac{1}{\delta}}$. Substituting from (12) into (9) gives

$$S(v_1,...,v_n) = \frac{\xi^{-1}\left(\prod_{i=1}^{n} \xi(m_i v_i)\right)}{\xi^{-1}\left(\prod_{i=1}^{n} \xi(m_i)\right)} = \frac{\left(1 - \prod_{i=1}^{n}(1 - m_i^\delta v_i^\delta)\right)^{\frac{1}{\delta}}}{\left(1 - \prod_{i=1}^{n}(1 - m_i^\delta)\right)^{\frac{1}{\delta}}}. \tag{12}$$

The functional form (13) is strictly increasing with each of its arguments when $0 < m_i < 1$, $i = 1,...,n$, and when $m_i = 1, i = 1,...,n$,

$$S(v_1,...,v_n) = \left(1 - \prod_{i=1}^{n}(1 - v_i^\delta)\right)^{\frac{1}{\delta}}, \tag{13}$$

which is equal to 1 if any of the arguments is equal to 1. Equation (13) also reduces to the multiplicative form when $\delta = 1$, giving

$$S(v_1,...,v_n) = \frac{\left(1 - \prod_{i=1}^{n}(1 - m_i v_i)\right)}{\left(1 - \prod_{i=1}^{n}(1 - m_i)\right)}. \tag{14}$$

The generating function (12) thus enables sensitivity to multiplicative utility independence using conditional utility assessments at the minimum values of the complement attributes.

29.4 ARCHIMEDEAN UTILITY COPULAS WITH IMPROPER GENERATING FUNCTIONS

We conclude our discussion of Archimedean utility copulas with the observation that other generating functions can also be used to construct utility copulas even if they are not normalized or restricted to the domain $[0,1]$. These generating functions expand the domain of definition of the utility copula, and we therefore call them *improper generating functions*.

To illustrate, let $\propto$ be a continuous and strictly decreasing function on the domain $[0,n]$, where n is the number of attributes, $\mu(0) = 1, \mu(1) = a, 0 < a < 1$, and $(\mu(1))^n \geq \mu(n)$. Consider a special case of $\propto$ where

$$\mu(v) = e^{-v}, v \in [0,n] \text{ with } \mu^{-1}(x) = \begin{cases} 1, & x < e^{-n} \\ -\ln x, & e^{-n} \leq x < 1 \end{cases}. \tag{15}$$

If we use the function $\propto$ with the scaled Archimedean form, we get

$$S(v_1,...,v_n) = \frac{\mu^{-1}\left(\prod_{i=1}^{n}\mu(m_i v_i)\right)}{\mu^{-1}\left(\prod_{i=1}^{n}\mu(m_i)\right)} = \frac{\sum_{i=1}^{n} m_i v_i}{\sum_{i=1}^{n} m_i} = \sum_{i=1}^{n} w_i v_i, \tag{16}$$

an additive utility copula with $w_i = \dfrac{m_i}{\displaystyle\sum_{i=1}^{n} m_i}, 0 " w_i " 1$, and $\displaystyle\sum_{i=1}^{n} w_i = 1$.

Example 29.4: Constructing a Class 0 utility copula with an improper generating function

Consider the improper generating function, $\mu(v) = e^{-v^{\delta}}, \delta > 0$. Substituting into (9) gives

$$S(v_1,...,v_n) = \frac{\mu^{-1}\left(\prod_{i=1}^{n}\mu(m_i v_i)\right)}{\mu^{-1}\left(\prod_{i=1}^{n}\mu(m_i)\right)} = \frac{\left(\sum_{i=1}^{n}(m_i v_i)^{\delta}\right)^{\frac{1}{\delta}}}{\left(\sum_{i=1}^{n}(m_i)^{\delta}\right)^{\frac{1}{\delta}}}, \tag{17}$$

which reduces to the additive utility copula of (17) when $\delta = 1$. For two attributes, X and Y, the mixed partial derivative of (18) can be determined directly as

$$\frac{\partial^2}{\partial v_x \partial v_y} S(v_x, v_y) = \delta\left(\frac{1}{\delta} - 1\right) \frac{\left((m_x v_x)^\delta + (m_y v_y)^\delta\right)^{\frac{1}{\delta} - 2}}{\left(m_x{}^\delta + m_y{}^\delta\right)^{\frac{1}{\delta}}} (m_x m_y)^\delta (v_x v_y)^{\delta - 1}. \tag{18}$$

Equation (19) shows that the mixed partial derivative is zero when either v_x, v_y is zero. At other instantiations, it is positive when $0 < \delta < 1$, and negative when $\delta > 1$. For example, when $\delta = 2$, (18) becomes

$$S(v_x, v_y) = \frac{\left((m_x v_x)^2 + (m_y v_y)^2\right)^{\frac{1}{2}}}{\left(m_x{}^2 + m_y{}^2\right)^{\frac{1}{2}}}. \tag{19}$$

If the decision-maker provides the assessments $U(x^*, y^0) = 0.6, U(x^0, y^*) = 0.8$, then from (11),

$$\frac{m_x}{\left(m_x{}^2 + m_y{}^2\right)^{\frac{1}{2}}} = 0.6, \frac{m_y}{\left(m_x{}^2 + m_y{}^2\right)^{\frac{1}{2}}} = 0.8. \tag{20}$$

Solving for the values of m_x, m_y in (21) gives $m_x = 0.3, m_y = 0.4$. Figure 29.3 shows the multiattribute utility function constructed using (20) and exponential marginal utility functions at the minimum values of the complement attributes, with $\gamma_x = 3, \gamma_y = 2$. Note that the boundary values $U(x, y^0)$ and $U(x^0, y)$ are preserved in the figure.

Using the generating function $\mu(v) = e^{-v^\delta}$, instead of $\mu(v) = e^{-v}$, also enables a sensitivity to additive utility independence to determine the percentage of incorrect decisions that could be made if the attributes have utility dependence. Figure 29.4 shows the simulation results for four attributes with $m_i = 1, i = 1, ..., n$. Appendix 29A shows the simulation procedure. The percentage of incorrect decisions is zero when $\delta \to 1$, since

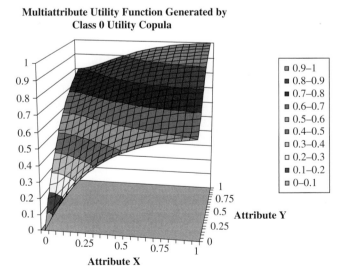

Figure 29.3. Multiattribute utility function generated by a Class 0 utility copula.

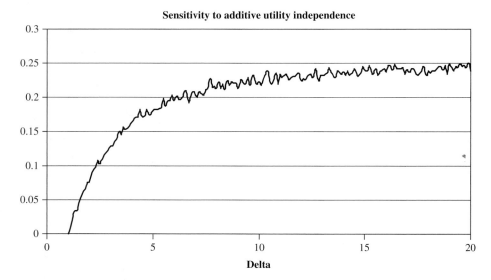

Figure 29.4. Sensitivity to additive utility independence.

$\lim_{\delta \to 1} e^{-\nu^\delta} = e^{-\nu}$. For larger values of δ, the percentage of incorrect decisions increases to about 25%. This result illustrates the need to verify additive independence before constructing an additive utility function, and also provides quantification on the percentage of incorrect decisions that could be made if additive independence is not satisfied but is used as an approximation.

29.5 SUMMARY

The Archimedean combination of individual utility functions

(i)   is the most general form of utility functions that are increasing with each arguments and have additive ordinal preferences for deterministic consequences (an additive preference function);

(ii) requires a single generating function;

(iii) enables both Class 1 and Class 0 utility copulas; and

(iv) enables sensitivity analysis to the additive and multiplicative forms of mutual utility independence.

ADDITIONAL READINGS

Abbas, A. E. 2004. Entropy Methods in Decision Analysis. *Ph.D. Dissertation Stanford University.*

Abbas, A. E. 2009. Multiattribute utility copulas. *Operations Research* 57(6): 1367–1383.

Abbas, A. E. and R. A. Howard. 2005. Attribute dominance utility. *Decision Analysis* 2(4): 185–206.

Abbas, A. E. and Z. Sun. 2015. A utility copula approach for preference functions in engineering design. *ASME Journal of Mechanical Design* 137(9): 1–8.

Abbas, A. E. and Z. Sun. 2017. Archimedean utility copulas with polynomial generating functions. *Working paper.*

Nelsen, R. B. 1998. *An Introduction to Copulas*. Springer-Verlag, New York.

Pratt, J. 1964. Risk aversion in the small and in the large. *Econometrica* 32: 122–136.

Richard, S. 1975. Multivariate risk aversion, utility independence and separable utility functions. *Management Science* 22: 12–21.

Sklar, A. 1959. Fonctions de répartition à n dimensions et leurs marges. *Publications de l'Institut de Statistique de L'Université de Paris* 8 : 229–231.

von Neumann, J. and O. Morgenstern. 1947. *Theory of Games and Economic Behavior*. 2nd Ed. Princeton University, Princeton, NJ.

APPENDIX 29A

Proof of Proposition 29.2: The continuity of ξ guarantees the continuity of the utility copula.

1. Monotonicity: Using similar analysis as that conducted in the outline of the Proof of Theorem 29.1, we can show that both ξ and ξ^{-1} are strictly decreasing and that

$\xi^{-1}\left(\delta_i \xi(m_i)\right)$ is a strictly increasing function of m_i when $0 < \delta_i \leq 1$. Let $\delta_i = \prod_{j \neq i}^{n} \xi(m_j v_j)$.

When $0 < m_i < 1, i = 1,...,n$, $0 < \delta_i \leq 1$ and (9) is strictly increasing with each of its arguments. If any $m_j = 1$, then function is non-decreasing with any of its arguments at the instantiation $v_j = 1$ (since $\delta_i = 0$), but even in this case, or the case where $m_j = 1, j = 1,...,n$

(9) reduces to $\xi^{-1}\left(\prod_{i=1}^{n} \xi(v_i)\right)$, which is strictly increasing with any argument v_i when the

remaining attributes are at their minimum values. Therefore, the monotonicity condition is satisfied.

2. Normalization: $S(0,...,0) = a\xi^{-1}\left(\prod_{i=1}^{n} \xi(0)\right) = a\xi^{-1}(1) = 0, S(1,...,1) = a\xi^{-1}\left(\prod_{i=1}^{n} \xi(m_i)\right) = 1.$

Furthermore, since $0 \text{ " } v_i \text{ " } 1$, and $0 < m_i \leq 1$, then $0 \text{ " } m_i \text{ " } 1$, and from (1) and (2), $C_\lambda : [0,1]^n \to [0,1]$.

3. When $v_j = 0, j \neq i$, we have $C_\lambda(0,...,0,v_i,0,...,0) = a\eta^{-1}\left(\eta(m_i)\right) = am_i v_i$, hence $a_i = am_i, b = 0$, and (9) is a class 0 utility copula. Furthermore, $C_\lambda(0,...,0,1,0,...,0) = a\xi^{-1}\left(\xi(m_i)\right) = am_i$. But from Proposition 29.2, $C_\lambda(0,...,0,1,0,...,0) = U(x_i^*, \overline{x}_i^0)$. This implies that $am_i = U(x_i^*, \overline{x}_i^0)$.

APPENDIX 29B STEPS OF MONTE CARLO SIMULATION

This appendix describes the sampling approach used to assess sensitivity to the multiplicative form of mutual utility independence. The idea is to generate joint probability distributions for two alternatives and calculate the expected utility of each alternative once using the multiplicative form and another using the copula form and see the percentage of time the calculation of the best decision alternative differs.

1. **Uniform sampling from the space of 3x3x3x3 probability distributions**

   Generate two 3x3x3x3 joint probability distributions to represent two decision alternatives (each having four variables discretized to three outcomes) by uniform sampling as follows:

- ◦ Generate $(3^4\text{-}1)$ independent samples, $x_1, x_2, ..., x_{3^4-1}$ from a uniform $[0,1]$ distribution.
- ◦ Sort the generated samples from highest to lowest to form an order statistic, $u_1 \leq u_2 \leq ..... \leq u_{3^4-1}$.
- ◦ Take the difference between each two successive elements of the order statistic $\{u_1 - 0, u_2 - u_1, ..., u_{3^4-1} - u_{3^4-2}, 1 - u_{3^4-1}\}$.
- ◦ The increments form a 3^4-outcome probability distribution that is uniformly sampled from the space of possible 3^4-outcome probability distributions.

2. **Generate four marginal utility values for each consequence in the tree from a uniform [0,1] distribution.**

   Since each consequence is characterized by four attributes, we generate four normalized utility values for each consequence.

3. **Evaluate the multiattribute utility function** for each prospect using two methods:

   (i) the additive form of multiattribute utility functions and
   (ii) the given utility copula form with $m_i = 0, i = 1, .., 4$.

4. **Calculate the expected utility** of both alternatives using (i) the product form and (ii) the copula form.

5. Repeat 10,000 times.

6. Calculate fraction of times a difference in the recommended decision alternatives occurs if we assume the product form when the actual utility function has a copula form.

7. Change the value of the dependence parameter, δ, in the copula form and repeat the simulation steps.

8. Repeat for different values of m_i.

CHAPTER 30

Assessing an Archimedean Utility Copula

Chapter Concepts

- An empirical method for constructing Archimedean utility copulas using indifference probability assessments
- The skewed diagonal curve of an Archimedean utility copula
- Functional equations relating curves on the surface of the Archimedean copula to its generating function

30.1 INTRODUCTION

By now, you should be quite familiar with the Archimedean form

$$C(u,v) = \varphi^{-1}\left(\varphi(u)\varphi(v)\right)$$

or its extension to multiple attributes

$$C(u_1,...,u_n) = \varphi^{-1}\left[\prod_{i=1}^{n}\varphi(u_i)\right], \tag{1}$$

where $u_i = U(x_i, \bar{x}_i^*)$ represents a boundary curve on the utility surface, and can be equivalently written as $u_i = U(x_i^0, \bar{x}_i^*) + (1 - U(x_i^0, \bar{x}_i^*))v_i$, where $v_i = U(x_i \mid \bar{x}_i^*)$.

In Chapter 12, we illustrated how this Archimedean form results from assigning a utility function over a value function that can be converted into an additive form using a monotone transformation. In Chapters 21 and 29, we illustrated how to construct such Archimedean utility copulas by assuming a functional form for the generating function. We also illustrated the use of Archimedean forms in conducting sensitivity analysis to the functional forms of mutual utility independence.

This chapter presents an empirical method for determining the generating function of an Archimedean form by direct utility elicitation. Using this approach, we shall characterize four curves on the surface of the Archimedean form and use these curves to construct the multiattribute utility surface.

30.2 ASSESSING AN ARCHIMEDEAN MULTIATTRIBUTE UTILITY FUNCTION

30.2.1 Verifying Additive Ordinal Preferences

As we discussed in earlier chapters, there is a correspondence between Archimedean forms and utility surfaces constructed by assigning a utility function over an additive value function. Therefore, the first step in the assessment process is to verify that ordinal preferences can be represented by an additive function,

$$V(x_1,...,x_n) = m\left(\sum_{i=1}^{n} f_i(x_i)\right),$$

and that more of an attribute is preferred to less.

This condition implies that the decision-maker's deterministic preferences over any of the attributes do not depend on the levels of the remaining attributes. The decision-maker can state a preference for each attribute independent of the others. In some cases, as we have seen with the peanut butter and jelly sandwich (Example 7.3), this need not be the case. For three or more attributes, this additive condition is known as *mutual preferential independence*.

30.2.2 Only Two Attributes Are Sufficient to Determine the Generating Function

Because a single generating function is sufficient to characterize the functional form of an Archimedean copula, it will suffice to assess this generating function using a two-attribute formulation, where any remaining attributes will be set at their maximum values.

To illustrate, note that if $C(v_1, v_2,...,v_n)$ is an Archimedean utility copula, with a generating function $\phi(1) = 1$ of the form

$$C(v_1,...,v_n) = \varphi^{-1}\left[\prod_{i=1}^{n} \varphi(U(x_i^0,\bar{x}_i^*) + (1 - U(x_i^0,\bar{x}_i^*))v_i)\right],$$

then the bivariate function

$$C_{12}(v_1,v_2) = C(v_1,v_2,1,1,...,1) = \varphi^{-1}\left[\prod_{i=1}^{2} \varphi(U(x_i^0,\bar{x}_i^*) + (1 - U(x_i^0,\bar{x}_i^*))v_i)\right]$$

is also an Archimedean functional form having the same generating function.

Therefore, an Archimedean utility copula of n attributes has the same generating function as a two-attribute Archimedean copula obtained by setting the remaining $(n\text{-}2)$ attributes at their maximum values. This observation allows us to use only two attributes in the assessment of the generating function and set the remaining attributes to their maximum values.

In principle, one can select from a library of functions to determine the generating function of the Archimedean form, and then conduct some utility assessments on the surface to estimate the parameters of the chosen functional form using a least-squares fit. This method of parameter estimation is widely used for utility functions, where the shape of the utility function is often assumed (such as an exponential function and the risk aversion coefficient is estimated to best match some utility assessments). As we have seen, however, the generating function of an Archimedean utility copula is strictly monotonic but does not need to be concave or convex on its entire domain. In fact, it can even be S-shaped to allow for further flexibility in the types of trade-offs that can be modeled. Therefore, the analyst must choose a functional form for the generating function that allows for a wide variety of shapes if the generating function is to accurately represent the assessments provided. One possibility is a polynomial generating function that provides various shapes and has a small number of parameters.

An alternate approach to determine the generating function is to assess a few points on the curve and then fit those points with a *smooth curve*. Unlike traditional utility function assessments, however, it is not possible to immediately assess points on the curve of the generating function of an Archimedean copula using lottery assessments, because there is no clear interpretation for the types of lottery questions one would ask to determine points on the generating function directly. To remedy this problem we need to first relate the generating function to some curves on the surface of the utility function that have a clear interpretation in terms of lottery assessments. This chapter explains an iterative approach to infer the generating function of the Archimedean form using direct utility assessments on specific curves on the domain of the attributes. The approach was proposed in Abbas and Sun (2015).

30.3 ASSESSMENTS NEEDED FOR CONSTRUCTING A TWO-ATTRIBUTE UTILITY FUNCTION USING AN ARCHIMEDEAN UTILITY COPULA

The utility assessments needed for constructing a two-attribute utility function using an Archimedean utility copula requires assessments of four curves, and can be divided into two steps:

Step 1: Assess the boundary utility functions for each attribute at the upper bound of the complement attributes, $U(x, y^*)$ and $U(x^*, y)$.

Step 2: Perform additional utility assessments to determine the generating function. These additional assessments include

(a) a utility assessment at the lower bound of the domain, and
(b) a utility assessment on a path on the domain of the copula, which we refer to as a *skewed diagonal* assessment, $S(t)$.

Define the corner values $k_x = U(x^*, y^0)$ and $k_y = U(x^0, y^*)$. Without loss of generality, we assume that $k_x \geq k_y$. If both corner values k_x, k_y are zero, then the assessment task would be simplified, and the lower bound, Step 2(a), would not be needed (the utility function would be grounded, and there would be no lower boundary assessments). We shall assume, therefore, that at least one of the corner values $k_x > 0$.

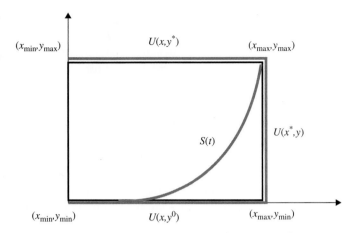

Figure 30.1. Paths for which utility assessments are needed to construct the utility surface.

Figure 30.1 shows the paths in the domain of the attributes for which the utility assessments are needed.

We now explain the assessments needed to construct the Archimedean utility function in more detail. First we derive convergence results to determine the generating function from the utility assessments of Steps 1 and 2.

Step 1: Assess Two Boundary Utility Functions $U(x, y^*)$ and $U(x^*, y)$

This step requires a utility assessment for each attribute at the upper bound of the complement attributes. For two attributes, we assess the two normalized conditional utility functions $U(x \mid y^*)$ and $U(x^* \mid y)$, and the two corner values $k_x = U(x^*, y^0)$ and $k_y = U(x^0, y^*)$. The normalized assessments, $U(x \mid y^*)$ and $U(y \mid x^*)$, can be determined by fitting the individual assessments to some of the widely used functional forms of utility functions or by assessing a few points and connecting them with a smooth path.

The utility function at the upper boundary values can then be determined from these normalized conditional assessments and corner values using the relations

$$U(x, y^*) = k_y + (1 - k_y)U(x \mid y^*)$$
$$U(x^*, y) = k_x + (1 - k_x)U(y \mid x^*). \tag{2}$$

Figure 30.2 plots an example of two boundary utility functions for attributes X and Y each on the domain $[0,1]$. The figure shows that $U(x, y^*)$ and $U(x^*, y)$ are strictly increasing and that $k_y = 0.1$ and $k_x = 0.5$.

Step 2: Additional Utility Assessments Needed to Determine the Generating Function

To determine the generating function when its functional form (or even shape) is not known, we need to make some additional utility assessments.

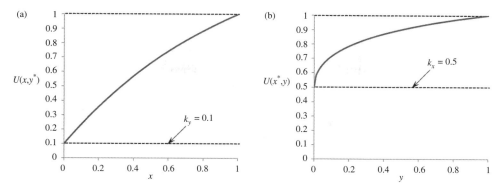

Figure 30.2. (a) The assessment $U(x,y^*)$ is strictly increasing from k_y to 1; (b) $U(x^*,y)$ is strictly increasing from k_x to 1.

STEP 2(A) UTILITY ASSESSMENTS AT A LOWER BOUND $U(x,y^0)$ Note that the lower bound of an Archimedean utility function is related to the upper bound using a transformation that depends on the generating function. We get by direct substitution into (1)

$$U(x,y^0) = C(v_1,0) = \varphi^{-1}\left[\{\varphi(U(x^0,y^*)) + (1 - U(x^0,y^*))U(x\,|\,y^*))\}\,\varphi(U(x^*,y^0))\right].$$

By assessing a lower boundary assessment, and comparing it to the upper boundary assessment, we can infer some information about the shape of the generating function. We have assumed that $k_x \geq k_y$, and so we assess the utility of the attribute with the higher corner value, X, at the lower bound of the attribute with the lower corner value, Y. In other words, we need to assess the curve $U(x,y^0)$.

STEP 2(B): UTILITY ASSESSMENT ON A SKEWED DIAGONAL CURVE The second assessment is conducted across a path on the domain of the attributes, which we refer to as a *skewed diagonal curve*. The intuition behind this name is that if both k_x, k_y were zero, this curve would be a straight (diagonal) line passing through the points $(0,0)$ and $(1,1)$ in the domain of the copula function. Because both k_x, k_y need not be zero, however, and they need not even be equal, the assessed curve in this case traces a skewed and offset path in the domain of the consequences, as we illustrate below.

Definition: Skewed Diagonal Path
The *skewed diagonal path* is determined by first defining a parameter $t \in [k_x, 1]$. The values of x and y that determine the skewed diagonal path are determined by the parametric equation

$$\begin{cases} U(x,y^*) = t \\ U(x^*,y) = t \end{cases}. \tag{3}$$

To determine the skewed diagonal path analytically, we define $x(t)$ as the inverse function of the curve $U(x,y^*)$ and $y(t)$ as the inverse function of the curve $U(x^*,y)$. The skewed diagonal path is traced by the points $(x(t),y(t))$ on the interval $t \in [k_x,1]$.

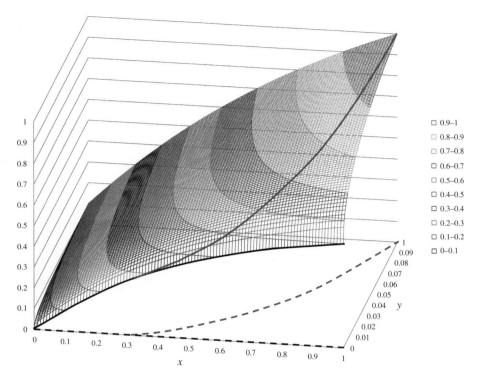

Figure 30.3. Assessments along the skewed diagonal curve and the lower bound.

Denote the utility values across the skewed diagonal path as $S(t)$, i.e.,

$$S(t) = U(x(t), y(t)), \quad t \in [k_x, 1]. \tag{4}$$

Figure 30.3 illustrates the utility assessments on the lower bound and the skewed diagonal path.

The following steps summarize the assessment procedure for the skewed diagonal curve:
1. Define the parameter t on the interval $[k_x, 1]$.
2. Define $x(t)$ using the equation $U(x(t), y^*) = t$.
3. Define $y(t)$ using the equation $U(x^*, y(t)) = t$.
4. Trace the path $(x(t), y(t))$, which is shown using the dashed line in Figure 30.3.
5. Conduct the utility assessments $S(t) = U(x(t), y(t))$ using indifference assessments.

The following example illustrates numerically the complete set of utility assessments needed to construct an Archimedean utility function.

EXAMPLE 30.1: Utility Assessments for the Archimedean Form

Step 1: Upper Boundary Assessments
The first step is to assess the upper boundary curves for each attribute. Once again, this can be done by identifying a functional form and assessing its parameters or by

assessing several points and fitting them. Here we assume a particular functional form. Suppose that the upper boundary utility functions are

$$U(x, y^*) = 1.52 - 1.42e^{-x}, x \in [0,1] \tag{5}$$

and

$$U(x^*, y) = 1.29 - 0.79e^{-\sqrt{y}}, y \in [0,1]. \tag{6}$$

By direct substitution, this implies that $k_x = U(x^*, y^0) = 0.5$ and $k_y = U(x^0, y^*) = 0.1$.

Step 2(a): Lower Boundary Assessment

Because the highest corner value is k_x, we need to assess the lower boundary curve $U(x, y^0)$. For this example, we use a Hyperbolic Absolute Risk Aversion (HARA) utility function at the lower bound because of its generality.

The analyst may assess utility values on this lower bound and then use these utility assessments to estimate the parameters of the HARA utility. Suppose that the resulting lower boundary assessment is

$$U(x, y^0) = 0.582 - (0.9884 - 0.041x)^{46.4}, x \in [0,1]. \tag{7}$$

Step 2(b): Skewed Diagonal Assessment

To assess the utility values along the skewed diagonal curve, we first define $t \in [k_x, 1]$. We then determine $x(t)$ and $y(t)$ for different values of t using (5) and (6), respectively. Table 30.1 shows the assessments. The first column shows discrete values of the parameter t. The second and third columns show the corresponding values of $x(t)$ and $y(t)$. Note that for $t = k_x$, $y(k_x) = y^0$ and for $t = 1$, $x(1) = x^*$ and $y(1) = y^*$.

Figure 30.4 plots this skewed diagonal path from Table 30.1, which is the x-y plane of Figure 30.2.

The last column in Table 30.1 shows the utility assessments for the points $x(t)$ and $y(t)$ defining the curve $S(t)$ obtained using indifference lottery assessments of $(x(t), y(t))$ for a binary gamble that gives either (x^*, y^*) with a probability $U(x(t), y(t))$ or (x^0, y^0) with a probability $1 - U(x(t), y(t))$.

Figure 30.5 illustrates the six assessments for $S(t)$ vs. t in Table 30.1.

We have now conducted all utility assessments needed to determine the utility surface.

Table 30.1. Determine the skewed diagonal path $(x(t), y(t))$ and the utility assessment $S(t)$.

T	$x(t)$	$y(t)$	$S(t) = U(x(t), y(t))$
$k_x = 0.5$	0.33	0	0.2765
0.6	0.43	0.02	0.3972
0.7	0.55	0.08	0.5253
0.8	0.68	0.23	0.6641
0.9	0.83	0.50	0.8189
1	1	1	1

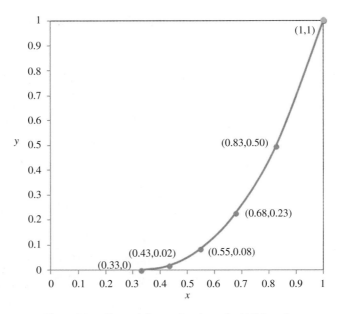

Figure 30.4. Skewed diagonal path on the X-Y domain.

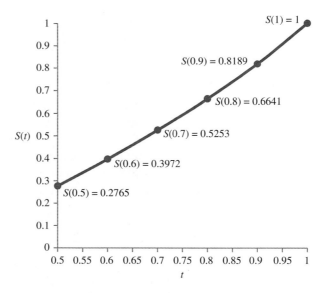

Figure 30.5. A utility assessment on the skewed diagonal curve $S(t)$.

Appendix 30.3 presents a method to connect finite ordered points using a differentiable smooth path. This method was proposed by Fritsch and Carlson (1980) and could be handy for fitting the curve $S(t)$ if needed.

The utility assessments across the skewed diagonal path can also be made by decomposing the assessment into multiple steps using the utility tree decomposition (Abbas 2011) where lotteries representing only one variation of each attribute can be incorporated. For example, consider the utility assessment at (x_1, y_1). The utility tree decomposition is

$$U(x_1, y_1) = U(x_1, y^*)U(y_1 \mid x_1) + U(x_1, y^0)\bar{U}(y_1 \mid x_1)) \tag{8}$$

where $\bar{U} = 1 - U$.

The assessment $U(x_1, y_1)$ can therefore be composed into $U(x_1, y^*), U(x_1, y^0)$, and $U(y_1 \mid x_1)$. Note that we have already assessed the utility function on the upper and lower bounds, $U(x, y^*)$ and $U(x, y^0)$. Therefore, $U(x_1, y^*)$ and $U(x_1, y^0)$ are already determined. Furthermore, the term $U(y_1 \mid x_1)$ is a single indifference assessment that can be obtained using indifference assessments of (x_1, y_1) for a binary gamble that gives either (x_1, y^*) or (x_1, y^0). This gamble keeps the level x_1 fixed and varies only y_1 from y^0 to y^*.

30.4 DETERMINING THE GENERATING FUNCTION

We shall now determine the generating function on the interval $[k_x, 1]$. Recall that the generating function of an Archimedean utility copula is strictly increasing on the interval $[0, 1]$, and so its derivative is positive (except possibly zero at some finite isolated points). We shall consider the case where the derivative of the generating function can be zero at finite points but assume it is strictly positive at the point $(1,1)$; i.e., we assume that $\varphi'(1) > 0$.

30.4.1 Relating $S(t)$ to the Generating Function on the Interval $[k_x, 1]$

The following proposition relates the assessment $S(t)$ to the generating function on the interval $[k_x, 1]$.

Proposition 30.1: Relating $S(t)$ to the generating function

$$S(t) = \varphi^{-1}\left((\varphi(t))^2\right), \qquad t \in [k_x, 1]$$

Proposition 30.1 shows that a portion of the generating function on the interval $t \in [k_x, 1]$ can be estimated if we solve the functional equation $\varphi(S(t)) = (\varphi(t))^2$. This functional equation is not easily solved for different curves $S(t)$, and so we shall use an iterative approach to determine the generating function using this functional equation.

Iterative Solution to the Functional Equation $S(t) = \varphi^{-1}\left((\varphi(t))^2\right)$, $t \in [k_x, 1]$

An iterative solution to this functional equation uses the following steps:

Step 1: Determine the inverse function S^{-1} on the interval $[k_x, 1]$.

Step 2: Determine the composite inverse function for any positive integer m as

$$S^{(-m)}(t) = S^{-1} \circ \cdots \circ S^{-1}(t), \forall t \in [k_x, 1].$$

Step 3: For any $S^{(-m)}(t)$, define the exponential function $\eta_m(t)$ such that

$$\eta_m(t) = e^{2^m\left(S^{(-m)}(t)-1\right)}.$$

Step 4: The iterations $\eta_m(t)$ converges to the generating function on the interval $[k_x, 1]$ as m increases.

The following example illustrates the steps needed to solve this functional equation numerically and to determine the generating function on the interval $[k_x, 1]$.

EXAMPLE 30.2: Determining the Generating Function on the Interval $[k_x, 1]$

Step 1: Determine the Inverse Function $S^{-1}(t)$

If we have fitted the assessments in Table 30.1, then we can determine the inverse function, $S^{-1}(t)$, by taking the inverse function of the fitted functional form. Alternatively, we can use the assessments of Table 30.1 to determine to determine assessments on the curve $S^{-1}(t)$.

To determine the inverse function $S^{-1}(t)$ from the assessments in Table 30.1, we interchange the order of the assessments of $(t, S(t))$ in the table to $(t, S^{-1}(t))$ by simply reversing the order of the columns. For example, the six assessments of $(t, S^{-1}(t))$ using the assessments of Table 30.1 are as follows: $(0.2765, 0.5)$, $(0.3972, 0.6)$, $(0.5253, 0.7)$, $(0.6641, 0.8)$, $(0.8189, 0.9)$, $(1, 1)$. The next step is to fit the assessments of $S^{-1}(t)$ using a smooth curve to help determine its composite functions.

The properties of $S(t)$ and its inverse $S^{-1}(t)$ are derived in Appendix 30.2 with an illustration of why the inverse function is clearly defined. Appendix 29.3 provides a procedure to determine a piecewise polynomial fit for $S^{-1}(t)$ that may be used in practice.

Step 2: Determine the Composite Functions $S^{-(m)}(t)$

Given $S^{-1}(t)$, the calculation of $S^{-2}(t)$ is obtained by iteration, where $S^{-(2)}(t) \triangleq S^{-1}\big(S^{-1}(t)\big)$, and similarly for higher orders to get $S^{-(m)}(t)$. Figure 30.6 plots the inverse function S^{-1}, and its composite functions, $S^{(-3)}$ and $S^{(-6)}$, on $[k_x, 1]$ as determined by the polynomial fit of S^{-1} using the approach in Appendix 29.3. Appendix 29.2 explains why $S^{(-m)}(t) \geq S^{(-(m-1))}(t)$, and therefore why the curves in Figure 30.6 are increasing with m.

Step 3: Determine the Iterations $\eta_m(t)$

For any $S^{(-m)}(t)$, define the exponential function

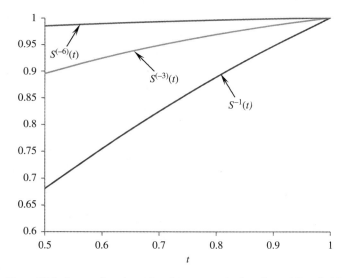

Figure 30.6. Inverse function S^{-1} and its composite functions $S^{(-3)}$ and $S^{(-6)}$.

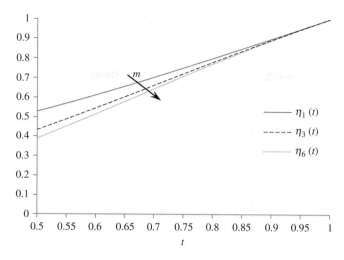

Figure 30.7. Plot of $\eta_m(t)$, $m = 1, 3, 6$ on the interval $[k_x, 1]$.

$$\eta_m(t) = e^{2^m\left(S^{(-m)}(t)-1\right)}. \tag{9}$$

The iterations $\eta_m(t)$ are obtained by direct substitution: for every t, determine the composite function $S^{(-m)}(t)$ and substitute into (9).

Figure 30.7 plots the curves $\eta_1(t) = e^{2(S^{-1}(t)-1)}$, $\eta_3(t) = e^{8(S^{(-3)}(t)-1)}$, $\eta_6(t) = e^{64(S^{(-6)}(t)-1)}$ computed directly from S^{-1}, $S^{(-3)}$, and $S^{(-6)}$. As we shall see below, the iterations $\eta_m(t)$ approximate the generating function on the interval $[k_x, 1]$.

30.4.2 Convergence of $\eta_m(t)$ to the Generating Function on the Interval $[k_x, 1]$

Observe that the generating function of an Archimedean copula is unique up to a power transformation. That is, if $C(v_1, ..., v_n)$ is an Archimedean utility copula with generating function φ, then it is also the copula formed by the generating function φ^α, $\alpha > 0$.

> **Lemma 30.2.** *If the derivative of the generating function at $t = 1$ is not equal to zero, i.e., $\varphi'(1) \neq 0$, then there always exists $\alpha > 0$ such that $\bar{\varphi}'(1) = 1$, where $\bar{\varphi} = \varphi^\alpha$.*

Lemma 30.2 implies that if $\varphi'(1) \neq 0$ (as we have assumed), then we can further assume without loss of generality that the generating function satisfies $\varphi'(1) = 1$.

> **Theorem 30.3. Determining the Generating Function on the Interval $[k_x, 1]$**
> *If the generating function of an Archimedean utility copula, $\varphi(t)$, satisfies $\varphi'(1) = 1$, then*
> $$\varphi(t) = \lim_{m \to \infty} \eta_m(t), \forall t \in [k_x, 1]. \tag{10}$$

Theorem 30.3 asserts that higher orders of $\eta_m(t)$ converge to the generating function, $\varphi(t)$, on the interval $t \in [k_x, 1]$. While any power of a generating function results in an equivalent Archimedean copula, the convergence of Theorem 30.3 results in the generating function that satisfies the condition $\varphi'(1) = 1$.

30.4.3 Determining the Generating Function on the Interval $[0, k_x)$

We have determined the generating function on the interval $[k_x, 1]$. We now show how to determine the generating function on the remaining interval using the estimated generating function on the interval $[k_x, 1]$ and a transformation function $g(r)$.

First, we define a general transformation function $g(r)$ that relates the upper and lower boundary assessments, $U(x, y^*)$ and $U(x, y^0)$, that we have already conducted, as

$$U(x, y^*) = g(U(x, y^0)). \tag{11}$$

As attribute X spans its minimum to maximum values, the domain of the function g spans $U(x^0, y^0) = 0$ to $U(x^*, y^0) = k_x$, and the range of g spans $U(x^0, y^*) = k_y$ to $U(x^*, y^*) = 1$. Therefore,

$$g : [0, k_x] \rightarrow [k_y, 1].$$

The following example illustrates the calculation of the transformation function, g, from the upper and lower boundary assessments.

EXAMPLE 30.3: Determining the Transformation Function $g(r)$

Suppose that $k_x \geq k_y$ and that the upper boundary utility function is

$$U(x, y^*) = 1.52 - 1.42e^{-x}, x \in [0,1]. \tag{12}$$

Suppose that the lower boundary utility assessment is

$$U(x, y^0) = 0.582 - (0.9884 - 0.041x)^{46.4}, x \in [0,1]. \tag{13}$$

From (11), (12), and (13), the transformation g satisfies

$$1.52 - 1.42e^{-x} = g\left(0.582 - (0.9884 - 0.041x)^{46.4}\right). \tag{14}$$

To determine the transformation g, define $r = U(x, y^0)$. From (13),

$$r = 0.582 - (0.9884 - 0.041x)^{46.4}.$$

Note that $r \in [0, k_x]$. Rearranging gives

$$x = 24.107 - 24.39(0.582 - r)^{0.0216}, r \in [0, k_x]. \tag{15}$$

Substituting for the value of x from (15) into (14) gives

$$g(r) = 1.52 - 1.42e^{-(24.107 - 24.39(0.582 - r)^{0.0216})}, r \in [0, k_x]. \tag{16}$$

Figure 30.8 plots the transformation $g(r)$ in (16). Note that $g : [0, k_x] \rightarrow [k_y, 1]$.

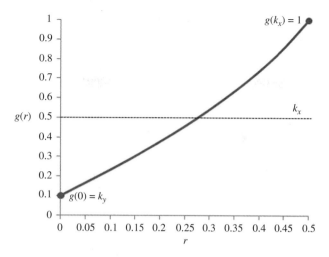

Figure 30.8. Transformation $g(r) : [0, k_x] \rightarrow [k_y, 1]$.

We discuss the properties of the transformation g in more detail in the next section and illustrate its use in determining the generating function.

30.4.4 Relating $g(r)$ to the Generating Function on the Interval $[k_x, 1]$

Define the pth composite function $g^{(p)}(r)$ as

$$g^{(p)}(r) = g \circ \cdots \circ g(r).$$

Before we relate $g(r)$ to the generating function, we need to provide some intuition about the shape of the compositions $g^{(p)}(r)$. Figure 30.9(a) plots the functions $g(r), g^{(2)}(r), g^{(3)}(r), g^{(4)}(r)$ for the assessments of Example 30.1.

Observe the following important features from Figure 30.9(a):

1. The domain of $g(r)$ is $[0, k_x]$ and its range is k_y to 1. Moreover, it satisfies the inequality $k_x \leq g(r) < 1$ on the interval $r \in [0.27, 0.5)$. No other composite function of g satisfies the inequality $k_x \leq g^{(p)}(r) < 1$ on the interval $r \in [0.27, 0.5)$.
2. The function $g^{(2)}(r)$ satisfies $k_x \leq g^{(2)}(r) < 1$ on the interval $r \in [0.13, 0.27)$. No other composite function of g satisfies $k_x \leq g^{(p)}(r) < 1$ on this interval.
3. The function $g^{(3)}(r)$ satisfies $k_x \leq g^{(3)}(r) < 1$ on the interval $r \in [0.03, 0.13)$. No other composite function of g satisfies $k_x \leq g^{(p)}(r) < 1$ on this interval.
4. The function $g^{(4)}(r)$ satisfies $k_x \leq g^{(4)}(r) < 1$ on the interval $r \in [0, 0.03)$. No other composite function of g satisfies $k_x \leq g^{(p)}(r) < 1$ on this interval. The function $g^{(4)}(r)$ has the property that its value at zero is greater than k_x. This is where we end the compositions of $g(r)$ for the purposes of estimating the generating function.

We can now define the integer-valued decreasing function $p(r)$ as the smallest integer, p, for any r, that satisfies $k_x \leq g^{(p)}(r) < 1$. From Figure 30.8(a), we have

$$p(r) = 1, \forall r \in [0.27, 0.5),$$

$$p(r) = 2, \forall r \in [0.13, 0.27),$$

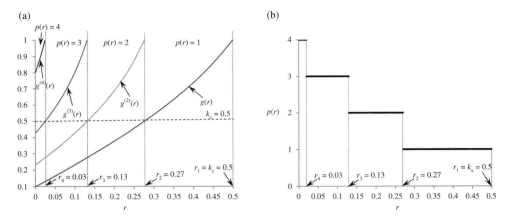

Figure 30.9. (a) Composite functions $g(r)$, $g^{(2)}(r)$, $g^{(3)}(r)$ and $g^{(4)}(r)$; (b) Integer function $p(r)$.

$$p(r) = 3, \forall r \in [0.03, 0.13),$$

$$p(r) = 4, \forall r \in [0, 0.03).$$

Figure 30.9(b) plots the function $p(r)$ and shows that it is a decreasing step function. The domain of $p(r)$ is the interval $(0, k_x)$ and is integer-valued.

The following proposition relates the functions $g^{(p)}(r)$ to the generating function.

Proposition 30.4: Relating $g^{(p)}(r)$ to the Generating Function on $r \in [0, k_x)$

$$\varphi(r) = (\varphi(k_x))^p \varphi(g^{(p)}(r)), \qquad r \in [0, k_x). \tag{17}$$

Equation (17) is a functional equation that shows that $\varphi(r)$ on the interval $[0, k_x)$ is equal to the product of the constant $(\varphi(k_x))^p$ and the composition $\varphi(g^{(p)}(r))$ for any value of p. This might suggest at first that we can determine the generating function on the interval $[0, k_x)$ by direct substitution for $r \in [0, k_x)$ into (17). The problem we encounter, however, is that we have only determined the value of $\varphi(t)$ on the interval $[k_x, 1]$. If the value of r in (17) is such that $g^{(p)}(r) \in [k_x, 1]$, then we can substitute directly into the right-hand side of (17) to determine the corresponding value of $\varphi(r)$. It might be possible, however, that the composition $g^{(p)}(r)$, for a given value of p, lies outside the interval $[k_x, 1]$, as we have seen in Figure 39(a). If this is the case, then we cannot determine $\varphi(r)$ by direct substitution into the right-hand side for that value of p.

The question that arises now is whether we can always find a value of p such that the composition $g^{(p)}(r)$ belongs to the interval $[k_x, 1]$ for any $r \in [0, k_x)$. If this were the case, then we can determine $\varphi(r)$ over the whole interval $[0, k_x)$ by direct substitution into the right-hand side of (17). The following lemma asserts this fact.

Lemma 30.5: Existence of the Integer p
For any given $r \in (0, k_x)$, there exists a composite function $g^{(p)}$ such that

$$k_x \leq g^{(p)}(r) < 1.$$

Due to Lemma 30.5, for any $r \in (0, k_x)$, we are guaranteed a value of p such that $k_x \leq g^{(p)}(r) < 1$. For ease of calculation of the composite functions, we shall use the lowest value of p that satisfies this condition, i.e., $p(r)$. This will enable us to determine the values of $\varphi(r)$ over $(0, k_x)$.

The Iteration $\tau_m(r)$

Define the function

$$\tau_m(r) = (\eta_m(k_x))^{p(r)} \eta_m(g^{(p(r))}(r)), \ \forall r \in (0, k_x), \tag{18}$$

where $\eta_m(t) = e^{2m[S^{(-m)}(t)-1]}$, $t \in [k_x, 1]$ is the same iteration defined earlier.

To better understand this function $\tau_m(r)$ note that the first term is simply a constant term $\eta_m(k_x)$ raised to the power of $p(r)$. The second term $\eta_m(g^{(p(r))}(r))$ is a composite function based on the iteration $\eta_m(t)$ and the $p(r)$ composition of $g(r)$.

We do not need to plot the full curves in Figure 30.9 every time we compute $\tau_m(r)$. To illustrate, suppose we wish to calculate $\tau_3(0.2)$. We first determine the value $g(0.2) = 0.377$. Because $g(0.2) < 0.5 = k_x$, we conduct another composition to get

$$g^{(2)}(0.2) = g(g(0.2)) = g(0.377) = 0.688 > 0.5 = k_x.$$

Hence, the integer valued function $p(0.2) = 2$, and we do not need higher compositions at $r = 0.2$. Now we calculate $\eta_3(g^{(2)}(0.2)) = \eta_3(0.688)$ as

$$\eta_3(g^{(2)}(0.2)) = e^{2^3(S^{(-3)}(0.688)-1))} = e^{8(0.946-1)} = 0.65. \tag{19}$$

Now for the constant term,

$$\eta_3(k_x) = e^{2^3(S^{(-3)}(0.5)-1))} = e^{8(0.8955-1)} = 0.43. \tag{20}$$

Substituting from (19) and (20) into (18) gives

$$\tau_3(0.2) = (\eta_3(k_x))^{p(0.2)} \eta_3(g^{(p(0.2))}(0.2)) = 0.43^2 \times 0.65 = 0.12.$$

Figure 30.10 plots the curve $\tau_m(r)$ for $m = 1, 3, 6$ for the skewed diagonal assessment of Table 30.1 and the boundary assessments of Example 30.1.

30.4.4.1 Determining the Generating Function on the Interval $(0, k_x)$

Theorem 30.6

If the generating function of an Archimedean utility copula, $\varphi(t)$, satisfies $\varphi'(1) = 1$, then

$$\varphi(r) = \lim_{m \to \infty} \tau_m(r), \ \forall r \in (0, k_x). \tag{21}$$

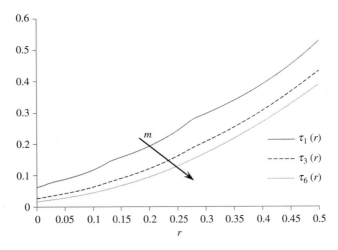

Figure 30.10. Convergence of $\tau_m(r)$, $m = 1,3,6$ on $(0,k_x)$.

Theorem 30.6 asserts that higher orders of τ_m converge to the generating function on the interval $(0,k_x)$. Once again, while any power of a generating function results in an equivalent Archimedean copula, the convergence of Theorem 30.6 results in a generating function that satisfies the condition $\varphi'(1) = 1$.

30.5 SUMMARY OF THE APPROACH

We now summarize the steps needed to determine the generating function from the utility assessments and then illustrate the deviation between consecutive iterations.

1. Assess two corner values, k_x and k_y. Assume that $k_x \geq k_y$.
2. Assess two utility functions at the upper bound, $U(x|y^*)$ and $U(y|x^*)$.
3. Assess the utility function at the lower bound, $U(x,y^0)$.
4. Assess the utility values along the skewed diagonal curve, $S(t)$.
5. Determine the inverse function $S^{-1}(t)$ and its composite functions $S^{(-m)}(t)$.
6. Determine $g(r)$ on the interval $[0,k_x]$ using (11) and its composites $g^{(p)}(r)$.
7. Determine the integer-valued function $p(r)$ from the functions $g^{(p)}(r)$.
8. Use Theorems 30.3 and 30.6 to determine the generating function $\varphi(t)$.

30.6 CONVERGENCE AND COMPARISON WITH OTHER APPROACHES

30.6.1 Convergences of Successive Iterations

To provide some insights into the rate of conversion of the results for examples, define the iteration $\varphi_m(t)$ as

$$\varphi_m(t) = \begin{cases} \eta_m(t), t \in [k_x,1] \\ \tau_m(t), t \in (0,k_x) \end{cases}. \tag{22}$$

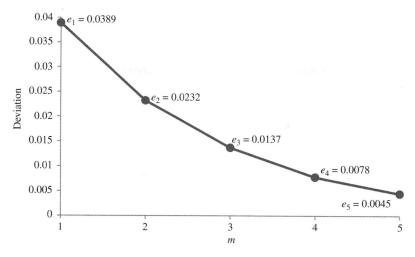

Figure 30.11. Deviation between $\varphi_{m+1}(t)$ and $\varphi_m(t)$.

Define the deviation between $\varphi_m(t)$ and $\varphi_{m+1}(t)$ over the interval $[0,1]$ as

$$e_m = \left(\int_0^1 (\varphi_{m+1}(t) - \varphi_m(t))^2 \, dt \right)^{\frac{1}{2}}.$$

Figure 30.11 shows the deviation, e_m, plotted versus m, where the iterations of $\varphi_m(t)$ are obtained from the assessments of Table 30.1 with the lower boundary assessment in (7). From the convergence results of Theorems 30.3 and 30.6, this deviation converges to 0. In other words,

$$\lim_{m \to \infty} e_m = 0.$$

The iterations can be conducted to reach the acquired accuracy of estimating the generating function as shown in Figure 30.11. Since the generating function $\varphi(t)$ is unknown in the approximation procedure, we terminate the iterations on $\varphi_m(t)$ when e_m is sufficiently small. Note that these iterations do not require additional cognitive effort from the decision-maker; they are simply computations used to calculate additional composite functions.

30.6.2 Comparison with Mutual Utility Independence

We now compare the estimates of the transformation g and the skewed diagonal assessment $S(t)$ to those obtained using the assumption of mutual utility independence. If two attributes are mutually utility independent, then $U(x \mid y^*) = U(x \mid y^0)$, and so

$$g(r) = k_y + \frac{1 - k_y}{k_x} r, \tag{23}$$

which is a linear function. It is natural to consider what $S(t)$ would look like if the attributes were mutually utility independent. The following proposition determines $S(t)$ in this case.

Proposition 30.7: *Two attributes are mutually utility independent with $k_x \geq k_y$ if and only if the utility function $U(x, y)$ has an Archimedean utility copula, and the following two statements hold:*

(i) $U(x \mid y^*) = U(x \mid y^0), x \in [x^0, x^*]$, *and*

(ii) $S(t) = kt^2 + 2(1-k)t + k - 1, t \in [k_x, 1]$, *where* $k = \dfrac{1 - k_x - k_y}{(1-k_x)(1-k_y)}$.

Proposition 30.7 shows a new method to verify mutual utility independence between two attributes. First, we verify that $U(x \mid y^*) = U(x \mid y^0)$, where X is the attribute with the greater corner value. Next, we assert that $S(t)$ is quadratic. If these conditions hold, then the attributes are mutually utility independent. For the special case where $k_x + k_y = 1$ (the case of an additive utility function), then $k = 0$ and $S(t)$ is a linear function. The following example compares the accuracy of the Archimedean utility copula obtained by the iterative approach for Example 30.1 to the utility function obtained assuming mutual utility independence.

EXAMPLE 30.4: Comparison with Mutual Utility Independence

Consider again the two-attribute utility function of Example 30.2, where

$$U(x, y^*) = 1.52 - 1.42e^{-x}, x \in [0, 1] \tag{24}$$

and

$$U(x^*, y) = 1.29 - 0.79e^{-\sqrt{y}}, y \in [0, 1]. \tag{25}$$

This implies that $k_x = 0.5, k_y = 0.1$. To compare our iterative assessment approach with that of mutual utility independence, we first compute the constant k

$$k = \frac{1 - k_x - k_y}{(1-k_x)(1-k_y)} = \frac{1 - 0.1 - 0.5}{(1-0.5)(1-0.1)} = 0.89.$$

Proposition 30.7 asserts that skewed diagonal assessment must be

$$S(t) = 0.89t^2 + 0.22t - 0.11, t \in [0.5, 1].$$

From (23), the transformation function has the form

$$g(r) = 0.1 + 1.8r, r \in [0, 0.5].$$

Figure 30.12 shows the skewed diagonal assessment $S(t)$ and the transformation function $g(r)$ for the utility function of Examples 30.2 and 30.3.

It is straightforward to see that a linear generating function of the form

$$\varphi(t) = kt + (1-k), t \in [0, 1] \tag{26}$$

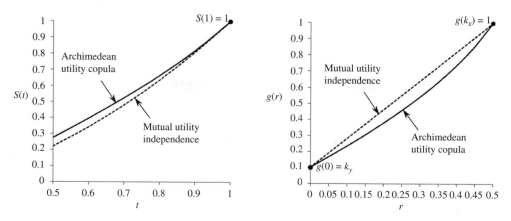

Figure 30.12. Archimedean copula versus mutual utility independence: (a) skewed diagonal assessment $S(t)$; (b) transformation function $g(r)$.

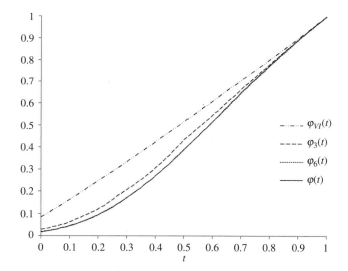

Figure 30.13. Generating functions $\varphi_{UI}(t)$, $\varphi(t)$, third-order iteration $\varphi_3(t)$ and sixth-order iteration $\varphi_6(t)$.

satisfies the condition of mutual utility independence. But to compare the generating function obtained from the iterative approach to that of mutual utility independence, we need take a power of (26) that makes $\varphi'(1) = 1$. Recall that the condition of $\varphi'(1) = 1$ applies to the generating functions obtained from the iterative approach. Direct substitution shows that the function $\varphi(t) = (kt + (1-k))^{\frac{1}{k}}$ satisfies the condition $\varphi'(1) = 1$. From here on we denote the generating function for mutual utility independence as

$$\varphi_{UI}(t) = [kt + (1-k)]^{\frac{1}{k}},$$

as it will be used in the comparison with the generating function obtained from the iterative approach.

Figure 30.13 plots the φ_3 and φ_6, $\varphi_{UI}(t)$ for Example 30.1. The figure also shows the actual generating function $\varphi(t)$ that we used to determine the numerical values of the examples in this chapter. Of course $\varphi(t)$ is not known to the analyst during this entire procedure. We simply included $\varphi(t)$ in the figure to illustrate the convergence of the results. The figure shows that $\varphi(t)$ and $\varphi_6(t)$ are indistinguishable over the entire interval $[0,1]$. The figure also shows the improvement that φ_3 offers over φ_{UI} in this example.

30.7 SUMMARY

If the assertion of an additive preference function is verified, then preferences over lotteries can be decomposed into two parts: (1) single-attribute assessments over each attribute at the upper bound of the complement attributes, and (2) a single generating function that combines these single-attribute assessments into an Archimedean form.

For additive ordinal preferences, the assumption of mutual utility independence focuses on the boundary assessments of each of the attributes but does not take into account the effect of the generating function.

The inclusion of the generating function into the Archimedean form allows for more general trade-offs and simultaneously satisfies the same utility values at the boundaries.

Once additive ordinal preferences are verified, we do not need to determine the actual values of the ordinal functions of the attributes when constructing the multiattribute utility function: assessing the boundary utility functions using indifference assessments and then constructing the generating function (also using indifference assessments) is sufficient to capture the whole structure of the utility function.

ADDITIONAL READINGS

Abbas A. E. 2009. Multiattribute utility copulas. *Operations Research* 57(6): 1367–1383.

Abbas, A. E. 2011. The multiattribute utility tree. *Decision Analysis* 8(3): 180–205.

Abbas, A. E. 2013. Utility copula functions matching all boundary assessments. *Operations Research* 61(2): 359–371.

Abbas, A. E. and D. E. Bell. 2011. One-switch independence for multiattribute utility functions. *Operations Research* 59(3): 764–771.

Abbas, A. E. and D. E. Bell. 2012. One-switch conditions for multiattribute utility functions. *Operations Research* 60(5): 1199–1212.

Abbas A. E. and R. A. Howard. 2005. Attribute dominance utility. *Decision Analysis* 2(4): 185–206.

Abbas, A. E. and Z. Sun. 2015. Multiattribute utility functions satisfying mutual preferential independence. *Operations Research* 62(2): 378–393.

Debreu, G. 1960. Topological methods in cardinal utility theory. In K. Arrow, S. Karlin, and P. Suppes, eds. *Mathematical Methods in the Social Sciences*, pp. 16–26. Stanford Univiversity Press, Stanford, CA.

Fritsch, F. N. and R. E. Carlson (1980) Monotone piecewise cubic interpolation. *SIAM Journal of Numerical Analytics* 17(2): 238–246.

Keeney, R. 1974. Multiplicative utility functions. *Operations Research* 22: 22–34.

Keeney, R. and L. H. Raiffa. 1976. *Decision with Multiple Objectives*. Wiley, New York.

Matheson, J. E. and A. E. Abbas. 2005. Utility transversality: a value-based approach. *Journal of Multi-Criteria Decision Analysis* 13: 229–238.

Nelsen, R. B. 1999. *An Introduction to Copulas*. Springer-Verlag, New York.

Sungur, A. E. and Y. Yang. 1996. Diagonal copulas of Archimedean class. *Communiques in Statistical-Theory Methodology* 25(7): 1659–1676.

von Neumann, J. and O. Morgenstern. 1947. *Theory of Games and Economic Behavior.* Princeton University Press, Princeton, NJ.

APPENDIX 30A

Proof of Proposition 30.1 for $S(t)$

By definition, $U(x(t), y^*) = t$ and $U(x^*, y(t)) = t$. Therefore,

$$\varphi(U(x(t), y^*)) = \varphi(U(x^*, y(t))) = \varphi(t).$$

Substituting into the definite of the Archimedean copula gives

$$S(t) = U(x(t), y(t)) = \varphi^{-1}(\varphi(U(x(t), y^*)\varphi(U(x^*, y(t)))) = \varphi^{-1}((\varphi(t))^2), \ \forall t \in [k_x, 1].$$

Proof of Lemma 30.2

Because the generating function φ is strictly increasing on the interval $[0,1]$, its derivative is non-negative, and so $\varphi'(1) \geq 0$. If $\varphi'(1) \neq 0$ (as we have assumed), then $\varphi'(1) > 0$. As shown in Appendix A at the end of the book, the generating function of an Archimedean copula is invariant to a power transformation, defined by the new generating function $\bar{\varphi} = \varphi^\alpha$, with $\alpha = \dfrac{1}{\varphi'(1)} > 0$. Note that $\bar{\varphi}(1) = 1$ and its derivative satisfies $\bar{\varphi}'(1) = \alpha \varphi'(1) \cdot [\varphi(1)] = 1$.

Proof of Theorem 30.3

Define $\psi(t) = -\ln(\varphi(t))$, $\forall t \in (0,1]$ and its inverse $\psi^{-1}(v) = \varphi^{-1}(e^{-v})$, $\forall v \in [0, \infty)$. Note that $\psi(t)$ is strict decreasing on the interval $[0,1]$ with $\psi(1) = -\ln(\varphi(1)) = 0$. Moreover, $\varphi(t) = e^{-\psi(t)}$, $\forall t \in (0,1]$, and so

$$S(t) \triangleq \varphi^{-1}\left((\varphi(t))^2\right) = \varphi^{-1}\left((e^{-\psi(t)})^2\right) = \varphi^{-1}\left(e^{-2\psi(t)}\right) = \psi^{-1}(2\psi(t)), \ \forall t \in [k_x, 1].$$

Define $q_m(t) = S^{(m)}(1 - 2^{-m}t)$ and its inverse function $q_m^{-1}(t) = 2^m\left(1 - S^{(-m)}(t)\right)$. Note that $\varphi'(1) = 1$, the derivative $\psi'(1) = -\dfrac{\varphi'(1)}{\varphi(1)} = -1 \neq 0$.

Sungur and Yang (1996) proved that for the equation $S(t) = \psi^{-1}(2\psi(t))$, the inverse function ψ^{-1} satisfies

$$\psi^{-1}(t) = \lim_{m \to \infty} S^{(m)}\left(1 + \frac{t}{2^m\,\psi'(1)}\right) = \lim_{m \to \infty} S^{(m)}(1 - 2^{-m}t) = \lim_{m \to \infty} q_m(t), \ \forall t \in [k_x, 1]. \quad (27)$$

Hence, the function ψ satisfies $\psi(t) = \lim_{m \to \infty} q_m^{-1}(t)$, $\forall t \in [k_x, 1]$. Therefore,

$$\varphi(t) = e^{-\psi(t)} = \lim_{m \to \infty} e^{-q_m^{-1}(t)} = \lim_{m \to \infty} e^{2^m(S^{(-m)}(t)-1)} = \lim_{m \to \infty} \eta_m(t).$$

Proof of Proposition 30.4

Substituting for $y = y^0$ in the Archimedean form $U(x, y) = \varphi^{-1}[\varphi(U(x, y^*)) \cdot \varphi(U(x^*, y))]$ gives

$$U(x, y^0) = \varphi^{-1}(\varphi(U(x, y^*)) \cdot \varphi(U(x^*, y^0))) = \varphi^{-1}(\varphi(U(x, y^*)) \cdot \varphi(k_x)). \qquad (28)$$

Denote $r = U(x, y^0)$, then $g(r) = g(U(x, y^0)) = U(x, y^*)$. Substituting $g(r)$ into (28) gives

$$r = \varphi^{-1}(\varphi(g(r)) \cdot \varphi(k_x)), \text{i.e., } \varphi(r) = \varphi(g(r)) \cdot \varphi(k_x). \qquad (29)$$

Applying (29) at $g(r)$ gives

$$\varphi(g(r)) = \varphi(g(g(r))) \cdot \varphi(k_x) = \varphi(g^{(2)}(r)) \cdot \varphi(k_x). \qquad (30)$$

Substituting from (30) into (29) gives

$$\varphi(r) = (\varphi(g(g(r))) \cdot \varphi(k_x)) \cdot \varphi(k_x) = \varphi(g^{(2)}(r)) \cdot (\varphi(k_x))^2. \qquad (31)$$

Repeating the above iteration with $g(r)$ gives $\varphi(r) = \varphi(g^{(p)}(r))(\varphi(k_x))^p$.

Proof of Lemma 30.5

For a given $r \in (0, k_x)$, $0 \le \varphi(0) < \varphi(r) < \varphi(k_x) < 1$, since φ is strictly increasing. Hence, $\lim_{p \to \infty}(\varphi(k_x))^p = 0 < \varphi(r)$. Therefore, $(\varphi(k_x))^p \le p(r)$ for sufficiently large integer $p > 1$. Define p_0 as the smallest such integer, i.e., $(\varphi(k_x))^{p_0} \le \varphi(r)$ and $(\varphi(k_x))^{p_0 - 1} > \varphi(r)$. Hence,

$$\varphi(k_x) \le \frac{\varphi(r)}{(\varphi(k_x))^{p_0}} \cdot \varphi(k_x) = \frac{\varphi(r)}{(\varphi(k_x))^{p_0 - 1}} < 1 = \varphi(1). \qquad (32)$$

Note that $\varphi(g^{(p_0 - 1)}(r)) = \dfrac{\varphi(r)}{(\varphi(k_x))^{p_0 - 1}}$ due to Proposition 30.3. Hence, inequality (32) becomes $\varphi(k_x) \le \varphi(g^{(p_0 - 1)}(r)) < \varphi(1)$. Since φ is strictly increasing, $\varphi(k_x) \le \varphi(g^{(p)}(r)) < \varphi(1)$, where $p = p_0 - 1$.

Proof of Theorem 30.6

Note that $g^{(p(r))}(r) \in [k_x, 1)$, $\forall r \in (0, k_x)$. Proposition 30.4 gives

$$\begin{aligned}
\varphi(r) &= [\varphi(k_x)]^{p(r)} \cdot \varphi(g^{(p(r))}(r)) \\
&= \lim_{m \to \infty}[\eta_m(k_x)]^{p(r)} \cdot \lim_{m \to \infty} \eta_m(g^{(p(r))}(r)) \\
&= \lim_{m \to \infty}[\eta_m(k_x)]^{p(r)} \cdot \eta_m(g^{(p(r))}(r)) = \lim_{m \to \infty} \tau_m(r).
\end{aligned}$$

Proof of Proposition 30.7

Necessity: If two attributes are mutually utility independent, then

$$U(x \mid y^*) = U(x \mid y^0), \forall x \in (x^0, x^*),$$

and

$$U(x, y) = k_x U(x \mid y^*) + k_y U(y \mid x^*) + (1 - k_x - k_y)U(x \mid y^*)U(y \mid x^*). \qquad (33)$$

With $u = U(x \mid y^*)$ and $v = U(y \mid x^*)$, we get its utility copula as

$$C(u, v) = k_x u + k_y v + (1 - k_x - k_y)uv. \qquad (34)$$

Substituting the definitions of $x(t)$ and $y(t)$ from (3) and (2) into (4) gives

$$S(t) = U(x(t), y(t))$$

$$= k_x \frac{t-k_y}{1-k_y} + k_y \frac{t-k_x}{1-k_x} + (1-k_x-k_y) \frac{t-k_y}{1-k_y} \frac{t-k_x}{1-k_x}$$

$$= kt^2 + 2(1-k)t + k - 1.$$

If $k_x + k_y = 1$, then $U(x,y) = k_x U(x \mid y^*) + k_y U(y \mid x^*)$, which has the additive value function as $k_x U(x \mid y^*) + k_y U(y \mid x^*)$. If $k_x + k_y \neq 1$, then

$$U(x,y) = (1-k_x-k_y)\left(U(x \mid y^*) - \frac{k_y}{(1-k_x-k_y)}\right)\left(U(x \mid y^*) - \frac{k_x}{(1-k_x-k_y)}\right) - \frac{k_x k_y}{(1-k_x-k_y)}$$

$$= (1-k_x-k_y)e^{\ln\left(U(x \mid y^*) - \frac{k_y}{(1-k_x-k_y)}\right) + \ln\left(U(x \mid y^*) - \frac{k_x}{(1-k_x-k_y)}\right)} - \frac{k_x k_y}{(1-k_x-k_y)},$$

which has the additive value function as

$$\ln\left(U(x \mid y^*) - \frac{k_y}{(1-k_x-k_y)}\right) + \ln\left(U(x \mid y^*) - \frac{k_x}{(1-k_x-k_y)}\right).$$

Hence, $U(x,y)$ has an Archimedean utility copula due to Proposition 30.1.

Sufficiency: Theorems 30.3 and 30.6 show that the generating function $\varphi(t)$ of the Archimedean utility copula in (34) satisfying $\varphi'(1) = 1$ is uniquely determined on the interval $(0,1]$ by the monotone transformation $g(r)$ and the skewed diagonal assessment $S(t)$. Hence, $g(r)$ and $S(t)$ uniquely determine the Archimedean utility copula $C(u,v)$ of (1). Note that if $U(x \mid y^*) = U(x \mid y^0)$, then $g(r) = k_y + \frac{1-k_y}{k_x} r$. Therefore, if (i) and (ii) holds, then the Archimedean utility copula $C(u,v)$ is in the form of (34) and the two attributes are mutually utility independent.

APPENDIX 30B

Properties of $S(t)$

1. $S(t): [k_x, 1] \rightarrow [S(k_x), 1].$
2. $S(t)$ is a continuous and strictly increasing function (refer to Lemma 30.5).
3. $S(t) < t$. This is because $(\varphi(t))^2 < \varphi(t)$, and so $\varphi^{-1}((\varphi(t))^2) < \varphi^{-1}(\varphi(t)) = t$.
4. The minimum value is $S(k_x) \leq k_x$.
5. The maximum value is $S(1) = U(x^*, y^*) = 1$.

Because $S(t)$ is continuous and strictly increasing, we can define the inverse function S^{-1} on the interval $[k_x, 1]$.

Properties of $S^{-1}(t)$

1. $S^{-1}: [k_x, 1] \rightarrow [S^{-1}(k_x), 1].$
2. $S^{-1}(t) > t, \forall t \in [k_x, 1)$ because $S(t) < t$.
3. Minimum value: $S^{-1}(k_x) > k_x$ because $S(k_x) < k_x, S^{-1}(k_x) > k_x$.

4.   Maximum value: $S^{-1}(1) = 1$ because $S(1) = 1$.
5. $S^{(-m)}(t) > S^{(-(m-1))}(t)$, $t \in [k_x, 1)$ because $S^{-1}(t) > t$.

Note that $S^{-1}(k_x) > k_x$, we know that the domain of S^{-1}, $[k_x, 1]$ contains its range, $[S^{-1}(k_x), 1]$. Therefore, the composites, $S^{(-m)}(t)$, $m = 1, 2, ...$, are well defined on $[k_x, 1]$.

APPENDIX 30C PIECEWISE CUBIC POLYNOMIAL INTERPOLATION

We applied a piecewise cubic polynomial interpolation for curve fitting of $S^{-1}(t)$ (see Fritsch and Carlson 1980 for further details). Denote the assessed points as (x_i, y_i), $i = 1, ..., n+1$. The approach assigns a cubic polynomial $S_i^{-1}(x)$ over each interval $[x_i, x_{i+1}]$, $i = 1, ..., n$ as the fitting curve, where

$$S_i^{-1}(x) = h_{00}(r)y_i + h_{10}(r)(x_{i+1} - x_i)m_i \\ + h_{01}(r)y_{i+1} + h_{11}(r)(x_{i+1} - x_i)m_{i+1}, \quad x \in [x_i, x_{i+1}], \tag{35}$$

where $r = \dfrac{x - x_i}{x_{i+1} - x_i}$, and where m_i, m_{i+1} are the derivatives at x_i and x_{i+1}, respectively. The functions h_{00}, h_{01}, h_{10}, and h_{11} are cubic polynomials defined as

$$h_{00}(t) = 2r^3 - 3r^2 + 1, h_{10}(r) = r^3 - 2r^2 + r, h_{01}(r) = -2r^3 + 3r^2, \text{ and } h_{11}(r) = r^3 - r^2.$$

The derivatives m_i, m_{i+1} are the only unknown parameters in (35) given the assessed points (x_i, y_i), $i = 1, ..., n+1$. We now show how to calculate these derivatives.

Denote the length of the ith interval as $l_i = x_{i+1} - x_i$ and the slope of the line connecting its two endpoints as $d_i = \dfrac{y_{i+1} - y_i}{x_{i+1} - x_i}$, $i = 1, ..., n-1$.

1.   The interior derivatives m_i, $i = 2, ..., n-1$ are

$$m_i = \begin{cases} \dfrac{d_{i-1}d_i}{w_1 d_{i-1} + w_2 d_i}, & \text{if } d_{i-1}d_i > 0; \\ 0, & \text{if } d_{i-1}d_i \leq 0, \end{cases} \tag{36}$$

where the weights satisfy $w_1 = \dfrac{2l_{i-1} + l_i}{3(l_{i-1} + l_i)}$ and $w_2 = 1 - w_1$.

2.   The derivatives at two endpoints of the whole domain, m_1 and m_n, are

$$m_1 = \begin{cases} 0, & \text{if } q_1 d_1 < 0; \\ 3d_1, & \text{if } q_1 d_1 \geq 0, d_1 d_2 < 0 \text{ and } |q_1| > 3|d_1|; \\ q_1, & \text{otherwise} \end{cases} \tag{37}$$

and

$$m_n = \begin{cases} 0, & \text{if } q_2 d_{n-1} < 0; \\ 3d_n, & \text{if } q_2 d_{n-1} \geq 0, d_{n-1} d_{n-2} < 0 \text{ and } |q_2| > 3|d_{n-1}|; \\ q_2, & \text{otherwise}, \end{cases} \tag{38}$$

where $q_1 = \dfrac{(2l_1 + l_2)d_1 - l_1 d_2}{l_1 + l_2}$ and $q_2 = \dfrac{(2l_{n-1} + l_{n-2})d_{n-1} - l_{n-1}d_{n-2}}{l_{n-1} + l_{n-2}}$.

Now we apply the curve-fitting approach to determine the inverse function $S^{-1}(t)$ from the assessments. From Table 30.1, we relabel t and $S(t)$ to get the six points:

$$(0.2765, 0.5), (0.3972, 0.6), (0.5253, 0.7), (0.6641, 0.8), (0.8189, 0.9), (1, 1).$$

We calculate the interior derivatives m_2, m_3, m_4, and m_5 from (36). For example,

$$l_1 = x_2 - x_1 = 0.3972 - 0.2765 = 0.1207 \text{ and } l_2 = x_3 - x_2 = 0.5253 - 0.3972 = 0.1281.$$

Furthermore, the slopes in these two intervals are

$$d_1 = \frac{y_2 - y_1}{x_2 - x_1} = \frac{0.6 - 0.5}{0.1207} = 0.8285 \text{ and } d_2 = \frac{y_3 - y_2}{x_3 - x_2} = \frac{0.7 - 0.6}{0.1281} = 0.7806.$$

From (36), the weights for m_2 are

$$w_1 = \frac{2l_1 + l_2}{3(l_1 + l_2)} = \frac{2 \times 0.1207 + 0.1381}{3 \times (0.1207 + 0.1381)} = 0.4950 \text{ and } w_2 = 1 - w_1 = 0.5050.$$

Substituting w_1, w_2, d_1, and d_2 into (36) gives

$$m_2 = \frac{d_1 d_2}{w_1 d_1 + w_2 d_2} = \frac{0.8285 \times 0.7806}{0.4950 \times 0.8285 + .0.5050 \times 0.7806} = 0.8041.$$

Similarly, $m_3 = 0.7497$, $m_4 = 0.6819$, $m_5 = 0.5966$.

To calculate the derivatives at the two endpoints $x_1 = 0.2765$ and $x_6 = 1$, note that $q_1 = \dfrac{(2l_1 + l_2)d_1 - l_1 d_2}{l_1 + l_2} = 0.8517 < 3d_1$ and $m_1 = q_1 = 0.8517$. Similarly, $m_6 = q_2 = 0.5016$.

Substituting the derivatives $m_i, i = 1,...,6$ into (35) and rearrange it gives the fitting curve of $S^{-1}(t)$ in the form of a piecewise polynomial over $[0.2765, 1]$ as

$$S^{-1}(t) = \begin{cases} -0.0814t^3 - 0.115t^2 + 0.934t + 0.2523, & t \in [0.24, 0.36]; \\ -0.4533t^3 - 2.4151t^2 + 0.6889t + 0.2893, & t \in [0.36, 0.5]; \\ -0.4828t^3 - 0.617t^2 + 0.5013t + 0.3364, & t \in [0.5, 0.65]; \\ -0.5626t^3 + 0.9761t^2 + 0.1297t + 0.4481, & t \in [0.65, 0.81]; \\ -0.1868t^3 + 0.2472t^2 + 0.5675t + 0.3721, & t \in [0.81, 1]. \end{cases}$$

The values of composite function $S^{(-2)}(t)$ are obtained by applying $S^{-1}(t)$ twice. For example, $S^{(-2)}(0.5) = S^{-1}(S^{-1}(0.5)) = S^{-1}(0.808) = 0.8114$.

CHAPTER 31

Utility Copula Functions Matching All
Boundary Assessments

Chapter Concepts

- Utility copula functions that have the flexibility to match multiple boundary assessments
- Examples of utility copula functions that match more than one boundary assessment

31.1 INTRODUCTION: THE NEED FOR COPULA FUNCTIONS MATCHING ALL BOUNDARY ASSESSMENTS

In Chapter 28, we introduced the idea of a utility copula function of the form

$$U(x,y) = C\left(U(x\,|\,y^*), U(y\,|\,x^*); k_x, k_y\right).$$

In Chapters 29 and 30, we discussed Archimedean utility copulas that correspond to additive ordinal preferences. We showed that this formulation requires only one utility function for each attribute as well as a generating function.

When additive ordinal preferences do not hold, it might be convenient to have a function that specifies two boundary curves for each attribute (for example, at the upper and lower boundaries of the complement attributes) as well as two corner assessments, k_x, k_y. This formulation would have the form

$$U(x,y) = C[U(x\,|\,y^0), U(x\,|\,y^*), U(y\,|\,x^0), U(y\,|\,x^*), k_x, k_y]. \tag{1}$$

Equation (1) expresses the utility function in terms of two boundary assessments for each attribute and some corner values. This chapter presents utility copula formulations that have the flexibility to match the conditional utility functions at more than just a single boundary value of the complement attributes. We refer to the function C in (1) as a ***double-sided utility copula*** for the case of two attributes, because it incorporates the upper and lower boundary assessments. For higher dimensions, we refer to this new function as an ***all-sided*** utility copula. The trade-offs between the attributes in this case are determined by (i) the conditional utility assessments at *all* boundary values, (ii) the normalizing constants, and (iii) the shape of the copula function, C.

612

31.2 ARCHIMEDEAN COPULAS MATCH ONLY ONE BOUNDARY ASSESSMENT FOR EACH ATTRIBUTE

In Chapter 29, we discussed Archimedean utility copulas. We showed that a single generating function was sufficient to combine individual utility functions over each attribute to determine the utility surface. As we shall see in this chapter, the use of an Archimedean utility copula asserts a functional relationship between the upper and lower boundary utility values of each attribute, and this functional relationship is determined by the generating function. The following example illustrates this idea.

EXAMPLE 31.1: Relating Boundary Assessments of an Archimedean Form Using the Generating Function

---

Consider the Archimedean form

$$C(u,v) = a\psi^{-1}[\psi(l_u + (1-l_u)u)\psi(l_v + (1-l_v)v)] + b, \tag{2}$$

where $0 \le l_u, l_v < 1$, $a = 1/(1 - \psi^{-1}[\psi(l_u)\psi(l_v)])$, $b = 1 - a$, and the generating function, ψ, is continuous and strictly increasing on the domain $v \in [0,1]$ with $\psi(1) = 1$.

By direct substitution into (2), we can determine the utility curves at the boundaries of the domain as

$$U(x,y^*) = a\psi^{-1}[\psi(l_u + (1-l_u)U(x\,|\,y^*))] + b = k_y + (1-k_y)U(x\,|\,y^*), \tag{3}$$

$$U(x^*,y) = a\psi^{-1}[\psi(l_v + (1-l_v)U(y\,|\,x^*))] + b = k_x + (1-k_x)U(y\,|\,x^*), \tag{4}$$

$$U(x,y^0) = a\psi^{-1}[\psi(l_v)\psi(l_u + (1-l_u)U(x\,|\,y^*))] + b, \tag{5}$$

$$U(x^0,y) = a\psi^{-1}[\psi(l_u)\psi(l_v + (1-l_v)U(y\,|\,x^*))] + b. \tag{6}$$

Equations (3) and (4) show that the utility surface is a positive linear transformations of the conditional utility functions $U(x\,|\,y^*)$ and $U(y\,|\,x^*)$ at the upper bounds, as expected for a Class 1 utility copula. Equations (5) and (6) show that the utility functions at the minimum margins are determined by some monotone transformation functions of the utility functions at the maximum margins. These monotone transformations are determined by the generating function. Therefore, we do not have an additional degree of freedom to specify the conditional utility functions at the minimum margins once the generating function and the utility functions at the upper margins are specified.

If the preference function is additive, we do not need to worry about an additional degree of freedom for the two boundary assessments, because the Archimedean combination is sufficient to characterize additive ordinal preferences. But if the ordinal preferences are not additive and the decision-maker would like to use a copula formulation that matches more than one boundary value, it would be convenient to have a utility copula formulation that matches more than one boundary assessment for each attribute.

31.3 DOUBLE-SIDED UTILITY COPULAS

31.3.1 Transformation Functions Relating the Boundary Assessments

Our focus will be two-attribute utility functions where each attribute has at least one conditional utility function that is strictly increasing at a boundary value of the complement attributes. Here, we discuss the case where $U(x\mid y^*), U(y\mid x^*)$ are strictly increasing, and discuss extensions to other boundary values in Appendix 31B. We assume that the utility function is normalized and that $U(x^*, y^*) = 1$ and $U(x^0, y^0) = 0$.

Proposition 31.1: *If $U(x\mid y^*), U(y\mid x^*)$ are continuous and strictly increasing, then there always exist unique transformations, g_1, g_2, defined on a domain $[0,1]$ such that for any continuous and bounded two-attribute utility function, the following equalities hold:*

$$U(x\mid y^0) = g_1\left(U(x\mid y^*)\right), \quad U(y\mid x^0) = g_2\left(U(y\mid x^*)\right).$$

Proposition 31.1 shows that we do not lose any generality by assuming that the conditional utility functions at the lower bound are functionally related to the conditional utility functions at the upper bound, provided those latter conditional utility functions are continuous and strictly increasing. If the utility functions at the upper and lower bounds are strictly increasing, then this formulation implies that $g_i(0) = 0$ and $g_i(1) = 1$.

Notation:

The following notation will be used throughout this chapter.

$$u = U(x\mid y^*), \ v = U(y\mid x^*), \ g_1(u) = U(x\mid y^0), \ g_2(v) = U(y\mid x^0).$$

We also use the same notation for corner values where $k_x = U(x^*, y^0), k_y = U(x^0, y^*)$.

Using the previous definitions of the transformation functions, we can express the utility surface at the upper and lower bounds as

$$U(x, y^*) = k_y + (1-k_y)U(x\mid y^*) = k_y + (1-k_y)u$$
$$U(x^*, y) = k_x + (1-k_x)U(y\mid x^*) = k_x + (1-k_x)v$$
$$U(x, y^0) = k_x U(x\mid y^0) = k_x g_1(u)$$
$$U(x^0, y) = k_y U(y\mid x^0) = k_y g_2(v).$$

31.3.2 Two-Attribute Double-Sided Class (1,1) Utility Copulas

---

Definition: Double-Sided Class (1,1) Utility Copula

*A **two-attribute double-sided Class (1,1) utility copula** is a continuous, bounded mapping, $C_{1,1}(u, v \,; g_1, g_2, k_x, k_y)$, with $0 " u, v " 1$, that satisfies the following conditions:*

1. *Linear transformation at maximum margins,*

$$C_{1,1}(u, 1; g_1, g_2, k_x, k_y) = k_y + (1-k_y)u, \quad C_{1,1}(1, v; g_1, g_2, k_x, k_y) = k_x + (1-k_x)v. \quad (7)$$

---

2. *Minimum margins defined by g-transformation of the maximum margins,*

$$C_{1,1}(u,0; g_1,g_2,k_x,k_y) = k_x g_1(u), \quad C_{1,1}(0,v; g_1,g_2,k_x,k_y) = k_y g_2(v). \tag{8}$$

Equations (7) and (8) imply by direct substitution that

$$C_{1,1}(0,0; g_1,g_2,k_x,k_y) = 0, \quad C_{1,1}(1,1; g_1,g_2,k_x,k_y) = 1$$

and that

$$C_{1,1}(1,0; g_1,g_2,k_x,k_y) = k_x, \quad C_{1,1}(0,1; g_1,g_2,k_x,k_y) = k_y.$$

The double-sided utility copula function is a linear function of u and v at the maximum margins. The minimum margins are the transformations g_1, g_2 scaled by k_x, k_y, respectively. Four corner assessments are also specified: the copula is zero at the origin; it is equal to one at the corner point $(1,1)$; it is equal to k_x at the corner $(1,0)$; and it is equal to k_y at the corner $(0,1)$. For shorthand, we may refer to the double-sided utility copula as simply $C(u,v)$. Figure 31.1 shows some examples of double-sided utility copulas and illustrates the generality of surfaces that can be modeled by this approach.

Note that we can also define double-sided utility copulas that are linear at different combinations of the boundary values of the attributes. For example, if a function is linear in X at the upper bound of Y and if it is linear in Y at the lower bound of X, then we refer to it as a double-sided Class $(1,0)$ utility copula. We discuss extensions to other boundary values such as Class $(0,1)$ and Class $(0,0)$ in Appendix 31B.

Recall the following about transformation functions on utility functions from Chapter 13:

1. If g_1 is increasing and concave, then the conditional utility function at the minimum boundary, $U(x \mid y^0)$, exhibits higher risk aversion with respect to x than the utility function at the upper boundary, $U(x \mid y^*)$. In Chapter 13, we referred to this as **concavifying** the utility function.
2. If g_1 is increasing and convex, then $U(x \mid y^0)$ exhibits lower risk aversion than $U(x \mid y^*)$.
3. If g_1 is S-shaped (or reverse S-shaped), then there will be a region on one boundary where the conditional utility function exhibits a higher risk aversion than the other boundary, and another region where it exhibits lower risk aversion.

Theorem 31.2

Any continuous, bounded two-attribute utility function that is strictly increasing with each argument at the maximum value of the complement, with transformation functions g_1, g_2, and corner assessments, $U(x^0,y^0) = 0$, $k_x = U(x^*,y^0)$, $k_y = U(x^0,y^*)$, $U(x^*,y^*) = 1$, can be constructed with a double-sided Class $(1,1)$ utility copula using conditional utility assessments $U(x \mid y^*), U(y \mid x^*)$ as

$$C(U(x \mid y^*), U(y \mid x^*); g_1,g_2,k_x,k_y).$$

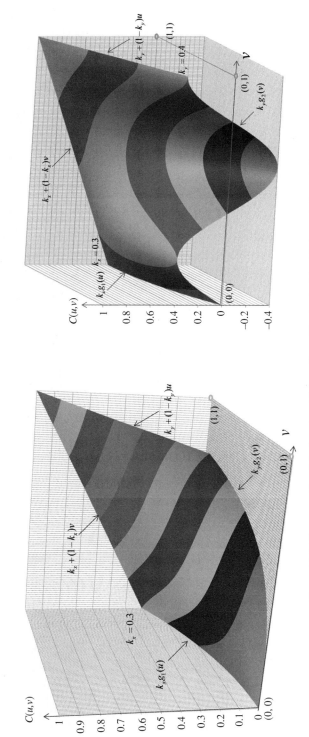

Figure 31.1. The generality of the functional forms allowed by the double-sided copula formulation.

Dialogue between Instructor (I) and Student (S)

I: So is the concept of a bidirectional copula for two attributes clear?

S: Yes, the motivation is that in general every attribute can have two boundary utility curves (at the minimum and maximum values of the other attribute). These new copula functions enable us to match all of these boundary assessments.

I: Correct, and we have the degree of freedom to assess those boundary assessments independently.

S: Yes, we have seen in Example 31.1 that for Archimedean forms, the generating function itself relates the upper and lower boundary assessments, so now we have more flexibility.

I: Correct, Archimedean forms correspond to additive ordinal preferences that need not be the case despite their ability to model very general trade-offs.

S: So Theorem 31.2 explains the importance of the double-sided utility copula definition.

I: Yes, it asserts that we can match all boundary assessments with any double-sided Class (1,1) utility copula. Moreover, if we change the parameters of a double-sided copula function, we will still preserve the boundary assessments, while varying the trade-off assessments. This provides an additional degree of freedom to model a wide range of multiattribute utility functions.

S: And how do we construct these new forms?

I: The next section illustrates some methods for constructing double-sided utility copula functions.

31.4 DOUBLE-SIDED UTILITY COPULAS USING THREE GROUNDED SURFACES

31.4.1 Defining a New Functional Form Using Grounded Surfaces

This section presents a method to construct double-sided utility copulas by decomposing the utility surface into grounded surfaces. This formulation is particularly useful when conducting sensitivity analysis to some of the widely used utility independence conditions for multiattribute utility functions. To start, consider the functional form,

$$C(u,v \; ; g_1,g_2,k_x,k_y) = C_1^d(u,v) + k_x C_2^d(g_1(u),1-v) + k_y C_3^d(1-u,g_2(v)), \qquad (9)$$

where $g_i(0) = 0, g_i(1) = 1, i = 1, 2,$ and $C_1^d(u,v), C_2^d(u,v), C_3^d(u,v)$ are grounded Class 1 utility copulas.

The utility function constructed by a grounded Class 1 utility copula is zero (hence grounded) when any attribute is at its minimum value. Because these functions will play an important role in our formulation, we give them a special symbol (a superscript, d). For example, we write $C^d(u,v)$ for the two-attribute case. A special case of grounded Class 1 utility copulas is the class of attribute dominance utility functions that we discussed in Chapter 21, where (in addition) more of any attribute is preferred to less.

The functional form (9) is a general expression of a weighted combination of three grounded functions. The weights are the corner assessments k_x, k_y provided by the decision-maker.

Proposition 31.3: *The functional form* (9) *is a double-sided Class* (1,1) *utility copula for all grounded Class* 1 *utility copula surfaces,* C_1^d, C_2^d, C_3^d.

The following theorem illustrates the importance of the functional form (9).

Theorem 31.4
Any continuous, bounded two-attribute utility function, $U(x,y)$, *that is (i) strictly increasing with each argument at the maximum values of the complement attributes with (ii) bounded (and well-defined) conditional utility functions can be expressed using* (9).

31.4.2 Important Special Cases

We now illustrate the connection of (9) with other approaches. The general expression for the conditional utility function of the copula implied by (9) is

$$C(u\,|\,v) = \frac{C(u,v)-C(0,v)}{C(1,v)-C(0,v)} = \frac{C_1^d(u,v)+k_x C_2^d\left(g_1(u),1-v\right)+k_y[C_3^d\left(1-u,g_2(v)\right)-g_2(v)]}{k_x+(1-k_x)v-k_y g_2(v)}, \quad (10)$$

and a similar expression can be derived for $C(v\,|\,u)$.

Because $C(u\,|\,v)$ is a function of both u and v, it is clear that (9) can model utility functions with utility dependence. Note that if $C(u,v)$ is grounded, then we get by direct substitution, $C(u\,|\,v) = C^d(u,v)/v$, and so $C(u\,|\,v=1) = u$. We now illustrate the variety of functional forms that may result from (9).

31.4.2.1 THE SUM PRODUCT DOUBLE-SIDED FAMILY If each of the grounded copulas in (9) is a simple product of its arguments, i.e.,

$$C_i^d\left(u,v\right) = uv, i = 1,2,3, \quad (11)$$

then we get the sum product form,

$$C(u,v\,;\,g_1,g_2,k_x,k_y) = uv + k_x g_1(u)(1-v) + k_y(1-u)g_2(v),$$

which can also be written as

$$C(u,v\,;\,g_1,g_2,k_x,k_y) = k_x g_1(u) + k_y g_2(v) + [uv - k_x g_1(u)v - k_y u g_2(v)]. \quad (12)$$

Equation (12) presents a family of multiattribute functions that matches all boundary assessments and all normalizing corner values provided by the decision-maker. It also incorporates utility dependence among the attributes using a sum of products of single-attribute utility assessments. The conditional utility functions implied by this functional form are

$$\begin{aligned} C(u\,|\,v) &= \frac{k_x g_1(u)+[uv-k_x v g_1(u)-k_y u g_2(v)]}{k_x+(1-k_x)v-k_y g_2(v)} \\ &= \left(\frac{v-k_y g_2(v)}{k_x(1-v)+v-k_y g_2(v)}\right)u + \left(\frac{k_x(1-v)}{k_x(1-v)+v-k_y g_2(v)}\right)g_1(u) \quad (13) \\ &= u\phi(v) + g_1(u)[1-\phi(v)], \end{aligned}$$

where $\phi(v) = \dfrac{v - k_y g_2(v)}{k_x(1-v) + v - k_y g_2(v)}$.

Equation (13) implies that u is interpolation independent of v. Using the same analysis, we can verify that v is interpolation independent of u. Hence, for two attributes, the functional form (12) is an assertion of mutual interpolation independence, which is also a special case of (9).

31.4.2.2 Hybrid Independence: Utility Independence and Interpolation Independence When each of the grounded copulas is a product and, in addition, one of the transformation functions is the identity function (for example $g_1(u) = u$), then the sum product family of (12) reduces to

$$C(u, v \; ; \; g_1, g_2, k_x, k_y) = k_x u + k_y g_2(v) + [uv - k_x uv - k_y u g_2(v)].$$

By direct substitution into (13), the conditional utility function $C(u \,|\, v)$ is

$$C(u \,|\, v) = u\phi(v) + u[1 - \phi(v)] = u. \qquad (14)$$

Therefore, this condition is an assertion that X is utility independent of Y and (as we already know) Y is interpolation independent of X. This is an example of the flexibility of this copula family to incorporate some forms of partial utility independence relations.

31.4.2.3 Mutual Utility Independence If in addition to (11), we have mutual boundary independence, i.e., $g_1(u) = u$ and $g_2(v) = v$, then

$$C(u \,|\, v) = u, \text{ and } C(v \,|\, u) = v,$$

which implies that X is utility independent of Y and Y is utility independent of X.

31.4.3 Matching All Boundary Assessments Using the Functional Form (9)

EXAMPLE 31.2: Double-Sided Utility Copula Using Archimedean Grounded Copulas

---

Suppose that a decision-maker faces a decision with two attributes, X and Y, defined on the normalized domain $[0,1] \times [0,1]$. He provides the following conditional utility functions at the four boundary values

$$U(x \,|\, y^0) = \frac{1 - e^{-\gamma_1 x}}{1 - e^{-\gamma_1}}, \quad U(x \,|\, y^*) = \frac{1 - e^{-\gamma_2 x}}{1 - e^{-\gamma_2}}, \quad U(y \,|\, x^0) = y^m, \quad U(y \,|\, x^*) = y^n,$$

where $\gamma_1 = 2, \gamma_2 = 3, m = 2, n = 4$. He also provides the corner utility values, $k_x = 0.4, k_y = 0.3$, using indifference probability assessments. It is clear the utility independence conditions do not exist.

To capture these boundary assessments into the functional form, a natural first step would be to use functional form of mutual interpolation independence because it matches the boundary assessments and does not require an additional interpolation function. If the corresponding isopreference contours do not match those of the decision-maker, we can use (9) to construct the utility surface by matching the boundary values and having some more flexibility in matching the trade-off assessments (and the shape of the isopreference contours).

Of particular interest, consider the Archimedean grounded utility copulas,

$$C(u,v \; ; g_1,g_2,k_x,k_y) = \psi^{-1}\big(\psi(u)\psi(v)\big) + k_x\psi^{-1}\big(\psi(g_1(u))\psi(1-v)\big)$$
$$+ k_y\psi^{-1}\big(\psi(1-u)\psi(g_2(v))\big), \tag{15}$$

where $\psi(t)$ is continuous and strictly increasing on the domain $t \in [0,1]$, with $\psi(0) = 0$ and $\psi(1) = 1$.

From Proposition 31.3, this must be a double-sided class (1,1) utility copula. Since $U(x\,|\,y^0) = g_1\big(U(x\,|\,y^*)\big)$ and $U(y\,|\,x^0) = g_2\big(U(y\,|\,x^*)\big)$, we have

$$g_1(t) = \frac{1-(1-(1-e^{-\gamma_2})t)^{\frac{\gamma_1}{\gamma_2}}}{1-e^{-\gamma_1}} \text{ and } g_2(t) = t^{\frac{m}{n}}. \tag{16}$$

Each copula function will match the boundary assessments provided but will result in a different utility surface. One form of generating functions that is widely used (because of its simplicity) is the exponential generating function, $\psi(t) = \dfrac{1-e^{-\delta t}}{1-e^{-\delta}}$. It is insightful to look at the diagonal assessments $C(u,u)$ that are implied by (15),

$$C(u,u \; ; g_1,g_2,k_x,k_y) = \psi^{-1}\big(\psi(u)^2\big) + k_x\psi^{-1}\big(\psi(g_1(u))\psi(1-u)\big) + k_y\psi^{-1}\big(\psi(1-u)\psi(g_2(u))\big).$$

Figure 31.2(a) plots this curve for different values of the parameter δ. As δ increases, the utility surface is higher, and as it decreases, the surface is lower. This is achieved while keeping the boundary values fixed. This observation highlights the additional degree of freedom that is provided by this copula function. As $\delta \to 0$, or $\psi(t) \to t$, we approach the case of mutual interpolation independence.

The diagonal assessments $C(u,u)$ are indifference assessments, $P_{(x_1,y_1)}$, for receiving a consequence (x_1,y_1), for certain (where $U(x_1\,|\,y^*) = u, U(y_1\,|\,x^*) = u$) or receiving a binary gamble of the form $< p_{(x_1,y_1)},(x^*,y^*); 1 - p_{(x_1,y_1)},(x^0,y^0) >$. The higher the value of $p_{(x_1,y_1)}$, the higher the value of δ, and the higher the corresponding utility surface. A parametric approach for estimating δ would repeat this assessment for different values of u and choose the diagonal assessment in Figure 31.2(a) that is closest to the assessments provided by the decision-maker. If the utility assessments do not match, then the analyst should use a different functional form for the generating function (or a grounded surface that need not be Archimedean). Figure 31.2(b) shows the effects of changing the value of δ on the shape of an isopreference contour and the corresponding trade-offs implied between the two attributes.

Figure 31.3 shows the utility copula surface constructed for $\delta = 2$. Note that the curves $C(u,1)$ and $C(1,v)$ in the top figure are linear, since this is a Class (1,1) utility copula. Moreover, since $g_1(t)$ is a convex function, $U(x\,|\,y^0)$ is less concave (exhibits lower risk aversion) than $U(x\,|\,y^*)$. On the other hand, since $g_1(t)$ is a concave function,

(a)

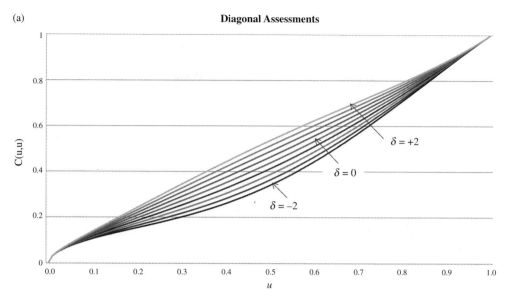

Figure 31.2a. Diagonal assessments, $C(u,u)$, for different values of δ.

(b)

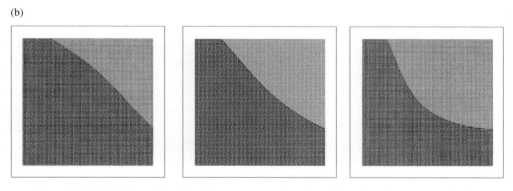

Figure 31.2b. The shape of an isopreference contour for value of $\delta = -2, 0.0001, +2$, respectively.

$U(y|x^0)$ is more concave (exhibits higher risk aversion) than $U(y|x^*)$. The corner assessments in the top figure in the copula space match the corner assessments provided by the decision-maker (as expected).

EXAMPLE 31.3: Changing the Height of the Utility Surface: A Monte Carlo Simulation

---

We now illustrate how the utility copula formulation enables sensitivity to the functional form of the multiattribute utility function. We use a Monte Carlo sensitivity analysis to determine the order of approximation provided by using $\delta = 0$ in this decision problem if the actual utility surface has the same boundary assessments but has a different value

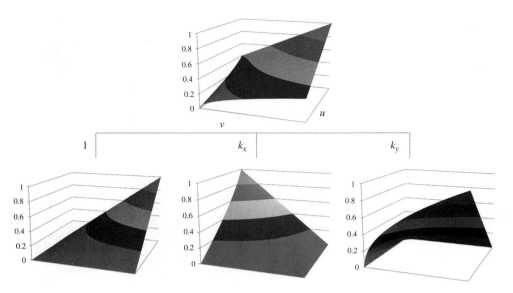

Figure 31.3. A double-sided utility copula (top figure) constructed from three grounded surfaces. The bottom left surface is $C^d(u,v)$, the middle figure is $C^d(g_1(u),1-v)$, and the third is $C^d(1-u,g_2(v))$.

of δ corresponding to a different height (or different trade-offs). The simulation was generated using a random decision tree with two alternatives. The joint distributions of the alternatives were generated uniformly from the space of all joint probability distributions as conducted in Chapters 21 and 29. The expected utility was then calculated once using the form (15) with an exponential generating function and a specific value of δ, and another time using $\delta = 0$.

Figure 31.4 shows the simulation results. The percentage of decisions where both functional forms gave different recommendations was recorded and plotted in the figure. As shown in the figure, if $\delta = 2$, then the percentage error obtained by using $\delta = 0$ is relatively small (about 5%). This implies that it could still be a reasonable approximation in some decision problems even if its conditions do not apply. However, if $\delta = 10$, then the percentage of incorrect decisions may be in the order of 15%, and more work would be needed to determine the functional form.

Note that this type of analysis does not require a particular functional form for the utility copula function. The same type of analysis can be generalized and the analyst can choose from a menu of double-sided copula functions that reduce to a particular form, by varying a parameter setting, to conduct this sensitivity analysis and determine whether more effort is needed to determine the functional form beyond the assumed functional form.

31.4.3 All-Sided Utility Copulas Extending (9)

We conclude this section by discussing the extensions of the functional form (9) to multiple attributes. Since the proof is similar to the two-attribute case, we merely state the result without proof to illustrate the basic idea behind the multiattribute extension.

For an n-attribute utility function, we have 2^{n-1} boundary instantiations of the complement for each attribute. We also have $2^n - 2$ corner assessments (since

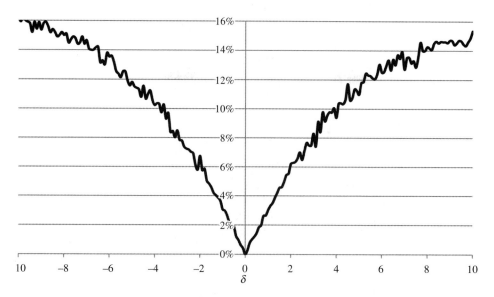

Figure 31.4. Percentage of incorrect decisions as we match the boundary assessments and raise the copula surface using the parameter δ.

$k_{00...0} \triangleq U(x_1^0,...,x_n^0) = 0$, $k_{11...1} \triangleq U(x_1^*,...,x_n^*) = 1)$. Once again, we require only one boundary assessment for each attribute to be continuous and strictly increasing. For each variable u_i in the copula function, define 2^{n-1} transformation functions, $g_{\bar{v}_i}^i(u_i)$, where $\bar{v}_i$ is the vector $(v_1,...,v_{i-1},v_{i+1},...,v_n)$, with $v_j \in \{0,1\}, j \neq i$. Depending on the type of copula function (which boundary assessment we will use as the reference), one of these functions $g_{\bar{v}_i}^i(u_i)$ will be the identity function $g(u) = u$.. Further define the variables

$$\tilde{u}_i = \begin{cases} g_{\bar{v}_i}^i(u_i), & v_i = 1 \\ 1 - g_{\bar{v}_i}^i(u_i), & v_i = 0. \end{cases}$$

Finally, define the 2^n corner assessments $k_{v_1,v_2,...,v_n}$; $v_1,v_2,...,v_n \in \{0,1\}$, with $k_{00...0} = 0, k_{11...1} = 1$.

The multivariate function consisting of 2^n grounded surfaces

$$C(u_1,...,u_n,g_{v_1v_2...v_n}^1,...,g_{v_1v_2...v_n}^n,k_{v_1,v_2,...,v_n}; v_1,v_2,...,v_n \in \{0,1\}) = \sum_{v_1,v_2,...,v_n \in \{0,1\}} k_{v_1,v_2,...,v_n} C_{v_1,v_2,...,v_n}^d(\tilde{u}_1,\tilde{u}_2,...\tilde{u}_n), \tag{17}$$

where $C_{v_1,v_2,...,v_n}^d(\tilde{u}_1,\tilde{u}_2,...,\tilde{u}_n)$ is a grounded Class 1 utility copula, is an all-sided utility copula function matching all boundary assessments regardless of the choice of the grounded surfaces.

31.5 UTILITY COPULA FUNCTIONS USING A SINGLE GROUNDED SURFACE

The utility copula function of the previous section allowed for numerous degrees of freedom for matching the boundary assessments while varying parameters and functional forms of the grounded surfaces. It also enabled sensitivity analysis to some widely used forms of utility independence. This section defines another class of copula functions that provides a sufficient condition to determine the whole utility surface using the boundary assessments.

31.5.1. Defining a New Functional Form

We start our analysis by rearranging the expression of the conditional utility function for the case of two attributes to get

$$U(x,y) = U(x \mid y)[k_x + (1-k_x)U(y \mid x^*) - k_y U(y \mid x^0)] + k_y U(y \mid x^0). \qquad (18)$$

Observe that the product $U(x \mid y)U(y \mid x^*)$ is zero if either x or y is minimum; it is equal to $U(x \mid y^*)$ and $U(y \mid x^*)$, respectively, when y and x are maximum, and it is equal to one when both x and y are maximum. Therefore, it can be expressed as a grounded Class 1 function,

$$C^d(U(x \mid y^*), U(y \mid x^*)) = U(x \mid y)U(y \mid x^*). \qquad (19)$$

Rearranging (19) gives

$$U(x \mid y) = \frac{C^d(U(x \mid y^*), U(y \mid x^*))}{U(y \mid x^*)} = \frac{C^d(u,v)}{v} = c(u \mid v), v \neq 0. \qquad (20)$$

Substituting into (18) and rearranging gives

$$\boxed{C(u,v) = \frac{C^d(u,v)}{v}[k_x + (1-k_x)v - k_y g_2(v)] + k_y g_2(v), \; v \neq 0}. \qquad (21)$$

The functional form (21) requires only three boundary assessments, $u, v, g_2(v)$, two normalizing constants, k_x, k_y, and one grounded copula function C^d to construct the functional form. The fourth boundary assessment is determined by the choice of the grounded copula, as we illustrate below.

Proposition 31.5: *If the limit,* $\lim\limits_{v \to 0} \dfrac{C^d(u,v)}{v}$ *exists, and is bounded, then the functional form (21) is a double-sided Class (1,1) utility copula with* $g_1(u) = \lim\limits_{v \to 0} \dfrac{C^d(u,v)}{v}$.

Since expansion (21) requires only the existence of a normalized conditional utility function, which was already assumed in Section 31.2, this formulation is very general and applies to a wide range of utility functions. We state without proof the following theorem based on the conditions used to define (21).

Theorem 31.6
Any two-attribute utility function that is continuous, bounded, strictly increasing with each argument at the maximum value of the complement attributes, with bounded conditional utility functions, can be expressed using (21).

By symmetry, we can also define another copula function,

$$C(u,v) = \frac{C_1^d(u,v)}{u}[k_y + (1-k_y)u - k_x g_1(u)] + k_y g_1(u), \; u \neq 0, \qquad (22)$$

where $g_2(v) = \lim_{u \to 0} \frac{C_1^d(u,v)}{u}$ and $C_1^d(U(x \mid y^*), U(y \mid x^*)) = U(y \mid x)U(x \mid y^*)$.

Theorem 31.6 shows that the choice of the grounded copula function $C^d(u,v)$ implicitly determines the transformation function, $g_1(u)$. But suppose that the decision-maker wishes to specify the function $g_1(u)$ and to further incorporate it into the functional form. It is natural to ask if *we can determine the grounded copula function, C^d, using the boundary assessments $u, v, g_1(u), g_2(v)$.* We answer this question below.

31.5.2. Characterizing the Grounded Copula Function If It Is Archimedean

Suppose that the grounded surface is Archimedean of the form $C^d(u,v) = \psi^{-1}(\psi(u)\psi(v))$. Archimedean functions provide a wide range of grounded surfaces that are used to model probability and utility functions in practice (see, for example, Sklar 1959; Abbas and Howard 2005). When used to model utility functions, there is even greater flexibility in the properties of the generating function we can choose since we merely require that the generating function, ψ, be strictly monotone with $\psi(0) = 0$ and $\psi(1) = 1$. Compare this condition to the additional requirement that $sign\left(\frac{d^n}{dt^n}\log\psi(t)\right) = (-1)^{n+1}$ for probability functions. The basic idea that will be used to determine the grounded copula function will be to use all the boundary assessments of the utility surface to identify its generating function. The convenience of requiring only one generating function to specify the Archimedean functional form allows us to specify the grounded surface using the boundary assessments.

Theorem 31.7: Identifying the Grounded Copula Function Using Boundary Assessments

If $C^d(u,v)$ in (21) is (i) Archimedean and (ii) has a differentiable generating function, $\psi(t)$, satisfying $\psi'(0) \neq 0$, then

$$\boxed{g_1(u) = \psi(u).} \qquad (23)$$

Theorem 31.7 enables us to characterize the generating function of an Archimedean copula uniquely by simply assessing the boundary assessments $U(x \mid y^0), U(x \mid y^*)$, and then calculating $g_1(u)$, from which we obtain $\psi(u)$.

EXAMPLE 31.4: Characterizing the Generating Function Using Boundary Assessments

---

Consider the functional form (21) with a grounded Archimedean copula function. The decision-maker provides the following boundary assessments:

$$U(x \mid y^*) = x, \quad U(x \mid y^0) = \frac{1 - e^{-\delta x}}{1 - e^{-\delta}}, \delta \neq 0.$$

We now calculate the transformation g_1 that satisfies

$$U(x \mid y^0) = g_1\big(U(x \mid y^*)\big),$$

which implies that $g_1(u) = \dfrac{1 - e^{-\delta u}}{1 - e^{-\delta}}, \delta \neq 0$. Note that this transformation is very useful in practice since it is either concave or convex, and therefore it either increases or decreases the risk aversion function from one boundary to the other. In particular, changing the value (and sign) of δ enables us to vary the concavity or convexity of $U(x \mid y^0)$ with respect to $U(x \mid y^*)$, thereby changing the relative risk aversion function between the two boundary curves. Since the derivative $g_1'(u) = \dfrac{\delta e^{-\delta u}}{1 - e^{-\delta}}$ is not zero at the origin, the generating function of the grounded copula function that can match this condition is

$$\psi(t) = \frac{1 - e^{-\delta t}}{1 - e^{-\delta}}.$$

We have now completely characterized the functional form of the grounded copula function based on the assumption of an Archimedean type.

We also have an interpretation for the parameter δ of the generating function in terms of the relative risk aversion between the two boundary values. To illustrate, we get, by direct substitution, the difference between the two risk aversion functions as

$$\gamma(x \mid y^0) - \gamma(x \mid y^*) = \delta \frac{\partial}{\partial x} U(x \mid y^*),$$

where γ is the Arrow-Pratt risk aversion function. Note that as $\delta \to 0$, the two risk aversion functions at the boundary values coincide, which is the condition of mutual utility independence.

If the transformation g_1 has zero-slope at the origin, but g_2 does not, then we can use the form (22) instead. To illustrate, let

$$U(y \mid x^0) = \frac{1 - e^{-\gamma y}}{1 - e^{-\gamma}}, \quad U(y \mid x^*) = y, \quad U(x \mid y^0) = x^m, \quad U(x \mid y^*) = x^n, \, m > n.$$

Here we have $g_2(v) = \dfrac{1 - e^{-\gamma v}}{1 - e^{-\gamma}}, \gamma \neq 0$, whose slope is non-zero at the origin, and $g_1(u) = u^{\frac{m}{n}}$, whose slope is zero. Substituting into (23) derives the functional form with $g_2(t) = \psi(t)$.

If both g_1 and g_2 have zero-slope at the origin, then we can either experiment with a variety of grounded copula surfaces or use the copula function of Section 31.4 to match the boundary assessments.

While Archimedean copulas could be an initial starting point for the choice of the grounded surface, they certainly need not apply to all decision situations. If the analyst starts with an Archimedean copula satisfying $\psi = g$, and determines that it is not a good fit (as determined by the assessed isopreference contours or the shape of the diagonal assessments, for example), then she can (i) choose from a variety of non-Archimedean grounded surfaces that better match the trade-off assessments between the attributes or (ii) assess a few points on the utility surface and try to find a non-Archimedean grounded surface that better matches the fit. We leave this topic as an area for future research.

31.5.3. Monotonicity of the Functional Form (21)

Many applications of multiattribute utility functions require that the utility surface be increasing with each of its arguments. This section presents sufficient conditions to illustrate how (21) can be used to construct these widely used functional forms. Consider the following conditions:

CONDITION (I): MONOTONICITY OF THE GROUNDED COPULA SURFACE WITH u Since $C^d(u,v)$ is grounded, we require only that it be strictly increasing with u when $v \neq 0$. This condition is satisfied, for example, by all Archimedean functions $C^d(u,v) = \psi^{-1}(\psi(u)\psi(v))$.

> **Proposition 31.8:** *If condition (i) holds, then the functional form* (21) *is strictly monotone with u when* $v \neq 0$

To analyze the monotonicity with v, we first observe that it merely requires monotonicity of the RHS of (21) with v. This can be achieved by various combinations of grounded surfaces and transformation functions. We now present the following (strong) sufficiency conditions that guarantee monotonicity.

CONDITION (II): MONOTONICITY OF THE TRANSFORMATION FUNCTION This condition requires that the function, $g_2(v)$, be strictly increasing with v.

CONDITION (III): MONOTONICITY OF THE DIFFERENCE IN BOUNDARY ASSESSMENTS We have already assumed strict inequality between $U(x^*,y)$ and $U(x^0,y)$. This monotonicity condition further requires that the difference $U(x^*,y) - U(x^0,y)$ be nondecreasing. Using our notation, this translates into the condition that the difference $k_x + (1-k_x)v - k_y g_2(v)$ be non-decreasing. For a differentiable transformation function, this condition simply implies that $g_2'(v) \leq \dfrac{1-k_x}{k_y}$. The smaller the values of k_x, k_y, the less constraining this condition becomes. In the limit, as $k_x, k_y \to 0$, this condition is always satisfied.

CONDITION (IV): MONOTONICITY OF THE CONDITIONAL UTILITY FUNCTION $C(u|v)$ WITH v Since $C(u|v) = \dfrac{C^d(u,v)}{v}$, this condition merely requires that $\dfrac{C^d(u,v)}{v}$ be nondecreasing with v. This condition implies higher conditional utility values for any level u as the level of v increases.

The requirement that $\dfrac{C^d(u,v)}{v} = C(u\,|\,v)$ is strictly increasing with v is a sufficient but not necessary condition. It also has a geometric interpretation: it implies that for any fixed u, the slope of the line connecting any point on the curve $C^d(u,v)$ to the origin (when plotted vs. v) must strictly increasing. Figure 31.5 illustrates this condition graphically for the generating function $\psi(t) = \dfrac{1-e^{-\delta t}}{1-e^{-\delta}}, \delta \neq 0$. The left curve plots for the curves for $\delta = 5$ and the right curve plots the case of $\delta = -5$. By inspection, the slope of the line connecting any point on the curves of the left figure to the origin is decreasing with v, while the slope connecting any point in the right figure to the origin is increasing with v.

> **Proposition 31.9:** *If conditions (ii)-(iv) hold, then the functional form* (21) *is strictly monotone with v.*

31.6 SUMMARY

The double-sided utility copula is a tool that the analyst may use to construct utility surfaces that incorporate utility dependence and match the conditional utility assessments at all boundary values of the domain of the attributes when preferential independence conditions do not exist.

We provided a systematic method for constructing such utility copula functions using grounded surfaces.

We presented two distinct formulations:

1. The first has the flexibility to match all boundary assessments regardless of the choice of the grounded surfaces themselves. This class allows us to vary the copula parameters while keeping the boundary assessments fixed.
2. The second class provides a convenient sufficient condition that enables the analyst to identify the grounded copula function that matches the boundary assessments when it is Archimedean. This result significantly simplifies the search for the appropriate grounded surface if it is of the Archimedean type.

> **Historical Note**
> When Sklar (1959) first introduced the idea of copula functions for joint cumulative probability distributions, it made sense to require that the copula match one boundary value for each variable. This is because, by definition, joint probability distributions have only one marginal probability distribution for each variable. In fact the term "marginal probability" was coined because it is the probability distribution that occurs at the margin (or the boundary value) of the joint cumulative probability distribution function. For attribute dominance utility functions, matching one boundary value for each attribute also makes sense because the utility function is grounded, and so each attribute has only one utility function at the boundary. For multiattribute utility functions that do not satisfy the attribute dominance condition, it makes sense to match more than one boundary value.

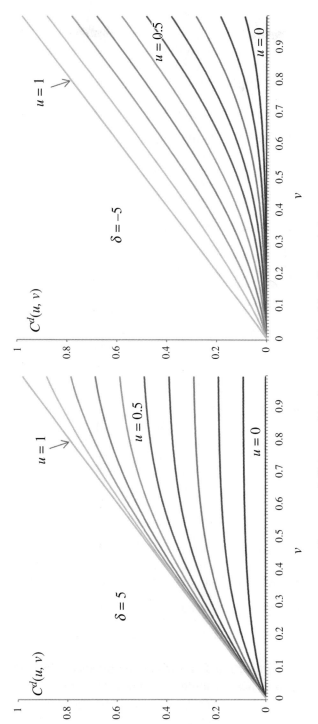

Figure 31.5. Sufficient condition for monotonicity of the utility copula function.

ADDITIONAL READINGS

Abbas, A. E. 2006. Entropy methods for joint distributions in decision analysis. *IEEE Transactions on Engineering. Management* 53: 146–159.

Abbas, A. E. 2009. Multiattribute utility copulas. *Operations Research* 57(6): 1367–1383.

Abbas, A. E. 2010. General decompositions of multiattribute utility functions. *Journal of Multicriteria Decision Analysis* 17(1, 2): 37–59.

Abbas, A. E. 2011. The multiattribute utility tree. *Decision Analysis* 8(3): 180–205.

Abbas, A. E. and D. E. Bell. 2011. One-switch independence for multiattribute utility functions. *Operations Research* 59(3): 764–771.

Abbas, A. E. and R. A. Howard. 2005. Attribute dominance utility. *Decision Analysis* 2(4): 185–206.

Bell, D. E. 1979. Multiattribute utility: Decomposition using interpolation. *Management Science* 25: 744–753.

Keeney, R. and H. Raiffa. 1976. *Decisions with Multiple Objectives*. Wiley, New York.

Matheson, J. E. and A. E. Abbas. 2005. Utility transversality: A value-based approach. *Journal of Multicriteria Decision Analysis* 13: 229–238.

Matheson, J. E. and R. A. Howard. 1968. An introduction to decision analysis. In *The Principles and Applications of Decision Analysis*, Vol I, R.A. Howard and J. E. Matheson (Eds). Strategic Decisions Group, Menlo Park, CA.

Nelsen, R. B. 1998. *An Introduction to Copulas*. Springer-Verlag, New York.

Sklar, A. 1959. Fonctions de répartition à n dimensions et leurs marges. *Publications de l'Institut de Statistique de l'Université de Paris* 8: 229–231.

von Neumann, J. and O. Morgenstern. 1947. *Theory of Games and Economic Behavior*. Princeton University Press, Princeton, NJ.

APPENDIX 31A

Proof of Proposition 31.1

If $U(x\,|\,y^*), U(y\,|\,x^*)$ are continuous, bounded, and strictly increasing, then their inverses exist and are normalized on the range $[0,1]$. Therefore, we can define $u = U(x\,|\,y^*), v = U(y\,|\,x^*), 0 \le u, v \le 1$, and further define $x = U_{y^*}^{-1}(u), y = U_{x^*}^{-1}(v)$. By direct substitution,

$$U(x\,|\,y^0) \triangleq U_{y^0}(x) = U_{y^0}\left(U_{y^*}^{-1}(u)\right) = U_{y^0}\left(U_{y^*}^{-1}(u)\right) = g_1(u) = g_1\left(U(x\,|\,y^*)\right),$$

$$U(y\,|\,x^0) \triangleq U_{x^0}(y) = U_{x^0}\left(U_{x^*}^{-1}(v)\right) = U_{x^0}\left(U_{x^*}^{-1}(v)\right) = g_2(v) = g_2\left(U(y\,|\,x^*)\right),$$

where $g_1(u) = U_{y^0}\left(U_{y^*}^{-1}(u)\right)$ and $g_2(v) = U_{x^0}\left(U_{x^*}^{-1}(v)\right)$. Therefore, there always exist unique transformations, g_1, g_2, defined on a domain $[0,1]$ that relate the upper and lower bounds.

Proof of Theorem 31.2

Since each $U(x\,|\,y^*), U(y\,|\,x^*)$ are continuous and strictly increasing, their inverse functions exist. Define $C(u,v) = U(U_x^{-1}(u), U_y^{-1}(v))$, where U_x^{-1}, U_y^{-1} are the inverse functions of $U(x\,|\,y^*), U(y\,|\,x^*)$ respectively. Therefore, $C(U_x(x), U_y(y)) = U(x,y)$.

We now show that this double-sided copula function matches all boundary assessments and corner values. By definition, since $C(u,1) = k_y + (1 - k_y)u$, then $U(x, y^*) = C(U(x\,|\,y^*), 1) = k_y + (1 - k_y)U(x\,|\,y^*)$, and therefore the resulting utility surface matches the utility function of X at the maximum margin. Further, since $C(u,0) = k_x g_1(u)$, then $U(x,y^0) = C(U(x \mid y^*),0) = k_x g_1(U(x \mid y^*)) = k_x U(x \mid y^0)$, and therefore the resulting utility surface matches the utility function of X at the minimum margin. The same applies to attribute Y. Moreover, in the resulting surface, we get by direct substitution, $U(x^*, y^0) = C(1,0) = k_x$ and $U(x^0, y^*) = C(0,1) = k_y$, and so it matches the corner assessments provided by the decision-maker.

Proof of Proposition 31.3

To show that (9) is a double-sided utility copula, we get by direct substitution,

$$C(u,0 \; ; \; g_1, g_2, k_x, k_y) = C_1^d(u,0) + k_x C_2^d(g_1(u),1) + k_y C_3^d(1-u,0).$$

From the properties of grounded utility copula surfaces, $C_1^d(u,0) = C_3^d(1-u,0) = 0$, and $C_2^d(g_1(u),1) = g_1(u)$. This implies that $C(u,0 \; ; \; g_1, g_2, k_x, k_y) = k_x g_1(u)$.

Similarly,

$$C(0,v \; ; \; g_1, g_2, k_x, k_y) = C_1^d(0,v) + k_x C_2^d(0,1-v) + k_y C_3^d(1,g_2(v)) = k_y g_2(v).$$

$$C(u,1 \; ; \; g_1, g_2, k_x, k_y) = C_1^d(u,1) + k_x C_2^d(g_1(u),0) + k_y C_3^d(1-u,1) = k_y + (1-k_y)u.$$

$$C(1,v \; ; \; g_1, g_2, k_x, k_y) = C_1^d(1,v) + k_x C_2^d(1,1-v) + k_y C_3^d(0,v) = k_x + (1-k_x)v.$$

Therefore, (9) is a double-sided utility copula. These conditions also imply that

$$C(0,0 \; ; \; g_1, g_2, k_x, k_y) = 0, C(1,1 \; ; \; g_1, g_2, k_x, k_y) = 1.$$

Continuity and boundedness of the resulting utility copula follows from continuity and boundedness of the individual grounded surfaces. Moreover, the domain of the resulting copula is $0 '' u, v '' 1$, as determined by the domain of the grounded copula surfaces.

Proof of Theorem 31.4

If the conditional utility function $U(x \mid y)$ is bounded and well defined, we use the basic expansion theorem to get

$$U(x,y) = U(x^*,y)U(x \mid y) + U(x^0,y)\bar{U}(x \mid y), \tag{24}$$

where $\bar{U}(x \mid y) = 1 - U(x \mid y)$. We further expand the two terms $U(x^*,y)$ and $U(x^0,y)$ with respect to Y, to get

$$U(x^*,y) = U(x^*,y^*)U(y \mid x^*) + U(x^*,y^0)\bar{U}(y \mid x^*) \tag{25}$$

and

$$U(x^0,y) = U(x^0,y^*)U(y \mid x^0) + U(x^0,y^0)\bar{U}(y \mid x^0). \tag{26}$$

Substituting from (25) and (26) into (24), with $U(x^*,y^*) = 1$ and (with no loss of generality) setting $U(x^0,y^0) = 0$, gives

$$U(x,y) = U(x \mid y)U(y \mid x^*) + k_x U(x \mid y)\bar{U}(y \mid x^*) + k_y \bar{U}(x \mid y)U(y \mid x^0). \tag{27}$$

The product term $U(x|y)U(y|x^*)$ in (27) is equal to zero if either X or Y is minimum. It is equal to $U(y|x^*)$ when X is maximum and equal to $U(x|y^*)$ when Y is maximum. Therefore, $U(x|y)U(y|x^*)$ can be expressed as a grounded Class 1 copula of the form $C_1^d(u,v)$, where $u = U(x|y^*)$ and $v = U(y|x^*)$.

Similarly, the term $U(x|y)\bar{U}(y|x^*)$ is equal to $U(x|y^0)$ when Y is minimum, y^0, and $\bar{U}(y|x^*)$ and X is maximum, x^*. It is zero when X is minimum or Y is maximum. Therefore, it can always be expressed as a grounded Class 1 copula of the form $C_2^d(g_1(u),1-v)$.

Finally, the term $\bar{U}(x|y)U(y|x^0)$ is equal to $U(y|x^0)$ when X is minimum and $\bar{U}(x|y^*)$ when Y is maximum, but is zero when Y is minimum or X is maximum. Therefore, it can be always expressed as a grounded Class 1 utility copula of the form $C_3^d(1-u,g_2(v))$.

Since (9) is derived from the general expression (27), a utility function can be expressed using (9) when it can be expressed using (27). This is possible when $U(x^0,y^0) = 0, U(x^*,y^*) = 1$, when $U(x|y^*)$ and $U(y|x^*)$ are strictly increasing, and when the conditional utility functions are bounded and well defined on the entire domain. It is also straightforward to see from (25) and (26) that if we had not assumed that $U(x^*,y^*) = 1$ and $U(x^0,y^0) = 0$, then we would simply add a fourth term in the expansion (27) of the form $U(x^0,y^0)\bar{U}(x|y)\bar{U}(y|x^0)$ and we would need to add a fourth term in (9) equal to $U(x^0,y^0)C_3^d(1-g_1(u),1-g_2(v))$. We would also scale the term $U(x|y)U(y|x^*)$ by $U(x^*,y^*)$, and therefore also scale the term $C_1^d(u,v)$ by $U(x^*,y^*)$.

Proof of Proposition 31.5

By construct of the grounded copula function and the transformation functions, the function $C(u,v) = \dfrac{C^d(u,v)}{v}[k_x + (1-k_x)v - k_y g_2(v)] + k_y g_2(v)$, $v \neq 0$ is continuous and bounded. We have

$$C(u,1) = \frac{C^d(u,1)}{1}[k_x + (1-k_x)1 - k_y g_2(1)] + k_y g_2(1) = u[1-k_y] + k_y \Rightarrow C(0,1) = k_y.$$

$$C(1,v) = \frac{C^d(1,v)}{v}[k_x + (1-k_x)v - k_y g_2(v)] + k_y g_2(v) = k_x + (1-k_x)v \Rightarrow C(1,0) = k_x.$$

$$C(0,v) = \frac{C^d(0,v)}{v}[k_x + (1-k_x)v - k_y g_2(v)] + k_y g_2(v) = k_y g_2(v).$$

$C(u,0) = \dfrac{0}{0}$. Taking the limit (if it exists and is not infinity) gives $C(u,0) = \lim\limits_{v \to 0} \dfrac{C^d(u,v)}{v} \triangleq g_1(u)$.

Therefore, it matches the minimum boundary of u for $g_1(u) = \lim\limits_{v \to 0} \dfrac{C^d(u,v)}{v}$.

Proof of Theorem 31.7

If the grounded copula function is differentiable, then by direct application of L'Hopital's rule, we get another expression for $g_1(u)$ as

$$\boxed{g_1(u) = \lim_{v \to 0} \frac{C^d(u,v)}{v} = \frac{\partial}{\partial v}C^d(u,v)|_{v=0}.}$$

For Archimedean copula functions, we have when $\psi'(0) \neq 0$,

$$\frac{\partial}{\partial v}C^d(u,v)\big|_{v=0}=\frac{\partial}{\partial v}\psi^{-1}\big(\psi(u)\psi(v)\big)\big|_{v=0}=\frac{\psi(u)\psi'(0)}{\psi'\big(\psi(u)\psi(0)\big)}=\psi(u).$$

Proof of Proposition 31.8

Since the functional form (21) depends on u only through the grounded surface $C^d(u,v)$, condition (i) is sufficient to guarantee monotonicity with u.

Proof of Proposition 31.9

The RHS of (21) has three terms that depend on v. The first is the conditional utility function $C(u\,|\,v)=\dfrac{C^d(u,v)}{v}$. The second is the difference $[k_x+(1-k_x)v-k_yg_2(v)]$, and the third is the term $g_2(v)$. If the first two terms are non-decreasing with v and the third term is strictly increasing with v, then the copula function is strictly increasing with v.

APPENDIX 31B DOUBLE-SIDED CLASSES (0,0), (1,0), AND (0,1) UTILITY COPULAS

We have considered situations where the functions $U(x\,|\,y^*),U(y\,|\,x^*)$ are strictly increasing and incorporated the other boundary assessments, $U(x\,|\,y^0),U(y\,|\,x^0)$, using the functions g_1,g_2. This appendix shows how to extend this formulation when other utility functions are continuous and strictly increasing. To formalize this procedure for other boundary values, we define three other types of double sided utility copulas.

Double-Sided Class (0,0) Utility Copulas

If the conditional utility functions are continuous at the minimum boundary values and/ or the decision-maker is more comfortable providing conditional utility assessments at these boundaries, we define

$$u=U(x\,|\,y^0),v=U(y\,|\,x^0),U(x\,|\,y^*)=g_1\big(U(x\,|\,y^0)\big),U(y\,|\,x^*)=g_2\big(U(y\,|\,x^0)\big).$$

Similar to the earlier development, there always exist functions g_1,g_2 to satisfy this condition. We will again require the functions g_1,g_2 to satisfy $g_i(0)=0,g_i(1)=1,i=1,2,$.

Definition: A *double-sided Class (0,0) utility copula* is a continuous bounded mapping, $C_{(0,0)}(u,v\,;\,g_1,g_2,k_x,k_y)$, with $0\text{ " }u,v\text{" }1$, that satisfies the following conditions:

1. Positive linear transformation at minimum margins,

$$C_{0,0}(u,0)=k_yu,\quad C_{0,0}(0,v)=k_xv.$$

2. Maximum margins defined by g-transformations of the minimum margins,

$$C_{0,0}(u,1)=k_x+(1-k_x)g_1(u),\quad C_{0,0}(1,v)=k_y+(1-k_y)g_2(v).$$

Note that this class also satisfies the conditions

$$C_{0,0}(0,0)=0,\quad C_{0,0}(1,1)=1,\quad C_{0,0}(1,0)=k_x,\quad C_{0,0}(0,1)=k_y.$$

Since the proof is similar to the previous development of Class (1,1), we state without proof that any two-attribute utility function that is continuous, bounded, and strictly increasing with each argument at the minimum boundary value of the complement attribute can be expressed in terms of a class (0,0) utility copula as

$$U(x,y) = C_{(0,0)}(U(x\,|\,y^0), U(y\,|\,x^0); g_1, g_2, k_x, k_d).$$

Define the functional form

$$C_{0,0}(u,v; g_1, g_2, k_x, k_d) = C^d(g_1(u), g_2(v)) + k_x C^d(u, \overline{g}_2(v)) + k_y C^d(\overline{g}_1(u), v), \quad (28)$$

where $\overline{g} = 1 - g$. We state without proof that (28) is a double-sided Class (0,0) utility copula.

Double-Sided Class (1,0) Utility Copulas

If $u = U(x\,|\,y^0), v = U(y\,|\,x^*)$ are strictly increasing, define

$$U(x\,|\,y^*) = g_1\big(U(x\,|\,y^0)\big), U(y\,|\,x^0) = g_2\big(U(y\,|\,x^*)\big).$$

Similar to the earlier development, there always exist functions g_1, g_2 to satisfy this condition. We will again require the functions g_1, g_2 to satisfy $g_i(0) = 0, g_i(1) = 1, i = 1, 2,$.

Definition: A *double-sided Class* **(1,0)** *utility copula* is a continuous bounded mapping, $C_{(1,0)}(u, v\,; g_1, g_2, k_x, k_y)$, with $0 '' u, v'' 1$, that satisfies the following conditions:

1. Positive linear transformation at minimum margins,

$$C_{1,0}(u,0) = k_y u, \quad C_{1,0}(1,v) = k_x + (1 - k_x)v.$$

2. Maximum margins defined by g-transformations of the minimum margins,

$$C_{1,0}(u,1) = k_x + (1 - k_x)g_1(u), \quad C_{0,1}(1,v) = k_y g_2(v).$$

This class also satisfies the conditions

$$C_{1,0}(0,0) = 0, \quad C_{1,0}(1,1) = 1, \quad C_{1,0}(1,0) = k_x, \quad C_{1,0}(0,1) = k_y.$$

Now we define the functional form

$$C_{1,0}(u,v; g_1, g_2, k_x, k_d) = C^d(g_1(u), v) + k_x C^d(u, \overline{v}) + k_y C^d(\overline{g}_1(u), g_2(v))$$

and state (without proof) that it is a class (1,0) utility copula.

Double-Sided Class (0,1) Utility Copulas

If $u = U(x\,|\,y^*), v = U(y\,|\,x^0)$, define $U(x\,|\,y^0) = g_1\big(U(x\,|\,y^*)\big), U(x\,|\,x^*) = g_2\big(U(y\,|\,x^0)\big)$.

The Class (0,1) requires $C_{0,1}(0,v) = k_x v, C_{0,1}(u,1) = k_y + (1 - k_y)u,$ and $C_{0,1}(1,v) = k_y g_2(v), C_{0,1}(u,1) = k_x + (1 - k_x)g_1(u)$. The functional form

$$C_{01}(u,v; g_1, g_2, k_x, k_d) = C^d(u, g_2(v)) + k_x C^d(g_1(u), \overline{g}_2(v)) + k_y C^d(\overline{u}, v)$$

is a Class (0,1) utility copula.

Using the previous definitions in this appendix, we now have the flexibility to construct utility surfaces using any continuous bounded and increasing boundary assessment for an attribute and express the other boundary using the transformation functions g_1, g_2.

CHAPTER 32

Preference and Value Copulas

Chapter Concepts

- Constructing preference and value functions that match boundary assessments
- Single-sided preference and value copulas
- Double-sided preference and value copulas
- The deterministic (ordinal) one-switch independence property (Abbas and Bell 2015)
- Ordinal one-switch preference and value copulas

32.1 INTRODUCTION

In Chapter 7, we discussed examples in which the preference function may be constructed by thinking about the shape of the trade-off contours or by thinking about our preferences such as in the peanut butter and jelly sandwich example. In Chapter 8, we discussed a special case of preference functions corresponding to additive ordinal preferences, and in Chapters 12 and 29, we showed how they correspond to Archimedean forms of multiattribute utility functions. While ordinal preferences described by an additive preference function may suffice in a variety of problems, it is natural to consider how to proceed with the construction of the utility function if it does not hold.

In Chapter 31, we extended the utility copula approach to account for multiple boundary assessments. We saw that this approach is particularly useful when the preference function is not additive. This chapter extends the double-sided copula formulation to construct preference and value functions that match multiple boundary assessments. This formulation enables more general preference and value functions that go beyond the additive forms. Our focus in this chapter will be on two-attribute preference and value copulas.

A second objective of this chapter is to revisit the condition of ordinal preferences that we referred to as "ordinal one-switch independence" in Chapter 7 (Abbas and Bell 2015) and illustrate its connection with double-sided copula formulations.

32.2 SINGLE-SIDED PREFERENCE AND VALUE COPULAS

32.2.1 Single-Sided Preference Copulas

Suppose we wish to construct a preference function that matches deterministic preferences of a decision-maker for the attributes. Unless preferential independence conditions exist, these preferences to be matched need to be specified at a reference value of the complement attributes. To illustrate, consider the preference function $P(x, y)$ and two curves on its surface, say $u = P(x, y_1)$ and $v = P(x_1, y)$ representing preferences for X at y_1 and preferences for Y at x_1, respectively.

If we wish to define a preference copula that matches these preferences for attributes X and Y, then we can write

$$P(x, y) = C_P(u, v) = C_P(P(x, y_1), P(x_1, y)),$$

where C_P is a preference copula.

If the resulting surface is to be consistent with the ordinal preferences used for its construction, then two consistency conditions at x_1 and y_1 are required with this preference copula, namely

1. Consistency with preferences for attribute X at y_1: $C_P(P(x, y_1), P(x_1, y_1)) = f(P(x, y_1))$ and
2. Consistency with preferences for attribute Y at x_1: $C_P(P(x_1, y_1), P(x_1, y)) = g(P(x_1, y))$,

where f and g are strictly monotone functions.

Note that the consistency condition requires the preference assessments to be preserved up to a monotone transformation at the reference values, and not the more restrictive linear transformation (as was the case with utility copulas). This is because preference functions are invariant up to a monotone transformation.

A sufficient condition for the existence of a preference copula that allows for this representation is that the preference function be strictly monotone with each argument for at least one reference value of the complement.

We can also normalize the curves $P(x, y_1), P(x_1, y)$, in which case the domain of the copula would be the unit square. For multiattribute utility functions, the normalization of the range of the utility surface, and the normalized conditional utility assessments, enabled a preference probability interpretation of the utility values. While not necessary, we can also normalize the surface of the preference function to range from zero to one.

There are many functions that satisfy the conditions of this preference function. Imposing additional conditions on the preference function leads to more specificity on the type of preference copula as we demonstrate below.

32.2.2 X is Preferentially Independent of Y

In Chapter 8, we defined the ordinal property where an attribute, X, is preferentially independent of another attribute, Y. The idea is that if prospects are characterized by two attributes, X and Y, and if you prefer level $x_1 \succ x_2$ of attribute X, for all values of Y and for all such x_1, x_2, then $P(x_1, y) - P(x_2, y)$ does not change sign for all y (this means that the sign is either always positive, always negative, or always zero for all y).

We discussed that X is preferentially independent of Y if and only if the preference function can be written as

$$P(x, y) = \phi(v(x), y),$$

where ϕ is either strictly monotonic or a constant function of its first argument for all y. We can now interpret this property using the preference copula as follows.

X is preferentially independent of Y if and only if the preference function can be written as

$$P(x, y) = C_P(P(x, y_1), y),$$

where the preference copula C_P is either strictly increasing, strictly decreasing, or constant in its first argument, and $P(x, y_1)$ is the preference for X at a fixed value of $Y = y_1$.

A consistency condition also requires that C_P satisfy the boundary condition

$$C_P(P(x, y_1), y_1) = f(P(x, y_1)),$$

where f is a monotone function.

Requiring preferential independence places further constraints on the function C_P, such as monotonicity with its first argument, than just the consistency condition.

32.2.3 Single-Sided Value Copulas

The construction of value functions using the copula formulation is also possible and should be clear. The difference is that the value function has an absolute scale and it is not invariant to a monotone transformation.

Consider the value function $V(x, y)$ and two curves on its surface, say $u = V(x, y_1), v = V(x_1, y)$. If we wish to define a value copula using this formulation, then we can write

$$V(x, y) = C_V(u, v) = C_V(V(x, y_1), V(x_1, y)),$$

where C_V is a value copula.

If the resulting surface is to be consistent with the values that are used for its construction, then we have

$$C_V(V(x, y_1), V(x_1, y_1)) = V(x, y_1) \text{ and } C_V(V(x_1, y_1), V(x_1, y)) = V(x_1, y).$$

Note that the consistency condition requires matching the exact values of the curves (and not a monotone transformation as was the case with preference copulas) because value functions have an absolute scale. We are matching two curves on a surface with an absolute scale.

32.3 DOUBLE-SIDED PREFERENCE AND VALUE COPULAS

What if the preference function is not additive? It is clear that we cannot rely on only one preference curve for each attribute. Therefore, it is natural to ask whether we can construct preference or value copulas matching two boundary values. For example, this preference or value copula would match $v_1(x) = v(x, y_1)$, $v_2 = v(x, y_2)$ at two values y_1, y_2 or it could match two curves for each attribute. The following example illustrates a bidirectional preference copula for one attribute.

EXAMPLE 32.1: A Double-Sided Ordinal Preference Copula

Consider the function

$$C(P_1, P_2, y) = S\big(yh(P_1) + (1-y)g(P_2)\big),\ 0 \le y \le 1, \tag{1}$$

where S, h, and g are strictly increasing (or strictly decreasing) functions.

It is clear that this function satisfies the boundary consistency conditions of a preference copula where the surface is a monotone transformation of its arguments at reference values of y equal to 0 and 1, where

$$C(P_1, P_2, 0) = S\big(g(P_2)\big) \text{ and } C(P_1, P_2, 1) = S\big(h(P_1)\big)$$

Therefore, if we assess $P_1(x) = P(x, 1)$ and $P_2 = P(x, 0)$, whose range lies in the domain of h and g, respectively, then we can substitute into (1), and the resulting surface will match the ordinal preferences of these assessments (preserving the assessments up to a monotone transformation).

Of course, the concept of a double-sided preference copula can also extend immediately to double-sided value copulas on an absolute scale.

EXAMPLE 32.2: A Double-Sided Ordinal Value Copula

Consider the function

$$C(v_1, v_2, y) = S^{-1}\big(yS(v_1) + (1-y)S(v_2)\big), \tag{2}$$

where S is a monotone function.

It is clear that this function satisfies the value copula boundary conditions at reference values of y equal to 0 and 1, where

$$C(v_1, v_2, 0) = S^{-1}\big(S(v_2)\big) = v_2 \text{ and } C(v_1, v_2, 1) = S^{-1}\big(S(v_1)\big) = v_1.$$

Therefore, if we assess $v_1(x) = v(x, 1)$ and $v_2 = v(x, 0)$, whose range lies in the domain of S, and substitute into (2), then the resulting surface will match these value assessments on their absolute scale.

Imposing additional conditions on the double-sided preference and value copulas also leads to more specificity as we illustrate below.

32.4 THE ORDINAL ONE-SWITCH PROPERTY (ABBAS AND BELL 2015)

What if preferences for deterministic prospects characterized by two levels of X, say x_1, x_2, can change, but only once, as we increase another attribute Y? We referred to this condition briefly in Chapter 8 as *ordinal one-switch independence*. This is also the "Food for Thought" question that we posed in Chapter 8, where we asked about the implications of the property that "$P(x_1, y) - P(x_2, y)$ can cross zero only once and at exactly one unique value of Y, and this relation is true for all x_1, x_2."

> **Definition**
>
> **X is ordinal one-switch independent of Y,** written **X 1S Y,** if for any pair of consequences in X, say $(x_1, y), (x_2, y)$, either one is preferred to the other for all y, or they are indifferent for all y, or there exists a unique level of y above which one of the consequences is preferred and below which the other is preferred.

Our definition of ordinal one-switch independence excludes the case where a pair of consequences is equally preferred only on an interval: if two X values are ever indifferent at two different values of Y, then they must be indifferent for all values of Y. This definition of ordinal one-switch independence is unchanged even if X is multidimensional. It also extends readily to "n-switch independence."

Observe that a function that satisfies ordinal zero-switch independence (preferential independence) of X from Y also satisfies the condition of ordinal one-switch independence of X from Y.

Example 32.3. Consider the function $v(x, y) = (x + y) - (x - y)^2$.

For any two values of X, say x_1, x_2, the difference

$$\Delta(y) = V(x_1, y) - V(x_2, y) = (x_1 - x_2) - (x_1 - x_2)(x_1 + x_2 - 2y).$$

This function satisfies X 1S Y, because the difference switches sign only once, when $y = (x_1 + x_2)/2$.

Example 32.4. Consider the function $V(x, y) = x^2 y - xy^2 + x$

The difference

$$\Delta(y) = V(x_1, y) - V(x_2, y) = (x_1 - x_2)(x_1 + x_2)y - (x_1 - x_2)y^2 + (x_1 - x_2)$$

is quadratic in y, and therefore it need not satisfy X 1S Y. For example, for $x_1 = 1$ and $x_2 = 0$, there are two switches as y varies.

32.4.1 Sufficient Conditions for Ordinal One-Switch Independence

Of course, we can always try the brute-force method to test for ordinal one-switch independence for particular values of X for simple functions, but more complex functions require additional tools. In this section we review well-known tests to determine whether a given function $V(x,y)$ satisfies the one-switch condition. In the next section, we illustrate how to construct functions satisfying this one-switch condition using double-sided value copulas.

Test 1: Monotone Differences:

$$X\,1S\,Y \text{ if } V(x_1,y)-V(x_2,y) \text{ is strictly monotonic in } y\,\forall x_1 \neq x_2.$$

If $\Delta(y) = v(x_1,y) - v(x_2,y)$ is strictly monotonic, then it can cross the x-axis at most once, thus satisfying the one-switch condition.

Monotonicity of $\mathcal{C}$ is a sufficient but not a necessary condition for satisfying the ordinal one-switch condition. To illustrate, the function $V(x,y) = e^{xy}$ satisfies the ordinal one-switch property. However, if we pick two values of X, say $x_1 = 2$ and $x_2 = 1$, the difference $\Delta(y) = e^{2y} - e^y$ is not monotone because the derivative $\varnothing(y)$ is positive when $e^y > 0.5$ and negative when $e^y < 0.5$.

Test 2: Constant Sign of the Cross-Derivative (for differentiable functions)

$$X\,1S\,Y \text{ if the cross-derivative } \frac{\partial^2 V(x,y)}{\partial x \partial y} \text{ does not change sign.}$$

If the cross-derivative has a constant sign (either positive or negative), then for any $x_1 > x_2, y_1 > y_2$, the difference

$$[v(x_1,y_1)-v(x_1,y_2)]-[v(x_2,y_1)-v(x_2,y_2)] = \int_{x_1}^{x_2}\int_{y_1}^{y_2} \frac{\partial^2 v(x,y)}{\partial x \partial y}\,dxdy$$

does not change sign. This implies that the difference $\Delta(y) = v(x_1,y) - v(x_2,y)$ is strictly monotone if the cross-derivative is either always positive or always negative. This condition is sufficient but not necessary. For example, $V(x,y) = x^2y$ satisfies $X\,1S\,Y$, but the cross-derivative is $2x$ that changes sign. Note that the condition on the sign of the cross-derivative is a symmetric condition. Therefore, it implies not only that $X\,1S\,Y$ but also that $Y\,1S\,X$.

Example 32.5 Consider the function $V(x,y) = \frac{x+y}{2} - k(x-y)^2$. The cross-derivative, $\frac{\partial^2 V(x,y)}{\partial x \partial y} = 2k$, has constant sign. This implies that V satisfies $X\,1S\,Y$ (and that $Y\,1S\,X$).

Test 3: Monotone Ratio

X 1S Y if the ratio $\dfrac{V(x_1, y)}{V(x_2, y)}$ is strictly monotone in y for $\forall x_1 \neq x_2$.

If $V(x_1, y)$ and $V(x_2, y)$ are equal at both y_1 and y_2, then their ratio is 1 at those two values, and thus the ratio cannot be strictly monotone in y. To illustrate, if $V(x, y) = e^{xy}$, then the ratio $\dfrac{V(x_1, y)}{V(x_2, y)} = e^{(x_1 - x_2)y}$ is strictly monotone, and therefore this function satisfies the ordinal one-switch condition.

32.4.2 Cardinal One-Switch Utility Independence Satisfies Ordinal One-Switch Independence

The one-switch utility independence idea that we discussed in Chapter 27 is

$$u(x, y) = g_0(y) + g_1(y)[f_1(x) + f_2(x)w(y)],$$

where $g_1(y)$ does not change sign and $w(y)$ is a monotone function, satisfies the condition of ordinal one-switch independence. This function necessarily satisfies ordinal one switch (because a sure thing is a special case of a gamble). To further illustrate, we have

$$\Delta(y) = V(x_1, y) - V(x_2, y) = g_1(y)[f_1(x_1) - f_1(x_2) + w(y)[f_2(x_1) - f_2(x_2)]],$$

which, since $g_1(y) > 0$, changes sign at most once when $w(y) = \dfrac{f_1(x_1) - f_1(x_2)}{f_2(x_2) - f_2(x_1)}$.

32.5 DOUBLE-SIDED VALUE COPULAS SATISFYING ORDINAL ONE-SWITCH INDEPENDENCE

Having defined the idea of ordinal one-switch independence, we now demonstrate that this ordinal condition can be constructed using a double-sided value copula formulation.

32.5.1 Existence of a Double-Sided Copula Representation

It will be convenient to adopt the notation $v_1(x) = V(x, y_1)$, $v_2(x) = V(x, y_2)$ where y_1, y_2 are the boundaries of an interval of interest.

Theorem 32.1: Necessary Condition for Ordinal One-Switch Functions (Abbas and Bell 2015)

If X is ordinally one-switch independent of Y, then there exists a function C such that
$$V(x, y) = C(v_1(x), v_2(x), y) \text{ for } y_1 \leq y \leq y_2,$$
where $v_1(x) = V(x, y_1), v_2(x) = V(x, y_2)$, and the function C satisfies the boundary conditions

$$C(v_1(x), v_2(x), y_1) = v_1(x), \quad C(v_1(x), v_2(x), y_2) = v_2(x).$$

Note: Theorem 32.1 shows that a value function satisfying ordinal one-switch independence can be expressed in terms of two curves on its surface and the attribute Y. Every value x_1 defines a unique pair of values $v_1(x_1), v_2(x_1)$, and it is also uniquely defined by them.

Example 32.6 Consider the cardinal one-switch form

$$U(x, y) = g_0(y) + g_1(y)[f_1(x) + f_2(x)w(y)],$$

where $g_1(y)$ does not change sign and $w(y)$ is a monotone function.

As we have seen, this function satisfies $X\,1S\,Y$ because it satisfies the cardinal condition, but note that it can also be written as $U(x, y) = \phi(v_1(x), v_2(x), y)$, where C is linear in both $v_1(x), v_2(x)$. Therefore, it satisfies the necessary condition for ordinal one-switch independence, as expected.

Example 32.7 Consider the function

$$V(x, y) = 2xy + \sin(x + y), \ 0 \le y \le \pi.$$

It is straightforward to see that this value function satisfies ordinal one-switch independence because the cross-derivative $\dfrac{\partial^2 V}{\partial x \partial y} = 2 - \sin(x + y)$ has constant sign. Theorem 32.1 asserts that it has a representation of the form $V(x, y) = C(v_1(x), v_2(x), y)$. Indeed, define

$$v_1(x) = V(x, 0) = \sin(x), \quad v_2(x) = V(x, \pi) = 2\pi x - \sin(x).$$

Therefore,

$$x = \frac{v_1 + v_2}{2\pi}.$$

We can therefore define

$$C(v_1, v_2, y) = 2\frac{v_1 + v_2}{2\pi} y + \sin\left(\frac{v_1 + v_2}{2\pi} + y\right)$$

and note that

$$V(x, y) = C(v_1(x), v_2(x), y).$$

Furthermore, C satisfies the boundary conditions

$$C(v_1, v_2, 0) = \sin\left(\frac{v_1 + v_2}{2\pi}\right) = \sin(x) = v_1(x)$$

$$C(v_1, v_2, \pi) = v_1 + v_2 - \sin\left(\frac{v_1 + v_2}{2\pi}\right) = v_2(x)$$

Dialogue between Instructor (I) and Student (S)

I: Is everyone clear on the notion of preference and value copulas?

S: Yes, the idea is that we can construct a preference or a value surface that matches input assessments.

I: Correct, and preference copulas would require consistency up to a monotone transformation of the input assessments while value copulas require absolute consistency with input value assessments because it is a value scale.

S: So these conditions are different than the utility copulas we discussed earlier? Because utility copulas required consistency up to a linear transformation.

I: Yes, in general conditions on ordinal preferences (preference functions) pose less restrctions and therefore do not lead to the same specificity in the corresponding functional form.

S: The extension to double-sided preference and value copulas is interesting. But how do we determine these copulas? And how do we know a double-sided copula representation exists?

I: As we have seen in Theorem 32.1, if we wish to satisfy the ordinal one-switch property, then there must exist a double-sided copula formulation that uses two curves on the surface of the value function to construct the utility surface. This is a necessary condition for ordinal one-switch value functions..

S: Yes, this is a very big insight, that we can express the utility surface using two curves if the one-switch ordinal property is satisfied.

I: Compare this formulation to that of preferential independence in that we discussed in Section 32.2.2, where only one univariate function of x was required. Theorem 32.1 shows that ordinal one-switch independence is a generalization of preferential independence requiring two functions $v_1(x), v_2(x)$.

S: Are there conditions that guarantee that a double-sided value copula will yield a surface that leads to the ordinal one-switch property?

I: Yes, we shall consider more conditions below.

32.5.2 Relation to the Single-Crossing Condition

It would be convenient to define some conditions such that a double-sided value copula, C, results in a value function that satisfies ordinal one-switch independence for arbitrary functions v_1, v_2 defined on its domain. Our focus will be on twice continuously differentiable functions C. For expositional simplicity, define $DR(v_1, v_2, y)$ as the ratio of partial derivatives $\partial C(v_1, v_2, y)/\partial v_1$ to $\partial C(v_1, v_2, y)/\partial v_2$, i.e.,

$$DR(v_1, v_2, y) = \frac{\partial C(v_1, v_2, y)/\partial v_1}{\partial C(v_1, v_2, y)/\partial v_2},$$

where we assume that $\partial C(v_1, v_2, y)/\partial v_2 \neq 0$.

Milgrom and Shannon's (1994) work derived conditions on a three-attribute function $U(x, y, z)$ to satisfy what is known as the single-crossing condition, where preferences for any pair $(x_1, y_1), (x_2, y_2)$ can switch only once as z increases, such that the difference $U(x_1, y_1, z) - U(x_2, y_2, z)$ crosses zero only once (and from below) when plotted as a function of z.

The single-crossing condition implies that the difference $C(v_1, v_2, y) - C(v_1^0, v_2^0, y)$ crosses zero only once (and from below) when plotted as a function of y. Milgrom and Shannon show that a function *satisfies the single-crossing condition for connected level sets if the ratio* $\dfrac{\dfrac{\partial U(x,y,z)}{\partial x}}{\dfrac{\partial U(x,y,z)}{\partial y}}$ *is non-deceasing (increasing) in y.* Building on their work, and on the representation of Theorem 32.1, we can place sufficient conditions on the double-sided value copula, C, so that the corresponding two-attribute function, $V(x, y)$, would satisfy X 1S Y. In particular, if C is monotone in v_1, v_2 and if $DR(v_1, v_2, y)$ is monotone in y, then the resulting value function will satisfy the conditions of ordinal one-switch independence.

32.6 SUMMARY

Preference and value copulas have single-sided and double-sided formulations.

Ordinal one-switch independence is a natural extension of preferential independence.

Ordinal one-switch preference functions can be represented in the form $V(x, y) = C(v_1(x), v_2(x), y)$.

Subject to some restrictions on C, a function of the form $C(v_1(x), v_2(x), y)$ can be used to construct an ordinal one-switch preference function.

ADDITIONAL READINGS

Abbas, A. E. 2009. Multiattribute utility copulas. *Operations Research* 57(6): 1367–1383.

Abbas, A. E. 2012. Utility copula functions matching all boundary assessments. *Operations Research* 61(2): 359–371.

Abbas, A. E. and D. E. Bell. 2011. One-switch independence for multiattribute utility functions. *Operations Research* 59(3): 764–771.

Abbas, A. E. and D. E. Bell. 2012. One-switch conditions for multiattribute utility functions. *Operations Research* 60(5): 1199–1212.

Abbas, A. E. and D. E. Bell. 2015. Ordinal one-switch utility functions. *Operations Research* 63(6): 1411–1419.

Bell, D. E. 1979. Multiattribute utility functions: Decompositions using interpolation. *Management Science* 25(2): 208–224.

Bell, D. E. 1988. One-switch utility functions and a measure of risk. *Management Science* 34(12): 1416–1424.

Debreu, G. 1960 Topological methods in cardinal utility theory. In K. Arrow, S. Karlin, and P. Suppes, eds. *Mathematical Methods in the Social Sciences*, pp. 16–26. Stanford University Press, Stanford, CA.

Farquhar, P. H. 1975. A fractional hypercube decomposition theorem for multiattribute utility functions. *Operations Research* 23: 941–967.

Fishburn, P. C. 1974. von Neumann-Morgenstern utility functions on two attributes. *Operations Research* 22: 35–45.

Fishburn, P. C. 1977. Approximations of two-attribute utility functions. *Mathematics of Operations Research* 2: 30–44.

Karni, E. and Z. Safra. 1998. Hexagon condition and additive representation of preferences: An
   algebraic approach. *Journal of Mathematical Psychology* 42: 393–399.
Keeney, R. L. and H. Raiffa. 1976. *Decisions with Multiple Objectives*. Wiley. New York.
Milgrom, P. and S. Shannon. 1994. Monotone comparative statics. *Econometrica* 62(1): 157–180.
von Neumann, J. and O. Morgenstern. 1947. *Theory of Games and Economic Behavior*. Princeton
   University Press, Princeton, NJ.

APPENDIX 32A

Proof of Theorem 32.1

We start with the following lemma, where we define $v_1(x) = V(x, y_1)$, $v_2(x) = V(x, y_2)$.

Lemma 32.1 *If X 1S Y and if there exist any two distinct values x_1, x_2 such that*

$$v_1(x_1) = v_1(x_2) \text{ and } v_2(x_1) = v_2(x_2),$$

then $V(x_1, y) = V(x_2, y)$ for all y.

Proof of Lemma. If there exist x_1, x_2 that satisfy $v_1(x_1) = v_1(x_2)$ and $v_2(x_1) = v_2(x_2)$,
then this means indifference of x_1, x_2 at two points y_1, y_2, which (by our definition)
is a violation of $X 1S Y$ unless $V(x_1, y) = V(x_2, y)$ for all y.

Geometrically speaking, Lemma 32.3 implies that if we pick any two curves
$V(x, y_1)$ and $V(x, y_2)$ on the surface of a function $V(x, y)$ that satisfies $X 1S Y$, then
there cannot be any *two distinct* x_1, x_2 whose values are equal on these curves unless
they are equal on the entire domain $V(x_1, y) = V(x_2, y)$. A corollary of this result
implies that there cannot be any regions of $v_1(x)$ and $v_2(x)$ that are flat for the same
values of x.

We now prove the theorem.

Proof of Theorem 32.1 Define C by the relation $C(v_1(x), v_2(x), y) = V(x, y)$. This
assignment converts a two-dimensional function $V(x, y)$ into a three-dimensional
function $C(v_1(x), v_2(x), y)$. To provide this mapping, every x must correspond to a
unique pair $(v_1(x), v_2(x))$, which it does if the one-switch independence relation
is satisfied. To illustrate in very simple terms, think of a two-dimensional table,
with x, y, and a vertical dimension $V(x, y)$ corresponding to the different values
of X and Y. For every value of x, we can define $v_1(x), v_2(x)$. If $X 1S Y$, then we
cannot have x_1, x_2 having equal values of $v_1(x_1), v_2(x_1)$ and $v_1(x_2), v_2(x_2)$ unless
$V(x_1, y) = V(x_2, y) \forall y$. Therefore, we can define $C(v_1(x_1), v_2(x_1), y)$ to correspond
to every $V(x_1, y)$. If $V(x_1, y) = V(x_2, y) \forall y$, then this will also be preserved with the
representation $C(v_1(x), v_2(x), y) = V(x, y)$.

CHAPTER 33

Reflections on What We Have Learned

Chapter Concepts

- Review of some of the concepts that we have discussed
- Good practices for modeling trade-offs

33.1 INTRODUCTION: REVIEW OF SOME IMPORTANT CONCEPTS

Dialogue between Instructor (I) and Student (S)

A recap of what we have learned in this book

I: First we made some definitions about a prospect: a decision alternative, and a decision. We emphasized that a prospect is deterministic. Then we defined the preference, value, and utility of a prospect.

S: It is clear why these distinctions were fundamental and why it was important to clearly define preference, value, and utility for deterministic prospects. There is much confusion about these terms in the literature.

I: Correct. Imagine if we do not get the correct meaning of utility as an indifference probability assessment and instead replace it with any measure, such as a measure of happiness?

S: This would lead us in a different direction from the start.

I: Correct. Then we showed how to use the preference, value, and utility of a prospect to make the best decision even when uncertainty is present using the expected utility criterion.

S: This was a result of the Five Rules. If there is no uncertainty, then preference or value statements are sufficient to determine the best decision alternative.

I: Correct. And we saw that there are many methods that are still widely used for decision-making under uncertainty, and are often motivated by simplicity, but they are inherently flawed.

S: It was also interesting to see that we could order deterministic prospects by visualization without much analysis. This was a big insight.

I: Yes, it is important to have distinct and clear preferences over deterministic prospects. The rules do not tell you what to order; they require you to have clear preferences according to the Order Rule.

S: And sometimes we do need analysis especially when the prospects are numerous. This also led to the idea of preference, value, and utility functions over direct value attributes.

I: It was also important to make the distinction between direct and indirect value attributes when characterizing prospects; direct value attributes play an important role in the rank order of deterministic prospects.

S: This concept was very insightful.

I: We illustrated that "probability of success" or "expectation of an attribute," for example, should not be used as arguments of preference, value, or utility functions. We also discussed the role of structural models in helping with the preference order and the choice of the attributes. Sometimes the structural model alone can determine the order.

S: Characterizing prospects by direct value attributes led to the idea of how to make trade-offs among those direct value attributes, and the idea of isopreference contours, as well as methods for their characterization, such as direct assessments, or using a functional form for the isopreference contours. We also discussed additive preference functions, their corresponding contours, and their representation using an Archimedean form.

I: Yes, then we talked about value functions for prospects characterized by multiple direct value attributes. We emphasized that value is an absolute scale. After that, we discussed the idea of a single-attribute utility function over a value measure or over an attribute.

S: Then we showed how a single attribute utility function – together with a preference function or a value function – was sufficient to construct a multiattribute utility function.

I: And we discussed the idea of utility transversality, using the risk aversion function over value and properties of the value function to relate the risk aversion functions over the individual attributes.

S: Yes, it was interesting to see that once the preference or value function are specified, we do not have an additional degree of freedom to assign utility functions over each of the attributes independently. The utility functions are related by the properties of the preference or value function.

I: We also discussed multivariate risk aversion and related it to risk aversion over value and to properties of the value function. We discussed its role in choice between bivariate lotteries and its role for deterministic prospects especially for ordinal one-switch independence.

S: This was an important insight. We also saw some issues with the implications of additive utility functions. They do not consider the correlation between the attributes. And we made a clear distinction between additive utility functions and the more general additive preference functions.

I: Then we discussed the special case of a utility function over an additive preference function to yield an Archimedean combination of individual utility functions. We showed that individual utility functions over each of the attributes could be combined into some Archimedean form to construct the multiattribute utility function.

S: This idea paved the way for constructing multiattribute utility functions by direct utility assessment over the attributes using marginal-conditional utility assessments.

I: And, as we have seen, there is a general structure for expanding the utility function into marginal-conditional assessments using the basic expansion theorem and the utility tree. The expansion is a summation of products of marginal-conditional utility assessments. To simplify the exposition, we started with the special case of attribute dominance utility functions where the expansion was a simple product.

S: It is clear now that the expansion was a product because the corner values were zero for attribute dominance utility functions.

I: And we saw the convenience of simplifying the assessments needed for the basic expansion theorem. This led to the idea of identifying independence assertions such as utility, boundary, interpolation, and one-switch independence. We saw how these conditions either reduce the number of assessments needed or simplify them by conditioning the assessments on fewer attributes.

S: And of course it was important to show how these independence assertions relate and their corresponding functional forms.

I: Correct. Then we discussed the idea of a utility copula that enables a mixture of matching both marginal utility assessments and trade-off assessments. We also discussed the use of utility copula functions in sensitivity analysis and methods to assess them empirically.

S: It was important to observe the linearity property for utility copulas at a reference value of the complement.

I: Yes. Then we saw double-sided copulas and preference and value copulas. Throughout this discussion, we saw the importance of single-attribute utility functions in many applications of multiattribute assessment. This is why we dedicated a whole section to single-attribute utility functions and we discussed their properties.

S: We saw how to characterize a utility function by the valuation of lotteries and by a change in the valuation of lotteries when the lottery outcomes are modified by monotone transformations. We also discussed the invariant transformation of a utility function and we extended this notion to one-switch independence and to one-switch utility functions.

I: At the beginning of the book, we also saw why some other methods, although simple, were not appropriate for making decisions. We discussed the difference between an approximate method and a fundamentally flawed method. It was important to make this distinction and to introduce some flawed methods to gain an appreciation of the process and the expected utility framework.

S: So what is next?

I: The field of multiattribute utility will continue to evolve. We have only studied the foundations. There will be new simplifying conditions and interpretations for properties of functional forms of utility functions. Some of them might be more complicated than others. New graphical representations and elicitation methods will be devised, and with automation, new algorithms for updating preferences will be developed.

S: We hope that current practitioners and researchers make more use of the methods we have studied already.

I: I hope so too. I hope it is clear why I do not necessarily advocate methods like "weight and rate" just because of simplicity.

S: Crystal clear. Do you think people still use it?

I: I am sure they do and they do so incorrectly. They use arbitrary attributes, with arbitrary metrics, and arbitrary constructed scales and also arbitrary weights.

S: Do you predict any other developments?

I: I would like to see more work on ordinal preference and value functions, like the seat in the movie theater example, but applied to large-scale situations, such as large-scale systems, and value of scientific discovery. The basic ideas are the same. I would also like to see more people from different fields incorporate the concepts that we have learned and think clearly about preference, value, and utility. Examples of value functions like the high-speed machining example can only be constructed when people with domain knowledge of different fields adopt this way of thinking.

S: We can see the importance of that especially for problems like climate change.

I: I would also like to see different domains learn what we have talked about and use this material appropriately in their fields. Or at least be aware of it, and request help of trained analysts when constructing multiattribute preference, value, or utility functions. They should not just assume that they know it without proper scholarship. I would also like to see more work on preferences for non-numerical attributes and methods for ordering them. There is a lot of room for developments there. Maybe one of you will do it?

C: Maybe.

I: Can you promise me that you will correct arbitrary methods of multi-objective decision-making if you see them being used?

S: Yes, I can promise that. I expect it will be hard to change an existing culture, but we have to start sometime. I have seen many misconceptions about the field by using wrong methods.

I: Another thing: you will not arbitrarily use "weight and rate" with arbitrary scales and arbitrary metrics in decision analysis.

S: I can promise that too. It is clear that using arbitrary constructed scales is a result of not defining the attributes appropriately, and not making a clear distinction between direct and indirect values. Why do we care about things like complexity or public perception and what do they really mean? We should relate these factors to the direct value attributes.

I: Thank you. Along those lines, be careful with the implications of additive utility functions that we discussed, and use them only when their conditions really apply. Also, promise me that you will not build preference, value, or utility functions with probabilities in their argument.

S: Yes. We saw in Chapter 10 how to make trade-offs between probability and consequences. We need a utility function and this trade-off depends on the risk aversion.

I: You will not set arbitrary threshold values as indicators of success and failure.

S: I promise, especially, if these thresholds will be used later for decision-making. Why should the state of the border change from orange to red, for example, for an infinitesimal change in probability? It is counterintuitive to continuity. How can we be indifferent between two events that have the same probability but different consequences and label them both as "orange"?

I: Okay, and you will also not set requirements independently on each attribute without specifying trade-offs among those requirements, and you will not set requirements in the first place without thinking about their implications to decision-making.

S: You have made this point very clear.

I: You will think about deterministic trade-offs and preference and value functions even when constructing a multiattribute utility function by eliciting conditional utility functions over the attributes.

S: This idea was also made clear. You can have the same marginal assessments at the boundary of the domain but have different trade-offs across the domain. Independence assumptions make assertions about trade-offs and they need to be verified.

I: If you construct a multiattribute utility function by direct assessments over the attributes, you will first consider the more general forms of the basic expansion theorem and you will not rush into assuming simplifying conditions if they do not exist.

S: I promise. Indeed, boundary, interpolation, and utility independence conditions are convenient when applicable, but they do need to be verified first. We have seen this clear with the sensitivity analysis to the product form.

I: You will check to see if preferential independence conditions really exist before using an Archimedean utility copula.

S: Sure. And I will think about the generating function either by assessment or by a functional form and will verify the implied trade-offs.

I: Great. Validation is an important step in what you do.

S: I do have a question. Given the recent advances in technology that we observe, how relevant will this material be in the next decades?

I: We will continue to observe advances in numerous fields, but the foundations of multiattribute utility will remain the same.

S: I can see that.

I: One more thing. Promise that you will maintain high ethical standards in research, practice, and work that you do.

S: I promise.

Ali's Pledge?

After reading this book, are you willing to take my pledge?

1. I will correct arbitrary methods of multi-objective decision-making if I see them being used.
2. I will not arbitrarily use "weight and rate" with arbitrary metrics and arbitrary scales.
3. I will not build preference or value or utility functions with probabilities in their argument.
4. I will not set arbitrary thresholds or targets as indicators of success or failure. Furthermore, in multiattribute problems, I will not set arbitrary thresholds independently for each attribute without specifying trade-offs among them. I will not set requirements or thresholds in general without thinking about their implications to decision-making.
5. I will think about deterministic trade-offs (and preference and value functions) even when I construct a multiattribute utility function by direct utility elicitation over the attributes.
6. If I do construct a multiattribute utility function by direct assessments over the attributes, I will consider the more general forms of the basic expansion theorem at first, and I will not rush into assuming simplifying conditions if they do not exist.
7. If I construct a multiattribute utility function using preference or value functions, I will verify the utility values of some points on the utility surface using indifference assessments.
8. I will check to see if preferential independence conditions really exist before using an Archimedean utility copula. I will not make arbitrary assumptions about the form of the generating function in the Archimedean form.
9. I will not value my life based on what was or what could have been.
10. I will maintain high ethical standards in research, practice, and work that I do.

APPENDIX A — Composite Functions, Inverse Functions, and Archimedean Combinations

A.1 THE COMPOSITE FUNCTION

The composite function is a function of some function of a variable. For example, if we have two functions $U(x)$ and $g(x)$, we may define a composite function as $g(U(x))$. The composite function maps every value of x to a new value obtained by first calculating $U(x)$ and then substituting for this new value into the argument of $g(x)$. This composition can also be written as $g \circ U$.

Example: If $U(x) = x^2$ and $g(x) = e^x$, then $g(U(x)) = g \circ U = e^{U(x)} = e^{x^2}$.

The composite operation can also occur multiple times. For example, we may write $S \circ g \circ U$, which implies $S(g(U(x)))$.

A.2 THE INVERSE OF A FUNCTION

The inverse of a function is a mirror image of that function around the axis $y=x$. Figure A.1 shows an example of a function and its inverse: mirror image around $y=x$

Given a function of a variable, say $U(x)$, the inverse function is usually written using the notation $U^{-1}(t)$. The composite function of both the inverse function and the function itself satisfies the relation

$$\boxed{U^{-1}(U(x)) = x} \text{ and } \boxed{U(U^{-1}(x)) = x}.$$

To determine the inverse of a function $U(x)$, set

$$U(x) = t$$

and solve for y in terms of t, i.e. $x = U^{-1}(t)$.

Example: The Inverse of a Function

Consider the function

$$U(x) = a + be^{-\gamma x}.$$

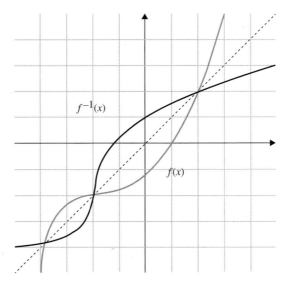

Figure A.1 A function and its inverse: mirror image around y=x.
"Inverse Function Graph" by Jim.belk at English Wikipedia – Transferred from en.wikipedia to Commons. Licensed under Public Domain via Wikimedia Commons – https://commons.wikimedia.org/wiki/ File:Inverse_Function_Graph.png#/media/File:Inverse_Function_Graph.png

To determine its inverse, set $U(x) = t$, to get

$$U(x) = t = a + be^{-\gamma x}.$$

Rearranging gives the relation between x and t as

$$x = \frac{-1}{\gamma} \ln \frac{t-a}{b} = U^{-1}(t).$$

Therefore, $U^{-1}(t) = \frac{-1}{\gamma} \ln \frac{t-a}{b}$ is the inverse of $U(x)$.

Note that, by direct substitution,

$$U^{-1}(U(x)) = \frac{-1}{\gamma} \ln\left(\frac{U(x)-a}{b}\right) = \frac{-1}{\gamma} \ln\left(\frac{(a+be^{-\gamma x})-a}{b}\right) = x.$$

Similarly, the inverse of the exponential function $y(x) = e^x$ is the logarithmic function, $x = \ln(y)$. Therefore, the composition of the exponential and the logarithmic functions must yield the identity function. Indeed,

$$e \circ \ln(x) = e^{\ln(x)} = x \text{ and } \ln \circ e^x = x.$$

This identity is used repeatedly throughout the book.

If a function is continuous and strictly increasing, then its inverse function exists and is also continuous and strictly increasing.

Even if the function does not have a closed-form expression, but it is continuous and strictly increasing, we can calculate its inverse function either graphically using a mirror image around $y = x$ or numerically by solving for $x = U^{-1}(t)$.

A.3 THE ARCHIMEDEAN COMBINATION OF TWO (OR MORE) FUNCTIONS

A.3.1 Archimedean Combinations of Univariate Functions (in One Dimension)

Consider the simple product of two functions for two variables, say $f_1(x)$ and $f_2(x)$, which is

$$f(x) = f_1(x) \times f_2(x).$$

Figure A.2(a) shows an example of two functions. Figure A.2(b) shows their product.

One step beyond the simple product is to define a function C and another function g, such that

$$g(C(x)) = g(f_1(x)) \times g(f_2(x))$$

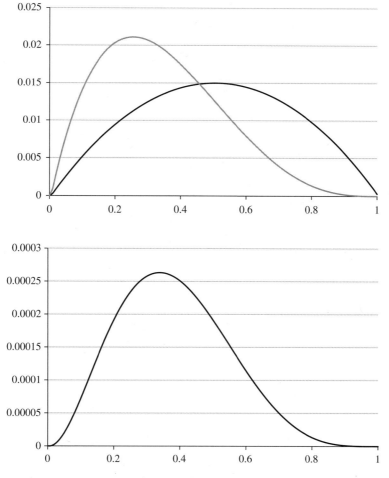

Figure A.2 (a) (Top) Two functions. (b) (Bottom) Product of two functions.

Note that for the special case in which $g(t) = t$ is the identity, the function $C(f_1(x), f_2(x))$ is the simple product $f_1(x) \times f_2(x)$, but for more general functions g, the previous definition need not be a product.

If the function g is continuous and strictly increasing, then we can take its inverse on both sides and write:

$$C(f_1(x), f_2(x)) = g^{-1}\big(g(f_1(x)) \times g(f_2(x))\big)$$

The function C is known as the Archimedean combination of two functions $f_1(x)$ and $f_2(x)$, and the function g is known as the generating function.

Consider, for example, $g(t) = e^{\lambda t}$. The Archimedean combination of two functions $f_1(x)$ and $f_2(x)$ of Figure A.2(a) is

$$C(f_1(x), f_2(x)) = \frac{1}{\lambda} \ln\big(e^{\lambda f_1(x)} \times e^{\lambda f_2(x)}\big) = f_1(x) + f_2(x)$$

The Archimedean combination of the two functions using a function, g, is shown in Figure A.3.

By changing the form of the generating function, we get different forms of the Archimedean combination.

Note that λ did not play a role in the Archimedean combination of the two functions. In general, raising the generating function to a power transformation does not affect the Archimedean combination as we illustrate below.

Example: Invariance of the Archimedean Combination to a Power Transformation on the Generating Function

---

Consider the generating function,

$$g(t) = t^m, \quad t, m > 0.$$

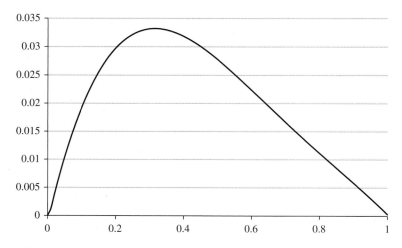

Figure A.3 Archimedean combination of the two functions using a function, g.

The Archimedean combination of two functions $f_1(x)$ and $f_2(x)$ is

$$C(f_1(x),f_2(x)) = \left(\left(f_1(x)\right)^m \times \left(f_2(x)\right)^m\right)^{\frac{1}{m}} = \left(f_1(x)\right) \times \left(f_2(x)\right),$$

which is the same Archimedean combination that results from $g(t) = t$. The Archimedean combination is invariant to a power transformation on the generating function.

Now we present a slightly more complicated example than the previous example.

Example: Exponential Generating Function

Consider the generating function

$$g(t) = \frac{1 - e^{\lambda t}}{1 - e^{\lambda}}.$$

Here the parameter λ does in fact play a role in the shape of the Archimedean combination because it is not merely a power transformation (it exists in the denominator).

To determine the inverse function, we set

$$\frac{1 - e^{\lambda t}}{1 - e^{\lambda}} = x \Rightarrow 1 - e^{\lambda t} = x(1 - e^{\lambda}) \Rightarrow t = \frac{1}{\lambda}\ln\left(1 - x(1 - e^{\lambda})\right) = g^{-1}(x)$$

The Archimedean combination of the two functions is now

$$C(f_1(x),f_2(x)) = \frac{1}{\lambda}\ln\left(1 - \frac{1 - e^{\lambda f_1(x)}}{1 - e^{\lambda}}\frac{1 - e^{\lambda f_2(x)}}{1 - e^{\lambda}}(1 - e^{\lambda})\right) = \frac{1}{\lambda}\ln\left(1 - \frac{(1 - e^{\lambda f_1(x)})(1 - e^{\lambda f_2(x)})}{1 - e^{\lambda}}\right)$$

This example shows the flexibility in the types of curves that can be generated using an Archimedean combination with the generating function $g(t) = \frac{1 - e^{\lambda t}}{1 - e^{\lambda}}$ when compared to a simple product of functions. Note the role that the function g plays in the shape of the Archimedean combination.

Common forms of Archimedean functions use normalized arguments where the functions f_1, f_2 are normalized to range from 0 to 1 (such as cumulative probability distributions or normalized utility functions). The generating function is also normalized using a normalized form of g where g is continuous, strictly increasing, and is normalized such that $g(0)=0$ and $g(1)=1$.

A.3.2 Archimedean Combinations in Two Dimensions

Now consider the simple product of two functions for two variables, say $U_1(x)$ and $U_2(y)$, which gives the two-dimensional function

$$U(x, y) = U_1(x) \times U_2(y).$$

One step beyond the simple product is to define a function C and another function g, such that

$$g\left(C(U_1(x),U_2(y))\right) = g\left(U_1(x)\right) \times g\left(U_2(y)\right)$$

For the special case where $g(t) = t$, the function $C(U_1(x), U_2(y))$ is the product $U_1(x) \times U_2(y)$, but for more general instantiations of g, it need not be.

> If the function g is continuous and strictly increasing, then we can take its inverse on both sides and write
>
> $$C(U_1(x),U_2(y)) = g^{-1}\left(g\left(U_1(x)\right) \times g\left(U_2(y)\right)\right)$$

The function C is known as the **_Archimedean combination of two functions $U_1(x)$ and $U_2(y)$_**, and the function g is known as the generating function.

> **Additive Forms of Archimedean Combinations**
> Define the composite transformation $g^{-1}(t) = \lambda^{-1}(-\ln(t))$. Therefore, $g(t) = e^{-\lambda(t)}$.
>   Note that if $g(0)=0$ and $g(1)=1$, then by direct substitution, $\lambda(0) = \infty$ and $\lambda(1) = 0$.
>   Substituting for the definition of the function λ into the Archimedean combination gives
>
> $$C(U_1(x),U_2(y)) = \lambda^{-1}\left(-\ln\left(e^{-\lambda(U_1(x))} \times e^{-\lambda(U_2(y))}\right)\right) = \lambda^{-1}\left(\lambda\left(U_1(x)\right) + \lambda\left(U_2(y)\right)\right)$$

The term $\lambda^{-1}\left(\lambda\left(U_1(x)\right) + \lambda\left(U_2(y)\right)\right)$ is an additive form of the Archimedean combination.

In the probability literature, the function $C(s,t) = \lambda^{-1}\left(\lambda(s) + \lambda(t)\right)$ is known as an Archimedean probability copula when the function λ satisfies certain conditions: λ is continuous, strictly decreasing, it maps $[0,1]$ to $[0,°]$, and it satisfies certain derivative conditions to guarantee that the joint probability density functions are non-negative. These conditions are $(-1)^k \dfrac{d^{(k)}}{dt^k}\lambda(t) \geq 0$, $k \in N$, the set of natural numbers. A function that satisfies this derivative condition is said to be completely monotone.

In the utility literature, the Archimedean combination defined in Abbas (2009) is more general as it does not require the constraints on the derivatives of λ (or the multiplicative form g) and it allows for more general definitions on the domain of g. Furthermore, the requirement that $g(1) = 1$ is associated with Class 1 utility copulas, and $g(1) = 0$ is associated with Class 0 utility copulas that are discussed in Chapters 28 and 29. Using an Archimedean combination for an n-attribute utility function implies that we can determine the surface using n curves and a generating function. This means that we construct a two-attribute utility function using two curves on the surface of the utility function and a utility copula.

Note: Similar to the univariate case, the Archimedean combination of two functions has several properties related to the generating function, one of which is that all positive powers of g yield the same Archimedean form.

To illustrate, consider the Archimedean form

$$C(x,y) = \varphi^{-1}\left[\varphi(x)\varphi(y)\right]$$

For a generating function, $\varphi(t) = t^m$ we have

$$C(x,y) = \left[x^m y^m\right]^{1/m} = xy$$

This is the same surface that would result from using the generating function $\varphi(t) = t$.

Using different generating functions results in different surfaces. The idea of utility copula functions in Chapter 28 and 29 constructs utility surfaces using the individual utility functions as arguments into the copula formulation.

Jensen's Equality and Inequality

Jensen's inequality determines the relative order of the expectation of a function of a random variable and the function of the expectation of the random variable, i.e., it determines the relation between $E[U(x)]$ and the utility of the expectation $U\big(E[(x)]\big)$. The interest is the relation

$$\sum_{i=1}^{n} p_i U(x_i) \quad \text{vs} \quad U\left(\sum_{i=1}^{n} p_i x_i\right)$$

Jensen's inequality considers three cases:

1. $\displaystyle\sum_{i=1}^{n} p_i U(x_i) = U\left(\sum_{i=1}^{n} p_i x_i\right)$ for all lotteries if and only if the function U is linear.

2. $\displaystyle\sum_{i=1}^{n} p_i U(x_i) < U\left(\sum_{i=1}^{n} p_i x_i\right)$ for all lotteries if and only if the function U is strictly concave.

3. If $\displaystyle\sum_{i=1}^{n} p_i U(x_i) > U\left(\sum_{i=1}^{n} p_i x_i\right)$ for all lotteries if and only if the function U is strictly convex.

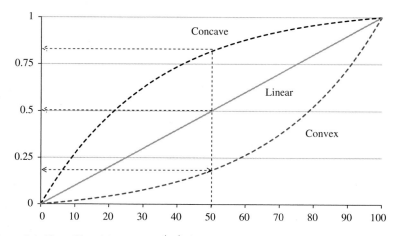

Figure B.1. The utility of the mean, $U(50)$, for concave, convex, and linear utility functions.

To memorize these inequalities, it will be useful to consider the following simple example.

Example
Consider the lottery

$$< 0.5, 0; \; 0.5, 100 >$$

and a utility function that is normalized such that $U(0) = 0$ and $U(100) = 1$.
The mean value of the lottery is $E[X] = 0.5 \times 0 + 0.5 \times 100 = 50$.
The expected utility of the lottery is $E[U(x)] = 0.5 \times 0 + 0.5 \times 1 = 0.5$.
Jensen's inequality is concerned with the relation between

$$E[U(x)] = 0.50 \text{ and } U\big(E[(x)]\big) = U(50).$$

Figure B.1 shows examples of concave convex and linear utility functions.
The figure shows immediately that if the utility function is concave, then $U(50) > 0.5$; if the utility function is convex, then $U(50) < 0.5$, and if the utility function is linear, then $U(50) = 0.5$.

If equality or inequality is not preserved for all lotteries, then the function U is neither linear nor concave nor convex. For example it could be S-shaped.

APPENDIX C Delta Property for a Discrete Number of Points

The delta property can be satisfied for a discrete number of shift amounts and would then allow for more general functions besides the linear and exponential functions. The type of analysis used to show this is part of the "***Mathematical Folklore***," and we shall discuss many examples below.

C.1 INVARIANCE TO A PARTICULAR SHIFT VALUE, δ_0

Consider a utility function that is invariant to a discrete shift amount δ_0, i.e., the wealth equivalent increases by an amount δ_0 when all lottery outcomes increase by δ_0.

As we have seen in Chapter 13, this implies that

$$U(y+\delta_0) = k(\delta_0)U(y) + d(\delta_0) \tag{1}$$

If the utility function is differentiable, we can take the derivative of both sides to get

$$u(y+\delta_0) = k(\delta_0)u(y), \tag{2}$$

where u is the derivative of the utility function.

To solve (2) for a particular value of δ_0, define $y = \delta_0 t$ and substitute to get

$$u(\delta_0(1+t)) = k(\delta_0)u(\delta_0 t) \tag{3}$$

Divide both sides of (3) by k^{t+1} to get

$$\frac{u(\delta_0(1+t))}{k^{1+t}} = \frac{u(\delta_0 t)}{k^t}$$

Further define

$$p(t) = k^{-t}u(\delta_0 t) \tag{4}$$

and substitute to get

$$p(t+1) = p(t), \tag{5}$$

which implies that $p(.)$ is a periodic function with period equal to one.

From (4), we have

$$u(y) = k^{y/\delta_0} p(y/\delta_0). \tag{6}$$

Making the substitution $\gamma = -\dfrac{1}{\delta_0}\ln(k)$ gives

$$u(x) = e^{-\gamma y} p(\dfrac{y}{\delta_0}). \qquad Q.E.D. \tag{7}$$

Consequently, it is not sufficient to test with a particular shift amount and assert constant absolute risk aversion.

Furthermore, it is not sufficient to test with even an infinite sequence of integer multiples of δ and assert that a decision-maker has constant absolute risk aversion, since the solution to the particular shift amount, δ_0, must also be invariant to all of its integer multiples:

$$\begin{aligned}
u(y + i\delta_0) &= e^{-\gamma(y + i\delta_0)} p(\dfrac{y + i\delta_0}{\delta_0}) \\
&= e^{-\gamma i\delta_0} e^{-\gamma y} p(\dfrac{y}{\delta_0} + i) \\
&= k'(\delta_0) e^{-\gamma y} p(\dfrac{y}{\delta_0})
\end{aligned} \tag{8}$$

where $k'(\delta_0) = e^{-\gamma i\delta_0}$ = constant.

Note: A decision-maker can thus satisfy transformation invariance for all shift amounts of $\delta_0 = \$100$ increments – i.e., $\delta_0 = \$100$, $2\delta_0 = \$200$, ..., $100\delta_0 = \$10,000$ – and still have neither an exponential nor a linear utility function over this interval.

C.2 INVARIANCE TO A PARTICULAR SCALE VALUE, C

Suppose all lottery wealth outcomes are scaled by a fixed amount $c > 0$ and the wealth equivalent of the modified lottery is equal to the wealth equivalent of the unmodified lottery multiplied by this scale amount. As we have seen in Chapter 15, if this occurs for values of c on an interval, then the utility function is either logarithmic or a power utility function.

If invariance occurs for only a particular scale value, $\delta = c$, then

$$U(cy) = k(c)U(y) + d(c).$$

Taking the derivatives with respect to y and dividing by c gives

$$u(cy) = \dfrac{k(c)}{c} u(y). \tag{9}$$

Since the value of c is fixed in (9), $k(c)$ must also be a fixed constant.

To solve (9) for a particular value of c, we can now make the substitution $y = c^t$ and divide both sides by $(\frac{c}{k})^{t+1}$ to get

$$\left(\frac{k}{c}\right)^{-(t+1)} u(c^{t+1}) = \left(\frac{k}{c}\right)^{-t} u(c^t). \tag{10}$$

Define

$$p(t) = (\frac{k}{c})^{-t} u(c^t). \tag{11}$$

Substituting (11) into (10) gives

$$p(t+1) = p(t). \tag{12}$$

Equation (12) implies that $p(t)$ is a periodic function with period equal to one. Rearranging (11) gives

$$u(c^t) = (\frac{c}{k})^t p(t). \tag{13}$$

Substituting back for $y = c^t$ gives

$$u(y) = (\frac{c}{k})^{\frac{\ln y}{\ln c}} p(\frac{\ln y}{\ln c}). \tag{14}$$

Using properties of logarithms gives

$$u(y) = y^{(\frac{\ln k}{\ln c} - 1)} p(\frac{\ln y}{\ln c}). \quad Q.E.D. \tag{15}$$

A decision-maker with a differentiable and strictly monotonic utility function is invariant to a particular scale amount, c, if and only if the first derivative of his utility function satisfies

$$u(x) = y^{(\frac{\ln k}{\ln c} - 1)} p(\frac{\ln y}{\ln c}), \tag{16}$$

where $p(.)$ is any periodic function with period equal to one.

Note: A differentiable utility function that satisfies transformation invariance for a scale amount, c, satisfies transformation invariance for all integer powers of c.

To see this note that if a decision-maker satisfies (16) and faces a modified lottery whose prospects are scaled by c^i, where i is any integer, we have

$$u(c^i y) = c^{i(\frac{\ln k}{\ln c} - 1)} y^{(\frac{\ln k}{\ln c} - 1)} p(\frac{\ln(c^i y)}{\ln c})$$

$$= k'(c) y^{(\frac{\ln k}{\ln c} - 1)} p(i + \frac{\ln(y)}{\ln c}) \tag{17}$$

$$= k'(c) y^{(\frac{\ln k}{\ln c} - 1)} p(\frac{\ln(y)}{\ln c}) = k'(c) u(y),$$

where $k'(c) = c^{i(\frac{\ln k}{\ln c} - 1)}$ is a constant. Equation (17) shows that $u(x)$ is invariant to the transformation $c^i y$ (i.e., all integer powers of c).

This implies that a decision-maker can satisfy scale invariance to the infinite sequence, $..., c^{-2}, c^{-1}, 1, c, c^2, c^3, ...etc$, and still not have a logarithmic or a power utility function. If we wish to assert constant relative risk aversion with transformation invariance, then we need to test the assertion with scale values other than those in this infinite sequence.

C.3 INVARIANCE WITH POWER TRANSFORMATIONS

For the power transformation, $g(y, \delta) = y^\delta$, we have $\frac{\partial}{\partial y} g(y, \delta) = \delta y^{\delta - 1}$. Invariance with power transformations requires

$$u(y^\delta) = \frac{k(\delta) u(y)}{\delta y^{\delta - 1}}. \tag{18}$$

The solution to (18) can be obtained using the substitution

$$y = e^t, \tag{19}$$

and multiplying both sides of the equation by $e^{\delta t}$ to get

$$e^{\delta t} u(e^{\delta t}) = \frac{k(\delta)}{\delta} e^t u(e^t). \tag{20}$$

Now we use the substitution

$$w(t) = e^t u(e^t). \tag{21}$$

Equation (20) reduces to

$$w(\delta t) = \frac{k(\delta)}{\delta} w(t). \tag{22}$$

When (22) is valid over an interval range for values of $\delta \in [\delta_{\min}, \delta_{\max}]$, we can define $z(\delta) = \frac{k(\delta)}{\delta}$ and substitute into (22) to get

$$w(\delta t) = z(\delta) w(t), \tag{23}$$

which is a Pexider equation, whose general solution has the form,

$$w(t) = at^b. \tag{24}$$

Substituting (24) into (22) shows the value of $b = \dfrac{\ln k}{\ln \delta} - 1$.

Now define $\mu = \dfrac{\ln k}{\ln \delta}$. From (24), (21), and (19), we have a general solution to (18) of the form

$$u(y) = \frac{a}{y}(\ln y)^{\mu-1}. \tag{25}$$

A decision-maker with a differentiable and strictly monotonic utility function, $U(y)$, satisfies transformation invariance to a monotonic power transformation $g(y, \delta) = y^\delta$, $\forall \delta \in [\delta_{\min}, \delta_{\max}], \delta_{\min} < \delta_{\max}$, if and only if the first derivative of his utility function satisfies

$$u(y) = \frac{a}{y}(\ln y)^{\mu-1}, \tag{26}$$

where $\propto$ and a are arbitrary constants.

By integration, we can write the utility function that corresponds to (26) as either

$$U(y) = \frac{a}{\mu}(\ln(y))^\mu + b, \ \mu \neq 0, \tag{27}$$

or

$$U(y) = a\ln(\ln(y)) + b, \ \mu = 0, \tag{28}$$

where $a, \mu \neq 0$ but otherwise arbitrary and b is an arbitrary constant.

C.4. INVARIANCE WITH A SINGLE POWER VALUE, M

Following the approach taken in the proofs of shift and scale for discrete amounts, it is relatively straightforward to show that when power invariance occurs for only a particular value of $\delta = m$, then (26) has an additional periodic component,

$$u(x) = \frac{(\ln x)^{\mu-1}}{x} p(\frac{\ln(\ln(x))}{\ln m}), \tag{29}$$

where $p(.)$ is a periodic function of period equal to one.

Proof: The proof is substantially similar to the prior two proofs, and hence omitted.

Decomposing Discrete Multiattribute
Indifference Probability Assessments Using
the Utility Tree

D.1 THE MEANING OF A MULTIATTRIBUTE UTILITY INDIFFERENCE ASSESSMENT

As we discussed, a multiattribute utility is an indifference probability assessment of a prospect characterized by multiple attributes according to the Equivalence Rule. Let X denote wealth level, with x^* denoting the maximum wealth and x^0 the minimum wealth. Further, let Y denote health as determined by equivalent life years, with y^* being the best and y^0 being the worst. We can think of three prospects:

- Prospect A with maximum wealth and health (x^*, y^*)
- Prospect C with minimum wealth and health (x^0, y^0)
- Prospect B in between characterized by (x, y), where $x^0 < x < x^*$ and $y^0 < y < y^*$.

The indifference probability of getting B for certain or a deal that gives A or C is the multiattribute utility value, $U(x, y)$. Figure D.1 shows an example of a two-attribute indifference probability assessment.

If more of an attribute is referred to less, then the utility of any prospect (point in the domain of definition) has a clear meaning as the indifference probability that makes us indifferent to either receiving that prospect or receiving a binary deal that provides the top right prospect (with a probability equal that indifference probability) or the bottom left prospect with a probability equal to one minus the indifference probability (Figure D.2). This indifference assessment requires us to think about multiple variations at once for each of the attributes.

> Note: Using a value function, we converted this multiattribute indifference assessment into a univariate indifference assessment over value. We can think of three prospects:
>
> Prospect A with a value V^* corresponding to maximum wealth and health (x^*, y^*)
> Prospect C with a value V^0 corresponding to minimum wealth and health (x^0, y^0)
> Prospect B with a value V, characterized by (x, y), where $V^0 < V < V^*$.
>
> The indifference probability of getting B for certain or a deal that gives A or C is the multiattribute utility value, $U_V(V)$. Figure D.3 shoes a multiattribute utility assessment using a value function.

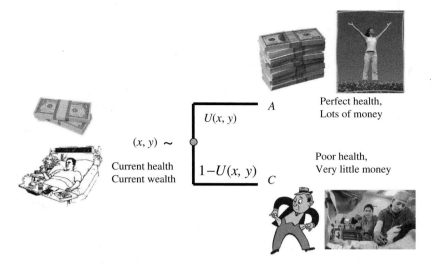

Figure D.1. Two-attribute indifference assessment.

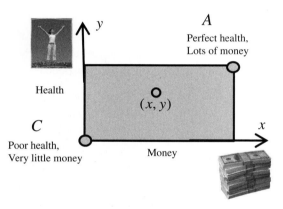

Figure D.2. Domain of a two-attribute utility function.

We can also decompose this multiattribute indifference assessment by considering variations over each attribute individually. Consider, for example, a situation with two attributes X and Y where more of an attribute is preferred to less. Suppose you would like to assess the indifference probability of the prospect (x_1, y_1) in terms of the best prospect (x^*, y^*) and the worst (x^0, y^0). Sometimes it is cognitively difficult to make this assessment because you need to think about two variations simultaneously: x_1 in terms of x^0 and x^*, and y_1 in terms of y^0 and y^*. It might be easier to first think about the indifference probability of (x_1, y_1) in terms of (x_1, y^*) and (x_1, y^0), where only one variation is considered with respect to y, and then think about (x_1, y^*) in terms of (x^*, y^*) and (x^0, y^*), and likewise (x_1, y^0) in terms of (x^*, y^0) and (x^0, y^0), focusing on variations with respect to x.

Below we illustrate how to decompose the joint indifference assessment of (x_1, y_1) in terms of indifference assessments that require only one variation at a time of each attribute and then combine them together using the utility tree to produce the required assessment.

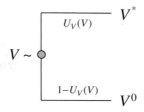

Figure D.3. Multiattribute utility assessment using a value function.

D.2 THE TWO-ATTRIBUTE UTILITY TREE FOR A SINGLE PROSPECT

When the number of consequences in a decision is small, the utility value of each prospect can be elicited using indifference probability assessments of binary gambles. Given three two-attribute consequences with strict indifference, $(x^0, y^0) \prec (x_1, y_1) \prec (x^*, y^*)$, the utility value $U(x_1, y_1)$ is the indifference probability, $p_{x_1 y_1}$, of receiving a binary gamble with a $p_{x_1 y_1}$ chance of getting (x^*, y^*) and $1 - p_{x_1 y_1}$ chance of getting (x^0, y^0) or receiving (x_1, y_1) for sure. By direct rollback analysis, we get

$$U(x_1, y_1) = p_{x_1 y_1} U(x^*, y^*) + (1 - p_{x_1 y_1}) U(x^0, y^0).$$

While this method of assessment is theoretically sound, observe that the levels of both attributes in the equivalent binary gamble are different than their levels for the consequence (x_1, y_1). It might be easier to use a binary gamble for the assessment in which the level of only one of the attributes changes. For example, suppose that $x^0 < x_1 < x^*, y^0 < y_1 < y^*$, and that more of any attribute is preferred to less. We may then ask about the indifference probability of receiving either (x_1, y_1) for certain or a binary gamble that provides either (x_1^*, y_1) or (x_1^0, y_1). Note that the level of y_1 is constant for both the consequence and the binary gamble. Denote this indifference probability as $p_{x_1 | y_1}$, where the subscript denotes that the level of y is held fixed at y_1 and the level of x will be expressed in terms of its best and worst values x_1^* and x_1^0.

In a sense, the indifference probability $p_{x_1 | y_1}$ can be regarded as a conditional indifference assessment over a gamble for a variation of only one of the attributes, while the probability $p_{x_1 y_1}$ can be viewed as an indifference probability assessment for the joint variation of the levels of the attributes. To relate the indifference probabilities, $p_{x_1 y_1}$ and $p_{x_1 | y_1}$, we now assess the indifference probability of (x_1^*, y_1) in terms of (x_1^*, y_1^*) and (x_1^*, y_1^0), which we refer to as $P_{y_1 | x_1^*}$ and the indifference probability of (x_1^0, y_1) in terms of (x_1^0, y_1^*) and (x_1^0, y_1^0), which we refer to as $P_{y_1 | x_1^0}$. By doing this, we can express the utility value of (x_1, y_1) in terms of the utility values of the four corner points in Figure D.4 using indifference assessments of a sequence of gambles where the level of only one attribute changes at each stage.

Denote the indifference probability of the corner point (x^*, y^0) in Figure D.4 as $k_x = P_{x^* y^0}$ and the indifference probability of the corner point (x^0, y^*) as $k_y = P_{x^0 y^*}$. Further let $P_{x^* y^*} = 1, P_{x^0 y^0} = 0$.

The top tree in Figure D.5 represents this sequence of gambles required. By rollback analysis, we have

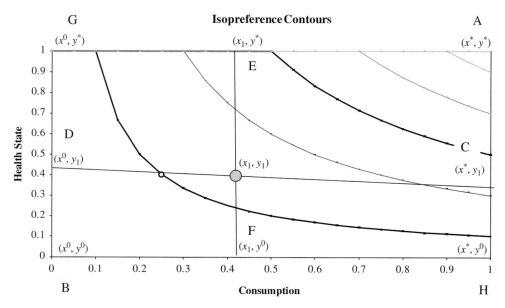

Figure D.4. Coordinates of nine important points in a two-attribute problem.

$$p_{x_1y_1} = p_{x_1|y_1} \times p_{y_1|x_1^*} \times 1 + p_{x_1|y_1} \times (1 - p_{y_1|x_1^*}) \times k_x$$
$$+ (1 - p_{x_1|y_1}) \times p_{y_1|x_1^0} \times k_y + (1 - p_{x_1|y_1}) \times (1 - p_{y_1|x_1^0}) \times 0. \tag{30}$$

Alternatively, we can think about the indifference probability of (x_1, y_1) in relation to a binary gamble that provides either (x_1, y^*) or (x_1, y^0), where the level of x_1 is held constant. Denote this indifference probability as $p_{y_1|x_1}$. To relate the indifference probabilities, $p_{x_1y_1}$ and $p_{y_1|x_1}$, we now assess the indifference probability of (x_1, y_1^*) in terms of (x_1^*, y_1^*) and (x_1^0, y_1^*), which we refer to as $P_{x_1|y_1^*}$ and the indifference probability of (x_1, y_1^0) in terms of (x_1^*, y_1^0) and (x_1^0, y_1^0), which we refer to as $P_{x_1|y_1^0}$.

Consistency requires that

$$p_{x_1y_1} = p_{y_1|x_1} \times p_{x_1|y_1^*} \times 1 + p_{y_1|x_1} \times (1 - p_{x_1|y_1^*}) \times k_y + (1 - p_{y_1|x_1})$$
$$\times p_{x_1|y_1^0} \times k_x + (1 - p_{y_1|x_1}) \times (1 - p_{x_1|y_1^0}) \times 0. \tag{31}$$

Figure D.5 shows the tree-representations for the assessments leading to (30) and (31). Note that we start with the level of the attributes of a consequence and, as we move from the left to the right, we replace a single attribute at each step with a binary gamble where its levels are at the minimum and maximum values.

Equating the two expressions (30) and (31) provides a consistency check on the assessments provided by the decision-maker in two different assessment orders, and it enables us to deduce some assessments that were not directly provided. The following examples present an application of utility trees illustrate their use in assessments and consistency checks.

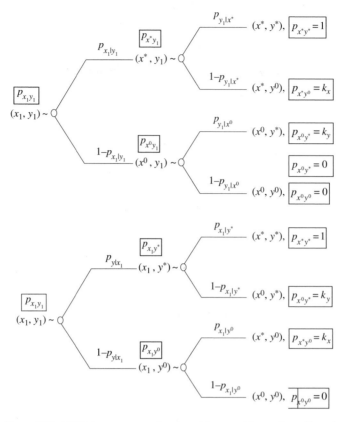

Figure D.5. Utility trees representing two decomposition orders of (x_1, y_1).

EXAMPLE: Assessing the Utility Value of Salary and Vacation Days

---

Suppose we wish to determine the utility of the consequence (Salary = $100k$, 24 vacation days) in terms of the best prospect ($200k$, 36 vacation days) and the worst prospect ($40k$, 12 vacation days), assuming that each of the two attributes is preferred to less.

To construct the utility tree, we need five assessments: two corner assessments, k_x, k_y, and three indifference probability assessments. At the vacation level of 24 days, the decision-maker thinks about the indifference probability of getting either $200k$ vs $40k$ instead of the $100k$ associated with this prospect. She then states an indifference probability $p_{100k|24days} = 0.8$. She is comfortable making this assessment since it requires only one variation of an attribute (salary) at the same vacation level. She then thinks about the indifference probability of getting either 36 vacation days or 12 vacation days in exchange for her current vacation level of 24 days, at salary levels of $200k and $40k. She realizes that she values 24 vacation days more when her salary is $200k than when it is $40k. She then provides $p_{24days|200k} = 0.7, p_{24days|40k} = 0.6$. The utility value of the consequence (Salary = $100k$, 24 vacation days) can therefore be expressed as

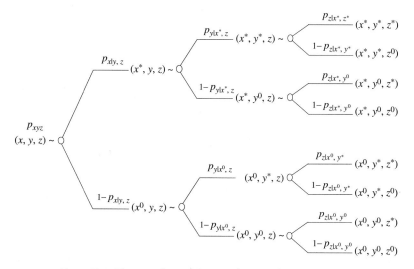

Figure D.6. Three-attribute utility tree for a single consequence.

$$p_{100k,24days} = 0.8 \times 0.7 \times 1 + 0.8 \times 0.3 \times k_x + 0.2 \times 0.6 \times k_y.$$

The corner assessments, k_x and k_y will be assessed once and can be used for all other consequences. The corner assessment, k_x, is the utility value of ($200k$, 12 vacation days) and the assessment k_y is the utility value of ($40k$, 36 vacation days). The decision-maker assigns $k_x = 0.9$ and $k_y = 0.5$. Therefore,

$$p_{100k,24days} = 0.8 \times 0.7 \times 1 + 0.8 \times 0.3 \times 0.9 + 0.2 \times 0.6 \times 0.5 = 0.836.$$

D.3 THE MULTIATTRIBUTE UTILITY TREE OF A SINGLE CONSEQUENCE

By induction, the utility tree of a three-attribute consequence can be represented as shown in Figure D.6, replacing an attribute by its minimum and maximum values at each step. The utility value of the consequence is a sum of the products of the indifference probabilities at each branch including the utility values of the corner assessments (the leaves of the tree). The number of assessments required in this tree is 13, namely 7 indifference probability assessments and 6 corner assessments (since we set one to unity and one to zero). The same methodology applies directly to n attributes. By induction, the number of assessments required for an n-attribute tree *is equal to the number of corner assessments plus the number of conditional assessments* $= (2^n - 2) + (2^n - 1) = 2^{n+1} - 3$.

Index